Uniform Distribution of Sequences

Uniform Distribution of Sequences

L. Kuipers

H. Niederreiter

Dover Publications, Inc.
Mineola, New York

Bibliographical Note

This Dover edition, first published in 2006, is an unabridged republication of the work originally published by John Wiley & Sons, Inc., New York, in 1974. The preface to the Dover Edition and Corrigenda were specially prepared for this edition by H. Niederreiter.

International Standard Book Number: 0-486-45019-8

Manufactured in the United States of America
Dover Publications, Inc., 31 East 2nd Street, Mineola, N.Y. 11501

To Francina and Gerlinde

PREFACE TO THE DOVER EDITION

When Dover Publications approached me with the proposal of bringing out a reprint edition of *Uniform Distribution of Sequences*, I gladly grasped the opportunity of making this book, which has been out of print for a long time, available again. It seems that this book, although over 30 years old, can still serve a useful purpose for students and researchers by providing a systematic introduction to the subject as well as a comprehensive coverage of the literature up to the early 1970s.

The development of the field of uniform and asymptotic distribution of sequences from 1974, the year of publication of the original version of the book, to the present has been marked by a tendency to move from the abstract to the concrete, which is in line with a general trend in mathematics during this period. In the context of the material in this book, this is reflected by the fact that comparatively little further work has been done on those parts of the theory that correspond to Chapters 3 and 4. On the other hand, research on the "down-to-earth" aspects of the area covered by Chapters 1 and 2 has flourished and exciting progress was achieved in the intervening years. These dynamic research activities on uniform distribution in concrete settings, and in particular on discrepancy theory, are driven by demands from applications. It is well known, and can be gleaned also from Chapter 2, Section 5, of the book, that the study of the discrepancy of sequences in unit cubes plays an important role in quasi-Monte Carlo methods for scientific computing. Other areas where discrepancy theory has made an impact are pseudorandom number generation, computer graphics, cryptology, and various parts of number theory.

A reprint edition does not allow adding new material, so I had to be content with listing misprints and updates for references. The reader who wants to gain access to research in the area since 1974 may want to consult the following works. Many aspects of the theory of uniform distribution of sequences are covered in the monograph of M. Drmota and R.F. Tichy, *Sequences, Discrepancies and Applications*, Lecture Notes in Mathematics, Vol. 1651, Springer-Verlag, Berlin, 1997. Uniformly distributed sequences are treated in the context of probabilistic number theory in G. Harman, *Metric Number Theory*, Oxford University Press, Oxford, 1998. The books of J. Beck and W.W.L. Chen, *Irregularities of Distribution*, Cambridge University Press, Cambridge, 1987, and J. Matoušek, *Geometric Discrepancy: An Illustrated Guide*, Springer-Verlag, Berlin, 1999, are devoted to discrepancy theory. A treatment of discrepancy theory from the perspective of theoretical computer science is provided in B. Chazelle, *The Discrepancy Method: Randomness and Complexity*, Cambridge University Press, Cambridge, 2000. Applications of discrepancy theory to quasi-Monte Carlo methods and pseudorandom number generation are discussed in the survey article by H. Niederreiter, Quasi-Monte Carlo methods and pseudo-random numbers, *Bull. Amer. Math. Soc.* **84**, 957–1041 (1978), as well as in the books of H. Niederreiter, *Random Number Generation and Quasi-Monte Carlo Methods*, SIAM, Philadelphia, 1992, and S. Tezuka, *Uniform Random Numbers: Theory and Practice*, Kluwer Academic Publishers, Boston, 1995. A series of biennial conferences on Monte Carlo and quasi-Monte Carlo methods was begun in 1994 and is still going strong. The proceedings of these conferences were all published by Springer-Verlag, with the first volume being *Monte Carlo and Quasi-Monte Carlo Methods in Scientific Computing* (H. Niederreiter and P.J.-S. Shiue, eds.), Lecture Notes in Statistics, Vol. 106, Springer-Verlag, New York, 1995, and the most recent one being *Monte Carlo and*

Quasi-Monte Carlo Methods 2004 (H. Niederreiter and D. Talay, eds.), Springer-Verlag, Berlin, 2006. A monograph updating the material in Chapter 5 of the book is that of W. Narkiewicz, *Uniform Distribution of Sequences of Integers in Residue Classes*, Lecture Notes in Mathematics, Vol. 1087, Springer-Verlag, Berlin, 1984.

I am dedicating this reprint edition to the memory of my coauthor Lauwerens Kuipers who passed away many years ago. Finally, I want to express my gratitude to John Grafton of Dover Publications for initiating the project of publishing this reprint edition.

Singapore, November 2005 Harald Niederreiter

PREFACE

The theory of uniform distribution modulo one (*Gleichverteilung modulo Eins*, *équirépartition modulo un*) is concerned, at least in its classical setting, with the distribution of fractional parts of real numbers in the unit interval (0, 1). The development of this theory started with Hermann Weyl's celebrated paper of 1916 titled: "Über die Gleichverteilung von Zahlen mod. Eins." Weyl's work was primarily intended as a refinement of Kronecker's approximation theorem, and, therefore, in its initial stage, the theory was deeply rooted in diophantine approximations. During the last decades the theory has unfolded beyond that framework. Today, the subject presents itself as a meeting ground for topics as diverse as number theory, probability theory, functional analysis, topological algebra, and so on. However, it must be said that the germs of various later developments can be found in Weyl's 1916 paper.

This book attempts to summarize the results of these investigations from the beginning to the present, with emphasis on the work done during the last 20 years. Because the literature on the subject is vast, it was inevitable that now and then choices had to be made and selection criteria had to be applied that reflect the personal taste of the authors. As a rule, we have endeavored to produce a comprehensive coverage of the methods used in the theory of uniform distribution. In some instances, we did not present a result in its most general version, but rather tried to describe the underlying principles and ideas, which might otherwise be shrouded in technicalities. The title we have chosen indicates that we have resolved the dilemma of "asymptotic distribution" versus "uniform distribution" in favor of the latter, since it is this aspect to which most of our exposition is devoted. We believe that this book should prove a useful introduction to the subject for students in number theory and analysis and a reference source for researchers in the field.

An important role in our presentation is played by the notes at the end of each section. These not only contain the pertinent bibliographical references, but also provide the reader with a brief survey of additional results relating to the material of that section. The exercises we include range from simple applications of theorems to proofs of propositions more general than, or shedding further light on, results in the text. The reader is encouraged to try his hand at many of these problems in order to increase his understanding of the theory.

The following is a brief outline of the contents of the book. The first chapter deals with the classical part of the theory. It assumes that the reader has a good background in real analysis. Important properties of uniformly distributed sequences of real numbers are developed in the early sections and specific examples of uniformly distributed sequences are described throughout. A central position in the theory is taken by the so-called Weyl criterion. As a means for investigating sequences with respect to uniform distribution, it caused in the early years of the development of the theory of uniform distribution a strong interest in exponential sums. One of the equivalents of the definition of a uniformly distributed sequence modulo 1 of real numbers is the functional definition. This form of the notion of uniform distribution reveals the measure-theoretic and topological character of the definition. To avoid repetition, we have, in the first chapter, concentrated on techniques and results that depend on the special structure of the real number system. Results that have analogues in a general setting (such as van der Corput's difference theorem and various metric theorems) will often be found in stronger versions in the more abstract Chapters 3 and 4. Some extensions of the theory are already touched on in Chapter 1, such as the multidimensional case in Section 6 and distribution functions and asymptotic distribution with respect to summation methods in Section 7. A study of normal numbers and their relation to uniform distribution is carried out in Section 8. Uniform distribution modulo 1 of measurable functions appears in Section 9.

In Chapter 2, the results of the preceding chapter are studied and complemented from a quantitative point of view. Here a modest background in number theory would be helpful. Many of the results presented here are current. The important results of K. F. Roth and W. M. Schmidt on irregularities of distribution, as well as the celebrated inequalities of Erdös-Turán and LeVeque, are proved in Section 2. The next section is concerned mainly with the sequences $(n\theta)$, $n = 1, 2 \ldots$, with θ irrational. This leads one back to the number-theoretic origins of the theory. Number-theoretic integration methods which derive from the theory of uniform distribution are treated in Section 5. This part of the book should be of interest to numerical analysts.

In Chapters 3 and 4, we develop in greater detail the theory of uniform distribution in compact Hausdorff spaces and in topological groups, respectively. A background in topology and measure theory is required. Furthermore, for Chapter 4 a knowledge of topological groups is desirable, but all the required results from structure theory and duality theory are stated. Many interesting relations to probability theory, ergodic theory, summability theory, and topological algebra emerge in these portions of the book. In Section 2 of Chapter 4 a detailed treatment of the method of correlation functions is given. Section 4 of Chapter 4 contains an in-depth discussion of the theory of monothetic groups.

The most recent branch of the theory is based on the notion of a uniformly distributed sequence of rational integers. It is not surprising that, subsequently, attention has been paid to the distribution of sequences of a more general type of integers, such as the p-adic ones, and of sequences of elements of the ring of polynomials over a finite field. These developments are described in Chapter 5. The reader should be familiar with the rudiments of p-adic number theory and the theory of finite fields.

Because of space limitations a number of topics could not be covered as fully as we wished or had to be omitted completely in our presentation. Topics in the latter category include metric quantitative results, relations between uniform distribution and harmonic analysis, and the theory of weak convergence of probability measures. However, the notes in the appropriate sections contain a survey of the literature on these aspects.

A graduate course in uniform distribution emphasizing the number-theoretic connections could be based on Chapters 1, 2, and 5 of the book and may be complemented by selections from the remaining chapters. The prerequisites for each chapter have been mentioned above.

It is with great pleasure and gratitude that we acknowledge helpful conversations and/or correspondence with I. D. Berg, D. L. Carlson, J. Cigler, S. Haber, J. H. Halton, W. J. LeVeque, H. G. Meijer, L. A. Rubel, W. M. Schmidt, R. Tijdeman, and S. K. Zaremba. Special thanks are due Professor Walter Philipp, who read drafts of the manuscript and made a number of valuable suggestions. The second author would like to express his indebtedness to Professor Edmund Hlawka, who guided his first steps into the theory of uniform distribution and whose supreme mastery of the subject was a constant source of enlightenment for his students. He also wants to thank the University of Illinois for its hospitality during a crucial period of the writing.

The enormous task of converting our drafts into a typewritten manuscript was mastered by Miss Pat Coombs, whose skill and unfailing patience we learned to admire. We extend our sincere gratitude to the staff of Wiley-Interscience for its care and professionalism in the production of the book.

October 1973

L. Kuipers
H. Niederreiter

CONTENTS

CORRIGENDA

Note. p. a^b, respectively p. a_b, means page a, line b from above, respectively from below.

p. vii^5: read [0,1)
p. xiii^{12}: read "*sequences* 197,"
p. xiv^{13}: read "*distributed sequences*"
p. 9^{16}: replace "$0 \le \theta(k) \le 1$" by "$|\theta(k)| \le 1$"
p. 15_{4-2}: replace these lines by: and therefore an application of integration by parts for Stieltjes integrals yields

$$|J| = \frac{1}{2\pi}\left|\frac{e^{2\pi i f(b)}}{f'(b)} - \frac{e^{2\pi i f(a)}}{f'(a)} - \int_a^b e^{2\pi i f(x)} d\left(\frac{1}{f'(x)}\right)\right|$$

p. 17^2: replace "the second mean value theorem" by "integration by parts"
p. 39_{22}: read "Cassels [3]"
p. 65^{11}: read $\sum_{k=1}^n f'(k)e^{2\pi i h f(k)} = o\left(\sum_{k=1}^n f'(k)\right)$
p. 66_{23}: read "Faĭnleĭb"
p. 75^7: read "is a nonconstant polynomial"
p. 85^9: replace "I. S. Kac [1]" by "Kac [3a]"
p. 97^{13}: read $\lambda(P) - \lambda(P_1) \le 2k\sqrt{k}/r$
p. 99, Exercise 1.2: read $\mathbb{R}^k$
p. 128^{5-6}: read $D_N(\omega) = O(|f(N)|/N + 1/|Nf'(N)|)$.
p. 132, Exercise 3.20: read $D_N(\omega) = O(N^{-1/3}(\log N)^{2/3})$.
p. 132, Exercise 3.21: read $D_N(\omega) = O(N^{-1/3}(\log N)^{1/3}(\log\log N)^{2/3})$.
p. 152, equation (5.23): replace "$f(x_n)$" by "$f(\mathbf{x}_n)$"
p. 159^{17}: read "Bass and Guilloud"
p. $224_{7,3}$: read "Gel'fand-Raikov"
p. $250_{4,2}$: in $\sum_{k\in\mathbb{Z}^+\setminus P}^{\infty}$ delete ∞ (three times)
p. 262, Theorem 3.2, condition 3: replace "$(y_n) * (x_n)$" by "$(x_n) * (y_n)$"
p. 266, above Exercise 3.1: insert headline **Exercises**
p. 281_1: replace "2c" by "2^c"
p. 297_1: delete "continuous"
p. 298^4: delete "continuous"
p. 334^{20}: read "Bass, J., and Guilloud, J."
p. 334, Bauer [1]: replace "to appear" by "**78**, 289–296 (1974)"
p. 341, Erdös-Koksma [2]: read "On the uniform distribution modulo 1 of sequences $(f(n,\theta))$"
p. 347: replace "Kac, I. S.: [1]" by "Kac, M.: [3a]"
p. 349_1: read "inequalities (Russian)"
p. 351^2: read "numbers, *Fibonacci*"
p. 354, Mück-Philipp [1]: replace "to appear" by "*Math. Z.* **142**, 195–202 (1975)"
p. 355, Niederreiter [17]: replace "to appear" by "*Proc. Amer. Math. Soc.* **47**, 305–310 (1975)"
p. 355, Niederreiter [18]: replace "to appear" by "*Math. Comp.* **28**, 1117–1132 (1974)"
p. 355, Niederreiter [19]: replace "to appear" by "*Acta Arith.* **28**, 321–339 (1975)"
p. 356, Philipp [10]: replace "to appear" by "*Acta Arith.* **26**, 241–251 (1975)"
p. 360, Schmidt [15]: replace "to appear" by "*Trans. Amer. Math. Soc.* **198**, 1–22 (1974)"

p. 362, Stoneham [3]: read "*Acta Arith.* **16**, 221–237 (1970)."
p. 362, Stoneham [4]: read "*Acta Arith.* **16**, 239–253 (1970)."
p. 366, Zaremba [9]: read "La méthode des "bons treillis" pour le calcul des intégrales multiples"
p. 366, Zaremba [11]: replace "to appear" by "**78**, 446–460 (1974)"
p. 376, second column: read "Guilloud, J."
p. 377, first column: delete "Kac, I. S., 85, 347"
p. 377, first column: under "Kac, M." insert "85" after "75"

Uniform Distribution of Sequences

1

UNIFORM DISTRIBUTION MOD 1

In this chapter, we develop the classical part of the theory of uniform distribution. We disregard quantitative aspects, which will be considered separately in Chapter 2.

1. DEFINITION

Uniform Distribution Modulo 1

For a real number x, let $[x]$ denote the *integral part* of x, that is, the greatest integer $\leq x$; let $\{x\} = x - [x]$ be the *fractional part* of x, or the residue of x modulo 1. We note that the fractional part of any real number is contained in the *unit interval* $I = [0, 1)$.

Let $\omega = (x_n)$, $n = 1, 2, \ldots$, be a given sequence of real numbers. For a positive integer N and a subset E of I, let the counting function $A(E; N; \omega)$ be defined as the number of terms x_n, $1 \leq n \leq N$, for which $\{x_n\} \in E$. If there is no risk of confusion, we shall often write $A(E; N)$ instead of $A(E; N; \omega)$. Here is our basic definition.

DEFINITION 1.1. The sequence $\omega = (x_n)$, $n = 1, 2, \ldots$, of real numbers is said to be *uniformly distributed modulo 1* (abbreviated u.d. mod 1) if for every pair a, b of real numbers with $0 \leq a < b \leq 1$ we have

$$\lim_{N\to\infty} \frac{A([a, b); N; \omega)}{N} = b - a. \tag{1.1}$$

Thus, in simple terms, the sequence (x_n) is u.d. mod 1 if every half-open subinterval of I eventually gets its "proper share" of fractional parts. In the course of developing the theory of uniform distribution modulo 1 (abbreviated u.d. mod 1), we shall encounter many examples of sequences that enjoy this property (for an easy example, see Exercise 1.13).

Let now $c_{[a,b)}$ be the characteristic function of the interval $[a, b) \subseteq I$. Then (1.1) can be written in the form

$$\lim_{N\to\infty} \frac{1}{N} \sum_{n=1}^{N} c_{[a,b)}(\{x_n\}) = \int_0^1 c_{[a,b)}(x)\, dx. \tag{1.2}$$

This observation, together with an important approximation technique, leads to the following criterion.

THEOREM 1.1. The sequence (x_n), $n = 1, 2, \ldots,$ of real numbers is u.d. mod 1 if and only if for every real-valued continuous function f defined on the closed unit interval $\bar{I} = [0, 1]$ we have

$$\lim_{N\to\infty} \frac{1}{N} \sum_{n=1}^{N} f(\{x_n\}) = \int_0^1 f(x)\, dx. \tag{1.3}$$

PROOF. Let (x_n) be u.d. mod 1, and let $f(x) = \sum_{i=0}^{k-1} d_i c_{[a_i, a_{i+1})}(x)$ be a step function on $\bar{I}$, where $0 = a_0 < a_1 < \cdots < a_k = 1$. Then it follows from (1.2) that for every such f equation (1.3) holds. We assume now that f is a real-valued continuous function defined on $\bar{I}$. Given any $\varepsilon > 0$, there exist, by the definition of the Riemann integral, two step functions, f_1 and f_2 say, such that $f_1(x) \leq f(x) \leq f_2(x)$ for all $x \in \bar{I}$ and $\int_0^1 (f_2(x) - f_1(x))\, dx \leq \varepsilon$. Then we have the following chain of inequalities:

$$\int_0^1 f(x)\, dx - \varepsilon \leq \int_0^1 f_1(x)\, dx = \lim_{N\to\infty} \frac{1}{N} \sum_{n=1}^{N} f_1(\{x_n\})$$
$$\leq \varliminf_{N\to\infty} \frac{1}{N} \sum_{n=1}^{N} f(\{x_n\}) \leq \varlimsup_{N\to\infty} \frac{1}{N} \sum_{n=1}^{N} f(\{x_n\})$$
$$\leq \lim_{N\to\infty} \frac{1}{N} \sum_{n=1}^{N} f_2(\{x_n\}) = \int_0^1 f_2(x)\, dx \leq \int_0^1 f(x)\, dx + \varepsilon,$$

so that in the case of a continuous function f the relation (1.3) holds.

Conversely, let a sequence (x_n) be given, and suppose that (1.3) holds for every real-valued continuous function f on $\bar{I}$. Let $[a, b)$ be an arbitrary subinterval of I. Given any $\varepsilon > 0$, there exist two continuous functions, g_1

and g_2 say, such that $g_1(x) \leq c_{[a,b)}(x) \leq g_2(x)$ for $x \in \bar{I}$ and at the same time $\int_0^1 (g_2(x) - g_1(x))\, dx \leq \varepsilon$. Then we have

$$b - a - \varepsilon \leq \int_0^1 g_2(x)\, dx - \varepsilon \leq \int_0^1 g_1(x)\, dx = \lim_{N\to\infty} \frac{1}{N} \sum_{n=1}^{N} g_1(\{x_n\})$$
$$\leq \varliminf_{N\to\infty} \frac{A([a,b);N)}{N} \leq \varlimsup_{N\to\infty} \frac{A([a,b);N)}{N} \leq \lim_{N\to\infty} \frac{1}{N} \sum_{n=1}^{N} g_2(\{x_n\})$$
$$= \int_0^1 g_2(x)\, dx \leq \int_0^1 g_1(x)\, dx + \varepsilon \leq b - a + \varepsilon.$$

Since ε is arbitrarily small, we have (1.1). ∎

COROLLARY 1.1. The sequence (x_n) is u.d. mod 1 if and only if for every Riemann-integrable function f on $\bar{I}$ equation (1.3) holds.

PROOF. The sufficiency is obvious, and the necessity follows as in the first part of the proof of Theorem 1.1. ∎

COROLLARY 1.2. The sequence (x_n) is u.d. mod 1 if and only if for every complex-valued continuous function f on $\mathbb{R}$ with period 1 we have

$$\lim_{N\to\infty} \frac{1}{N} \sum_{n=1}^{N} f(x_n) = \int_0^1 f(x)\, dx. \tag{1.4}$$

PROOF. By applying Theorem 1.1 to the real and imaginary part of f, one shows first that (1.3) also holds for complex-valued f. But the periodicity condition implies $f(\{x_n\}) = f(x_n)$, and so we arrive at (1.4). As to the sufficiency of (1.4), we need only note that in the second part of the proof of Theorem 1.1 the functions g_1 and g_2 can be chosen in such a way that they satisfy the additional requirements $g_1(0) = g_1(1)$ and $g_2(0) = g_2(1)$, so that (1.4) can be applied to the periodic extensions of g_1 and g_2 to $\mathbb{R}$. ∎

Some simple but useful properties may be deduced easily from Definition 1.1. We mention the following results.

LEMMA 1.1. If the sequence (x_n), $n = 1, 2, \ldots$, is u.d. mod 1, then the sequence $(x_n + \alpha)$, $n = 1, 2, \ldots$, where α is a real constant, is u.d. mod 1.

PROOF. This follows immediately from Definition 1.1. ∎

THEOREM 1.2. If the sequence (x_n), $n = 1, 2, \ldots$, is u.d. mod 1, and if (y_n) is a sequence with the property $\lim_{n\to\infty}(x_n - y_n) = \alpha$, a real constant, then (y_n) is u.d. mod 1.

PROOF. Because of Lemma 1.1 it suffices to consider the case $\alpha = 0$. Set $\varepsilon_n = x_n - y_n$ for $n \geq 1$. Let $0 < a < b < 1$, and choose ε such that

$$0 < \varepsilon < \min\left(a, 1 - b, \frac{b-a}{2}\right).$$

There exists an $N_0 = N_0(\varepsilon)$ such that $-\varepsilon \leq \varepsilon_n \leq \varepsilon$ for $n \geq N_0$. Let $n \geq N_0$, then $a + \varepsilon \leq \{x_n\} < b - \varepsilon$ implies $a \leq \{y_n\} < b$, and on the other hand $a \leq \{y_n\} < b$ implies $a - \varepsilon \leq \{x_n\} < b + \varepsilon$. Hence, if $\sigma = (x_n)$ and $\omega = (y_n)$, then

$$\begin{aligned} b - a - 2\varepsilon &= \lim_{N\to\infty} \frac{A([a+\varepsilon, b-\varepsilon); N; \sigma)}{N} \leq \varliminf_{N\to\infty} \frac{A([a, b); N; \omega)}{N} \\ &\leq \varlimsup_{N\to\infty} \frac{A([a, b); N; \omega)}{N} \leq \lim_{N\to\infty} \frac{A([a-\varepsilon, b+\varepsilon); N; \sigma)}{N} \\ &= b - a + 2\varepsilon. \end{aligned}$$

Since ε can be taken arbitrarily small, the sequence ω satisfies (1.1) for all a and b with $0 < a < b < 1$. To complete the proof, one uses the result in Exercise 1.2. ■

Uniform Distribution Modulo a Subdivision

We mention briefly one of the many variants of the definition of u.d. mod 1. Let Δ: $0 = z_0 < z_1 < z_2 < \cdots$ be a subdivision of the interval $[0, \infty)$ with $\lim_{k\to\infty} z_k = \infty$. For $z_{k-1} \leq x < z_k$ put

$$[x]_\Delta = z_{k-1} \qquad \text{and} \qquad \{x\}_\Delta = \frac{x - z_{k-1}}{z_k - z_{k-1}},$$

so that $0 \leq \{x\}_\Delta < 1$.

DEFINITION 1.2. The sequence (x_n), $n = 1, 2, \ldots$, of nonnegative real numbers is said to be *uniformly distributed modulo* Δ (abbreviated u.d. mod Δ) if the sequence $(\{x_n\}_\Delta)$, $n = 1, 2, \ldots$, is u.d. mod 1.

If Δ is the subdivision Δ_0 for which $z_k = k$, this concept reduces to that of u.d. mod 1. An interesting case occurs if (x_n) is an increasing sequence of nonnegative numbers with $\lim_{n\to\infty} x_n = \infty$. Then we let $A(x, \alpha)$ be the number of $x_n < x$ with $\{x_n\}_\Delta < \alpha$, and we set $A(x) = A(x, 1)$. Clearly, the sequence (x_n) is u.d. mod Δ if and only if for each $\alpha \in (0, 1)$,

$$\lim_{x\to\infty} \frac{A(x, \alpha)}{A(x)} = \alpha. \tag{1.5}$$

The following remarkable result can be shown.

THEOREM 1.3. Let (x_n) be an increasing sequence of nonnegative numbers with $\lim_{n\to\infty} x_n = \infty$. A necessary condition for (x_n) to be u.d. mod Δ is that

$$\lim_{k\to\infty} \frac{A(z_{k+1})}{A(z_k)} = 1. \tag{1.6}$$

PROOF. Suppose (x_n) is u.d. mod Δ. Since

$$A(\tfrac{1}{2}(z_k + z_{k+1}), \tfrac{1}{2}) - A(z_k, \tfrac{1}{2}) = A(\tfrac{1}{2}(z_k + z_{k+1})) - A(z_k),$$

we have

$$\begin{aligned}\frac{A(\frac{1}{2}(z_k + z_{k+1}), \frac{1}{2})}{A(\frac{1}{2}(z_k + z_{k+1}))} &= \frac{A(z_k, \frac{1}{2})}{A(\frac{1}{2}(z_k + z_{k+1}))} + \frac{A(\frac{1}{2}(z_k + z_{k+1})) - A(z_k)}{A(\frac{1}{2}(z_k + z_{k+1}))} \\ &= \frac{A(z_k, \frac{1}{2})}{A(z_k)} \cdot \frac{A(z_k)}{A(\frac{1}{2}(z_k + z_{k+1}))} + 1 - \frac{A(z_k)}{A(\frac{1}{2}(z_k + z_{k+1}))} \\ &= 1 + \frac{A(z_k)}{A(\frac{1}{2}(z_k + z_{k+1}))}\left(\frac{A(z_k, \frac{1}{2})}{A(z_k)} - 1\right). \qquad (1.7)\end{aligned}$$

Now the extreme left member of (1.7) goes to $\frac{1}{2}$ as $k \to \infty$, according to (1.5) and the assumption about the sequence (x_n). Similarly, the second factor of the second term of the extreme right member of (1.7) goes to $-\frac{1}{2}$ as $k \to \infty$. Hence, (1.7) implies

$$\lim_{k\to\infty} \frac{A(z_k)}{A(\frac{1}{2}(z_k + z_{k+1}))} = 1. \qquad (1.8)$$

In a similar way, it can be shown that

$$\lim_{k\to\infty} \frac{A(z_{k+1})}{A(\frac{1}{2}(z_k + z_{k+1}))} = 1. \qquad (1.9)$$

From (1.8) and (1.9) we obtain (1.6). ■

Notes

The formal definition of u.d. mod 1 was given by Weyl [2, 4]. The distribution mod 1 of special sequences was already investigated earlier (see the notes in Section 2). Theorem 1.1 and its corollaries also come from Weyl [2, 4]. The notion of u.d. mod Δ was introduced by LeVeque [4] and was studied further by Cigler [1], Davenport and LeVeque [1], Erdös and Davenport [1], W. M. Schmidt [10], and Burkard [1, 2].

A detailed survey of the results on u.d. mod 1 prior to 1936 can be found in Koksma [4, Kap. 8, 9]. The period from 1936 to 1961 is covered in the survey article of Cigler and Helmberg [1]. An expository treatment of some of the classical results is given in Cassels [9, Chapter 4]. The survey article of Koksma [16] touches upon some of the interesting aspects of the theory.

Let λ be the Lebesgue measure in I. If (x_n) is u.d. mod 1, the limit relation

$$\lim_{N\to\infty} \frac{1}{N} A(E; N) = \lambda(E)$$

will still hold for all Jordan-measurable (or λ-continuity) sets E in I (see Chapter 3, Theorem 1.2) but not for all Lebesgue-measurable sets E in I (see Exercise 1.9). See also Section 1 of Chapter 3 and Rimkevičiūtė [1]. Similarly, the limit relation (1.3) cannot hold for all

Lebesgue-integrable functions f on $\bar{I}$. See Koksma and Salem [1] for strong negative results. The following converse of Theorem 1.1 was shown by de Bruijn and Post [1]: if f is defined on $\bar{I}$ and if $\lim_{N\to\infty}(1/N)\sum_{n=1}^{N} f(\{x_n\})$ exists for every (x_n) u.d. mod 1, then f is Riemann-integrable on $\bar{I}$. Binder [1] gives an alternative proof and a generalization. See also Bass and Couot [1]. Rudin [2] discusses a related question.

Elementary criteria for u.d. mod 1 have been given by O'Neil [1] (see also the notes in Section 3 of Chapter 2) and Niederreiter [15]. Sequences of rationals of the type considered in Exercise 1.13 were studied by Knapowski [1] using elementary methods.

In the sequel, we shall discuss many variants of the definition of u.d. mod 1. One rather special variant was introduced by Erdös and Lorentz [1] in the context of a problem from probabilistic number theory. A sequence (x_n) is called *homogeneously equidistributed mod 1* if $((1/d)x_{nd})$, $n = 1, 2, \ldots$, is u.d. mod 1 for every positive integer d. This notion was also studied by Schnabl [1].

For the result in Exercise 1.14, see Pólya and Szegö [1, II. Abschn., Aufg. 163].

Exercises

1.1. A definition equivalent to Definition 1.1 is the following: A sequence (x_n) of real numbers is u.d. mod 1 if $\lim_{N\to\infty} A([0, c); N)/N = c$ for each real number c with $0 \le c \le 1$.

1.2. If (1.1) holds for all a, b with $0 < a < b < 1$, then it holds for all a, b with $0 \le a < b \le 1$.

1.3. A sequence (x_n) is u.d. mod 1 if and only if (1.1) is satisfied for every interval $[a, b) \subseteq I$ with rational end points.

1.4. A sequence (x_n) is u.d. mod 1 if and only if $\lim_{N\to\infty} A([a, b]; N)/N = b - a$ for all a, b with $0 \le a \le b \le 1$.

1.5. A sequence (x_n) is u.d. mod 1 if and only if $\lim_{N\to\infty} A((a, b); N)/N = b - a$ for all a, b with $0 \le a < b \le 1$.

1.6. If (x_n) is u.d. mod 1, then the sequence $(\{x_n\})$ of fractional parts is everywhere dense in $\bar{I}$.

1.7. If we leave out finitely many terms from a sequence that is u.d. mod 1, the resulting sequence is still u.d. mod 1. What additional condition is needed if "finitely" is replaced by "infinitely"?

1.8. If finitely many terms of a sequence that is u.d. mod 1 are changed in an arbitrary manner, the resulting sequence is still u.d. mod 1. Generalize as in Exercise 1.7.

1.9. Let (x_n) be an arbitrary sequence of real numbers. Construct a Lebesgue-measurable subset E of I with $\lambda(E) = 1$ and $\lim_{N\to\infty} A(E; N)/N = 0$. *Hint:* Consider the complement in I of the set determined by the range of the sequence $(\{x_n\})$.

1.10. Let (x_n) be u.d. mod 1. Then the relation (1.3) is not valid for every Lebesgue-integrable function f on $\bar{I}$.

1.11. Let (x_n) and (y_n) be u.d. mod 1. Then the sequence $x_1, y_1, x_2, y_2, \ldots, x_n, y_n, \ldots$ is u.d. mod 1.

1.12. If r is a rational number, then the sequence (nr), $n = 1, 2, \ldots$, is not u.d. mod 1. Is there a nonempty proper subinterval $[a, b)$ of I for which (1.1) holds?

1.13. Prove that the sequence $0/1, 0/2, 1/2, 0/3, 1/3, 2/3, \ldots, 0/k, 1/k, \ldots, (k-1)/k, \ldots$ is u.d. mod 1.

1.14. Let (x_n) be a sequence in I. For a subinterval $[a, b)$ of I and $N \geq 1$, let $S([a, b); N)$ be the sum of the elements from $x_1, x_2, \ldots, x_N$ that are in $[a, b)$. Then (x_n) is u.d. mod 1 if and only if

$$\lim_{N\to\infty} \frac{S([a, b); N)}{N} = \tfrac{1}{2}(b^2 - a^2)$$

for all subintervals $[a, b)$ of I.

2. THE WEYL CRITERION

The Criterion

The functions f of the form $f(x) = e^{2\pi i h x}$, where h is a nonzero integer, satisfy the conditions of Corollary 1.2. Thus, if (x_n) is u.d. mod 1, the relation (1.4) will be satisfied for those f. It is one of the most important facts of the theory of u.d. mod 1 that these functions already suffice to determine the u.d. mod 1 of a sequence.

THEOREM 2.1: Weyl Criterion. The sequence (x_n), $n = 1, 2, \ldots$, is u.d. mod 1 if and only if

$$\lim_{N\to\infty} \frac{1}{N} \sum_{n=1}^{N} e^{2\pi i h x_n} = 0 \qquad \text{for all integers } h \neq 0. \tag{2.1}$$

PROOF. The necessity follows from Corollary 1.2. Now suppose that (x_n) possesses property (2.1). Then we shall show that (1.4) is valid for every complex-valued continuous function f on $\mathbb{R}$ with period 1. Let ε be an arbitrary positive number. By the Weierstrass approximation theorem, there exists a trigonometric polynomial $\Psi(x)$, that is, a finite linear combination of functions of the type $e^{2\pi i h x}$, $h \in \mathbb{Z}$, with complex coefficients, such that

$$\sup_{0\leq x\leq 1} |f(x) - \Psi(x)| \leq \varepsilon. \tag{2.2}$$

Now we have

$$\begin{aligned}\left|\int_0^1 f(x)\,dx - \frac{1}{N}\sum_{n=1}^{N} f(x_n)\right| \leq{}& \left|\int_0^1 (f(x) - \Psi(x))\,dx\right| \\ &+ \left|\int_0^1 \Psi(x)\,dx - \frac{1}{N}\sum_{n=1}^{N}\Psi(x_n)\right| \\ &+ \left|\frac{1}{N}\sum_{n=1}^{N}(f(x_n) - \Psi(x_n))\right|.\end{aligned}$$

The first and the third terms on the right are both $\leq\varepsilon$ whatever the value of N, because of (2.2). By taking N sufficiently large, the second term on the right is $\leq\varepsilon$ because of (2.1). ■

Applications to Special Sequences

EXAMPLE 2.1. Let θ be an irrational number. Then the sequence $(n\theta)$, $n = 1, 2, \ldots$, is u.d. mod 1. This follows from Theorem 2.1 and the inequality

$$\left|\frac{1}{N}\sum_{n=1}^{N} e^{2\pi i h n\theta}\right| = \frac{|e^{2\pi i h N\theta} - 1|}{N\,|e^{2\pi i h\theta} - 1|} \leq \frac{1}{N\,|\sin \pi h\theta|}$$

for integers $h \neq 0$. ■

EXAMPLE 2.2. Let $\theta = 0.12345678910111213 14 \cdots$ in decimal notation. Then θ is irrational. Therefore, the sequence $(n\theta)$ is u.d. mod 1 by Example 2.1. It follows that the sequence $(\{n\theta\})$ is dense in $\bar{I}$ (see Exercise 1.6). One can even show that the subsequence $(\{10^n\theta\})$ is dense in $\bar{I}$. For let $\alpha = 0\,.\,a_1a_2\cdots a_k$ be a decimal fraction in I. One chooses n such that $\{10^n\theta\}$ begins with the digits $a_1, a_2, \ldots, a_k$ followed by r zeros. Then we have $0 < \{10^n\theta\} - \alpha < 10^{-k-r}$. ■

EXAMPLE 2.3. The sequence $(\{ne\})$, $n = 1, 2, \ldots$, is u.d. mod 1 according to Example 2.1. However, the subsequence $(\{n!e\})$ has 0 as the only limit point. We have

$$e = 1 + \frac{1}{1!} + \frac{1}{2!} + \cdots + \frac{1}{n!} + \frac{e^\alpha}{(n+1)!}, \qquad 0 < \alpha < 1,$$

so that $n!e = k + e^\alpha/(n + 1)$, where k is a positive integer. Hence, for $n \geq 2$ we get $\{n!e\} = e^\alpha/(n + 1) < e/(n + 1)$. ■

EXAMPLE 2.4. The sequence $(\log n)$, $n = 1, 2, \ldots$, is not u.d. mod 1. In order to show this we use the Euler summation formula. If $F(t)$ is a complex-valued function with a continuous derivative on $1 \leq t \leq N$, where $N \geq 1$ is an integer, then

$$\sum_{n=1}^{N} F(n) = \int_1^N F(t)\,dt + \tfrac{1}{2}(F(1) + F(N)) + \int_1^N (\{t\} - \tfrac{1}{2})F'(t)\,dt. \quad (2.3)$$

Let $F(t) = e^{2\pi i \log t}$, and divide both sides of (2.3) by N. Then the first term on the right of (2.3) is equal to

$$\frac{Ne^{2\pi i \log N} - 1}{N(2\pi i + 1)}$$

and this expression does not converge as $N \to \infty$. The second term on the right of (2.3), divided by N, tends to zero as $N \to \infty$, as does the third term on the right of (2.3) divided by N, as follows from

$$\left| \int_1^N (\{t\} - \tfrac{1}{2})F'(t)\, dt \right| \le \pi \int_1^N \frac{dt}{t}.$$

Hence, (2.1) with $x_n = \log n$ and $h = 1$ is not satisfied. ■

EXAMPLE 2.5. In the previous example it was shown that $(\log n)$ is not u.d. mod 1. This general statement, however, does not describe the behavior of $(\log n)$ mod 1 with regard to a particular interval $[a, b) \subseteq I$. In the following we show in an elementary way that for every proper nonempty subinterval $[a, b)$ of I the sequence $(\log n)$ fails to satisfy (1.1). Consider first an interval $[a, b)$ with $0 \le a < b < 1$. Choose a sequence of integers (N_m), $m \ge m_0$, such that $e^{m+b} < N_m < e^{m+1}$ for $m \ge m_0$. Now the number of indices $n = 1, 2, \ldots, N_m$ for which $a \le \{\log n\} < b$, or $k + a \le \log n < k + b$, or $e^{k+a} \le n < e^{k+b}$, $k = 0, 1, \ldots, m$, is given by the expression

$$\sum_{k=0}^{m}(e^{k+b} - e^{k+a} + \theta(k)) = (e^b - e^a)\frac{e^{m+1} - 1}{e - 1} + \sum_{k=0}^{m}\theta(k),$$

with $0 \le \theta(k) \le 1$.

Now it is clear that the fraction obtained by dividing this expression by N_m, which was chosen between e^{m+b} and e^{m+1}, is not convergent as $m \to \infty$ for every choice of the sequence (N_m). If $0 < a < b = 1$, one works in a similar way with a sequence (N_m) satisfying $e^m < N_m < e^{m+a}$ for $m \ge m_0$. With a slight modification of the calculation, one may show the same with respect to the sequence $(c \log n)$, $c \in \mathbb{R}$, $n = 1, 2, \ldots$. ■

EXAMPLE 2.6. Suppose we are given an infinitely large table of the Brigg logarithms $({}^{10}\log n)$, $n = 1, 2, \ldots$, in decimal representation, and consider the sequence of the consecutive digits in the kth column after the decimal point for some fixed $k \ge 1$. Let $c = 10^{k-1}(\log 10)^{-1}$; then

$$c \log n = 10^{k-1}\, {}^{10}\log n.$$

If for some n we have $\{{}^{10}\log n\} = 0.b_1 b_2 \cdots b_k \cdots$, then $\{c \log n\} = 0.b_k b_{k+1} \cdots$. Thus, we observe that the digit b_k at the kth decimal place of ${}^{10}\log n$ is equal to g, $g = 0, 1, \ldots, 9$, if and only if $g/10 \le \{c \log n\} < (g + 1)/10$. Now the sequence $(c \log n)$ has the property described in Example 2.5. This implies that the relative frequency with which the digit g appears in the kth column of the table is not equal to $1/10$. ■

Applications to Power Series

In the following, some interesting results in the theory of power series are deduced from the fact that the sequences $(n\alpha)$ with irrational α are u.d. mod 1 (see Example 2.1).

THEOREM 2.2. Let α be a real number, and let g be a polynomial over $\mathbb{C}$ of positive degree. Define

$$G(x) = \sum_{n=0}^{\infty} g([n\alpha])x^n.$$

Then $G(x)$ is a rational function if and only if α is a rational number.

PROOF. The proof is based on the following auxiliary result: Let α be an irrational number, and let S be a finite set of nonintegral real numbers. Then there are infinitely many positive integers m such that

$$[\{m\alpha\} + \eta] = [\eta] \qquad \text{for all } \eta \in S, \tag{2.4}$$

and also infinitely many positive integers n such that

$$[\{n\alpha\} + \eta] = 1 + [\eta] \qquad \text{for all } \eta \in S. \tag{2.5}$$

Observe that (2.4) is equivalent to

$$0 \leq \{m\alpha\} + \{\eta\} < 1 \qquad \text{for all } \eta \in S,$$

and that (2.5) is equivalent to

$$0 \leq \{n\alpha\} + \{\eta\} - 1 < 1 \qquad \text{for all } \eta \in S,$$

and also that these relations follow easily from the fact that the sequence $(n\alpha)$, $n = 1, 2, \ldots,$ is u.d. mod 1—or in fact from the property that the sequence $(\{n\alpha\})$ is everywhere dense in $\bar{I}$.

Now we turn to the proof of the theorem. Let α be irrational. If $G(x)$ were rational, then polynomials $A(x)$ and $B(x)$, of degrees $a \geq 1$ and b, respectively, would exist such that $G(x) = B(x)/A(x)$. Assume that

$$A(x) = x^a - c_1x^{a-1} - \cdots - c_{a-1}x - c_a.$$

From $A(x)G(x) = B(x)$ it follows, by equating corresponding coefficients, that

$$g([n\alpha]) = \sum_{r=1}^{a} g([n\alpha + r\alpha])c_r \qquad \text{for } n \geq \max(0, b - a + 1). \tag{2.6}$$

Since g is a polynomial of degree $p \geq 1$, we have

$$\lim_{n\to\infty} \frac{g([n\alpha + r\alpha])}{g([n\alpha])} = \lim_{n\to\infty} \frac{[n\alpha + r\alpha]^p}{[n\alpha]^p} = 1,$$

so that (2.6) implies

$$c_1 + c_2 + \cdots + c_a = 1. \tag{2.7}$$

Moreover, (2.6) and (2.7) imply

$$\sum_{r=1}^{a} (g([n\alpha + r\alpha]) - g([n\alpha]))c_r = 0. \tag{2.8}$$

We have $[n\alpha + r\alpha] = [\{n\alpha\} + r\alpha] + [n\alpha]$, and so

$$g([n\alpha + r\alpha]) - g([n\alpha]) = \sum_{k=1}^{p} \frac{g^{(k)}([n\alpha])}{k!} [\{n\alpha\} + r\alpha]^k.$$

Therefore, after multiplying both sides of this last equality by c_r and summing from $r = 1$ to $r = a$, for large n one obtains using (2.8),

$$\sum_{r=1}^{a} [\{n\alpha\} + r\alpha]c_r + \sum_{r=1}^{a} \sum_{k=2}^{p} \frac{g^{(k)}([n\alpha])}{k!\, g'([n\alpha])} [\{n\alpha\} + r\alpha]^k c_r = 0. \tag{2.9}$$

For $p = 1$ the last sum on the left of (2.9) is empty, and if $p \geq 2$, we have

$$\lim_{n\to\infty} \frac{g^{(k)}([n\alpha])}{g'([n\alpha])} [\{n\alpha\} + r\alpha]^k = 0 \quad \text{for } 2 \leq k \leq p \text{ and } 1 \leq r \leq a.$$

So we arrive at

$$\lim_{n\to\infty} \sum_{r=1}^{a} [\{n\alpha\} + r\alpha]c_r = 0. \tag{2.10}$$

The numbers $r\alpha$ in (2.10) are not integers. Thus, according to the auxiliary result and (2.10) we can find integers m and n such that the expressions

$$\sum_{r=1}^{a} [\{m\alpha\} + r\alpha]c_r = \sum_{r=1}^{a} [r\alpha]c_r$$

and

$$\sum_{r=1}^{a} [\{n\alpha\} + r\alpha]c_r = \sum_{r=1}^{a} (1 + [r\alpha])c_r$$

differ from 0 as little as we please, which contradicts (2.7). In this way, it is shown that if α is irrational, $G(x)$ is not a rational function.

Now assume that α is rational. Set $\alpha = c/d$, where c and d are integers with $d > 0$. Applying the division algorithm, we have $n = md + r$ with $0 \leq r \leq d - 1$, and so

$$n\alpha = \frac{nc}{d} = \frac{(md + r)c}{d} = mc + \frac{rc}{d},$$

so that $[n\alpha] = mc + \left[\frac{rc}{d}\right]$. Then

$$G(x) = \sum_{n=0}^{\infty} g([n\alpha])x^n = \sum_{r=0}^{d-1}\sum_{m=0}^{\infty} g\left(mc + \left[\frac{rc}{d}\right]\right)x^{md+r}$$

$$= \sum_{r=0}^{d-1}\sum_{m=0}^{\infty}\sum_{k=0}^{p}\frac{g^{(k)}([rc/d])}{k!}(mc)^k x^{md+r}$$

$$= \sum_{r=0}^{d-1}\sum_{k=0}^{p}\frac{g^{(k)}([rc/d])}{k!}c^k x^r \sum_{m=0}^{\infty} m^k x^{md}.$$

Now

$$\sum_{m=0}^{\infty} m^k x^m = \left(x\frac{d}{dx}\right)^k (1-x)^{-1}$$

is rational, and so it is shown that $G(x)$ is rational. ■

THEOREM 2.3. Let α be a positive number and let

$$F(x) = \sum_{t=1}^{\infty} x^{[t\alpha]}.$$

Then $F(x)$ is a rational function if and only if α is rational.

PROOF. Suppose that α is irrational. Let $X(n)$ be the number of solutions of $n = [t\alpha]$ in positive integers t. Then

$$F(x) = \sum_{n=0}^{\infty} X(n)x^n.$$

Now $X(n)$ is the number of integers t satisfying $n < t\alpha < n + 1$; hence,

$$X(n) = \left[\frac{n+1}{\alpha}\right] - \left[\frac{n}{\alpha}\right],$$

and therefore, $F(x) = (1 - x)\sum_{n=1}^{\infty} [n/\alpha]x^{n-1}$. According to Theorem 2.2, $F(x)$ is not a rational function.

Suppose that α is rational. Write $\alpha = c/d$ with positive integers c and d. Then, using $t = md + r$ with $0 \le r \le d - 1$, we have

$$1 + F(x) = \sum_{t=0}^{\infty} x^{[t\alpha]} = \sum_{r=0}^{d-1}\sum_{m=0}^{\infty} x^{mc+[rc/d]} = (1 - x^c)^{-1}\sum_{r=0}^{d-1} x^{[rc/d]},$$

so that $F(x)$ is rational. ■

THEOREM 2.4. Let f be a Riemann-integrable function on $[0, 1]$ for which $\int_0^1 f(x)e^{2\pi imx}\,dx \ne 0$ for all but finitely many positive integers m. Let α be irrational. Then the power series

$$G(z) = \sum_{n=1}^{\infty} f(\{n\alpha\})z^n$$

cannot be analytically extended across the unit circle $\{z \in \mathbf{C} : |z| = 1\}$.

PROOF. We recall the following theorem of Frobenius (see Knopp [1, pp. 507–508] and G. M. Petersen [2, pp. 48–49]). Let (c_n), $n = 1, 2, \ldots,$ be a sequence of complex numbers with the property that

$$\lim_{N\to\infty} (1/N)(c_1 + \cdots + c_N)$$

exists and equals c. Set $\Phi(r) = \sum_{n=1}^{\infty} c_n r^n$; then we have $\lim_{r\to 1-0} (1 - r)\Phi(r) = c$.

Now we turn to the proof of our theorem. We have, according to the definition of $G(z)$,

$$G(re^{2\pi i m\alpha}) = \sum_{n=1}^{\infty} f(\{n\alpha\})e^{2\pi i m n\alpha} r^n \qquad \text{for} \qquad -1 < r < 1 \quad \text{and} \quad m \in \mathbb{Z}.$$

Since $(n\alpha)$ is u.d. mod 1, one obtains from Corollary 1.1 the relation

$$\lim_{N\to\infty} \frac{1}{N} \sum_{n=1}^{N} f(\{n\alpha\})e^{2\pi i m n\alpha} = \int_0^1 f(x)e^{2\pi i m x}\, dx = d_m, \quad \text{say.}$$

Using the theorem of Frobenius, we get

$$\lim_{r\to 1-0} (1 - r)G(re^{2\pi i m\alpha}) = d_m. \tag{2.11}$$

For $m \geq m_0$ we have $d_m \neq 0$. So (2.11) implies that if $z \to e^{2\pi i m\alpha}$ along the radius, the function value $G(z)$ tends to ∞, and therefore, G has singularities on an everywhere dense subset of $\{z \in \mathbb{C}: |z| = 1\}$. ■

Fejér's Theorem

As another consequence of the Weyl criterion, we obtain a theorem that will provide many more examples of sequences that are u.d. mod 1.

THEOREM 2.5. Let $(f(n))$, $n = 1, 2, \ldots,$ be a sequence of real numbers such that $\Delta f(n) = f(n + 1) - f(n)$ is monotone as n increases. Let, furthermore,

$$\lim_{n\to\infty} \Delta f(n) = 0 \qquad \text{and} \qquad \lim_{n\to\infty} n\,|\Delta f(n)| = \infty. \tag{2.12}$$

Then the sequence $(f(n))$ is u.d. mod 1.

PROOF. For every pair of real numbers u and v we have

$$\begin{aligned} |e^{2\pi i u} - e^{2\pi i v} - 2\pi i(u - v)e^{2\pi i v}| &= |e^{2\pi i(u-v)} - 1 - 2\pi i(u - v)| \\ &= 4\pi^2 \left| \int_0^{u-v} (u - v - w)e^{2\pi i w}\, dw \right| \\ &\leq 4\pi^2 \left| \int_0^{u-v} (u - v - w)\, dw \right| \\ &= 2\pi^2(u - v)^2. \end{aligned} \tag{2.13}$$

Now set $u = hf(n+1)$ and $v = hf(n)$, where h is a nonzero integer. Then, according to (2.13),

$$\left|\frac{e^{2\pi i h f(n+1)}}{\Delta f(n)} - \frac{e^{2\pi i h f(n)}}{\Delta f(n)} - 2\pi i h e^{2\pi i h f(n)}\right| \leq 2\pi^2 h^2 |\Delta f(n)| \qquad \text{for } n \geqq 1;$$

hence,

$$\left|\frac{e^{2\pi i h f(n+1)}}{\Delta f(n+1)} - \frac{e^{2\pi i h f(n)}}{\Delta f(n)} - 2\pi i h e^{2\pi i h f(n)}\right|$$
$$\leq \left|\frac{1}{\Delta f(n)} - \frac{1}{\Delta f(n+1)}\right| + 2\pi^2 h^2 |\Delta f(n)| \qquad \text{for } n \geqq 1. \quad (2.14)$$

Then,

$$\left|2\pi i h \sum_{n=1}^{N-1} e^{2\pi i h f(n)}\right|$$
$$= \left|\sum_{n=1}^{N-1}\left(2\pi i h e^{2\pi i h f(n)} - \frac{e^{2\pi i h f(n+1)}}{\Delta f(n+1)} + \frac{e^{2\pi i h f(n)}}{\Delta f(n)}\right) + \frac{e^{2\pi i h f(N)}}{\Delta f(N)} - \frac{e^{2\pi i h f(1)}}{\Delta f(1)}\right|$$
$$\leq \sum_{n=1}^{N-1}\left|2\pi i h e^{2\pi i h f(n)} - \frac{e^{2\pi i h f(n+1)}}{\Delta f(n+1)} + \frac{e^{2\pi i h f(n)}}{\Delta f(n)}\right| + \frac{1}{|\Delta f(N)|} + \frac{1}{|\Delta f(1)|}$$
$$\leq \sum_{n=1}^{N-1}\left|\frac{1}{\Delta f(n)} - \frac{1}{\Delta f(n+1)}\right| + 2\pi^2 h^2 \sum_{n=1}^{N-1}|\Delta f(n)| + \frac{1}{|\Delta f(N)|} + \frac{1}{|\Delta f(1)|},$$

where we used (2.14) in the last step. Because of the monotonicity of $\Delta f(n)$, we get

$$\left|\frac{1}{N}\sum_{n=1}^{N-1} e^{2\pi i h f(n)}\right| \leq \frac{1}{\pi |h|}\left(\frac{1}{N|\Delta f(1)|} + \frac{1}{N|\Delta f(N)|}\right) + \frac{\pi |h|}{N}\sum_{n=1}^{N-1}|\Delta f(n)|,$$

and therefore, in view of (2.12),

$$\lim_{N\to\infty}\frac{1}{N}\sum_{n=1}^{N-1} e^{2\pi i h f(n)} = 0. \quad \blacksquare$$

COROLLARY 2.1: Fejér's Theorem. Let $f(x)$ be a function defined for $x \geq 1$ that is differentiable for $x \geq x_0$. If $f'(x)$ tends monotonically to 0 as $x \to \infty$ and if $\lim_{x\to\infty} x|f'(x)| = \infty$, then the sequence $(f(n))$, $n = 1, 2, \ldots,$ is u.d. mod 1.

PROOF. The mean value theorem shows that $\Delta f(n)$ satisfies the conditions of Theorem 2.5, at least for sufficiently large n. The finitely many exceptional terms do not influence the u.d. mod 1 of the sequence. $\blacksquare$

EXAMPLE 2.7. Fejér's theorem immediately implies the u.d. mod 1 of the following types of sequences: (i) $(\alpha n^\sigma \log^\tau n)$, $n = 2, 3, \ldots,$ with $\alpha \neq 0$,

$0 < \sigma < 1$, and arbitrary τ; (ii) $(\alpha \log^\tau n)$, $n = 1, 2, \ldots$, with $\alpha \neq 0$ and $\tau > 1$; (iii) $(\alpha n \log^\tau n)$, $n = 2, 3, \ldots$, with $\alpha \neq 0$ and $\tau < 0$. ■

The following simple result shows that the second condition in (2.12) cannot be relaxed too much.

THEOREM 2.6. If a sequence $(f(n))$, $n = 1, 2, \ldots$, is u.d. mod 1, then necessarily $\overline{\lim}_{n\to\infty} n\,|\Delta f(n)| = \infty$.

PROOF. Suppose that $(f(n))$ is u.d. mod 1 and that $\overline{\lim}_{n\to\infty} n\,|\Delta f(n)| < \infty$. For any two real numbers u and v, we have

$$|e^{2\pi i u} - e^{2\pi i v}| \leq 2\pi\,|u - v|, \tag{2.15}$$

and so,

$$|e^{2\pi i f(n+1)} - e^{2\pi i f(n)}| \leq 2\pi\,|\Delta f(n)| = O\left(\frac{1}{n}\right).$$

On the other hand,

$$\lim_{N\to\infty} \frac{1}{N} \sum_{n=1}^{N} e^{2\pi i f(n)} = 0.$$

By a well-known Tauberian theorem (Hardy [2, p. 121], G. M. Petersen [2, p. 51]) it follows that $\lim_{n\to\infty} e^{2\pi i f(n)} = 0$, an obvious absurdity. ■

An Estimate for Exponential Sums

Although Corollary 2.1 is a very powerful result, there are various interesting sequences to which it does not apply. For instance, the question whether $(n \log n)$, $n = 1, 2, \ldots$, is u.d. mod 1 cannot be settled by appealing to Fejér's theorem. In such cases, the following estimate may prove to be useful. We first need some technical lemmas. The values of the absolute constants will not be important in these estimates.

LEMMA 2.1. Suppose the real-valued function f has a monotone derivative f' on $[a, b]$ with $f'(x) \geq \lambda > 0$ or $f'(x) \leq -\lambda < 0$ for $x \in [a, b]$. Then, if $J = \int_a^b e^{2\pi i f(x)}\,dx$, we have $|J| < 1/\lambda$.

PROOF. We have

$$J = \frac{1}{2\pi i} \int_a^b \frac{de^{2\pi i f(x)}}{f'(x)},$$

and therefore, an application of the second mean value theorem yields, with some $x_0 \in [a, b]$,

$$\begin{aligned} |J| &= \left|\frac{1}{2\pi i}\left(\frac{1}{f'(a)} \int_a^{x_0} de^{2\pi i f(x)} + \frac{1}{f'(b)} \int_{x_0}^{b} de^{2\pi i f(x)}\right)\right| \\ &\leq \frac{1}{2\pi}\left(\frac{2}{|f'(a)|} + \frac{2}{|f'(b)|}\right) \leq \frac{2}{\pi\lambda} < \frac{1}{\lambda}. \end{aligned}$$ ■

LEMMA 2.2. Let f be twice differentiable on $[a, b]$ with $f''(x) \geq \rho > 0$ or $f''(x) \leq -\rho < 0$ for $x \in [a, b]$. Then the integral J from Lemma 2.1 satisfies $|J| < 4/\sqrt{\rho}$.

PROOF. We may suppose that $f''(x) \geq \rho$ for $x \in [a, b]$; otherwise, we replace f by $-f$. We note that f' is increasing. Suppose for the moment that f' is of constant sign in $[a, b]$, say $f' \geq 0$. If $a < c < b$, then $f'(x) \geq (c - a)\rho$ for $c \leq x \leq b$ by the mean value theorem. Therefore, by Lemma 2.1,

$$|J| \leq \left|\int_a^c e^{2\pi i f(x)}\,dx\right| + \left|\int_c^b e^{2\pi i f(x)}\,dx\right| < (c - a) + \frac{1}{(c - a)\rho},$$

and choosing c so as to make the last sum a minimum, we obtain $|J| < 2/\sqrt{\rho}$. In the general case, $[a, b]$ is the union of two intervals in each of which f' is of constant sign, and the desired inequality follows by adding the inequalities for these two intervals. ■

LEMMA 2.3. Let f' be monotone on $[a, b]$ with $|f'(x)| \leq \frac{1}{2}$ for $x \in [a, b]$. Then, if $J_1 = \int_a^b (\{x\} - \frac{1}{2})\,de^{2\pi i f(x)}$, we have

$$|J_1| \leq 2. \tag{2.16}$$

PROOF. We start from the Fourier-series expansion

$$\{x\} - \tfrac{1}{2} = -\sum_{h=1}^{\infty} (\sin 2\pi hx)/\pi h,$$

valid for all $x \notin \mathbb{Z}$. For $m \geq 1$, let $\chi_m(x) = -\sum_{h=1}^{m} (\sin 2\pi hx)/\pi h$, $x \in \mathbb{R}$, be the mth partial sum. The functions χ_m, $m = 1, 2, \ldots$, are uniformly bounded as is seen easily after summation by parts. Therefore,

$$J_1 = \lim_{m\to\infty} \int_a^b \chi_m(x)\,de^{2\pi i f(x)}. \tag{2.17}$$

Now for $m \geq 1$ we have

$$\begin{aligned}
&\int_a^b \chi_m(x)\,de^{2\pi i f(x)} \\
&\quad = \sum_{h=1}^{m} \frac{1}{h} \int_a^b (-2i \sin 2\pi hx) e^{2\pi i f(x)} f'(x)\,dx \\
&\quad = \sum_{h=1}^{m} \frac{1}{h} \int_a^b (e^{-2\pi i hx} - e^{2\pi i hx}) e^{2\pi i f(x)} f'(x)\,dx \\
&\quad = \frac{1}{2\pi i} \sum_{h=1}^{m} \frac{1}{h} \left(\int_a^b \frac{f'(x)}{f'(x) - h}\,de^{2\pi i(f(x) - hx)} - \int_a^b \frac{f'(x)}{f'(x) + h}\,de^{2\pi i(f(x) + hx)} \right).
\end{aligned}$$

Since the functions $f'/(f' \pm h)$ are monotone and $|f'| \leq \frac{1}{2}$, an application of the second mean value theorem shows that

$$\left|\int_a^b \frac{f'(x)}{f'(x) \pm h}\, de^{2\pi i(f(x) \pm hx)}\right| \leq \frac{2}{h - \frac{1}{2}},$$

and so

$$\left|\int_a^b \chi_m(x)\, de^{2\pi i f(x)}\right| \leq \frac{2}{\pi} \sum_{h=1}^{m} \frac{1}{h(h - \frac{1}{2})} < 2. \tag{2.18}$$

Equations (2.17) and (2.18) imply (2.16). ■

THEOREM 2.7. Let a and b be integers with $a < b$, and let f be twice differentiable on $[a, b]$ with $f''(x) \geq \rho > 0$ or $f''(x) \leq -\rho < 0$ for $x \in [a, b]$. Then,

$$\left|\sum_{n=a}^{b} e^{2\pi i f(n)}\right| \leq (|f'(b) - f'(a)| + 2)\left(\frac{4}{\sqrt{\rho}} + 3\right). \tag{2.19}$$

PROOF. We write

$$\sum_{n=a}^{b} e^{2\pi i f(n)} = \sum_{p=-\infty}^{\infty} S_p \tag{2.20}$$

with

$$S_p = \sum_{\substack{a \leq n \leq b \\ p-1/2 \leq f'(n) < p+1/2}} e^{2\pi i f(n)}. \tag{2.21}$$

The sum over p in (2.20) is in reality just a finite sum. Let p be an integer for which the sum in (2.21) is nonvoid. Since f' is monotone, this sum is over consecutive values of n, say from $n = a_p$ to $n = b_p$. With $F_p(x) = f(x) - px$, we get

$$\begin{aligned} S_p &= \sum_{n=a_p}^{b_p} e^{2\pi i f(n)} = \sum_{n=a_p}^{b_p} e^{2\pi i F_p(n)} \\ &= \int_{a_p}^{b_p} e^{2\pi i F_p(x)}\, dx + \tfrac{1}{2}(e^{2\pi i F_p(a_p)} + e^{2\pi i F_p(b_p)}) \\ &\quad + \int_{a_p}^{b_p} (\{x\} - \tfrac{1}{2})\, de^{2\pi i F_p(x)} \end{aligned} \tag{2.22}$$

by the Euler summation formula; compare with (2.3). Now the first integral in (2.22) is in absolute value less than $4/\sqrt{\rho}$ by Lemma 2.2. The second integral in (2.22) is in absolute value at most 2 because of $|F_p'(x)| \leq \frac{1}{2}$ for $x \in [a_p, b_p]$ and Lemma 2.3. Therefore, $|S_p| < (4/\sqrt{\rho}) + 3$. Since there are at most $|f'(b) - f'(a)| + 2$ values of p for which S_p is a nonvoid sum, we arrive at (2.19). ■

EXAMPLE 2.8. From Theorem 2.7 we infer that

$$\left|\frac{1}{N}\sum_{n=1}^{N} e^{2\pi i h\, n \log n}\right| \leq \frac{1}{N}(|h| \log N + 2)\left(4\sqrt{\frac{N}{|h|}} + 3\right)$$

for all nonzero integers h, and so, the sequence $(n \log n)$, $n = 1, 2, \ldots$, is u.d. mod 1 by the Weyl criterion. More generally, the method yields that $(\alpha n \log^{\tau} n)$, $n = 1, 2, \ldots$, $\alpha \neq 0$, $\tau > 0$, is u.d. mod 1. In the same way, the sequence $(n \log \log n)$, $n = 2, 3, \ldots$, can be shown to be u.d. mod 1. Compare also with Exercises 2.23–2.26. ■

Uniform Distribution of Double Sequences

DEFINITION 2.1. A double sequence (s_{jk}), $j = 1, 2, \ldots$, $k = 1, 2, \ldots$, of real numbers is said to be u.d. mod 1 if for any a and b such that $0 \leq a < b \leq 1$,

$$\lim_{M,N\to\infty} \frac{A([a, b); M, N)}{MN} = b - a, \tag{2.23}$$

where $A([a, b); M, N)$ is the number of s_{jk}, $1 \leq j \leq M$, $1 \leq k \leq N$, for which $a \leq \{s_{jk}\} < b$.

THEOREM 2.8. The double sequence (s_{jk}) is u.d. mod 1 if and only if for every Riemann-integrable function f on $\bar{I}$ we have

$$\lim_{M,N\to\infty} \frac{1}{MN}\sum_{j=1}^{M}\sum_{k=1}^{N} f(\{s_{jk}\}) = \int_0^1 f(x)\, dx.$$

THEOREM 2.9. The double sequence (s_{jk}) is u.d. mod 1 if and only if

$$\lim_{M,N\to\infty} \frac{1}{MN}\sum_{j=1}^{M}\sum_{k=1}^{N} e^{2\pi i h s_{jk}} = 0 \qquad \text{for all integers } h \neq 0.$$

The proofs of these theorems can be given along the same lines as those of Corollary 1.1 and Theorem 2.1, respectively.

EXAMPLE 2.9. Let θ be irrational and α an arbitrary real number. Then $(j\theta + k\alpha)$, $j = 1, 2, \ldots$, $k = 1, 2, \ldots$, is u.d. mod 1, as follows easily from Theorem 2.9. ■

Any u.d. double sequence (s_{jk}) can be arranged into a u.d. single sequence (s_n), say. One just has to choose (s_n) in such a way that its first M^2 terms consist exactly of all numbers s_{jk} with $1 \leq j, k \leq M$. For instance, the arrangement $s_{11}, s_{21}, s_{22}, s_{12}, s_{31}, s_{32}, s_{33}, s_{23}, s_{13}, s_{41}, \ldots, s_{14}, \ldots$ satisfies this property. To prove our assertion, we note that for a given positive integer m there exists a unique integer M with $(M - 1)^2 \leq m < M^2$, which implies

$M^2 - m \leq 2M - 1$. Set $A(m) = A([a, b); m; (s_n))$ and $A(M, M) = A([a, b); M, M)$. Let ε be an arbitrary positive number. Then there exists an integer $N_0 = N_0(\varepsilon)$ such that for $M > N_0$ and $M^2A(m) \geq mA(M, M)$,

$$\left|\frac{A(m)}{m} - \frac{A(M, M)}{M^2}\right| = \frac{M^2A(m) - mA(M, M)}{M^2m}$$
$$\leq \frac{(M^2 - m)A(m)}{M^2m} \leq \frac{2M - 1}{M^2} < \varepsilon,$$

and for $M > N_0$ and $M^2A(m) < mA(M, M)$,

$$\left|\frac{A(m)}{m} - \frac{A(M, M)}{M^2}\right| = \frac{mA(M, M) - M^2A(m)}{M^2m}$$
$$\leq \frac{m(A(M, M) - A(m))}{M^2m} \leq \frac{2M - 1}{M^2} < \varepsilon.$$

From these inequalities it follows that $\lim_{m\to\infty} A(m)/m = b - a$, since we have $\lim_{M\to\infty} A(M, M)/M^2 = b - a$ by (2.23).

A weaker concept of u.d. for double sequences, for which the above argument holds as well, is the following one: We call the double sequence (s_{jk}) *u.d. mod 1 in the squares* $1 \leq j,\ k \leq N$ *as* $N \to \infty$ if $\lim_{N\to\infty} A([a, b); N, N)/N^2 = b - a$ for $0 \leq a < b \leq 1$. As is easily seen, this is equivalent to

$$\lim_{N\to\infty} \frac{1}{N^2} \sum_{j=1}^{N} \sum_{k=1}^{N} e^{2\pi i h s_{jk}} = 0 \qquad \text{for all integers } h \neq 0. \tag{2.24}$$

It is interesting that an analogue of Fejér's theorem can be shown for this type of u.d. We first need the following simple generalization of the Euler summation formula.

LEMMA 2.4. For integers $M \geq 2$ and $N \geq 2$, let $g = g(x, y)$ be defined for $1 \leq x \leq M$, $1 \leq y \leq N$, with g_{xy} continuous in this region. Then,

$$\sum_{j=2}^{M} \sum_{k=2}^{N} g(j, k)$$
$$= \int_1^N \int_1^M g(x, y)\, dx\, dy + \int_1^N \int_1^M \{y\} g_y(x, y)\, dx\, dy$$
$$+ \int_1^N \int_1^M \{x\} g_x(x, y)\, dx\, dy + \int_1^N \int_1^M \{x\}\{y\} g_{xy}(x, y)\, dx\, dy. \tag{2.25}$$

PROOF. We may write (2.3) in the form

$$\sum_{k=2}^{N} F(k) = \int_1^N F(t)\, dt + \int_1^N \{t\} F'(t)\, dt. \tag{2.26}$$

For fixed j, $2 \leq j \leq M$, apply (2.26) with $F(y) = g(j, y)$, and we get

$$\sum_{k=2}^{N} g(j, k) = \int_1^N g(j, y)\, dy + \int_1^N \{y\} g_y(j, y)\, dy. \tag{2.27}$$

Summing up the equations (2.27) from $j = 2$ to M, and applying again (2.26), we arrive at (2.25). ■

THEOREM 2.10. Let the real-valued function $f = f(x, y)$ be defined for $x \geq 1$, $y \geq 1$, with f_{xy} continuous in this region. Let f be increasing in x and y, and let f_x be nonincreasing in x and y. Assume that $\lim_{x\to\infty} f_x(x, 1) = 0$, $\lim_{y\to\infty} f_y(1, y) = 0$, and $\lim_{x\to\infty} f(x, x)/x^2 = 0$. Furthermore, suppose that the integrals $\int_1^N \int_1^N f_x(x, y) f_y(x, y)\, dx\, dy$ and $\int_1^N dy/f_x(N, y)$ are both $o(N^2)$. Then the double sequence $(f(j, k))$, $j = 1, 2, \ldots$, $k = 1, 2, \ldots$, is u.d. mod 1 in the squares $1 \leq j, k \leq N$ as $N \to \infty$.

PROOF. We verify (2.24) for all $h \neq 0$. To this end, we use (2.25) with $g(x, y) = e^{2\pi i h f(x,y)}$. This yields

$$\begin{aligned}
\sum_{j=1}^{N} \sum_{k=1}^{N} e^{2\pi i h f(j,k)}
&= \int_1^N \int_1^N e^{2\pi i h f(x,y)}\, dx\, dy \\
&\quad + 2\pi i h \int_1^N \int_1^N \{y\} f_y(x, y) e^{2\pi i h f(x,y)}\, dx\, dy \\
&\quad + 2\pi i h \int_1^N \int_1^N \{x\} f_x(x, y) e^{2\pi i h f(x,y)}\, dx\, dy \\
&\quad + 2\pi i h \int_1^N \int_1^N \{x\}\{y\} f_{xy}(x, y) e^{2\pi i h f(x,y)}\, dx\, dy \\
&\quad + (2\pi i h)^2 \int_1^N \int_1^N \{x\}\{y\} f_x(x, y) f_y(x, y) e^{2\pi i h f(x,y)}\, dx\, dy + o(N^2).
\end{aligned}$$

The first double integral is $o(N^2)$ because of Lemma 2.1, the condition on f_x, and

$$\int_1^N \frac{dy}{f_x(N, y)} = o(N^2).$$

The last double integral is $o(N^2)$ because of

$$\int_1^N \int_1^N f_x(x, y) f_y(x, y)\, dx\, dy = o(N^2).$$

As to the third double integral, note that

$$\left|\int_1^N \int_1^N \{x\} f_x(x, y) e^{2\pi i h f(x,y)}\, dx\, dy\right|$$

$$\leq \int_1^N \int_1^N f_x(x, y)\, dx\, dy \leq (N-1)\int_1^N f_x(x, 1)\, dx$$

$$= (N-1)(f(N, 1) - f(1, 1)) = o(N^2).$$

The second double integral is treated in a similar fashion (note that our conditions imply that f_y is nonincreasing in x). Finally, for the fourth double integral we have

$$\left|\int_1^N \int_1^N \{x\}\{y\} f_{xy}(x, y) e^{2\pi i h f(x,y)}\, dx\, dy\right|$$

$$\leq \int_1^N \int_1^N (-f_{xy}(x, y))\, dx\, dy$$

$$= -f(N, N) + f(1, N) + f(N, 1) - f(1, 1) = o(N^2). \quad \blacksquare$$

EXAMPLE 2.10. Let α and β be positive numbers. Let $f(t)$ be increasing for $t > 0$, $f'(t) \to 0$ (monotonically), $tf'(t) \to \infty$ as $t \to \infty$, and $f''(t)$ continuous for $t > 0$. Then the double sequence $(f(\alpha j + \beta k))$, $j = 1, 2, \ldots$, $k = 1, 2, \ldots$, is u.d. mod 1 in the squares $1 \leq j, k \leq N$ as $N \to \infty$, as follows from Theorem 2.10. See also Exercises 2.28–2.30. $\blacksquare$

Notes

The fundamental Theorem 2.1 was first shown by Weyl [2, 4]. Many proofs of this result can be found in the literature, most of them proceeding along the same lines as Weyl's original proof, which we reproduced here. See Artémiadis and Kuipers [1], Cassels [9, Chapter 4], Chandrasekharan [1, Chapter 8], Niven [1, Chapter 6], and Exercises 2.6 and 2.7 of Chapter 2. We shall encounter various generalizations of the Weyl criterion in Sections 6 and 7 of this chapter and in Chapters 3 and 4. See also Exercises 2.1, 2.2, and 2.3, as well as the papers of Blum and Mizel [2], Brown and Duncan [1], Holewijn [3], Kuipers and Stam [1], Loynes [1], and Robbins [1].

Because of the Weyl criterion, there are intimate connections between the theory of u.d. mod 1 and the estimation of exponential sums. Concerning the latter, we refer to the monographs and articles of Hua [1, 2], Koksma [4, Kap. 9], Teghem [2], Vinogradov [5], and Walfisz [1].

The fact that $(n\theta)$, $n = 1, 2, \ldots$, is u.d. mod 1 for irrational θ was established independently by Bohl [1], Sierpiński [1, 2], and Weyl [1] in 1909–1910. Their proofs were elementary. The proof in Example 2.1 comes from Weyl [2, 4]. For other elementary proofs, see Callahan [1], Hardy and Wright [1, Chapter 23], Jacobs [1], Miklavc [1], Niven [1, Chapter 6; 3, Chapter 3], O'Neil [1], and Weyl [4]. The problem of the distribution of $(n\theta)$ mod 1 had its origin in the theory of secular perturbations in astronomy. Weyl [3] discusses this connection in detail. For a survey of the early literature, see Koksma [4, Kap.

8]. We remark that Example 2.1 improves Kronecker's theorem, which says that $(\{n\theta\})$ is everywhere dense in $[0, 1]$.

The distribution of $(n\theta)$ mod 1 has been studied extensively, especially in connection with certain conjectures of Steinhaus. Given a real θ, a positive integer N, and $0 < b \leq 1$, it turns out that the "gaps" between the successive values of n, $0 \leq n \leq N$, for which $\{n\theta\} < b$ can have at most three lengths, and if there are three, one is the sum of the other two (Slater [1], Florek [1]). Also, if N and θ are as above, and if one arranges the points $0, \{\theta\}, \ldots, \{N\theta\}$ in ascending order, the "steps" between consecutive points can have at most three lengths, and if there are three, one will be the sum of the other two (Sós [2], Surányi [1], Swierczkowski [1]). Further investigations along these lines have been carried out by Hartman [3], Sós [1], Halton [2], and Graham and van Lint [1]. Slater [3] gives a summary of these results with simple proofs. For quantitative results on $(n\theta)$ with θ irrational, see Section 3 of Chapter 2.

Interesting relations between $(n\theta)$, θ irrational, and ergodic theory are discussed in Hartman, Marczewski, and Ryll-Nardzewski [1] and Postnikov [8, Chapter 2]. In this context, the remarkable work of Veech [1, 2, 4] should also be mentioned. See also Hlawka [27]. In a different direction, Hardy and Littlewood [6] investigate for which θ one has $\lim_{N\to\infty}(1/N)\sum_{n=1}^{N} f(\{n\theta\}) = \int_0^1 f(x)\,dx$ for a class of functions f having infinities at $x = 0$ and 1. Mostly, questions of this type have been considered from the metric viewpoint (see the notes in Section 4 of this chapter). Erdös [7, pp. 52–53] states an open problem concerning $(n\theta)$.

Various subsequences of $(n\theta)$, θ irrational, have been studied. For instance, it is known that $(p_n\theta)$, $n = 1, 2, \ldots$, is u.d. mod 1, where $p_1 = 2, p_2 = 3, \ldots, p_n, \ldots$ is the sequence of primes arranged in ascending order (Vinogradov [3, 4; 5, Chapter 11], Hua [1]). If $\omega(n)$ is the number of prime divisors of n, then $(\omega(n)\theta)$ is u.d. mod 1 (Erdös [3], Delange [4]). Let (q_n) be an increasing sequence of integers >1 and set $r_n = q_1 \cdots q_n$ for $n \geq 1$; then Korobov [5] characterized the numbers θ for which $(r_n\theta)$ is u.d. mod 1. Some improvements were obtained by Šalát [2]. For other results on subsequences of $(n\theta)$, see Sections 4 and 8 of this chapter and Section 3 of the next. In particular, it will turn out that for the number θ in Example 2.2 the sequence $(10^n\theta)$ is u.d. mod 1.

Concerning the sequences $(c \log n)$ considered in Examples 2.4, 2.5, and 2.6, we refer to Franel [3], Pólya and Szegö [1, II. Abschn., Aufg. 179–181], and Thorp and Whitley [1]. We shall return to these sequences in Section 7. Wintner [1] showed that $(\log p_n)$ is not u.d. mod 1, where (p_n) is the sequence of primes. The sequences of logarithms of natural numbers are related to an amusing problem in elementary number theory. See Bird [1] and Exercises 2.14 to 2.17.

Theorems 2.2 and 2.3 are from Newman [1], and Theorem 2.4 which is essentially due to Mordell [3, 4] generalizes a result of Hecke [1]. More general results are known, and they can be found in Cantor [3], Carroll [1, 2], Carroll and Kemperman [1], Davenport [3], Meijer [1], Mordell [1, 2, 3, 4], Popken [1], Salem [1], Schwarz [1, 2], Schwarz and Wallisser [1], and Wallisser [1, 2]. Most of the proofs depend on results from the theory of u.d. mod 1. Some of these papers generalize, or are connected with, the classical theorem of Carlson-Pólya to the effect that a power series $\sum_{n=0}^{\infty} a_n z^n$ with integer coefficients a_n that converges for $|z| < 1$ either represents a rational function or has the unit circle as its natural boundary. Quantitative versions of Hecke's result are in Hlawka [28] and Niederreiter [19].

Theorem 2.5 is from van der Corput [5]. The special case enunciated in Corollary 2.1 was already known earlier (Pólya and Szegö [1, II. Abschn., Aufg. 174]). For other approaches to Fejér's theorem, see Tsuji [2] and Niederreiter [2]. Kuipers [3] proves some analogues of Fejér's theorem. See also Theorem 9.8 and Koksma [2; 4, Kap. 8]. Theorem 2.6 is essentially from Kennedy [1]. A special case was shown earlier by Kuipers [3].

Our proof follows Kano [1]. In a certain sense, the result is best possible (Kennedy [1]). For a thorough investigation of slowly growing sequences, see Kemperman [4].

The estimate in Theorem 2.7 is a result of van der Corput [1]. For various generalizations, see van der Corput [2, 3, 4] and Koksma [4, Kap. 9].

The presentation of u.d. of double sequences is based on the work of Cigler [1] and Hiergeist [1]. Using the same method as in the proof of Theorem 2.10, a result concerning the u.d. mod 1 of $(f(j, k))$ can be shown (Cigler [1]).

For the result in Exercise 2.27, see Koksma and Salem [1].

Exercises

2.1. The sequence (x_n) is u.d. mod 1 if and only if (2.1) holds for all positive integers h.

2.2. Let $f_1(x), f_2(x), \ldots, f_h(x), \ldots$ be a sequence of continuous functions on $\bar{I}$ that is dense in the space of all continuous functions on $\bar{I}$ in the sense of uniform convergence. Prove that (x_n) is u.d. mod 1 if and only if

$$\lim_{N\to\infty} \frac{1}{N} \sum_{n=1}^{N} f_h(\{x_n\}) = \int_0^1 f_h(x)\, dx \qquad \text{for all } h = 1, 2, \ldots.$$

2.3. Prove that the sequence (x_n) is u.d. mod 1 if and only if

$$\lim_{N\to\infty} \frac{1}{N} \sum_{n=1}^{N} \{x_n\}^h = \frac{1}{h+1} \qquad \text{for all } h = 1, 2, \ldots.$$

2.4. If (x_n) is u.d. mod 1, then (mx_n) is u.d. mod 1 for every nonzero integer m.

2.5. Let θ be an irrational number. Then the sequence $(a\theta)$, $a = 0, 1, -1, 2, -2, \ldots$, is u.d. mod 1.

2.6. Let θ be an irrational number, and let a and d be integers with $a \geq 0$ and $d > 0$. For $n \geq 1$, one sets $\varepsilon_n = 1$ if the integer closest to $n\theta$ is to the left of $n\theta$; otherwise, $\varepsilon_n = 0$. Then we have

$$\lim_{N\to\infty} \frac{1}{N} \sum_{n=1}^{N} \varepsilon_{a+nd} = \tfrac{1}{2}.$$

2.7. Prove that the sequence $(\{\sin n\})$, $n = 1, 2, \ldots$, is dense in $\bar{I}$ but not u.d. mod 1.

2.8. Prove the first part of Exercise 1.12 using the Weyl criterion.

2.9. Prove Lemma 1.1 using the Weyl criterion.

2.10. Prove Theorem 1.2 using the Weyl criterion.

2.11. If (x_n) is u.d. mod 1 and (y_n) satisfies

$$\lim_{N\to\infty} \frac{1}{N} \sum_{n=1}^{N} |y_n| = 0,$$

then $(x_n + y_n)$ is u.d. mod 1.

2.12. Let (x_n) be a sequence of real numbers and Q a natural number. If each of the Q sequences $(y_n^{(q)}) = (x_{Qn+q})$, $0 \leq q \leq Q - 1$, is u.d. mod 1, then so is (x_n).

2.13. The sequence $(c \log n)$, $n = 1, 2, \ldots$, where c is a constant, is not u.d. mod 1.

2.14. Let $b \geq 2$ be an integer. A set S of positive integers is called *extendable in base b* if for every finite string of b-adic digits, there exists an $s \in S$ whose initial digits in b-adic representation coincide with the given string of digits. Prove that $S = \{s_1, s_2, \ldots\}$ is extendable in base b if and only if the sequence $(\{{}^b\log s_n\})$, $n = 1, 2, \ldots$, of fractional parts is dense in $\bar{I}$.

2.15. For a given positive integer k, prove that the set $\{n^k: n = 1, 2, \ldots\}$ is extendable in any base (see Exercise 2.14).

2.16. Let $b \geq 2$ and k a positive integer that is not a rational power of b. Prove that the set $\{k^n: n = 1, 2, \ldots\}$ is extendable in base b (see Exercise 2.14).

2.17. The set $\{n^n: n = 1, 2, \ldots\}$ is extendable in any base (see Exercise 2.14).

2.18. Let β be an arbitrary real number. Prove Theorem 2.4 with $\{n\alpha\}$ replaced by $\{n\alpha + \beta\}$.

2.19. Let α be irrational and $\beta \in \mathbb{R}$ arbitrary. Then the power series $\sum_{n=1}^{\infty} \{n\alpha + \beta\} z^n$ has the unit circle as its natural boundary. What happens if α is rational?

2.20. Let the Bernoulli polynomials $B_k(x)$ be defined recursively by $B_1(x) = x - \frac{1}{2}$ and $B'_{k+1}(x) = (k + 1)B_k(x)$ and $\int_0^1 B_{k+1}(x)\, dx = 0$ for $k \geq 1$. Prove that the power series $\sum_{n=1}^{\infty} B_k(\{n\alpha + \beta\}) z^n$ has the unit circle as its natural boundary, whenever α is irrational, $\beta \in \mathbb{R}$, and $k \geq 1$.

2.21. Prove the Euler summation formula (2.3). *Hint:* Start from the identity $\int_n^{n+1} F(t)\, dt = \frac{1}{2}(F(n) + F(n + 1)) - \int_n^{n+1} (\{t\} - \frac{1}{2})F'(t)\, dt$ for integers n.

2.22. Use the Weyl criterion and the Euler summation formula to prove the following version of Fejér's theorem: Let $f(x)$ have a continuous derivative for sufficiently large x with $f'(x)$ tending monotonically to 0 as $x \to \infty$ and $\lim_{x\to\infty} x\,|f'(x)| = \infty$; then $(f(n))$, $n = 1, 2, \ldots$, is u.d. mod 1.

2.23. Use Theorem 2.7 to show that the sequence (αn^σ), $n = 1, 2, \ldots$, $\alpha \neq 0$, $1 < \sigma < 2$, is u.d. mod 1.

2.24. Prove that the sequence $(n \log \log n)$, $n = 2, 3, \ldots$, is u.d. mod 1.

2.25. Define $\log_k x$ recursively by $\log_1 x = \log x$ and $\log_k x = \log_{k-1}(\log x)$ for $k \geq 2$. For each $k \geq 1$, prove that $(n \log_k n)$, $n = n_0(k), n_0(k) + 1, \ldots$, is u.d. mod 1, where $n_0(k)$ is the least integer in the domain of $\log_k x$.

2.26. Deduce the following result from Theorem 2.7: Let $f(x)$ be defined for $x \geq 1$ and twice differentiable for sufficiently large x with $f''(x)$ tending monotonically to 0 as $x \to \infty$. Suppose also

$$\lim_{x\to\infty} f'(x) = \pm\infty \qquad \text{and} \qquad \lim_{x\to\infty} \frac{(f'(x))^2}{x^2\,|f''(x)|} = 0.$$

Then $(f(n))$, $n = 1, 2, \ldots$, is u.d. mod 1.

2.27. Let $f \in L^2[0, 1]$ be a function with period 1 and with $\int_0^1 f(x)\,dx = 0$. Then for any sequence (x_n) that is u.d. mod 1, one has

$$\lim_{N\to\infty} \int_0^1 \left| \frac{1}{N} \sum_{n=1}^{N} f(t + x_n) \right|^2 dt = 0.$$

Hint: Expand f into a Fourier series.

2.28. Consider the double sequence (s_{jk}) with $s_{jk} = j/k$ if $j \leq k$ and $s_{jk} = k/j$ if $j > k$. Then (s_{jk}) is not u.d. mod 1, but (s_{jk}) is u.d. mod 1 in the squares $1 \leq j, k \leq N$ as $N \to \infty$.

2.29. Let (s_k), $k = 1, 2, \ldots$, be u.d. mod 1 and define $s_{jk} = s_k$ for all $j, k = 1, 2, \ldots$. Then the double sequence (s_{jk}) is u.d. mod 1.

2.30. Let α and β be positive numbers. Then the double sequences $((\alpha j + \beta k)^\sigma)$, $j, k = 1, 2, \ldots$, $0 < \sigma < 1$, and $(\log^\tau (\alpha j + \beta k))$, $j, k = 1, 2, \ldots$, $\tau > 1$, are u.d. mod 1 in the squares $1 \leq j, k \leq N$ as $N \to \infty$.

2.31. See the proof of Theorem 2.2. Show in detail that $A(x)G(x) = B(x)$ implies (2.6).

3. DIFFERENCE THEOREMS

Van der Corput's Difference Theorem

LEMMA 3.1: Van der Corput's Fundamental Inequality. Let $u_1, \ldots, u_N$ be complex numbers, and let H be an integer with $1 \leq H \leq N$. Then

$$H^2 \left| \sum_{n=1}^{N} u_n \right|^2 \leq H(N + H - 1) \sum_{n=1}^{N} |u_n|^2 + 2(N + H - 1) \sum_{h=1}^{H-1} (H - h) \operatorname{Re} \sum_{n=1}^{N-h} u_n \bar{u}_{n+h},$$

where Re z denotes the real part of $z \in \mathbb{C}$.

PROOF. Define $u_n = 0$ for all $n \leq 0$ and all $n > N$. Then we have

$$H \sum_{n=1}^{N} u_n = \sum_{p=1}^{N+H-1} \sum_{h=0}^{H-1} u_{p-h}. \tag{3.1}$$

Using the Cauchy-Schwarz inequality, one obtains

$$\begin{aligned} H^2\left|\sum_{n=1}^{N} u_n\right|^2 &\le (N+H-1)\sum_{p=1}^{N+H-1}\left|\sum_{h=0}^{H-1} u_{p-h}\right|^2 \\ &= (N+H-1)\sum_{p=1}^{N+H-1}\left(\sum_{r=0}^{H-1} u_{p-r}\right)\left(\sum_{s=0}^{H-1} \bar{u}_{p-s}\right) \\ &= (N+H-1)\sum_{p=1}^{N+H-1}\sum_{h=0}^{H-1} |u_{p-h}|^2 \\ &\quad + 2(N+H-1)\mathrm{Re}\sum_{p=1}^{N+H-1}\sum_{\substack{r,s=0\\ s<r}}^{H-1} u_{p-r}\bar{u}_{p-s} \\ &= (N+H-1)(\Sigma_1 + 2\mathrm{Re}\,\Sigma_2). \end{aligned}$$

By (3.1) we see that Σ_1 is equal to $H\sum_{n=1}^{N}|u_n|^2$. The sum Σ_2 contains terms of the form $u_n\bar{u}_{n+h}$ with $n = 1, 2, \ldots, N$ and $h = r - s = 1, 2, \ldots, H-1$. For fixed n, $1 \le n \le N$, and h, $1 \le h \le H-1$, the possible choices for (r, s) yielding the term $u_n\bar{u}_{n+h}$ can be enumerated explicitly, namely, $(h, 0)$, $(h+1, 1), \ldots, (H-1, H-h-1)$. For each such choice, the value of p is uniquely determined. Thus, we have precisely $H - h$ occurrences of $u_n\bar{u}_{n+h}$ in Σ_2. Thus, we can write

$$\Sigma_2 = \sum_{h=1}^{H-1}(H-h)\sum_{n=1}^{N} u_n\bar{u}_{n+h}.$$

Now $u_n = 0$ for $n > N$; hence the summation over n can be restricted to $1 \le n \le N - h$. ■

THEOREM 3.1: Van der Corput's Difference Theorem. Let (x_n) be a given sequence of real numbers. If for every positive integer h the sequence $(x_{n+h} - x_n)$, $n = 1, 2, \ldots$, is u.d. mod 1, then (x_n) is u.d. mod 1.

PROOF. Let m be a fixed nonzero integer. We apply Lemma 3.1 with $u_n = e^{2\pi i m x_n}$ and, dividing by H^2N^2, we get

$$\left|\frac{1}{N}\sum_{n=1}^{N} e^{2\pi i m x_n}\right|^2 \le \frac{N+H-1}{HN} + 2\sum_{h=1}^{H-1}\frac{(N+H-1)(H-h)(N-h)}{H^2N^2} \cdot\left|\frac{1}{N-h}\sum_{n=1}^{N-h} e^{2\pi i m(x_n - x_{n+h})}\right|. \tag{3.2}$$

The sequence $(x_n - x_{n+h})$ is u.d. mod 1 for every $h \ge 1$; hence,

$$\lim_{N\to\infty}\frac{1}{N-h}\sum_{n=1}^{N-h} e^{2\pi i m(x_n - x_{n+h})} = 0 \qquad \text{for every } h \ge 1. \tag{3.3}$$

By (3.2) and (3.3) we obtain

$$\overline{\lim_{N\to\infty}} \left| \frac{1}{N} \sum_{n=1}^{N} e^{2\pi i m x_n} \right|^2 \leq \frac{1}{H}, \tag{3.4}$$

and since (3.4) holds for every positive integer H, we have

$$\lim_{N\to\infty} \frac{1}{N} \sum_{n=1}^{N} e^{2\pi i m x_n} = 0. \quad \blacksquare$$

This theorem yields an important sufficient condition for u.d. mod 1, but not a necessary one, as is seen by considering the sequence $(n\theta)$ with θ irrational. One of the many applications of Theorem 3.1 is to sequences of polynomial values.

THEOREM 3.2. Let $p(x) = \alpha_m x^m + \alpha_{m-1}x^{m-1} + \cdots + \alpha_0$, $m \geq 1$, be a polynomial with real coefficients and let at least one of the coefficients α_j with $j > 0$ be irrational. Then the sequence $(p(n))$, $n = 1, 2, \ldots$, is u.d. mod 1.

PROOF. The case $m = 1$ is settled as in Example 2.1, and so, we may suppose $m \geq 2$. Let first $\alpha_2, \ldots, \alpha_m$ be rational and α_1 irrational. Then write $p(x) = P(x) + \alpha_1 x + \alpha_0$. Let D be the least common multiple of the denominators of $\alpha_2, \ldots, \alpha_m$. We have $\{P(Dk + d)\} = \{P(d)\}$ for $k \geq 0$ and $d \geq 1$. Therefore, for every nonzero integer h,

$$\begin{aligned}
\frac{1}{N}\sum_{n=1}^{N} e^{2\pi i h p(n)} &= \frac{1}{N} \sum_{n=[N/D]D+1}^{N} e^{2\pi i h p(n)} \\
&\quad + \frac{1}{N} \sum_{d=1}^{D} \sum_{k=0}^{[N/D]-1} e^{2\pi i h(P(Dk+d)+\alpha_1(Dk+d)+\alpha_0)} \\
&= \frac{1}{N} \sum_{n=[N/D]D+1}^{N} e^{2\pi i h p(n)} \\
&\quad + \left(\sum_{d=1}^{D} e^{2\pi i h(P(d)+\alpha_1 d+\alpha_0)}\right)\left(\frac{1}{N}\sum_{k=0}^{[N/D]-1} e^{2\pi i h \alpha_1 D k}\right).
\end{aligned} \tag{3.5}$$

Since α_1 is irrational, the sequence $(\alpha_1 Dk)$, $k = 0, 1, \ldots$, is u.d. mod 1, so that the second term on the right of (3.5) tends to 0 as $N \to \infty$. The first term on the right of (3.5) is in absolute value less than D/N. According to Theorem 2.1 the sequence $(p(n))$ is u.d. mod 1 in this special case.

In order to show Theorem 3.2 we use induction. Let $p(x)$ be a polynomial whose highest degree term with irrational coefficient is $\alpha_q x^q$. Because of the above argument, the theorem is true for $q = 1$. Now let q be any integer with $m \geq q > 1$. Then every polynomial

$$p_h(x) = p(x + h) - p(x), \qquad h = 1, 2, \ldots,$$

has the property that the term of largest degree with an irrational coefficient is the term containing the factor x^{q-1}. Because of the induction hypothesis the sequence $(p_h(n))$, $n = 1, 2, \ldots$, is u.d. mod 1. Finally, applying Theorem 3.1, we conclude that the sequence $(p(n))$ is u.d. mod 1. ∎

EXAMPLE 3.1. As a sample result exhibiting the usefulness of Theorem 3.2 in diophantine problems, we shall prove that the system of inequalities

$$2x^{20} + 7x^4y^2 < y^4 - y + 1 < 2x^{20} + x^{10}y - 2x^2y^2 \tag{3.6}$$

has infinitely many solutions in positive integers x and y. We set $y = x^5 \sqrt[4]{2} + \delta$ with $|\delta| < 1$ and substitute in (3.6). Upon substitution we divide the members of (3.6) by $4x^{15} \sqrt[4]{8}$. The system (3.6) can then be written in the form

$$\varepsilon_1(x) - \tfrac{1}{8}\sqrt{2} < x^5 \sqrt[4]{2} - y < \varepsilon_2(x), \tag{3.7}$$

where $\varepsilon_1(x)$ and $\varepsilon_2(x)$ have the property that they tend to 0 as $x \to \infty$. Let η be a positive number $<\frac{1}{16}\sqrt{2}$. Then from a certain $x = x_0$ on, we have $-\eta < \varepsilon_1(x) < \eta$ and $-\eta < \varepsilon_2(x) < \eta$. Instead of (3.7) we consider the system $\eta - \frac{1}{8}\sqrt{2} < x^5 \sqrt[4]{2} - y < -\eta$, or

$$1 + \eta - \tfrac{1}{8}\sqrt{2} < x^5 \sqrt[4]{2} < 1 - \eta \pmod 1. \tag{3.8}$$

We note that the sequence $\omega = (n^5 \sqrt[4]{2})$, $n = 1, 2, \ldots$, is u.d. mod 1 by Theorem 3.2, and so,

$$\frac{1}{N} A((1 + \eta - \tfrac{1}{8}\sqrt{2}, 1 - \eta); N; \omega) \to \tfrac{1}{8}\sqrt{2} - 2\eta \qquad \text{as } N \to \infty.$$

Hence, there are infinitely many positive integers x satisfying (3.8) and therefore infinitely many pairs (x, y) of positive integers satisfying (3.7). ∎

Other Difference Theorems

THEOREM 3.3. If a sequence (x_n), $n = 1, 2, \ldots$, has the property

$$\Delta x_n = x_{n+1} - x_n \to \theta \text{ (irrational)} \qquad \text{as } n \to \infty, \tag{3.9}$$

then the sequence (x_n) is u.d. mod 1.

PROOF. If q is a positive integer, then by (3.9) there exists an integer $g_0 = g_0(q)$ such that for any integers $n \geq g \geq g_0$,

$$|x_n - x_g - (n - g)\theta| = \left| \sum_{j=g}^{n-1} (\Delta x_j - \theta) \right| \leq \frac{n - g}{q^2}. \tag{3.10}$$

Hence, if h is an integer $\neq 0$, then (3.10) and (2.15) yield

$$|e^{2\pi i h x_n} - e^{2\pi i h (x_g + (n-g)\theta)}| \leq \frac{2\pi |h| (n - g)}{q^2}, \tag{3.11}$$

and from (3.11) and the triangle inequality

$$\left|\sum_{n=g}^{g+q-1} e^{2\pi i h x_n}\right| \leq \left|\sum_{n=g}^{g+q-1} e^{2\pi i h(x_g+(n-g)\theta)}\right| + \frac{2\pi\,|h|}{q^2}\sum_{n=g}^{g+q-1}(n-g) \leq K, \quad (3.12)$$

where K stands for $|\sin \pi h\theta|^{-1} + \pi\,|h|$. Thus, by (3.12), for every positive integer H,

$$\left|\sum_{n=g}^{g-1+Hq} e^{2\pi i h x_n}\right| \leq HK. \quad (3.13)$$

So for every integer $N \geq g$ we have from (3.13),

$$\left|\sum_{n=1}^{N} e^{2\pi i h x_n}\right| \leq g - 1 + \frac{N-g}{q} K + q.$$

Keeping q fixed, we obtain

$$\overline{\lim_{N\to\infty}} \left|\frac{1}{N}\sum_{n=1}^{N} e^{2\pi i h x_n}\right| \leq \frac{K}{q}.$$

Since q can be as large as we please, the theorem follows. ■

We define recursively the difference operator Δ^k on a sequence (x_n) by $\Delta^1 x_n = \Delta x_n = x_{n+1} - x_n$ and $\Delta^k x_n = \Delta(\Delta^{k-1} x_n)$ for $k \geq 2$. With this notation, the following generalization of Theorem 2.5 can be established by means of Theorem 3.1.

THEOREM 3.4. Let $(f(n))$, $n = 1, 2, \ldots$, be a sequence of real numbers, and let k be a positive integer. If $\Delta^k f(n)$ is monotone in n, if $\Delta^k f(n) \to 0$ and $n\,|\Delta^k f(n)| \to \infty$ as $n \to \infty$, then the sequence $(f(n))$ is u.d. mod 1.

PROOF. We proceed by induction on k. For $k = 1$, the theorem reduces to Theorem 2.5. Suppose the theorem is true for the positive integer k, and let $(f(n))$ be a sequence with $\Delta^{k+1} f(n)$ monotone in n, $\lim_{n\to\infty} \Delta^{k+1} f(n) = 0$, and $\lim_{n\to\infty} n\,|\Delta^{k+1} f(n)| = \infty$. For a fixed positive integer h, we have $f(n+h) - f(n) = \sum_{j=0}^{h-1} \Delta f(n+j)$, and so $\Delta^k(f(n+h) - f(n)) = \sum_{j=0}^{h-1} \Delta^{k+1} f(n+j)$. In view of the assumptions, this leads to the following properties: $\Delta^k(f(n+h) - f(n))$ is monotone in n, $\lim_{n\to\infty} \Delta^k(f(n+h) - f(n)) = 0$, and $\lim_{n\to\infty} n\,|\Delta^k(f(n+h) - f(n))| = \infty$ (here we use that $\Delta^{k+1} f(n)$ has constant sign). By the induction hypothesis we obtain that the sequence $(f(n+h) - f(n))$, $n = 1, 2, \ldots$, is u.d. mod 1, and this for each $h \geq 1$. Thus, according to Theorem 3.1, the sequence $(f(n))$ is u.d. mod 1. ■

By using similar ideas, we may extend Fejér's theorem to the following result.

THEOREM 3.5. Let k be a positive integer, and let $f(x)$ be a function defined for $x \geq 1$, which is k times differentiable for $x \geq x_0$. If $f^{(k)}(x)$ tends monotonically to 0 as $x \to \infty$ and if $\lim_{x\to\infty} x\,|f^{(k)}(x)| = \infty$, then the sequence $(f(n))$, $n = 1, 2, \ldots$, is u.d. mod 1.

PROOF. We use induction on k. For $k = 1$, the assertion was shown in Corollary 2.1. Let f be a function satisfying the conditions of the theorem with k replaced by $k + 1$. For a positive integer h, set $g_h(x) = f(x + h) - f(x)$ for $x \geq 1$. Then $g_h^{(k)}(x) = f^{(k)}(x + h) - f^{(k)}(x)$ for $x \geq x_0$, and it is thus easily seen that the induction hypothesis can be applied to g_h. Hence, $(g_h(n))$, $n = 1, 2, \ldots$, is u.d. mod 1, and by Theorem 3.1 we are done. ∎

The above theorem leads to many interesting new classes of sequences that are u.d. mod 1. We refer to Exercises 3.9, 3.10, and 3.11.

Notes

Lemma 3.1 is from van der Corput [4, 5], who also showed Theorem 3.1 (van der Corput [5]). Theorem 3.2 was already proved earlier by Weyl [2, 4], who used more complicated methods. See also Hardy and Littlewood [1, 2] for weaker results. For an exposition of Weyl's method, see Titchmarsh [1, Chapter 5] and Walfisz [1]. Lemma 3.1 may also be found in Titchmarsh [1, Chapter 5] and Cassels [9, Chapter 4]. In the latter monograph one also finds a slightly different proof of Theorem 3.2. An interesting approach to Theorem 3.2 is possible by means of ergodic theory. See Furstenberg [1, 2], F. J. Hahn [1], and Cigler [14]. For applications of Theorem 3.2 to ergodic theory, see Akuliničev [1] and Postnikov [8, Chapter 2]. The statement of Theorem 3.2 holds also if n only runs through the sequence of primes (Vinogradov [4, 5], Hua [1], Rhin [4]).

For Example 3.1 and related problems, see van der Corput [5]. The same author [6] also developed other methods for the study of diophantine inequalities. Theorems 3.3 and 3.4 are also from van der Corput [5]. For an application of Theorem 3.3, see Exercise 3.3 and the papers of Brown and Duncan [2, 4], Duncan [1], and Kuipers [11]. The sequences considered in Exercise 3.9 were first shown to be u.d. mod 1 by Csillag [1]. For u.d. mod 1 of sequences of values of entire functions, see Rauzy [3]. Karacuba [1] proves the u.d. mod 1 of $(f(n))$ with f growing somewhat faster than a polynomial, e.g., $f(x) = e^{c(\log x)^\gamma}$ with $c > 0$ and $1 < \gamma < \frac{3}{2}$. Brezin [1] proves the u.d. mod 1 of a specific sequence arising from nilmanifold theory. Elliott [2] investigates sequences arising from zeros of the Riemann zeta-function. Blanchard [1, Chapter 6] studies sequences connected with prime number theory for Gaussian integers.

Important generalizations of the difference theorem (Theorem 3.1) will be shown in Chapter 4, Section 2. The following result of Korobov and Postnikov [1] may already be mentioned here: namely, if (x_n) satisfies the assumptions of Theorem 3.1, then (x_{qn+r}), $n = 1, 2, \ldots$, is u.d. mod 1, where $q > 0$ and $r \geq 0$ are integers. A detailed study of various difference theorems was carried out by Hlawka [8]. See Section 6 of Chapter 2 for quantitative versions of the difference theorem.

Exercises

3.1. See Theorem 3.2. If all the coefficients α_j with $j > 0$ are rational, then the sequence $(p(n))$ is not u.d. mod 1.

3.2. The sequence $(n\theta + \sin(2\pi\sqrt{n}))$, $n = 1, 2, \ldots$, θ irrational, is u.d. mod 1.

3.3. Consider the sequence (F_n) of Fibonacci numbers defined by $F_1 = F_2 = 1$ and $F_n = F_{n-1} + F_{n-2}$ for $n \geq 3$. Prove that the sequence $(\log F_n)$ is u.d. mod 1. *Hint:* Show first that $\lim_{n\to\infty} F_{n+1}/F_n = (1 + \sqrt{5})/2$.

3.4. The set $\{F_n: n = 1, 2, \ldots\}$ of Fibonacci numbers is extendable in any base (see Exercise 2.14).

3.5. Let $f(x)$ be a function defined for $x \geq 1$ that is differentiable for sufficiently large x. If $\lim_{x\to\infty} f'(x) = \theta$ (irrational), then the sequence $(f(n))$, $n = 1, 2, \ldots$, is u.d. mod 1.

3.6. Let (x_n) be a sequence of real numbers with the property $\lim_{n\to\infty} \Delta^k x_n = \theta$ (irrational) for some integer $k \geq 1$. Then (x_n) is u.d. mod 1.

3.7. Let k be a positive integer, and let $f(x)$ be a function defined for $x \geq 1$ that is k times differentiable for sufficiently large x. If

$$\lim_{x\to\infty} f^{(k)}(x) = \theta \quad \text{(irrational)},$$

then the sequence $(f(n))$, $n = 1, 2, \ldots$, is u.d. mod 1.

3.8. The sequence $(n^2\theta + \sin(2\pi\sqrt{n}))$, $n = 1, 2, \ldots$, with θ irrational, is u.d. mod 1.

3.9. Let $\alpha \neq 0$ and $\sigma > 0$ with σ not an integer. Then the sequence (αn^σ), $n = 1, 2, \ldots$, is u.d. mod 1.

3.10. Let α and σ be as in Exercise 3.9, and let τ be arbitrary. Then the sequence $(\alpha n^\sigma \log^\tau n)$, $n = 2, 3, \ldots$, is u.d. mod 1.

3.11. Let k be a positive integer, let $\alpha \neq 0$ and $\tau < 0$. Then the sequence $(\alpha n^k \log^\tau n)$, $n = 2, 3, \ldots$, is u.d. mod 1. The same is true for $\tau > 1$.

3.12. Prove that $(n^2 \log n)$ is u.d. mod 1. *Hint:* Use the difference theorem and Theorem 2.7.

3.13. Prove that $(n^2 \log\log n)$, $n = 2, 3, \ldots$, is u.d. mod 1.

3.14. For $\alpha \neq 0$ and $0 < \tau \leq 1$, prove that $(\alpha n \log^\tau n)$ and $(\alpha n^2 \log^\tau n)$ are u.d. mod 1.

3.15. Let $\sigma > 0$ and let $g(x)$ be a nonconstant linear combination of arbitrary powers of x. Prove that the sequence $(n^\sigma g(\log n))$, $n = 2, 3, \ldots$, is u.d. mod 1. *Hint:* Distinguish between $\sigma \in \mathbb{Z}$ and $\sigma \notin \mathbb{Z}$.

3.16. For an arbitrary sequence (x_n) of real numbers, prove that

$$\Delta^k x_n = \sum_{i=0}^{k} (-1)^i \binom{k}{i} x_{n+k-i}$$

for $n \geq 1$ and $k \geq 1$.

3.17. For any $\varepsilon > 0$, there exist arbitrarily large x with $\cos x^2 > 1 - \varepsilon$ and $\cos (x + 1)^2 < -1 + \varepsilon$.

3.18. Prove that $\overline{\lim}_{x\to\infty} |x\int_x^{x+1} \sin t^2\, dt| = 1$. *Hint:* Use integration by parts and Exercise 3.17.

4. METRIC THEOREMS

Some Basic Results

Let $(u_n(x))$, $n = 1, 2, \ldots$, for every x lying in some given bounded or unbounded interval J, be a sequence of real numbers. The sequence $(u_n(x))$ is said to be *u.d. mod 1 for almost all* x if for every $x \in J$, apart from a set which has Lebesgue measure 0, the sequence $(u_n(x))$ is u.d. mod 1. An early result of this type is the following one.

THEOREM 4.1. Let (a_n), $n = 1, 2, \ldots$, be a given sequence of distinct integers. Then the sequence $(a_n x)$, $n = 1, 2, \ldots$, is u.d. mod 1 for almost all real numbers x.

PROOF. It suffices to prove that $(a_n x)$ is u.d. mod 1 for almost all $x \in I = [0, 1)$. For if k is an integer, then $\{a_n(k + x)\} = \{a_n x\}$ implies that the set of $y \in [k, k + 1)$ for which $(a_n y)$ fails to be u.d. mod 1 is just the translate by k of the set of $x \in I$ for which $(a_n x)$ fails to be u.d. mod 1, and so also a null set. Since the countable union of null sets is again a null set, we are then done. For a fixed nonzero integer h, define

$$S(N, x) = \frac{1}{N}\sum_{n=1}^{N} e^{2\pi i h a_n x} \qquad \text{for } N \geq 1 \quad \text{and} \quad 0 \leq x \leq 1.$$

Then

$$|S(N, x)|^2 = S(N, x)\overline{S(N, x)} = \frac{1}{N^2}\sum_{m,n=1}^{N} e^{2\pi i h(a_m - a_n)x},$$

and so

$$\int_0^1 |S(N, x)|^2\, dx = \frac{1}{N^2}\sum_{m,n=1}^{N}\int_0^1 e^{2\pi i h(a_m - a_n)x}\, dx = \frac{1}{N}, \tag{4.1}$$

since the only contribution to the double sum comes from the terms with $m = n$. Now (4.1) implies

$$\sum_{N=1}^{\infty}\int_0^1 |S(N^2, x)|^2\, dx = \sum_{N=1}^{\infty}\frac{1}{N^2} < \infty,$$

and then Fatou's lemma yields

$$\int_0^1 \sum_{N=1}^{\infty} |S(N^2, x)|^2\, dx < \infty,$$

so that $\sum_{N=1}^{\infty} |S(N^2, x)|^2 < \infty$ for almost all $x \in [0, 1]$. Consequently, $\lim_{N\to\infty} S(N^2, x) = 0$ for almost all $x \in [0, 1]$. Now, given $N \geq 1$, there exists a positive integer m such that $m^2 \leq N < (m + 1)^2$. Then, by trivial estimates,

$$|S(N, x)| \leq |S(m^2, x)| + \frac{2m}{N} \leq |S(m^2, x)| + \frac{2}{\sqrt{N}}.$$

It follows that $\lim_{N\to\infty} S(N, x) = 0$ holds for almost all $x \in [0, 1]$, the exceptional set depending on the integer h chosen earlier. Forming the countable union of all the exceptional sets corresponding to $h = \pm 1, \pm 2, \ldots$, we arrive at a null set B. The Weyl criterion shows that $(a_n x)$ is u.d. mod 1 for all $x \in [0, 1]\backslash B$. ∎

In the proof of Theorem 4.1 a more general principle is hidden. Let $(u_n(x))$, $n = 1, 2, \ldots$, be a sequence depending on a real parameter x, each u_n being a real-valued Lebesgue-measurable function on the interval $[a, b]$. For integers $h \neq 0$ and $N \geq 1$, we set

$$S_h(N, x) = \frac{1}{N} \sum_{n=1}^{N} e^{2\pi i h u_n(x)} \qquad \text{for } a \leq x \leq b,$$

and also

$$I_h(N) = \int_a^b |S_h(N, x)|^2 \, dx.$$

Then the following general result holds.

THEOREM 4.2. If the series $\sum_{N=1}^{\infty} I_h(N)/N$ converges for each integer $h \neq 0$, then the sequence $(u_n(x))$ is u.d. mod 1 for almost all $x \in [a, b]$.

PROOF. We keep h fixed, and so drop the reference to h. Since $\sum_{N=1}^{\infty} I(N)/N$ converges, there exists an increasing sequence $(\lambda(N))$, $N = 1, 2, \ldots$, of real numbers >1 with $\lambda(N) \to \infty$ such that $\sum_{N=1}^{\infty} I(N)\lambda(N)/N$ still converges (see Exercise 4.9). Let $M_1 < M_2 < \cdots$ be positive integers such that

$$M_{r+1} = \left[\frac{\lambda(M_r)}{\lambda(M_r) - 1} M_r\right] + 1 \qquad \text{for } r \geq 1. \tag{4.2}$$

Let N_r be an integer in the range $M_r < N \leq M_{r+1}$ for which $I(N)$ attains its least value. Then

$$I(N_r) \leq \frac{1}{M_{r+1} - M_r} \sum_{N=M_r+1}^{M_{r+1}} I(N) \leq \frac{M_{r+1}}{M_{r+1} - M_r} \sum_{N=M_r+1}^{M_{r+1}} I(N)/N.$$

Since, according to (4.2),

$$\frac{M_{r+1}}{M_{r+1} - M_r} < \lambda(M_r),$$

we have

$$I(N_r) \leq \sum_{N=M_r+1}^{M_{r+1}} \frac{I(N)\lambda(N)}{N},$$

so that $\sum_{r=1}^{\infty} I(N_r) < \infty$. As in the proof of Theorem 4.1, one shows that then $\lim_{r\to\infty} S(N_r, x) = 0$ for almost all $x \in [a, b]$. We note that $M_{r+1}/M_r \to 1$ as $r \to \infty$ according to (4.2), and so, we have $N_{r+1}/N_r \to 1$ as $r \to \infty$. Now if $N_r < N \leq N_{r+1}$, then

$$|S(N, x)| \leq |S(N_r, x)| + \frac{N_{r+1} - N_r}{N_r},$$

whence $\lim_{N\to\infty} S(N, x) = 0$ for almost all $x \in [a, b]$. The proof is then completed as in Theorem 4.1. ■

EXAMPLE 4.1. The above theorem yields immediately the following generalization of Theorem 4.1. Let ψ be a positive function with

$$\sum_{n=1}^{\infty} \psi(n)n^{-3} < \infty.$$

Let (a_n), $n = 1, 2, \ldots$, be a sequence of integers for which $a_m = a_n$ for at most $\psi(N)$ ordered pairs (m, n) with $1 \leq m, n \leq N$. Then the sequence $(a_n x)$, $n = 1, 2, \ldots$, is u.d. mod 1 for almost all real numbers x. ■

Koksma's General Metric Theorem

THEOREM 4.3. Let $(u_n(x))$, $n = 1, 2, \ldots$, be a sequence of real numbers defined for every x in an interval $[a, b]$. For every $n \geq 1$, let $u_n(x)$ be continuously differentiable on $[a, b]$. Suppose that, for any two positive integers $m \neq n$, the function $u'_m(x) - u'_n(x)$ is monotone with respect to x and that $|u'_m(x) - u'_n(x)| \geq K > 0$, where K does not depend on x, m, and n. Then the sequence $(u_n(x))$ is u.d. mod 1 for almost all x in $[a, b]$.

PROOF. For a fixed integer $h \neq 0$, we set again

$$S_h(N, x) = \frac{1}{N} \sum_{n=1}^{N} e^{2\pi i h u_n(x)} \qquad \text{for } a \leq x \leq b.$$

Then,

$$\begin{aligned} I_h(N) &= \int_a^b |S_h(N, x)|^2 \, dx = \frac{1}{N^2} \sum_{m,n=1}^{N} \int_a^b e^{2\pi i h(u_m(x) - u_n(x))} \, dx \\ &\leq \frac{1}{N^2} \sum_{m,n=1}^{N} \left| \int_a^b e^{2\pi i h(u_m(x) - u_n(x))} \, dx \right| \\ &= \frac{b-a}{N} + \frac{2}{N^2} \sum_{m=2}^{N} \sum_{n=1}^{m-1} \left| \int_a^b e^{2\pi i h(u_m(x) - u_n(x))} \, dx \right|. \end{aligned}$$

To each of the integrals appearing in the last expression, we apply Lemma 2.1. Since $|u'_m(x) - u'_n(x)|$ attains its minimum at one of the end points of $[a, b]$, we obtain

$$\begin{aligned} I_h(N) &\le \frac{b-a}{N} + \frac{2}{|h|\,N^2} \sum_{m=2}^{N} \sum_{n=1}^{m-1} \max\left(\frac{1}{|u'_m(a) - u'_n(a)|}, \frac{1}{|u'_m(b) - u'_n(b)|}\right) \\ &\le \frac{b-a}{N} + \frac{2}{|h|\,N^2} \sum_{m=2}^{N} \sum_{n=1}^{m-1} \left(\frac{1}{|u'_m(a) - u'_n(a)|} + \frac{1}{|u'_m(b) - u'_n(b)|}\right). \end{aligned} \tag{4.3}$$

For fixed $x \in [a, b]$ and $2 \le m \le N$, we can order the numbers $u'_1(x)$, $u'_2(x), \ldots, u'_m(x)$ according to their magnitude. In the new ordering, the difference of any two consecutive numbers will be $\ge K$. Therefore,

$$\sum_{n=1}^{m-1} \frac{1}{|u'_m(x) - u'_n(x)|} < 2\sum_{j=1}^{N} \frac{1}{jK} < \frac{2}{K} \log (3N). \tag{4.4}$$

Combining (4.3) and (4.4), we obtain

$$I_h(N) \le \frac{b-a}{N} + \frac{8}{|h|\,K} \cdot \frac{\log(3N)}{N}.$$

The rest follows from Theorem 4.2. ∎

The above theorem contains many interesting special cases. We mention a few of them.

COROLLARY 4.1. Let δ be a positive constant, and let $(F(n))$, $n = 1, 2, \ldots$, be a sequence of numbers ≥ 1 with $|F(m) - F(n)| \ge \delta$ for $m \ne n$. Then the sequence $(\lambda x^{F(n)})$, $\lambda \ne 0$, $n = 1, 2, \ldots$, is u.d. mod 1 for almost all $x \ge 1$.

PROOF. Let k be a positive integer, and set $u_n(x) = \lambda x^{F(n)}$ for $k \le x \le k + 1$. For $m \ne n$, the function $u'_m(x) - u'_n(x) = \lambda(F(m)x^{F(m)-1} - F(n)x^{F(n)-1})$ is monotone in x since $u''_m(x) - u''_n(x)$ has the sign of $\lambda(F(m) - F(n))$. Moreover, for $m \ne n$ and $k \le x \le k + 1$, one has

$$|u'_m(x) - u'_n(x)| = |\lambda|\, x^{p-1}(qx^{q-p} - p) \ge |\lambda|\,(q - p) \ge |\lambda|\,\delta,$$

where $p = \min(F(m), F(n))$ and $q = \max(F(m), F(n))$. Theorem 4.3 implies that $(\lambda x^{F(n)})$ is u.d. mod 1 for almost all $x \in [k, k + 1]$. The countable union of these exceptional sets leads to a null set in $[1, \infty)$. ∎

COROLLARY 4.2. The sequence (x^n), $n = 1, 2, \ldots$, is u.d. mod 1 for almost all $x \ge 1$.

PROOF. Immediate from Corollary 4.1. ∎

COROLLARY 4.3. Let δ be a positive constant, and let (λ_n), $n = 1, 2, \ldots$, be a sequence of real numbers with $|\lambda_m - \lambda_n| \ge \delta$ for $m \ne n$. Then the sequence $(\lambda_n x)$, $n = 1, 2, \ldots$, is u.d. mod 1 for almost all real numbers x.

PROOF. For a fixed integer k, Theorem 4.3 yields that $(\lambda_n x)$ is u.d. mod 1 for almost all $x \in [k, k+1]$. The countable union of these exceptional sets leads to a null set in $\mathbb{R}$. ■

Concerning Corollary 4.2, it is interesting to note that one does not know whether sequences such as (e^n), (π^n), or even $((3/2)^n)$ are u.d. mod 1 or not. However, as we shall see below, one can say something about the exceptional set.

EXAMPLE 4.2. A real number $\alpha > 1$ is called a *Pisot-Vijayaraghavan number* (abbreviated P.V. number) if α is an algebraic integer all of whose conjugates (apart from α itself) lie in the open unit circle $\{z \in \mathbb{C}: |z| < 1\}$. In other words, $\alpha > 1$ is a P.V. number if there exists a polynomial $f(x) = x^m + a_{m-1}x^{m-1} + \cdots + a_0$, $a_i \in \mathbb{Z}$, irreducible over $\mathbb{Q}$ such that $f(x) = (x - \alpha_1) \cdots (x - \alpha_m)$ with $\alpha_1 = \alpha$ and $|\alpha_j| < 1$ for $2 \leq j \leq m$. Trivially, every rational integer >1 is a P.V. number. A less trivial example is $\alpha = (1 + \sqrt{5})/2$. We claim that, for a P.V. number α, the only possible limit points of the sequence $(\{\alpha^n\})$, $n = 1, 2, \ldots$, are 0 and 1, so that the sequence (α^n) can obviously not be u.d. mod 1. With the above notation, we set $T_n(\alpha) = \alpha_1{}^n + \cdots + \alpha_m{}^n$ for $n \geq 1$. Since $T_n(\alpha)$ is defined as a symmetric function of $\alpha_1, \ldots, \alpha_m$ with integral coefficients, $T_n(\alpha)$ is in fact a rational integer. Now $|\alpha^n - T_n(\alpha)| = |\alpha_1{}^n - T_n(\alpha)| \leq |\alpha_2|^n + \cdots + |\alpha_m|^n$, and so $\lim_{n\to\infty} |\alpha^n - T_n(\alpha)| = 0$. This proves the assertion. ■

Trigonometric Sequences

There are some interesting classes of sequences for which the conditions of Theorem 4.3 are not satisfied, among them, trigonometric sequences. We indicate how a more refined method can lead to a metric result for this case.

THEOREM 4.4. Let (a_n), $n = 1, 2, \ldots$, be an increasing sequence of positive integers. Then the sequence $(a_n \cos a_n x)$ is u.d. mod 1 for almost all real numbers x.

PROOF. It clearly suffices to prove that the sequence is u.d. mod 1 for almost all $x \in [0, 2\pi]$. For an integer $h \neq 0$ we have

$$\begin{aligned} I_h(N) &= \int_0^{2\pi} \left| \frac{1}{N} \sum_{n=1}^{N} e^{2\pi i h a_n \cos a_n x} \right|^2 dx \\ &\leq \frac{1}{N^2} \sum_{m,n=1}^{N} \left| \int_0^{2\pi} e^{2\pi i h (a_m \cos a_m x - a_n \cos a_n x)}\, dx \right| \\ &= \frac{2\pi}{N} + \frac{2}{N^2} \sum_{\substack{m,n=1\\ m<n}}^{N} \left| \int_0^{2\pi} e^{2\pi i h (a_m \cos a_m x - a_n \cos a_n x)}\, dx \right|. \end{aligned} \tag{4.5}$$

We estimate the integral occurring in the last expression for fixed m and n with $1 \leq m < n$. Set $G(x) = a_m \cos a_m x - a_n \cos a_n x$ and $g(x) = G'(x) = a_n^2 \sin a_n x - a_m^2 \sin a_m x$. We divide $[0, 2\pi]$ into three parts. Let E_1 be the set of $x \in [0, 2\pi]$ for which $|g(x)| \geq \frac{1}{2}a_n(a_n - a_m)^{1/2}$. Let E_2 be the set of $x \in [0, 2\pi]$ for which $|g(x)| < \frac{1}{2}a_n(a_n - a_m)^{1/2}$ and $|\sin a_n x| \geq (a_n - a_m)^{-1/2}$. Let E_3 be the set of $x \in [0, 2\pi]$ for which $|g(x)| < \frac{1}{2}a_n(a_n - a_m)^{1/2}$ and $|\sin a_n x| < (a_n - a_m)^{-1/2}$. The Lebesgue measure of the set of $x \in [0, 2\pi]$ for which $|\sin a_n x| < (a_n - a_m)^{-1/2}$ is clearly $O((a_n - a_m)^{-1/2})$, and since the integrand is of absolute value 1, we get

$$\int_{E_3} e^{2\pi i h G(x)}\, dx = O((a_n - a_m)^{-1/2}). \tag{4.6}$$

Since $g(x) = \pm\frac{1}{2}a_n(a_n - a_m)^{1/2}$ for $O(a_n)$ values of $x \in [0, 2\pi]$, it follows that E_1 is the union of $O(a_n)$ intervals. Furthermore, since $g'(x)$ has $O(a_n)$ zeros in $[0, 2\pi]$, we see that E_1 can be decomposed into $O(a_n)$ intervals J in each of which $g(x)$ is monotone. For such an interval J, Lemma 2.1 implies

$$\int_J e^{2\pi i h G(x)}\, dx = O(a_n^{-1}(a_n - a_m)^{-1/2}).$$

It follows that

$$\int_{E_1} e^{2\pi i h G(x)}\, dx = O((a_n - a_m)^{-1/2}). \tag{4.7}$$

For $x \in E_2$, we have $|g(x)| < \frac{1}{2}a_n(a_n - a_m)|\sin a_n x|$. We claim that this inequality implies

$$|g'(x)| \geq c(a_n^4 - a_m^4)^{1/2}(a_n^2 - a_m^2)^{1/2} \tag{4.8}$$

with some absolute constant $c > 0$. First of all, since both hypothesis and conclusion remain unchanged when replacing x by $-x$, we can suppose that $\sin a_n x > 0$. Then,

$$g(x) = a_n^2 \sin a_n x - a_m^2 \sin a_m x = (1 - t)a_n^2 \sin a_n x$$

with $\frac{1}{2} + a_m/2a_n \leq t \leq 3/2 - a_m/2a_n$. Thus, we can write

$$t a_n^2 \sin a_n x = a_m^2 \sin a_m x. \tag{4.9}$$

We also have

$$a_n^3 \cos a_n x = \eta + a_m^3 \cos a_m x, \tag{4.10}$$

where $\eta = g'(x)$. By squaring the equations (4.9) and (4.10) and using $\sin^2 a_n x + \cos^2 a_n x = 1$, we obtain

$$(a_m^4 a_n^2 - t^2 a_m^6)\cos^2 a_m x - 2t^2 \eta a_m^3 \cos a_m x + t^2 a_n^6 - a_m^4 a_n^2 - t^2\eta^2 = 0.$$

Hence, the discriminant of this quadratic equation in $\cos a_m x$ must be nonnegative; after simplifying, we arrive at

$$\eta^2 \geq a_m^6 + a_n^6 - a_m^2 a_n^4 t^2 - \frac{a_m^4 a_n^2}{t^2}.$$

Call the expression on the right-hand side of this inequality $s(t)$. The following statements are easily verified: $s(t)$ is concave downward for $t > 0$, $s(a_m/a_n) = s(1) = (a_n^4 - a_m^4)(a_n^2 - a_m^2)$, and $s(a_n/a_m) = 0$. The first and second of these facts imply $|\eta| \geq (a_n^4 - a_m^4)^{1/2}(a_n^2 - a_m^2)^{1/2}$ when $\frac{1}{2} + a_m/2a_n \leq t \leq 1$, and the three together imply that

$$s\left(\frac{1 + a_n/a_m}{2}\right) \geq \tfrac{1}{2}s(1),$$

so that $|\eta| \geq (\frac{1}{2}(a_n^4 - a_m^4)(a_n^2 - a_m^2))^{1/2}$ for $1 \leq t \leq \frac{1}{2} + a_n/2a_m$. Since $3/2 - a_m/2a_n \leq \frac{1}{2} + a_n/2a_m$, the inequality (4.8) is shown. We shall only use the weaker inequality

$$|g'(x)| \geq c a_n^2(a_n - a_m) \qquad \text{for } x \in E_2. \tag{4.11}$$

We note that E_2 is the union of $O(a_n)$ intervals K. Moreover, by the mean value theorem and (4.11) the length of such an interval K is

$$\sup_{x_1, x_2 \in K} |x_1 - x_2| \leq \sup_{x, x_1, x_2 \in K} \frac{|g(x_1)| + |g(x_2)|}{|g'(x)|} \leq \frac{a_n(a_n - a_m)^{1/2}}{c a_n^2(a_n - a_m)},$$

and so the Lebesgue measure of E_2 is $O((a_n - a_m)^{-1/2})$. Consequently, we have

$$\int_{E_2} e^{2\pi i h G(x)}\, dx = O((a_n - a_m)^{-1/2}), \tag{4.12}$$

and (4.6), (4.7), and (4.12) imply

$$\int_0^{2\pi} e^{2\pi i h(a_m \cos a_m x - a_n \cos a_n x)}\, dx = O((a_n - a_m)^{-1/2}) \qquad \text{for } 1 \leq m < n.$$

Combining this with (4.5), we arrive at

$$I_h(N) = O\left(\frac{1}{N} + \frac{1}{N^2}\sum_{n=2}^{N}\sum_{m=1}^{n-1}(a_n - a_m)^{-1/2}\right).$$

But

$$\sum_{n=2}^{N}\sum_{m=1}^{n-1}(a_n - a_m)^{-1/2} \leq \sum_{n=2}^{N}\sum_{m=1}^{n-1}(n - m)^{-1/2} = O(N^{3/2}),$$

and so $I_h(N) = O(N^{-1/2})$. The rest follows from Theorem 4.2. ∎

Notes

Theorem 4.1 is a result of Weyl [2, 4]. The case $a_n = b^n$ for some integer $b \geq 2$ was studied earlier by Hardy and Littlewood [1]. Sequences $(a_n x)$ with integers a_n have been investigated extensively from the metric point of view. See R. C. Baker [5], Erdös [4, 7], Erdös and Taylor [1], Hardy and Littlewood [6], Kahane and Salem [1], Khintchine [4], and Koksma [12, 14]. For the special case $a_n = b^n$, see also Section 8. A growth condition (see e.g. Exercise 4.5) is needed in these results: Dress [1] shows that if (a_n) is nondecreasing with $a_n = o(\log n)$, then for no real x is $(a_n x)$ u.d. mod 1. Some metric theorems arise from the individual ergodic theorem (Khintchine [6], Riesz [1], Franklin [1]). The following was a long-standing conjecture of Khintchine [1]: Let E be a Lebesgue-measurable subset of I; then for almost all θ the sequence $(n\theta)$ satisfies $\lim_{N\to\infty} A(E; N)/N = \lambda(E)$. This was disproved by Marstrand [1]. The sequences $(\lambda_n x)$ with arbitrary real numbers λ_n have also been studied, especially in the case where $\lambda_{n+1}/\lambda_n \geqq c > 1$ for all $n \geq 1$ (the so-called lacunary case). See Erdös [4], Furstenberg [3], Helson and Kahane [1], Kac, Salem, and Zygmund [1], and Koksma [10, 15]. A related subject is that of normal sets (see the notes in Section 8). For quantitative refinements, we refer to the notes in Section 3 of Chapter 2. A survey of results on the above classes of sequences can be found in Erdös [7] and Koksma [16]. For the early literature on the subject, see Koksma [4, Kap. 9].

The important Theorem 4.2 comes from Davenport, Erdös, and LeVeque [1]. See also Kuipers and van der Steen [1] and Philipp [2]. Holewijn [2] applies this theorem to u.d. of random variables. See also Loynes [1] and Lacaze [2] for related applications of metric theorems to stochastic processes. In a certain sense, Theorem 4.2 is best possible (Davenport, Erdös, and LeVeque [1]).

Theorem 4.3 was shown by Koksma [3]. Some variations of this theme, and some more consequences, can be found in LeVeque [1], Franklin [2], and Kuipers and van der Steen [1]. See Koksma [10] for a related question, and Erdös and Koksma [2] and Cassels [1] for quantitative versions. Generalizations of Theorem 4.3 to more abstract settings have been given by F. Bertrandias [1] and de Mathan [3].

For expository accounts of the theory of P. V. numbers introduced in Example 4.2, we refer to Cassels [9, Chapter 8] and Salem [3]. A good bibliography is given in the paper of Pisot [1]. For recent work, see Amara [1], Boyd [1], Cantor [1], Halter-Koch [1], Mahler [7], Pathiaux [1], Pisot and Salem [1], Senge [1], Senge and Straus [1], and Zlebov [1]. There are interesting relations between P.V. numbers and questions in harmonic analysis. See Salem [3] and Meyer [3, 5] and the literature given there. A well-known problem is connected with the sequences (r^n), $n = 1, 2, \ldots$, where $r > 1$ is a nonintegral rational (Mahler [1, 2, 4, 7], Tijdeman [1]). Supnick, Cohen, and Keston [1] study multiplicities in the sequences $(\{\theta^n\})$ with $\theta > 1$. Forman and Shapiro [1] discuss arithmetic properties of the sequences $([(4/3)^n])$ and $([(3/2)^n])$ of integral parts. See also Shapiro and Sparer [1].

Theorem 4.4 derives from LeVeque [3], who also has more general results. Dudley [1] improves some of LeVeque's theorems.

By identifying sequences in I with elements of the infinite-dimensional unit cube $I^\infty = \prod_{j=1}^{\infty} I_j$, $I_j = I$ for $j \geq 1$, it can be shown that "almost all" sequences in I are u.d. mod 1 (see Chapter 3, Theorem 2.2). Here "almost all" has to be taken in the sense of the product measure in I^∞ induced by Lebesgue measure in I.

We remark that metric results are also available for u.d. mod Δ. See LeVeque [4], Davenport and LeVeque [1], Erdös and Davenport [1], and W. M. Schmidt [10]. In a different direction, Petersen and McGregor [2] prove that a sequence is u.d. mod 1 if and only if "almost all" subsequences are u.d. mod 1. For a study of u.d. mod 1 subsequences of a given sequence, see Mendès France [8] and Dupain and Lesca [1].

Exercises

4.1. In Corollary 4.1 the condition "$F(n) \geq 1$" can be replaced by "$(F(n))$ is bounded from below".

4.2. Let $(F(n))$ be a sequence of positive integers with $F(p) \neq F(q)$ for $p \neq q$. Then the sequence $(\lambda x^{F(n)})$, $\lambda \neq 0$, is u.d. mod 1 for almost all $x \geq 1$, and also for almost all $x \leq -1$.

4.3. For real θ with $|\theta| > 1$, the sequence $(\theta^n x)$ is u.d. mod 1 for almost all x.

4.4. Using results of this section, prove that for $\lambda \neq 0$ the sequence (λn^x) is u.d. mod 1 for almost all $x \geq 0$. (Note: Using results of earlier sections, one can in fact describe the exceptional set explicitly).

4.5. Let (a_n) be a sequence of integers for which there exist positive constants ε and c such that $a_m \neq a_n$ whenever $|m - n| \geq cn/(\log n)^{1+\varepsilon}$. Prove that $(a_n x)$ is u.d. mod 1 for almost all x.

4.6. Let (λ_n) be a sequence of real numbers for which there exist positive constants ε and δ such that $|\lambda_m - \lambda_n| \geq \delta$ whenever $|m - n| \geq n/(\log n)^{1+\varepsilon}$ (Weyl's growth condition). Then $(\lambda_n x)$ is u.d. mod 1 for almost all x.

4.7. Prove that $\alpha = (1 + \sqrt{5})/2$ is a P.V. number.

4.8. Prove that the unique root >1 of the polynomial $x^3 - x - 1$ is a P.V. number.

4.9. Let $\sum_{n=1}^{\infty} u_n$ be a convergent series of positive terms. Set $r(n) = \sum_{j=n}^{\infty} u_j$ and $\lambda(n) = (\sqrt{r(n)} + \sqrt{r(n+1)})^{-1}$ for $n \geq 1$. Show that $\lambda(n) \to \infty$ monotonically as $n \to \infty$ and that $\sum_{n=1}^{\infty} u_n \lambda(n)$ is convergent.

5. WELL-DISTRIBUTED SEQUENCES MOD 1

Definition and Weyl Criteria

Let (x_n), $n = 1, 2, \ldots$, be a sequence of real numbers. For integers $N \geq 1$ and $k \geq 0$ and a subset E of I, let $A(E; N, k)$ be the number of terms among $\{x_{k+1}\}, \{x_{k+2}\}, \ldots, \{x_{k+N}\}$ that are lying in E.

DEFINITION 5.1. The sequence (x_n), $n = 1, 2, \ldots$, is said to be *well-distributed mod 1* (abbreviated w.d. mod 1) if for all pairs a, b of real numbers with $0 \leq a < b \leq 1$ we have

$$\lim_{N\to\infty} \frac{A([a, b); N, k)}{N} = b - a \qquad \text{uniformly in } k = 0, 1, 2, \ldots.$$

Evidently, a sequence that is w.d. mod 1 is also u.d. mod 1. The converse is not true, as is shown by the following counterexample.

EXAMPLE 5.1. Let $\omega = (x_n)$, $n = 1, 2, \ldots$, be u.d. mod 1. Now form a sequence $\sigma = (y_n)$, $n = 1, 2, \ldots$, by setting $y_n = 0$ if $m^3 + 1 \leq n \leq m^3 + m$, $m = 1, 2, \ldots$, and $y_n = x_n$ for all other values of n. The sequence σ is u.d. mod 1, since it is obtained from ω by replacing a sufficiently small number of terms x_n by zeros. For a detailed proof, we note that for each integer $N \geq 1$ there is an integer $p \geq 1$ such that $p^3 \leq N < (p+1)^3$. Then for any subinterval $[a, b)$ of I,

$$|A([a, b); N; \omega) - A([a, b); N; \sigma)| \leq 1 + 2 + \cdots + p \leq N^{2/3},$$

and so

$$\lim_{N\to\infty} \frac{A([a, b); N; \sigma)}{N} = \lim_{N\to\infty} \frac{A([a, b); N; \omega)}{N} = b - a.$$

On the other hand, σ is not w.d. mod 1. To see this, choose ε with $0 < \varepsilon < \frac{1}{2}$. If σ were w.d. mod 1, there would exist a positive integer $N_0 = N_0(\varepsilon)$, independent of k, such that for $N \geq N_0$ and for the sequence σ,

$$\left|\frac{A([0, \frac{1}{2}); N, k)}{N} - \frac{1}{2}\right| < \varepsilon \qquad \text{for all } k \geq 0,$$

and therefore also

$$\left|\frac{A([0, \frac{1}{2}); N_0, N_0^3)}{N_0} - \frac{1}{2}\right| < \varepsilon. \tag{5.1}$$

However, $A([0, 1/2); N, N^3) = N$ for every $N \geq 1$, since all elements y_n with $N^3 + 1 \leq n \leq N^3 + N$ are 0. Hence, (5.1) gives $\varepsilon > 1/2$, which contradicts the assumption concerning ε. ■

The Weyl criteria for w.d. mod 1 are those for u.d. mod 1 with an additional uniformity condition in k. We omit the proofs. The interested reader is encouraged to consult Section 3 of Chapter 3.

THEOREM 5.1. The sequence (x_n), $n = 1, 2, \ldots$, is w.d. mod 1 if and only if for all continuous functions f on $\bar{I}$ we have

$$\lim_{N\to\infty} \frac{1}{N} \sum_{n=k+1}^{k+N} f(\{x_n\}) = \int_0^1 f(x)\, dx \qquad \text{uniformly in } k = 0, 1, 2, \ldots.$$

THEOREM 5.2. The sequence (x_n), $n = 1, 2, \ldots$, is w.d. mod 1 if and only if for all integers $h \neq 0$ we have

$$\lim_{N\to\infty} \frac{1}{N} \sum_{n=k+1}^{k+N} e^{2\pi i h x_n} = 0 \qquad \text{uniformly in } k = 0, 1, 2, \ldots.$$

EXAMPLE 5.2. Using the same argument as in Example 2.1, Theorem 5.2 shows that the sequence $(n\theta)$, $n = 1, 2, \ldots$, with θ irrational is w.d. mod 1. ∎

Since w.d. mod 1 is a comparatively rare phenomenon (see the notes for a rigorous interpretation of this statement), there are many negative results about w.d. mod 1. One such result is as follows.

THEOREM 5.3. The sequence $(p^n\theta)$, $n = 1, 2, \ldots$, where p is some integer and θ is real, is not w.d. mod 1.

PROOF. Let $|p| \geq 2$ and $\theta \neq 0$ (all other cases are trivial). Set $x_n = p^n\theta$ and consider the expression

$$\sum_{n=k+1}^{k+N} e^{2\pi i x_n} = \sum_{n=1}^{N} e^{2\pi i x_{n+k}} = \sum_{n=1}^{N} e^{2\pi i p^n x_k}.$$

We have

$$\left|\sum_{n=k+1}^{k+N} e^{2\pi i x_n}\right| \geq \left|\sum_{n=1}^{N} \cos 2\pi p^n x_k\right| = \left|\sum_{n=1}^{N} \cos 2\pi\, |p|^n \{x_k\}\right|. \tag{5.2}$$

If we assume that the sequence $(p^n\theta)$ is w.d. mod 1, then 0 is a limit point of the sequence $(\{x_n\})$. Hence, to any $N \geq 1$ there is a $k(N)$ such that

$$\{x_{k(N)}\} < \frac{1}{6\,|p|^N}.$$

With $k = k(N)$, the arguments in the last expression in (5.2) satisfy

$$0 \leq 2\pi\, |p|^n \{x_{k(N)}\} \leq 2\pi\, |p|^N \{x_{k(N)}\} < \frac{\pi}{3} \qquad \text{for } 1 \leq n \leq N,$$

and so the cosine of each of these arguments is $> \frac{1}{2}$. Hence, for every $N \geq 1$,

$$\left|\frac{1}{N} \sum_{n=k(N)+1}^{k(N)+N} e^{2\pi i x_n}\right| > \frac{1}{2},$$

and this contradicts Theorem 5.2. ∎

This theorem should be compared with an immediate consequence of Theorem 4.1, namely, that for an integer p with $|p| \geq 2$ the sequence $(p^n\theta)$, $n = 1, 2, \ldots$, is u.d. mod 1 for almost all real numbers θ.

Admissible Sequences

DEFINITION 5.2. A sequence (s_n) of real numbers is said to be *admissible* if whenever the sequence (x_n) is w.d. mod 1, the sequence $(x_n + s_n)$ is also w.d. mod 1.

An obvious example of an admissible sequence is a constant sequence (see Exercise 5.4). More generally, every convergent sequence is admissible (see Exercise 5.5). The following is a sufficient condition for admissible sequences going beyond these simple cases.

THEOREM 5.4. Let (t_n) be a sequence of real numbers, and let $s_n = \sum_{j=1}^{n} t_j$. Then the sequence (s_n) is admissible if

$$\lim_{N\to\infty} \frac{1}{N} \sum_{m=k+1}^{k+N} |t_m| = 0 \qquad \text{uniformly in } k = 0, 1, 2, \ldots. \tag{5.3}$$

PROOF. Let the sequence (x_n) be w.d. mod 1. Choose $\varepsilon > 0$ and a nonzero integer h. According to the assumptions there exist positive integers P and Q such that

$$\left| \frac{1}{p} \sum_{n=k+1}^{k+p} e^{2\pi i h x_n} \right| < \frac{\varepsilon}{3} \qquad \text{for all } p \geq P \text{ and all } k \geq 0, \tag{5.4}$$

$$\frac{1}{q} \sum_{n=k+1}^{k+q} |t_n| < \frac{\varepsilon}{3AP} \qquad \text{for all } q \geq Q \text{ and all } k \geq 0, \tag{5.5}$$

where $A = 2\pi\,|h|$. Suppose N is an integer $\geq \max(P+1, Q, 3P/\varepsilon)$. Then, for any $k \geq 0$,

$$\left| \frac{1}{N} \sum_{n=k+1}^{k+N} e^{2\pi i h(x_n+s_n)} \right| = \frac{1}{N} \left| \sum_{r=1}^{v} b_r + \sum_{n=k+vP+1}^{k+N} e^{2\pi i h(x_n+s_n)} \right|,$$

where

$$b_r = \sum_{n=k+(r-1)P+1}^{k+rP} e^{2\pi i h(x_n+s_n)} \qquad \text{and} \qquad v = \left[\frac{N-1}{P} \right].$$

Now

$$\left| \sum_{n=k+vP+1}^{k+N} e^{2\pi i h(x_n+s_n)} \right| \leq P.$$

Furthermore,

$$b_r = \sum_{n=k+(r-1)P+1}^{k+rP} e^{2\pi i h(x_n+s_{k+(r-1)P})} + \sum_{n=k+(r-1)P+1}^{k+rP} e^{2\pi i h x_n}\left(e^{2\pi i h s_n} - e^{2\pi i h s_{k+(r-1)P}}\right),$$

so that, using (5.4) for the first and (2.15) for the second sum,

$$|b_r| \leq \frac{\varepsilon P}{3} + A \sum_{n=k+(r-1)P+1}^{k+rP} |s_n - s_{k+(r-1)P}|$$

$$\leq \frac{\varepsilon P}{3} + A \sum_{n=1}^{P} (|t_{k+(r-1)P+1}| + \cdots + |t_{k+(r-1)P+n}|).$$

Using (5.5) we obtain

$$\left|\sum_{r=1}^{v} b_r\right| \le \frac{v\varepsilon P}{3} + A \sum_{n=k+1}^{k+N} P\,|t_n| \le \frac{v\varepsilon P}{3} + AP\frac{N\varepsilon}{3AP} \le \frac{2N\varepsilon}{3}.$$

Hence,

$$\left|\frac{1}{N}\sum_{n=k+1}^{k+N} e^{2\pi i h(x_n+s_n)}\right| \le \frac{2\varepsilon}{3} + \frac{P}{N} \le \varepsilon,$$

and $(x_n + s_n)$ is w.d. mod 1 by Theorem 5.2. ■

Since (s_n) is only relevant modulo 1, we may suppose without loss of generality that $|s_{n+1} - s_n| = |t_{n+1}| \le \frac{1}{2}$ for $n \ge 1$. With this additional hypothesis, the condition (5.3) is also necessary for the sequence (s_n) to be admissible, but this is by no means easy to show. In connection with (5.3) we remark also that a sequence (u_n) is called *almost convergent* to the value u if $\lim_{N\to\infty} (1/N)\sum_{m=k+1}^{k+N} u_m = u$ uniformly in $k = 0, 1, 2, \ldots$.

Metric Theorems

THEOREM 5.5. If p and q are positive integers, then for almost all real numbers x the sequence $((p/q)^n x)$, $n = 1, 2, \ldots$, is not w.d. mod 1.

PROOF. Evidently, we may assume $p > q$. For a fixed integer $N \ge 1$, let E_N be the set of $x \in \mathbb{R}$ for which $(p^n x/q^{n+N})$, $n = 1, 2, \ldots$, is not u.d. mod 1. Then, according to Corollary 4.3, we have $\lambda(E_N) = 0$, where λ denotes Lebesgue measure. Hence, if $E = \bigcup_{N=1}^{\infty} E_N$, then $\lambda(E) = 0$. We note that for a sequence that is u.d. mod 1 the sequence of fractional parts is everywhere dense in I. Thus, if $x \notin E$, then for every $N \ge 1$ we can find a $k = k(N)$ such that

$$\left\{\frac{p^k x}{q^{k+N}}\right\} < \frac{1}{6p^N q^N}. \tag{5.6}$$

Now we consider the expression

$$\sum_{n=k+1}^{k+N} e^{2\pi i p^n x/q^n} = \sum_{n=k+1}^{k+N} e^{2\pi i q^N (p^n x/q^{n+N})}$$
$$= \sum_{n=1}^{N} e^{2\pi i q^N (p^n/q^n)(p^k x/q^{k+N})} = \sum_{n=1}^{N} e^{2\pi i q^{N-n} p^n \{p^k x/q^{k+N}\}}.$$

If $x \notin E$ and $k = k(N)$, then from (5.6) we have for $1 \le n \le N$,

$$0 \le q^{N-n} p^n \left\{\frac{p^k x}{q^{k+N}}\right\} < \frac{q^{N-n} p^n}{6p^N q^N} \le \frac{1}{6}.$$

This implies

$$\left|\frac{1}{N}\sum_{n=k(N)+1}^{k(N)+N} e^{2\pi i p^n x/q^n}\right| > \frac{1}{2}$$

as in the proof of Theorem 5.3, and so, the sequences $((p/q)^n x)$ with $x \notin E$ are not w.d. mod 1 by Theorem 5.2. ∎

For the proof of the following lemma, we need some results and notions from Section 6. This lemma, which is highly useful in establishing that a sequence is not w.d. mod 1, is also closely related to the material on complete uniform distribution in Section 3 of Chapter 3 and is in fact a special case of Theorem 3.12 of Chapter 3.

LEMMA 5.1. Let (x_n) be a sequence of real numbers such that for every $p \geq 1$ and every p-tuple $(h_1, \ldots, h_p) \neq (0, \ldots, 0)$ of integers, the sequence $(h_1 x_n + h_2 x_{n+1} + \cdots + h_p x_{n+p-1})$, $n = 1, 2, \ldots$, is u.d. mod 1. Then (x_n) is not w.d. mod 1.

PROOF. If (x_n) is w.d. mod 1, then it is easily seen from Definition 5.1 that there exists an integer $p \geq 1$ such that at least one of any p consecutive terms of $(\{x_n\})$ lies in $[0, \frac{1}{2})$ (see Exercise 5.7). On the other hand, the sequence $((x_n, x_{n+1}, \ldots, x_{n+p-1}))$, $n = 1, 2, \ldots$, is u.d. mod 1 in $\mathbb{R}^p$ according to Theorem 6.3. In particular, there is an $N \geq 1$ such that

$$(\{x_N\}, \{x_{N+1}\}, \ldots, \{x_{N+p-1}\}) \in [\tfrac{1}{2}, 1) \times \cdots \times [\tfrac{1}{2}, 1),$$

the Cartesian product of p copies of $[\frac{1}{2}, 1)$. This results in a contradiction. ∎

THEOREM 5.6. Let (λ_n) be a sequence of nonzero real numbers such that $\lim_{n\to\infty} |\lambda_{n+1}/\lambda_n| = \infty$. Then for almost all real numbers x the sequence $(\lambda_n x)$, $n = 1, 2, \ldots$, is not w.d. mod 1.

PROOF. We prove that the condition of Lemma 5.1 is satisfied for almost all sequences $(\lambda_n x)$. Consider a fixed $p \geq 1$ and a fixed p-tuple $(h_1, \ldots, h_p) \neq (0, \ldots, 0)$ of integers. Clearly, we may suppose $h_p \neq 0$. We write

$$h_1 \lambda_n x + \cdots + h_p \lambda_{n+p-1} x = b_n x$$

with $b_n = h_1 \lambda_n + \cdots + h_p \lambda_{n+p-1}$ for $n \geq 1$, and we set $H = \max_{1 \leq j \leq p} |h_j|$. By hypothesis, there exists a positive integer N such that

$$\frac{|\lambda_{r+1}|}{|\lambda_r|} \geq \frac{(2p-1)H}{|h_p|} + 1 \qquad \text{for all } r \geq N.$$

Then for $m > n \geq N$ we have

$$\begin{aligned} |b_m - b_n| &= |h_p \lambda_{m+p-1} + \cdots + h_1 \lambda_m - h_p \lambda_{n+p-1} - \cdots - h_1 \lambda_n| \\ &\geq |h_p|\,|\lambda_{m+p-1}| - (2p-1)H\,|\lambda_{m+p-2}| \geq |h_p|\,|\lambda_{m+p-2}| \geq |h_p|\,|\lambda_N|, \end{aligned}$$

and so Corollary 4.3 implies that the sequence $(b_n x)$ is u.d. mod 1 for almost all real numbers x. Letting p run through all positive integers and considering all p-tuples of integers that are permitted, we arrive at a countable union of exceptional sets of x that is still a null set. ∎

Notes

The definition of a w.d. sequence mod 1 is from Hlawka [1] and G. M. Petersen [1]. The criteria in Theorems 5.1 and 5.2 were also given by these authors. Example 5.1 is from G. M. Petersen [1]. Theorem 5.3 derives from Dowidar and Petersen [1], who corrected an erroneous statement in Keogh, Lawton, and Petersen [1, Theorem 5]. Admissible sequences were introduced and studied by Petersen and Zame [1]. Theorem 5.5 is a result of Petersen and McGregor [2]. Murdoch [1] proved that if α is any given real number, then for almost all x the sequence $(\alpha^n x)$ is not w.d. mod 1. Theorem 5.6 is from Dowidar and Petersen [1]. This result was improved by G. M. Petersen [4] and Zame [2], who showed that lacunary sequences (λ_n) satisfy the same property, thereby settling a conjecture of Petersen and McGregor [3]. For further improvements, see Zame [4]. Related results can be found in Cigler [12], Erdös [7], Gerl [5], G. M. Petersen [3], and Petersen and McGregor [1]. The construction of w.d. sequences mod 1 is discussed in Gerl [4] and Keogh, Lawton, and Petersen [1].

An analogue of van der Corput's difference theorem can be shown for w.d. sequences mod 1. See Theorem 2.2 of Chapter 4 and Hlawka [1, 8]. It follows that the sequences $(p(n))$ from Theorem 3.2 are even w.d. mod 1. See Hlawka [8] and Lawton [1]. This result may also be shown using ergodic theory (Furstenberg [2], Cigler [12, 14, 15]). Cigler [13] shows that the sequences $(f(n))$ in Theorem 3.5 are not w.d. mod 1. Burkard [2] discusses sequences that are w.d. mod Δ (compare with Definition 1.2).

Dowidar and Petersen [1] show that in a certain sense almost no sequence is w.d. mod 1 (see Exercise 5.15). Also, in the sense of the product measure on I^∞ (see the notes in Section 4) almost no sequence is w.d. mod 1. See Chapter 3, Theorem 3.8, for a proof. Further results on w.d. sequences can be found in Chapter 1, Section 6; Chapter 3, Sections 3 and 4; and Chapter 4, Sections 1, 2, and 4.

Exercises

5.1. Prove Theorem 5.1.

5.2. Prove Theorem 5.2.

5.3. The sequence (x_n), $n = 1, 2, \ldots$, is w.d. mod 1 if and only if, for all Riemann-integrable functions f on $\bar{I}$, we have

$$\lim_{N\to\infty} \frac{1}{N} \sum_{n=k+1}^{k+N} f(\{x_n\}) = \int_0^1 f(x)\, dx \qquad \text{uniformly in } k = 0, 1, 2, \ldots.$$

5.4. If the sequence (x_n) is w.d. mod 1, then the sequence $(x_n + c)$ is w.d. mod 1, where c is a real constant.

5.5. If the sequence (x_n) is w.d. mod 1 and if (y_n) is a sequence with the property $\lim_{n\to\infty} (x_n - y_n) = c$, a real constant, then (y_n) is w.d. mod 1.

5.6. Prove in detail that the sequence $(n\theta)$, $n = 1, 2, \ldots$, with θ irrational, is w.d. mod 1.

5.7. Let $J = [a, b)$ be a subinterval of I of positive length, and let (x_n) be w.d. mod 1. Show that there exists a positive integer Q such that at least one of any Q consecutive terms of $(\{x_n\})$ lies in J.

5.8. Let (x_n) be a sequence of real numbers with $\lim_{n\to\infty}(x_{n+1} - x_n) = 0$. Prove that (x_n) is not w.d. mod 1. *Hint:* Use Exercise 5.7.

5.9. If (x_n) and (y_n) are w.d. mod 1, then the sequence $x_1, y_1, x_2, y_2, \ldots, x_n, y_n, \ldots$ is w.d. mod 1. Generalize.

5.10. Let (r_n) and (s_n) be admissible sequences, and let a and b be integers. Prove that the sequence $(ar_n + bs_n)$ is admissible.

5.11. If (s_n) is admissible, then (Δs_n) is admissible.

5.12. If α is transcendental, then for almost all x the sequence $(\alpha^n x)$, $n = 1, 2, \ldots$, is not w.d. mod 1. *Hint:* Use Lemma 5.1.

5.13. Let (x_n) be a sequence in I. For $n \geq 1$, define $a_n = [nx_n]$. Prove that the sequence (x_n) is w.d. mod 1 if and only if the sequence (a_n/n) is w.d. mod 1.

5.14. With the same notation as in Exercise 5.13, introduce the number $\alpha = \sum_{n=1}^{\infty} a_n/n!$. Prove that (x_n) is w.d. mod 1 if and only if $(n!\alpha)$ is w.d. mod 1.

5.15. Using Exercise 5.14, prove that "almost no" sequence in I is w.d. mod 1 and specify in which sense this is meant.

5.16. Carry out the arguments in Exercises 5.13, 5.14, and 5.15 with "w.d. mod 1" replaced by "u.d. mod 1". Of course, in Exercise 5.15 "almost no" has to be replaced by "almost all."

5.17. Is the sequence in Exercise 1.13 w.d. mod 1?

6. THE MULTIDIMENSIONAL CASE

Definition and Basic Results

Let s be an integer with $s \geq 2$. Let $\mathbf{a} = (a_1, \ldots, a_s)$ and $\mathbf{b} = (b_1, \ldots, b_s)$ be two vectors with real components; that is, let $\mathbf{a}, \mathbf{b} \in \mathbb{R}^s$. We say that $\mathbf{a} < \mathbf{b}$ $(\mathbf{a} \leq \mathbf{b})$ if $a_j < b_j$ $(a_j \leq b_j)$ for $j = 1, 2, \ldots, s$. The set of points $\mathbf{x} \in \mathbb{R}^s$ such that $\mathbf{a} \leq \mathbf{x} < \mathbf{b}$ will be denoted by $[\mathbf{a}, \mathbf{b})$. The other s-dimensional intervals such as $[\mathbf{a}, \mathbf{b}]$ have similar meanings. The *s-dimensional unit cube* I^s is the interval $[\mathbf{0}, \mathbf{1})$, where $\mathbf{0} = (0, \ldots, 0)$ and $\mathbf{1} = (1, \ldots, 1)$. The *integral part* of $\mathbf{x} = (x_1, \ldots, x_s)$ is $[\mathbf{x}] = ([x_1], \ldots, [x_s])$ and the *fractional part* of $\mathbf{x}$ is $\{\mathbf{x}\} = (\{x_1\}, \ldots, \{x_s\})$.

Let $(\mathbf{x}_n)$, $n = 1, 2, \ldots$, be a sequence of vectors in $\mathbb{R}^s$. For a subset E of I^s, let $A(E; N)$ denote the number of points $\{\mathbf{x}_n\}$, $1 \leq n \leq N$, that lie in E.

DEFINITION 6.1. The sequence $(\mathbf{x}_n)$, $n = 1, 2, \ldots$, is said to be *u.d. mod 1 in* $\mathbb{R}^s$ if

$$\lim_{N\to\infty} \frac{A([\mathbf{a}, \mathbf{b}); N)}{N} = \prod_{j=1}^{s} (b_j - a_j)$$

for all intervals $[\mathbf{a}, \mathbf{b}) \subseteq I^s$.

DEFINITION 6.2. The sequence $(\mathbf{x}_n)$, $n = 1, 2, \ldots$, is said to be *w.d. mod 1 in* $\mathbb{R}^s$ if, uniformly in $k = 0, 1, \ldots$, and for all intervals $[\mathbf{a}, \mathbf{b}) \subseteq I^s$, we have

$$\lim_{N\to\infty} \frac{A([\mathbf{a}, \mathbf{b}); N, k)}{N} = \prod_{j=1}^{s} (b_j - a_j),$$

where $A([\mathbf{a}, \mathbf{b}); N, k)$ denotes the number of points $\{\mathbf{x}_n\}$, $k + 1 \leq n \leq k + N$, that lie in $[\mathbf{a}, \mathbf{b})$.

DEFINITION 6.3. Let (z_n), $n = 1, 2, \ldots$, be a sequence of complex numbers. Let $\operatorname{Re} z_n = x_n$ and $\operatorname{Im} z_n = y_n$. Then the sequence (z_n) is said to be *u.d. mod 1 in* $\mathbb{C}$ if the sequence $((x_n, y_n))$, $n = 1, 2, \ldots$, is u.d. mod 1 in $\mathbb{R}^2$.

The *closed s-dimensional unit cube* $\bar{I}^s$ is the interval $[\mathbf{0}, \mathbf{1}]$. Also, for $\mathbf{x} = (x_1, \ldots, x_s)$ and $\mathbf{y} = (y_1, \ldots, y_s)$ in $\mathbb{R}^s$, let $\langle \mathbf{x}, \mathbf{y} \rangle$ be the standard inner product in $\mathbb{R}^s$; that is, $\langle \mathbf{x}, \mathbf{y} \rangle = x_1 y_1 + \cdots + x_s y_s$. Then we have the following analogues of one-dimensional results.

THEOREM 6.1. A sequence $(\mathbf{x}_n)$, $n = 1, 2, \ldots$, is u.d. mod 1 in $\mathbb{R}^s$ if and only if for every continuous complex-valued function f on $\bar{I}^s$ the following relation holds:

$$\lim_{N\to\infty} \frac{1}{N} \sum_{n=1}^{N} f(\{\mathbf{x}_n\}) = \int_{\bar{I}^s} f(\mathbf{x})\, d\mathbf{x}. \tag{6.1}$$

THEOREM 6.2: Weyl Criterion. A sequence $(\mathbf{x}_n)$, $n = 1, 2, \ldots$, is u.d. mod 1 in $\mathbb{R}^s$ if and only if for every lattice point $\mathbf{h} \in \mathbb{Z}^s$, $\mathbf{h} \neq \mathbf{0}$,

$$\lim_{N\to\infty} \frac{1}{N} \sum_{n=1}^{N} e^{2\pi i \langle \mathbf{h}, \mathbf{x}_n \rangle} = 0. \tag{6.2}$$

The proof of Theorem 6.1 is similar to that of Theorem 1.1. The proof of Theorem 6.2 follows from the fact that the finite linear combinations of the functions $e^{2\pi i \langle \mathbf{h}, \mathbf{x} \rangle}$, $\mathbf{h} \in \mathbb{Z}^s$, with complex coefficients are dense with respect to the uniform norm in the space of all continuous complex-valued functions on $\bar{I}^s$ with period 1 in each variable (see the proof of Theorem 2.1).

THEOREM 6.3. A sequence $(\mathbf{x}_n)$, $n = 1, 2, \ldots$, is u.d. mod 1 in $\mathbb{R}^s$ if and only if for every lattice point $\mathbf{h} \in \mathbb{Z}^s$, $\mathbf{h} \neq \mathbf{0}$, the sequence of real numbers $(\langle \mathbf{h}, \mathbf{x}_n \rangle)$, $n = 1, 2, \ldots$, is u.d. mod 1.

PROOF. This follows immediately from Theorems 6.2 and 2.1. ∎

EXAMPLE 6.1. If the vector $\boldsymbol{\theta} = (\theta_1, \ldots, \theta_s)$ has the property that the real numbers $1, \theta_1, \ldots, \theta_s$ are linearly independent over the rationals, then the sequence $(n\boldsymbol{\theta}) = ((n\theta_1, \ldots, n\theta_s))$, $n = 1, 2, \ldots$, is u.d. mod 1 in $\mathbb{R}^s$.

For the proof, we simply note that for every lattice point $\mathbf{h} \in \mathbb{Z}^s$, $\mathbf{h} \neq \mathbf{0}$, the real number $\langle \mathbf{h}, \boldsymbol{\theta} \rangle$ is irrational, and so Theorem 6.3 can be applied. ∎

EXAMPLE 6.2. For any lattice point $\mathbf{h} = (h_1, \ldots, h_s)$ in $\mathbb{Z}^s$, let $\|\mathbf{h}\| = \max_{1 \le j \le s} |h_j|$. If $\mathbf{p} = (\alpha_1, \ldots, \alpha_s)$ is an arbitrary vector in $\mathbb{R}^s$, define the vector $\mathbf{ph}$ by $\mathbf{ph} = (\alpha_1 h_1, \ldots, \alpha_s h_s)$. We claim that if the set of all lattice points in $\mathbb{Z}^s$ is ordered as a sequence $\mathbf{h}_1, \mathbf{h}_2, \ldots$ in such a way that $\|\mathbf{h}_m\| < \|\mathbf{h}_n\|$ implies $m < n$, then the sequence $\mathbf{ph}_1, \mathbf{ph}_2, \ldots$ is u.d. mod 1 in $\mathbb{R}^s$ for a vector $\mathbf{p}$ with irrational coordinates. For the proof, let $J = [\mathbf{a}, \mathbf{b}) \subseteq I^s$ with $\mathbf{a} = (a_1, \ldots, a_s)$ and $\mathbf{b} = (b_1, \ldots, b_s)$, and set $J_k = [a_k, b_k)$ for $1 \le k \le s$. We note that it suffices to show

$$\lim_{L \to \infty} \frac{A(J; (2L+1)^s)}{(2L+1)^s} = \prod_{k=1}^{s} (b_k - a_k). \tag{6.3}$$

By the construction of the sequence $\mathbf{h}_1, \mathbf{h}_2, \ldots$, the first $(2L+1)^s$ terms of this sequence are exactly all the lattice points $\mathbf{h}$ with $\|\mathbf{h}\| \le L$. Therefore,

$$A(J; (2L+1)^s) = \prod_{k=1}^{s} A([a_k, b_k); 2L+1; \omega_k), \tag{6.4}$$

where $\omega_k = (\alpha_k h)$, $h = 0, 1, -1, 2, -2, \ldots$. Now each of the sequences ω_k is u.d. mod 1 by Exercise 2.5. It follows that

$$\lim_{L \to \infty} \frac{A([a_k, b_k); 2L+1; \omega_k)}{2L+1} = b_k - a_k \qquad \text{for } 1 \le k \le s.$$

Together with (6.4) we obtain (6.3). ∎

THEOREM 6.4. Let $\mathbf{p}(x) = (p_1(x), \ldots, p_s(x))$, where all $p_i(x)$ are real polynomials, and suppose $\mathbf{p}(x)$ has the property that for each lattice point $\mathbf{h} \in \mathbb{Z}^s$, $\mathbf{h} \neq \mathbf{0}$, the polynomial $\langle \mathbf{h}, \mathbf{p}(x) \rangle$ has at least one nonconstant term with irrational coefficient. Then the sequence $(\mathbf{p}(n))$, $n = 1, 2, \ldots$, is u.d. mod 1 in $\mathbb{R}^s$.

PROOF. According to Theorem 3.2 the sequence $(p(n))$, $n = 1, 2, \ldots$, where $p(x)$ is a real polynomial that has a nonconstant term with an irrational coefficient, is u.d. mod 1. Hence, Theorem 6.4 is true in view of Theorem 6.3. ∎

For the sake of completeness we mention without proof the following general theorem.

THEOREM 6.5. Let Ω denote a sequence of intervals Q of the form $Q = [\mathbf{a}, \mathbf{b})$ with lattice points $\mathbf{a} < \mathbf{b}$ in $\mathbb{Z}^s$. To each $Q \in \Omega$ there corresponds a positive integer n and, furthermore, $2n$ numbers α_ν, β_ν $(1 \le \nu \le n)$ satisfying $\alpha_\nu < \beta_\nu \le \alpha_\nu + 1$ for $1 \le \nu \le n$, and n real-valued functions $f_\nu(\mathbf{x})$,

$1 \leq \nu \leq n$, defined for each lattice point $\mathbf{x} \in Q$. For each $Q \in \Omega$ and $c > 0$ we put

$$T(Q; c) = \sum_{\mathbf{h}}' \left| \frac{1}{N(Q)} \sum_{\mathbf{x} \in Q} e^{2\pi i (h_1 f_1(\mathbf{x}) + \cdots + h_n f_n(\mathbf{x}))} \right|,$$

where $N(Q)$ denotes the number of lattice points $\mathbf{x} \in Q$, and where $\sum_{\mathbf{h}}'$ is to be extended over all lattice points $\mathbf{h} = (h_1, \ldots, h_n) \neq (0, \ldots, 0)$ that satisfy

$$|h_\nu| \leq \frac{cn}{\beta_\nu - \alpha_\nu} \log \frac{2n}{\beta_\nu - \alpha_\nu} \qquad \text{for } 1 \leq \nu \leq n.$$

Assume that for each fixed value of c we have $T(Q; c) \to 0$ as Q runs through Ω. Then

$$\frac{N_p(Q)}{N(Q) \prod_{\nu=1}^{n} (\beta_\nu - \alpha_\nu)} \to 1 \quad \text{as } Q \text{ runs through } \Omega,$$

where $N_p(Q)$ denotes the number of lattice points $\mathbf{x} \in Q$ for which

$$\alpha_\nu \leq f_\nu(\mathbf{x}) < \beta_\nu \ (\mathrm{mod}\ 1) \qquad \text{for } 1 \leq \nu \leq n.$$

Applications

THEOREM 6.6. Let f be Riemann-integrable on $[0, 1]$, and let $1, \theta, \alpha$ be linearly independent over the rationals. Then the power series $G(z) = \sum_{n=1}^{\infty} f(\{n\alpha\}) z^n$ has the property that

$$\lim_{r \to 1-0} (1 - r) G(re^{2\pi i \theta}) = 0. \tag{6.5}$$

PROOF. By what we have seen in the proof of Theorem 2.4, the identity

$$\lim_{r \to 1-0} (1 - r) G(re^{2\pi i \theta}) = \lim_{N \to \infty} \frac{1}{N} \sum_{n=1}^{N} f(\{n\alpha\}) e^{2\pi i n \theta}$$

holds whenever the limit on the right-hand side exists. Since it is assumed that 1, θ, and α are linearly independent over the rationals, the sequence of vectors $((n\alpha, n\theta))$, $n = 1, 2, \ldots$, is u.d. mod 1 in $\mathbb{R}^2$. So for all Riemann-integrable functions g on $\bar{I}^2$ we have

$$\lim_{N \to \infty} \frac{1}{N} \sum_{n=1}^{N} g(\{n\alpha\}, \{n\theta\}) = \int_0^1 \int_0^1 g(x, y)\, dx\, dy$$

according to the two-dimensional analogue of Corollary 1.1 (see Exercise 6.3), and therefore

$$\lim_{N \to \infty} \frac{1}{N} \sum_{n=1}^{N} f(\{n\alpha\}) e^{2\pi i n \theta} = \int_0^1 \int_0^1 f(x) e^{2\pi i y}\, dx\, dy = 0. \qquad \blacksquare$$

It should be remarked that, in case $\theta = m\alpha + p$ with integers m and p, the proof of Theorem 2.4 shows that the limit in (6.5) is equal to the integral d_m occurring in that proof.

For other types of applications, see, for instance, Chapter 4, Example 4.1, and the proof of Theorem 1.8 of Chapter 5.

Notes

U.d. mod 1 in $\mathbb{R}^s$ was first considered by Weyl [2, 4], who proved Theorems 6.1–6.4 as well as the result in Example 6.1. A discussion of the exceptional cases in this example was also carried out by Weyl [4]. For further remarks concerning this example, see Bergström [1], Jacobs [1], Rizzi [1], and Slater [2]. We refer also to the notes in Chapter 2, Sections 3 and 5. Example 6.2 is from Volkmann [3, 5], who applied the result to additive number theory. In connection with Theorem 6.4, see also Kovalevskaja [1] and the notes in Chapter 2, Section 3. Results on other special classes of sequences can be found in Delange [5], Karimov [1], Korobov [12], and Polosuev [2, 4]. For the many investigations concerning the sequences $(\mathbf{A}^n\mathbf{x})$, where $\mathbf{A}$ is a given real $s \times s$ matrix and $\mathbf{x}$ a given point in $\mathbb{R}^s$, see the notes in Section 8 of this chapter. Sequences in $\mathbb{R}^s$ of the form

$$((x_n, x_{n+1}, \ldots, x_{n+s-1})),$$

$n = 1, 2, \ldots$, where (x_n) is a given sequence in $\mathbb{R}$, have been studied by Carroll [2], Cigler [2], Franklin [2], Hlawka [8], Kemperman [4], and Knuth [2, Chapter 3] (compare also with Chapter 3, Section 3). Cigler [14, 15] shows that the sequences in Exercise 6.10 are w.d. mod 1 in $\mathbb{R}^s$. For further results on w.d. mod 1 in $\mathbb{R}^s$, see Gerl [4, 6]. Koksma's metric theorem (Theorem 4.3) has been extended to the multidimensional case by LeVeque [2] and Gerl [2], and Philipp [1, 3] has shown an analogue of Theorem 4.1. See also Carroll [2] for another metric result. Definition 6.3 goes back to LeVeque [3], who proved that for almost all $z \in \mathbb{C}$ with $|z| \geq 1$ the sequence (z^n), $n = 1, 2, \ldots$, is u.d. mod 1 in $\mathbb{C}$.

A general definition of *u.d. mod 1 in sequences of intervals*, which is the basis for Theorem 6.5, was given by van der Corput [5] (see also Koksma [4, Kap. 8]). A proof of Theorem 6.5 (also due to van der Corput) has never been published, although the theorem was applied a number of times (see, e.g., Koksma [1], Teghem [1]). A refined version of Theorem 6.5 was shown by Koksma [11]. Van der Corput [5] extends many of Weyl's results to u.d. mod 1 in sequences of intervals and also proves a difference theorem. For some further results in the one-dimensional case, see Kuipers [4].

U.d. of sequences on curves and surfaces was studied by Gerl [1, 3]. For the case of a sphere, see Arnol'd and Krylov [1] and Gerl [9]. We refer also to a related investigation by Každan [1]. Jessen [1] proves a Weyl criterion for u.d. in I^∞, the Cartesian product of denumerably many copies of I. An even more general setting for u.d. will be developed in Chapters 3 and 4.

For literature on Theorem 6.6 and related results, see the notes in Section 2. Luthar [1] applies u.d. mod 1 in $\mathbb{R}^s$ to a lattice point problem, Pham Phu Hien [1] applies it to distribution of functions, and Slater [2] discusses applications to gas theory (see also Hlawka [19, 21, 23]). Further applications are presented in Chapter 2, Section 5.

Exercises

6.1. Prove Theorem 6.1 in detail.

6.2. Prove Theorem 6.2 in detail.

6.3. A sequence $(\mathbf{x}_n)$ is u.d. mod 1 in $\mathbb{R}^s$ if and only if the relation (6.1) holds for every Riemann-integrable function f on $\bar{I}^s$.

6.4. A sequence $(\mathbf{x}_n)$, $n = 1, 2, \ldots$, in $\mathbb{R}^s$ with $\mathbf{x}_n = (x_{1n}, x_{2n}, \ldots, x_{sn})$, $n = 1, 2, \ldots$, is u.d. mod 1 in $\mathbb{R}^s$ if and only if

$$\lim_{N\to\infty} \frac{1}{N} \sum_{n=1}^{N} \{x_{1n}\}^{h_1} \cdots \{x_{sn}\}^{h_s} = \prod_{j=1}^{s} (h_j + 1)^{-1}$$

for all nonnegative integers $h_1, \ldots, h_s$.

6.5. If a sequence is u.d. mod 1 in $\mathbb{R}^s$, then all s coordinate sequences are u.d. mod 1.

6.6. If a sequence $(\mathbf{x}_n)$ is u.d. mod 1 in $\mathbb{R}^s$, then the sequence $(\{\mathbf{x}_n\})$ of fractional parts is dense in $\bar{I}^s$.

6.7. If the real numbers $1, \theta_1, \ldots, \theta_s$ are linearly dependent over the rationals, then the sequence $((n\theta_1, \ldots, n\theta_s))$, $n = 1, 2, \ldots$, is not u.d. mod 1 in $\mathbb{R}^s$.

6.8. Let $\theta_1, \ldots, \theta_s$ be irrational numbers. Then the sequence

$$((\theta_1 n^s, \theta_2 n^{s-1}, \ldots, \theta_s n)),$$

$n = 1, 2, \ldots$, is u.d. mod 1 in $\mathbb{R}^s$.

6.9. Let $\alpha_1, \ldots, \alpha_s$ be nonzero real numbers, and let $\tau_1, \ldots, \tau_s$ be distinct positive numbers not in $\mathbb{Z}$. Then the sequence $((\alpha_1 n^{\tau_1}, \ldots, \alpha_s n^{\tau_s}))$, $n = 1, 2, \ldots$, is u.d. mod 1 in $\mathbb{R}^s$.

6.10. Let $p(x)$ be a polynomial of degree $s \geq 1$ with irrational leading coefficient. Then the sequence $((p(n), p(n+1), \ldots, p(n+s-1)))$, $n = 1, 2, \ldots$, is u.d. mod 1 in $\mathbb{R}^s$.

6.11. Prove that the sequence $((n^2 \log n, n \log n))$, $n = 1, 2, \ldots$, is u.d. mod 1 in $\mathbb{R}^2$.

6.12. Let (a_n), $n = 1, 2, \ldots$, be a sequence of distinct integers. Prove that for almost all $(\alpha_1, \ldots, \alpha_s) \in \mathbb{R}^s$ (in the sense of Lebesgue measure) the sequence $((a_n\alpha_1, \ldots, a_n\alpha_s))$, $n = 1, 2, \ldots$, is u.d. mod 1 in $\mathbb{R}^s$.

6.13. The sequence $(\mathbf{x}_n)$, $n = 1, 2, \ldots$, is w.d. mod 1 in $\mathbb{R}^s$ if and only if for every lattice point $\mathbf{h} \in \mathbb{Z}^s$, $\mathbf{h} \neq \mathbf{0}$, we have

$$\lim_{N\to\infty} \frac{1}{N} \sum_{n=k+1}^{k+N} e^{2\pi i \langle \mathbf{h}, \mathbf{x}_n \rangle} = 0 \qquad \text{uniformly in } k = 0, 1, \ldots.$$

6.14. Let the vector $\boldsymbol{\theta}$ be as in Example 6.1. Then the sequence $(n\boldsymbol{\theta})$, $n = 1, 2, \ldots$, is w.d. mod 1 in $\mathbb{R}^s$.

6.15. If $z \in \mathbb{C}$ and $|z| \leq 1$, then (z^n), $n = 1, 2, \ldots$, is not u.d. mod 1 in $\mathbb{C}$.

6.16. The sequence $(n\xi)$, $n = 1, 2, \ldots$, is u.d. mod 1 in $\mathbb{C}$ where ξ is a primitive fifth root of unity.

6.17. The sequence (n^i), $n = 1, 2, \ldots$, is not u.d. mod 1 in $\mathbb{C}$.

6.18. The power series $F(z) = \sum_{n=1}^{\infty} \{n\alpha\} z^n$, where α is irrational, has the property that $\lim_{r \to 1-0} (1 - r)F(re^{2\pi i\theta}) = (2\pi iq)^{-1}$, where $\theta = p + q\alpha$, $p, q \in \mathbb{Z}$, $q \neq 0$.

7. DISTRIBUTION FUNCTIONS

Various Types of Distribution Functions

Let (x_n), $n = 1, 2, \ldots$, be a sequence of real numbers, and let $A([a, b); N)$ have the same meaning as in Section 1.

Definition 7.1. The sequence (x_n) is said to have the *asymptotic distribution function mod 1* (abbreviated a.d.f. (mod 1)) $g(x)$ if

$$\lim_{N\to\infty} \frac{A([0, x); N)}{N} = g(x) \qquad \text{for } 0 \leq x \leq 1. \tag{7.1}$$

Evidently, the function g on $[0, 1]$ is nondecreasing with $g(0) = 0$ and $g(1) = 1$. It will be shown later on (see Chapter 2, Theorem 4.3) that for any function g satisfying these conditions there exists a sequence (x_n) having g as its a.d.f. (mod 1). An arbitrary sequence (x_n) need not have an a.d.f. (mod 1). But in any case we can consider the limits

$$\underline{\lim}_{N\to\infty} \frac{A([0, x); N)}{N} = \varphi(x) \qquad \text{for } 0 \leq x \leq 1,$$

$$\overline{\lim}_{N\to\infty} \frac{A([0, x); N)}{N} = \Phi(x) \qquad \text{for } 0 \leq x \leq 1.$$

The functions φ and Φ are nondecreasing with $\varphi(0) = \Phi(0) = 0$ and $\varphi(1) = \Phi(1) = 1$, while $0 \leq \varphi(x) \leq \Phi(x) \leq 1$ for $0 \leq x \leq 1$. The functions φ and Φ may be called the *lower* resp. the *upper d.f. (mod 1)* of (x_n). If $\varphi \equiv \Phi$, then the sequence (x_n) under consideration has the a.d.f. (mod 1) φ. If $\varphi(x) = \Phi(x) = x$ for $0 \leq x \leq 1$, the sequence (x_n) is u.d. mod 1.

Definition 7.2. Let (x_n) be a sequence of real numbers. If there exists an increasing sequence of natural numbers $N_1, N_2, \ldots$ such that

$$\lim_{i\to\infty} \frac{A([0, x); N_i)}{N_i} = z(x) \qquad \text{for } 0 \leq x \leq 1, \tag{7.2}$$

then $z(x)$ is called a *distribution function mod 1* (abbreviated d.f. (mod 1)) of (x_n). If (7.2) holds with $z(x) = x$, then the sequence (x_n) is called *almost u.d. mod 1*.

THEOREM 7.1. A sequence (x_n) of real numbers has at least one d.f. (mod 1).

PROOF. The functions F_N, defined by $F_N(x) = A([0, x); N)/N$ for $0 \leq x \leq 1$, $F_N(x) = 0$ for $x < 0$, and $F_N(x) = 1$ for $x > 1$, are distribution functions in the sense of probability theory; that is, they are nondecreasing and left continuous on $\mathbb{R}$ with $\lim_{x\to-\infty} F_N(x) = 0$ and $\lim_{x\to\infty} F_N(x) = 1$. By the Helly selection principle (Loève [1, p. 179]), there exists a distribution function z on $\mathbb{R}$ and a subsequence (N_j) of the natural numbers such that $\lim_{j\to\infty} F_{N_j}(x) = z(x)$ for all continuity points x of z. By passing to a suitable subsequence of (N_j), one can also guarantee the existence of the limit at the countably many points of discontinuity of z (compare with Exercise 7.7). ∎

Criteria

THEOREM 7.2. A sequence (x_n) has the continuous a.d.f. (mod 1) $g(x)$ if and only if for every real-valued continuous function f on [0, 1],

$$\lim_{N\to\infty} \frac{1}{N} \sum_{n=1}^{N} f(\{x_n\}) = \int_0^1 f(x)\, dg(x). \tag{7.3}$$

PROOF. The necessity of (7.3) can be shown by using the definition of the Riemann-Stieltjes integral. The line of proof is the same as that of the first part of the proof of Theorem 1.1. In an alternative method, one uses the functions F_N from the proof of Theorem 7.1, notes that $(1/N)\sum_{n=1}^{N} f(\{x_n\}) = \int_0^1 f(x)\, dF_N(x)$, and applies the Helly-Bray lemma (Loève [1, p. 180]). The sufficiency of (7.3) follows as in the second part of the proof of Theorem 1.1. ∎

THEOREM 7.3. A sequence (x_n) has the continuous a.d.f. (mod 1) $g(x)$ if and only if

$$\lim_{N\to\infty} \frac{1}{N} \sum_{n=1}^{N} e^{2\pi i h x_n} = \int_0^1 e^{2\pi i h x}\, dg(x) \qquad \text{for all integers } h \neq 0. \tag{7.4}$$

PROOF. The necessity of (7.4) follows from Theorem 7.2. For the sufficiency, one notes that (7.4) holds for $h = 0$ as well, and then one proceeds as in the proof of Theorem 2.1. ∎

Let g_1 and g_2 be two nondecreasing functions on [0, 1] with $g_1(0) = g_2(0) = 0$ and $g_1(1) = g_2(1) = 1$. The functions g_1 and g_2 are considered to be equivalent, denoted by $g_1 \sim g_2$, if $g_1(x) = g_2(x)$ for every x where both g_1 and g_2 are continuous. Using the fact that the numbers x of the latter type are dense in [0, 1], it follows that $g_1(x + 0) = g_2(x + 0)$ and $g_1(x - 0) = g_2(x - 0)$ for all $x \in (0, 1)$. In particular, we have $g_1(x) = g_2(x)$ for every x where either g_1 or g_2 is continuous. We conclude also that $\sim$ is an equivalence relation. The equivalence class containing the function g will be denoted by $\tilde{g}$. If g_1 and g_2 belong to $\tilde{g}$ and f is continuous on $[0, 1]$, then

$$\int_0^1 f(x)\, dg_1(x) = \int_0^1 f(x)\, dg_2(x),$$

and so the common value of these integrals may be denoted by $\int_0^1 f(x)\, d\tilde{g}(x)$. Now we have the following result.

THEOREM 7.4. The sequence (x_n) has a d.f. (mod 1) belonging to the equivalence class $\tilde{g}$ if and only if there exists a subsequence (N_i) of the natural numbers such that

$$\lim_{i\to\infty} \frac{1}{N_i} \sum_{n=1}^{N_i} f(\{x_n\}) = \int_0^1 f(x)\, d\tilde{g}(x) \tag{7.5}$$

for all real-valued continuous functions f on $[0, 1]$.

PROOF. The necessity of (7.5) follows as in Theorem 7.2 by using the functions F_{N_i} and applying the Helly-Bray lemma. In order to show the sufficiency of (7.5) one observes that (7.5) implies in the usual way (see the second part of the proof of Theorem 1.1) that

$$\lim_{i\to\infty} \frac{A([0, x); N_i)}{N_i} = g(x)$$

holds for every x where g is continuous. By selecting a suitable subsequence of (N_i) one can guarantee the existence of the limit at the points of the countable set where g is discontinuous. The resulting d.f. (mod 1) of (x_n) is obviously equivalent to g. ■

Miscellaneous Results

THEOREM 7.5: Wiener-Schoenberg Theorem. The sequence (x_n) has a continuous a.d.f. (mod 1) if and only if for every positive integer h the limit

$$\omega_h = \lim_{N\to\infty} \frac{1}{N} \sum_{n=1}^{N} e^{2\pi i h x_n} \tag{7.6}$$

exists and, in addition,

$$\lim_{H\to\infty} \frac{1}{H} \sum_{h=1}^{H} |\omega_h|^2 = 0. \tag{7.7}$$

PROOF. The existence of the limits (7.6) is certainly necessary. Next we show that if for (x_n) we have

$$\omega_h = \lim_{N\to\infty} \frac{1}{N} \sum_{n=1}^{N} e^{2\pi i h x_n} = \int_0^1 e^{2\pi i h x}\, dg(x)$$

for all positive integers h, then $g(x)$ is continuous if and only if (7.7) holds. For we have

$$\begin{aligned}\lim_{N\to\infty}\frac{1}{H}\sum_{h=1}^{H}|\omega_h|^2 &= \lim_{H\to\infty}\frac{1}{H}\sum_{h=1}^{H}\omega_h\bar{\omega}_h\\ &= \lim_{H\to\infty}\frac{1}{H}\sum_{h=1}^{H}\int_0^1\int_0^1 e^{2\pi ih(x-y)}\,dg(x)\,dg(y)\\ &= \int_0^1\int_0^1\left(\lim_{H\to\infty}\frac{1}{H}\sum_{h=1}^{H}e^{2\pi ih(x-y)}\right)dg(x)\,dg(y)\\ &= \iint\limits_{\{(x,y)\in\bar{I}^2:\,x-y\in\mathbb{Z}\}} dg(x)\,dg(y),\end{aligned}$$

and the last integral is zero if and only if g is continuous. In particular, if (x_n) has a continuous a.d.f. (mod 1), then (7.7) follows. Finally, suppose that the limits (7.6) exist and that (7.7) holds. By the usual approximation method, it follows that the limit

$$L(f) = \lim_{N\to\infty}\frac{1}{N}\sum_{n=1}^{N}f(\{x_n\})$$

exists for every continuous function f on $[0, 1]$ with $f(0) = f(1)$. If the space of these functions is equipped with the supremum norm, then L is a bounded linear functional on it with $L(f) \geq 0$ whenever $f \geq 0$. Thus, by the Riesz representation theorem,

$$L(f) = \int_0^1 f(x)\,dg(x)$$

with a nondecreasing function g on $[0, 1]$. Without loss of generality, we may assume $g(0) = 0$. Then, choosing $f \equiv 1$, we conclude $g(1) = 1$. By what we have already shown, $g(x)$ is continuous. The rest follows from Theorem 7.3. ∎

THEOREM 7.6. Suppose the sequence (x_n) with $x_n \notin \mathbb{Z}$ has the continuous a.d.f. (mod 1) $g(x)$. Then the sequence $(1/\{x_n\})$ has the a.d.f. (mod 1) given by

$$g^*(x) = \sum_{n=1}^{\infty}\left(g\left(\frac{1}{n}\right) - g\left(\frac{1}{n+x}\right)\right) \qquad \text{for } 0 \leq x \leq 1.$$

PROOF. Let (u_n) be a sequence of real numbers. For $N \geq 1$ and $\xi \in \mathbb{R}$, let $N^*(\xi)$ be the number of u_n, $1 \leq n \leq N$, such that $u_n < \xi$. If $\lim_{N\to\infty} N^*(\xi)/N = f(\xi)$ exists for all ξ and if in addition $\lim_{\xi\to\infty} f(\xi) = 1$ and $\lim_{\xi\to-\infty} f(\xi) = 0$, then f is called the *asymptotic d.f.* of (u_n). Suppose (u_n) has the continuous asymptotic d.f. $f(\xi)$. Then, by the Helly-Bray theorem (Loève [1, p. 182]),

$$\lim_{N\to\infty} \frac{1}{N} \sum_{n=1}^{N} e^{2\pi i h u_n} = \lim_{N\to\infty} \int_{-\infty}^{\infty} e^{2\pi i h \xi}\, d\left(\frac{N^*(\xi)}{N}\right)$$
$$= \int_{-\infty}^{\infty} e^{2\pi i h \xi}\, df(\xi) \qquad \text{for } h \in \mathbb{Z}. \tag{7.8}$$

On the other hand, we have

$$\int_{-\infty}^{\infty} e^{2\pi i h \xi}\, df(\xi) = \sum_{n=-\infty}^{\infty} \int_{n}^{n+1} e^{2\pi i h \xi}\, df(\xi)$$
$$= \sum_{n=-\infty}^{\infty} \int_{0}^{1} e^{2\pi i h x}\, df(n + x)$$
$$= \sum_{n=-\infty}^{\infty} \int_{0}^{1} e^{2\pi i h x}\, d(f(n + x) - f(n))$$
$$= \int_{0}^{1} e^{2\pi i h x}\, d\left(\sum_{n=-\infty}^{\infty} (f(n + x) - f(n))\right), \tag{7.9}$$

where the interchange of summation and integration can be justified by integration by parts, using the fact that the series $\sum_{n=-\infty}^{\infty} (f(n + x) - f(n))$ is uniformly convergent in $0 \le x \le 1$ since it has the convergent majorant $\sum_{n=-\infty}^{\infty} (f(n + 1) - f(n)) = 1$. Combining (7.8) and (7.9) with Theorem 7.3, we have shown that whenever (u_n) has the continuous asymptotic d.f. $f(\xi)$, then (u_n) has a continuous a.d.f. (mod 1) given by $\sum_{n=-\infty}^{\infty} (f(n + x) - f(n))$ for $0 \le x \le 1$.

Now consider the given sequence (x_n). Then by elementary reasoning one shows that $(1/\{x_n\})$ has the continuous asymptotic d.f. $f(\xi)$ given by $f(\xi) = 1 - g(\xi^{-1})$ for $\xi > 1$ and $f(\xi) = 0$ for $\xi \le 1$ (use Exercise 7.5). By what we have already shown, it follows that $(1/\{x_n\})$ has the a.d.f. (mod 1) given by

$$g^*(x) = \sum_{n=-\infty}^{\infty} (f(n + x) - f(n))$$
$$= \sum_{n=1}^{\infty} \left(g\left(\frac{1}{n}\right) - g\left(\frac{1}{n + x}\right)\right) \qquad \text{for } 0 \le x \le 1. \quad ■$$

An Elementary Method

Remarkable results can already be obtained by applying elementary methods. In the following a property of the sequence $(\log n)$, $n = 1, 2, \ldots$, is shown. Let x be a number with $0 \le x \le 1$. Then, as in Example 2.5, we have

$$A([0, x); n) = \sum_{k=0}^{N-1} (e^{k+x} - e^{k}) + e^{N+\lambda_n} - e^{N} + O(N),$$

where $N = [\log n]$ and $\lambda_n = \min(x, \{\log n\})$. It follows from Example 2.5 that $(\{\log n\})$, $n = 1, 2, \ldots$, is dense in $\bar{I}$. We choose a subsequence (n_i), $i = 1, 2, \ldots$, of the natural numbers such that $\{\log n_i\} \to \xi$ as $i \to \infty$, where ξ is a (fixed) number with $0 \leq \xi \leq 1$. Then, if n runs through this subsequence, we have

$$\frac{A([0, x); n)}{n} = \frac{A([0, x); n)}{e^{N+\{\log n\}}} \to \frac{e^x - 1}{e - 1} e^{-\xi} + (e^\lambda - 1)e^{-\xi},$$

$$\text{where} \qquad \lambda = \min(x, \xi).$$

Let us denote the resulting d.f. (mod 1) by $z(x, \xi)$. We note that $z(x, 0) = z(x, 1) = (e^x - 1)/(e - 1)$. By Theorem 7.4 we have for every continuous function f on $[0, 1]$,

$$\frac{1}{n}\sum_{k=1}^{n} f(\{\log k\}) \to \int_0^1 f(x)\, dz(x, \xi) \tag{7.10}$$

if n runs through a sequence for which $\{\log n\} \to \xi$. By differentiating the function $z(x, \xi)$ with respect to x, one obtains an alternate form for (7.10):

$$\frac{1}{n}\sum_{k=1}^{n} f(\{\log k\}) \to \int_0^1 f(x)K(x, \xi)\, dx, \tag{7.11}$$

where

$$K(x, \xi) = \frac{e^{x-\xi+1}}{e - 1} \qquad \text{for } 0 \leq x < \xi,$$

$$K(x, \xi) = \frac{e^{x-\xi}}{e - 1} \qquad \text{for } \xi \leq x \leq 1.$$

For fixed f, consider the right-hand side of (7.11) and denote it by $h(\xi)$. Now $h(\xi)$ is continuous in $0 \leq \xi \leq 1$ with $h(0) = h(1)$, while $h(\xi)$ is in general not a constant. Hence, the limit points of the set of values

$$(1/n) \sum_{k=1}^{n} f(\{\log k\}),$$

$n = 1, 2, \ldots$, may fill an entire interval.

The above elementary method can be applied to functions not necessarily differentiable. We mention the following result without proof.

THEOREM 7.7. Let $f(t)$, $t \geq 1$, be a continuous increasing function with $\lim_{t\to\infty} f(t) = \infty$. Let $F(u)$ denote the inverse function of $f(t)$. Assume that $\Delta F(n) = F(n + 1) - F(n) \to \infty$ as n runs through the positive integers. Moreover, it is assumed that for every x, $0 \leq x \leq 1$, the limit

$$\lim_{n\to\infty} \frac{F(n + x) - F(n)}{\Delta F(n)} = \psi(x)$$

exists. Then, if

$$\lim_{n\to\infty} \frac{F(n)}{F(n + x)} = \chi(x) \qquad \text{for } 0 \leq x \leq 1,$$

the functions $\varphi(x) = \psi(x)$ and $\Phi(x) = 1 - \chi(x)(1 - \psi(x))$, $0 \leq x \leq 1$, are respectively the lower and upper d.f. (mod 1) of the sequence $(f(n))$, $n = 1, 2, \ldots$.

Application of Theorem 7.7 to the function ${}^b\log t$, $b > 1$, gives immediately the result that $\varphi(x) = (b^x - 1)/(b - 1)$ and $\Phi(x) = (1 - b^{-x})/(1 - b^{-1})$ are respectively the lower and upper d.f. (mod 1) of the sequence $({}^b\log n)$, $n = 1, 2, \ldots$.

Metric Theorems

THEOREM 7.8. Let (x_n) be a sequence of real numbers, and let A be the set of all real numbers a such that (ax_n), $n = 1, 2, \ldots$, converges mod 1 to 0 (i.e., 0 and 1 are the only limit points of $(\{ax_n\})$). Furthermore, suppose that (x_n) does not converge to 0 as $n \to \infty$. Then the set A has Lebesgue measure 0.

PROOF. If $\lim_{n\to\infty} x_n = 0$, then (ax_n) converges mod 1 to 0 for every $a \in \mathbb{R}$. Hence, this case is excluded. Assume the sequence (x_n) has a finite limit point $c \neq 0$. Then there is a subsequence (x_{n_i}) converging to c as $i \to \infty$. If $a \in A$, then (ax_{n_i}), $i = 1, 2, \ldots$, converges mod 1 to 0, but also $\lim_{i\to\infty} ax_{n_i} = ac$, and so $ac \in \mathbb{Z}$. Then $A \subseteq \{0, \pm 1/c, \pm 2/c, \ldots\}$, so that the Lebesgue measure $\lambda(A)$ of A is 0. It remains to consider the case $\lim_{n\to\infty} |x_n| = \infty$. We note that

$$\lim_{n\to\infty} e^{2\pi i a x_n} = 1 \qquad \text{if } a \in A. \tag{7.12}$$

Let $A_z = A \cap [-z, z]$ for $z > 0$. Then (7.12) implies by the dominated convergence theorem

$$\lim_{n\to\infty} \int_{A_z} e^{2\pi i a x_n}\, da = \lambda(A_z). \tag{7.13}$$

On the other hand, we conclude from $\lim_{n\to\infty} |x_n| = \infty$ and the Riemann-Lebesgue lemma that the limit on the left-hand side of (7.13) is equal to 0, or $\lambda(A_z) = 0$, and finally $\lambda(A) = 0$. ∎

THEOREM 7.9. Let (x_n) be a sequence of real numbers satisfying $\lim_{n\to\infty} |x_n| = \infty$. Let B be the set of positive numbers b such that the sequence (bx_n), $n = 1, 2, \ldots$, has an a.d.f. (mod 1). If $\lambda(B) > 0$, then for almost all $b \in B$ the sequence (bx_n) is u.d. mod 1.

PROOF. For $b \in B$, let $g_b(x)$, $0 \leq x \leq 1$, be the a.d.f. (mod 1) of (bx_n). By the necessary part of Theorem 7.3, which holds also for discontinuous $g(x)$ (see the proof of Theorem 7.2), we have

$$\omega_h(b) = \lim_{n\to\infty} \frac{1}{n} \sum_{k=1}^{n} e^{2\pi i h b x_k} = \int_0^1 e^{2\pi i h x}\, dg_b(x) \qquad \text{for } h \in \mathbb{Z}.$$

For $z > 0$, set $B_z = B \cap [0, z]$. Then, because of $\lim_{n\to\infty} |x_n| = \infty$ and the Riemann-Lebesgue lemma, we have for $h \neq 0$,

$$\lim_{n\to\infty} \int_{B_z} e^{2\pi i h b x_n}\, db = 0.$$

Then by Cauchy's theorem

$$\lim_{n\to\infty} \frac{1}{n} \sum_{k=1}^{n} \int_{B_z} e^{2\pi i h b x_n}\, db = 0,$$

and from the dominated convergence theorem

$$\int_{B_z} \omega_h(b)\, db = 0.$$

Since this holds for all $z > 0$, we have $\omega_h(b) = 0$ for all $b \in B$ with the exception of a set S_h with $\lambda(S_h) = 0$. But $\lambda(\bigcup_{h\neq 0} S_h) = 0$, and so, for almost all $b \in B$, we have $\omega_h(b) = 0$ for all $h \neq 0$. This property characterizes u.d. mod 1 of the sequence (bx_n). ∎

Summation Methods

In general, a summation method S is some notion of convergence for sequences of complex numbers. We call a sequence (z_n) *summable by S to the value z* if (z_n) "converges" to z under this notion of convergence S. The most common types of summation methods are introduced via an infinite real matrix $\mathbf{A} = (a_{nk})$, $n = 1, 2, \ldots$, $k = 1, 2, \ldots$. Such summation methods are called *matrix methods*, and by an abuse of language we speak of the matrix method $\mathbf{A}$. We think of a given sequence (z_n) of complex numbers being transformed by means of the matrix $\mathbf{A}$ into the sequence (z_n') of so-called **A**-*transforms*, where $z_n' = \sum_{k=1}^{\infty} a_{nk} z_k$. Then (z_n) is said to be *summable by* **A** *to the value* z if $\lim_{n\to\infty} z_n' = z$. Here we shall only be interested in a special class of matrix methods, namely, positive Toeplitz matrices (for a more general class, see Chapter 3, Section 4). The matrix $\mathbf{A} = (a_{nk})$, $n = 1, 2, \ldots$, $k = 1, 2, \ldots$, is called a *positive Toeplitz matrix* if $a_{nk} \geq 0$ for all n and k and if $\lim_{n\to\infty} \sum_{k=1}^{\infty} a_{nk} = 1$.

DEFINITION 7.3. Let $\mathbf{A} = (a_{nk})$ be a positive Toeplitz matrix, and let (x_n), $n = 1, 2, \ldots$, be a sequence of real numbers. For $0 \leq x \leq 1$, let c_x be the characteristic function of the interval $[0, x)$. The function $g(x)$, $0 \leq x \leq 1$, is the **A**-*asymptotic distribution function mod 1* (abbreviated **A**-a.d.f. (mod 1)) of (x_n) if the sequence $(c_x(\{x_n\}))$, $n = 1, 2, \ldots$, is summable by **A** to the value $g(x)$ for $0 \leq x \leq 1$; that is, if

$$\lim_{n\to\infty} \sum_{k=1}^{\infty} a_{nk} c_x(\{x_k\}) = g(x) \qquad \text{for } 0 \leq x \leq 1.$$

The function $g(x)$ is nondecreasing on $[0, 1]$ with $g(0) = 0$ and $g(1) = 1$. In the case $g(x) = x$ for $0 \leq x \leq 1$, the sequence (x_n) is called **A**-*u.d. mod 1*. If we choose for **A** the matrix method of arithmetic means, defined by $a_{nk} = 1/n$ for $1 \leq k \leq n$ and $a_{nk} = 0$ for $k > n$, then we see that Definition 7.3 contains Definition 7.1 as a special case. Theorems 7.2 and 7.3 can be generalized as follows.

THEOREM 7.10. Let $\mathbf{A} = (a_{nk})$ be a positive Toeplitz matrix. A sequence (x_n) has the continuous **A**-a.d.f. (mod 1) $g(x)$ if and only if for every real-valued continuous function f on $[0, 1]$,

$$\lim_{n\to\infty} \sum_{k=1}^{\infty} a_{nk} f(\{x_k\}) = \int_0^1 f(x)\, dg(x).$$

PROOF. One proceeds as in Theorem 1.1, but replaces arithmetic means by **A**-transforms and Riemann integrals by Riemann-Stieltjes integrals with respect to the function $g(x)$. ■

THEOREM 7.11. Let $\mathbf{A} = (a_{nk})$ be a positive Toeplitz matrix. A sequence (x_n) has the continuous **A**-a.d.f. (mod 1) $g(x)$ if and only if

$$\lim_{n\to\infty} \sum_{k=1}^{\infty} a_{nk} e^{2\pi i h x_k} = \int_0^1 e^{2\pi i h x}\, dg(x) \qquad \text{for all integers } h \neq 0. \tag{7.14}$$

PROOF. See the proof of Theorem 7.3. Compare also with the proof of Theorem 4.1 of Chapter 3. ■

If $g(x) = x$, (7.14) becomes

$$\lim_{n\to\infty} \sum_{k=1}^{\infty} a_{nk} e^{2\pi i h x_k} = 0 \qquad \text{for all integers } h \neq 0. \tag{7.15}$$

Since the theory of summation methods does not lie in the mainstream of our investigation, we shall refer more frequently to the literature. We make use of the following concepts and results.

DEFINITION 7.4. A summation method S_1 *includes* a summation method S_2 if every sequence that is summable by S_2 (to the value z, say) is summable by S_1 to the same value z. S_1 and S_2 are called *equivalent* if S_1 includes S_2 and S_2 includes S_1.

DEFINITION 7.5. A summation method S is called *regular* if every convergent sequence (converging to z, say) is summable by S to the same value z.

THEOREM 7.12: Silverman-Toeplitz Theorem. An arbitrary matrix method $\mathbf{A} = (a_{nk})$ is regular if and only if the following conditions hold:

i. $\sup_{n} \sum_{k=1}^{\infty} |a_{nk}| < \infty$

ii. $\lim_{n\to\infty} \sum_{k=1}^{\infty} a_{nk} = 1$

iii. $\lim_{n\to\infty} a_{nk} = 0 \qquad \text{for } k = 1, 2, \ldots.$

In particular, a positive Toeplitz matrix is regular if and only if condition (iii) holds.

EXAMPLE 7.1. For a real number $r \geq 0$, define the matrix of the *Cesàro means* $(\mathbf{C}, r)$ by $a_{nk} = A_{n-k}^{(r-1)}/A_{n-1}^{(r)}$ for $1 \leq k \leq n$ and $a_{nk} = 0$ for $k > n$, where

$$A_i^{(j)} = \frac{(j+1)(j+2)\cdots(j+i)}{i!}$$

for $i \geq 1$ and $A_0^{(j)} = 1$. We note that $(\mathbf{C}, 1)$ is just the matrix method of arithmetic means. It can be shown that $(\mathbf{C}, r)$ is a positive regular Toeplitz matrix (Cooke [1, pp. 68–69]), and that the Cesàro means $(\mathbf{C}, r)$ with $r \geq 1$ include $(\mathbf{C}, 1)$ (Zeller and Beekmann [1, p. 104], Peyerimhoff [1, p. 15]). For positive integers r, the Cesàro means $(\mathbf{C}, r)$ are equivalent to the so-called *Hölder means* $(\mathbf{H}, r)$ defined recursively as follows. The Hölder means $(\mathbf{H}, 1)$ is the same as $(\mathbf{C}, 1)$; if $(\mathbf{H}, r)$ is already defined and $(z_n^{(r)})$, $n = 1, 2, \ldots$, is the sequence of $(\mathbf{H}, r)$-transforms of a given sequence (z_n), then the sequence of $(\mathbf{H}, r+1)$-transforms of (z_n) is defined as the sequence of $(\mathbf{C}, 1)$-transforms of $(z_n^{(r)})$. Hence, Hölder means are iterated $(\mathbf{C}, 1)$ means. For a proof of the equivalence of $(\mathbf{C}, r)$ and $(\mathbf{H}, r)$, see Hardy [2, p. 103]. ∎

It is clear from the definitions that the concepts of an a.d.f. (mod 1) and a $(\mathbf{C}, 1)$-a.d.f. (mod 1) are identical. For $(\mathbf{C}, r)$ with $r > 1$, we have the following less trivial result.

THEOREM 7.13. Let $r > 1$. Then a sequence (x_n) has the $(\mathbf{C}, r)$-a.d.f. (mod 1) $g(x)$ if and only if it has the a.d.f. (mod 1) $g(x)$.

PROOF. The sufficiency is clear since $(\mathbf{C}, r)$ includes $(\mathbf{C}, 1)$ (see Example 7.1). For the converse, we use the following well-known Tauberian theorem (Hardy [2, Theorem 92], Peyerimhoff [1, Theorem III.2]): If (z_n) is bounded and summable by $(\mathbf{C}, r)$ to the value z, then (z_n) is summable by $(\mathbf{C}, 1)$ to the value z. Now, by hypothesis, for $0 \leq x \leq 1$, the bounded sequence $(c_x(\{x_n\}))$, $n = 1, 2, \ldots$, is summable by $(\mathbf{C}, r)$ to the value $g(x)$, and an application of the Tauberian theorem yields the desired conclusion. ∎

EXAMPLE 7.2. For a real number $s > 0$, the *Euler method* $(\mathbf{E}, s)$ is defined by $a_{nk} = \binom{n}{k} s^{n-k}(s+1)^{-n}$ for $1 \leq k \leq n$ and $a_{nk} = 0$ for $k > n$. It is easily seen that $(\mathbf{E}, s)$ is a positive regular Toeplitz matrix. The method $(\mathbf{E}, s)$ includes $(\mathbf{E}, s')$ whenever $s > s'$ (Hardy [2, Theorem 118]). However, the method $(\mathbf{E}, s)$ does not include any of the Cesàro means $(\mathbf{C}, r)$ with $r > 0$ (Hardy [2, p. 213]). ■

THEOREM 7.14. If a sequence (x_n) has the $(\mathbf{E}, s)$-a.d.f. (mod 1) $g(x)$, then it has the a.d.f. (mod 1) $g(x)$.

PROOF. We use the following Tauberian theorem (Hardy [2, Theorems 128 and 147]): If (z_n) is summable by $(\mathbf{E}, s)$ to the value z and if $z_{n+1} - z_n = o(n^\rho)$ for some $\rho \geq -\frac{1}{2}$, then (z_n) is summable by $(\mathbf{C}, 2\rho + 1)$ to the value z. Now, by hypothesis, for $0 \leq x \leq 1$ the sequence $(c_x(\{x_n\}))$, $n = 1, 2, \ldots$, is summable by $(\mathbf{E}, s)$ to the value $g(x)$, and the sequence satisfies also the additional condition in the Tauberian theorem with $\rho = 1$. Hence, the sequence is summable by $(\mathbf{C}, 3)$ to the value $g(x)$, and so (x_n) has the $(\mathbf{C}, 3)$-a.d.f. (mod 1) $g(x)$. An application of Theorem 7.13 completes the proof. ■

EXAMPLE 7.3. Let (p_n), $n = 1, 2, \ldots$, be a given sequence of non-negative real numbers with $p_1 > 0$. For $n \geq 1$, put $P_n = p_1 + \cdots + p_n$. Then the *simple Riesz* (or *weighted arithmetic*) *means* $(\mathbf{R}, p_n)$ is given by $a_{nk} = p_k/P_n$ for $1 \leq k \leq n$ and $a_{nk} = 0$ for $k > n$. It is obvious that $(\mathbf{R}, p_n)$ is a positive Toeplitz matrix and that $(\mathbf{R}, p_n)$ is regular if and only if $\lim_{n\to\infty} P_n = \infty$. The Cesàro means $(\mathbf{C}, 1)$ is identical with the simple Riesz means $(\mathbf{R}, 1)$. ■

In the following, we give some sufficient conditions for a simple Riesz means to include another simple Riesz means.

LEMMA 7.1. If $p_n > 0$ and $q_n > 0$ for $n \geq 1$, if $\lim_{n\to\infty} P_n = \lim_{n\to\infty} Q_n = \infty$, and if either

$$\frac{q_{n+1}}{q_n} \leq \frac{p_{n+1}}{p_n} \quad \text{for } n \geq 1 \tag{7.16}$$

or

$$\frac{q_{n+1}}{q_n} \geq \frac{p_{n+1}}{p_n} \quad \text{and} \quad \frac{P_n}{p_n} \leq H\frac{Q_n}{q_n} \quad \text{for some } H \text{ and } n \geq 1, \tag{7.17}$$

then $(\mathbf{R}, q_n)$ includes $(\mathbf{R}, p_n)$.

PROOF. We show that the $(\mathbf{R}, q_n)$-transform of a sequence (z_n) can be expressed by means of the $(\mathbf{R}, p_n)$-transform and that the corresponding "transition matrix" is regular. Set $v_n = (p_1 z_1 + \cdots + p_n z_n)/P_n$ and $w_n = (q_1 z_1 + \cdots + q_n z_n)/Q_n$. Then $p_1 z_1 = P_1 v_1$ and $p_n z_n = P_n v_n - P_{n-1} v_{n-1}$ for $n \geq 2$, and therefore,

$$w_n = \frac{1}{Q_n}\left(\frac{q_1}{p_1} P_1 v_1 + \frac{q_2}{p_2}(P_2 v_2 - P_1 v_1) + \cdots + \frac{q_n}{p_n}(P_n v_n - P_{n-1} v_{n-1})\right)$$

$$= \sum_{k=1}^{\infty} c_{nk} v_k,$$

where

$$c_{nk} = \left(\frac{q_k}{p_k} - \frac{q_{k+1}}{p_{k+1}}\right)\frac{P_k}{Q_n} \qquad \text{for } 1 \leq k < n,$$

$$c_{nn} = \frac{q_n}{p_n} \cdot \frac{P_n}{Q_n}, \qquad c_{nk} = 0 \qquad \text{for } k > n.$$

It remains to show that the matrix (c_{nk}) satisfies the three conditions of Theorem 7.12. Property (iii) follows immediately from $\lim_{n\to\infty} Q_n = \infty$. If we choose $z_n = 1$ for all n, then $v_n = w_n = 1$ for all n, and so $\sum_{k=1}^{\infty} c_{nk} = 1$ for all n, which yields (ii). If (7.16) holds, then $c_{nk} \geq 0$, and (i) follows. If (7.17) holds, then $c_{nk} \leq 0$ for $1 \leq k < n$; hence,

$$\sum_{k=1}^{\infty} |c_{nk}| = -\sum_{k=1}^{n-1} c_{nk} + \frac{q_n P_n}{p_n Q_n} = -\sum_{k=1}^{\infty} c_{nk} + 2\frac{q_n P_n}{p_n Q_n} \leq -1 + 2H$$

for all n, and (i) follows again. ■

THEOREM 7.15. If $\sigma > -1$, then a sequence (x_n) has the $(\mathbf{R}, n^\sigma)$-a.d.f. (mod 1) $g(x)$ if and only if it has the a.d.f. (mod 1) $g(x)$.

PROOF. It suffices to show that the summation methods $(\mathbf{R}, n^\sigma)$ and $(\mathbf{R}, 1)$ are equivalent. Let $\sigma \geq 0$; then $1 \leq (n+1)^\sigma/n^\sigma$, so that according to (7.16) $(\mathbf{R}, 1)$ includes $(\mathbf{R}, n^\sigma)$. Furthermore, we observe that

$$\frac{P_n}{p_n} = n \leq H \frac{1 + 2^\sigma + \cdots + n^\sigma}{n^\sigma} = H \frac{Q_n}{q_n} \qquad \text{for } n \geq 1,$$

with a suitable H, implies according to (7.17) that $(\mathbf{R}, n^\sigma)$ includes $(\mathbf{R}, 1)$. If $-1 < \sigma < 0$, the proof runs in a similar way. ■

THEOREM 7.16. Let $f(x)$, $x \geq 1$, be an increasing function having a continuous second derivative, let $\lim_{x\to\infty} f(x) = \infty$, and let $f'(x)$ tend monotonically to 0 as $x \to \infty$. Then the sequence $(f(n))$, $n = 1, 2, \ldots$, is $(\mathbf{R}, f'(n))$-u.d. mod 1.

PROOF. For a nonzero integer h, we get by the Euler summation formula (see Example 2.4),

$$\begin{aligned}\sum_{k=1}^{n} f'(k)e^{2\pi i h f(k)} &= \int_1^n f'(x)e^{2\pi i h f(x)}\,dx \\ &\quad + \tfrac{1}{2}(f'(1)e^{2\pi i h f(1)} + f'(n)e^{2\pi i h f(n)}) \\ &\quad + \int_1^n (\{x\} - \tfrac{1}{2})(f'(x)e^{2\pi i h f(x)})'\,dx \\ &= \int_{f(1)}^{f(n)} e^{2\pi i h x}\,dx + \mathrm{O}(1) \\ &\quad + \mathrm{O}\left(\int_1^n |f''(x)|\,dx\right) + \mathrm{O}\left(\int_1^n (f'(x))^2\,dx\right) \\ &= \mathrm{O}(1) + \mathrm{O}(1) + \mathrm{O}(1) + \mathrm{o}(f(n)) = \mathrm{o}(f(n)).\end{aligned}$$

On the other hand, an application of the Euler summation formula yields also $f(n) = \mathrm{O}(\sum_{k=1}^{n} f'(k))$, and therefore,

$$\sum_{k=1}^{n} f'(k)e^{2\pi i h f(k)} = \mathrm{o}\left(\sum_{k=1}^{n} f'(k)\right) \qquad \text{for all integers } h \neq 0.$$

This property characterizes the $(\mathbf{R}, f'(n))$-u.d. mod 1 of $(f(n))$ by (7.15). ■

COROLLARY 7.1. If $f(x)$ satisfies the conditions of Theorem 7.16, then the sequence $(f(n))$, $n = 1, 2, \ldots$, either has no a.d.f. (mod 1) or is u.d. mod 1.

PROOF. Suppose $(f(n))$ has the a.d.f.(mod 1) $g(x)$. Now $f'(n+1)/f'(n) \leq 1$ for all $n \geq 1$, so, according to (7.16), $(\mathbf{R}, f'(n))$ includes $(\mathbf{C}, 1)$. It follows that $g(x)$ is also the $(\mathbf{R}, f'(n))$-a.d.f. (mod 1) of $(f(n))$. On the other hand, $(f(n))$ is $(\mathbf{R}, f'(n))$-u.d. mod 1 by Theorem 7.16, and so, $g(x) = x$; that is, $(f(n))$ is u.d. mod 1. ■

EXAMPLE 7.4. An important example of a summation method that is not a matrix method is the *Abel method.* Let (z_n) be a sequence such that the radius of convergence of the power series $\alpha(x) = (1 - x)\sum_{n=1}^{\infty} z_n x^n$ is at least 1. Then the sequence (z_n) is summable by the Abel method to the value z if $\lim_{x \to 1-0} \alpha(x) = z$. By the theorem of Frobenius quoted in the proof of Theorem 2.4, the Abel methods includes $(\mathbf{C}, 1)$ and is therefore regular. In fact, the Abel method includes all Cesàro means $(\mathbf{C}, r)$ with $r \geq 0$ (Hardy [2, Theorem 55]). Theorem 7.13 holds with $(\mathbf{C}, r)$ replaced by the Abel method, since the Tauberian theorem used in the proof is true for the Abel method as well (Hardy [2, Theorem 92]). ■

Notes

The study of asymptotic distribution functions mod 1 was initiated by Schoenberg [1], who proved Theorems 7.2 and 7.3. Other proofs or generalizations of these results are contained in Kuipers [10], Kuipers and Stam [1], and Brown and Duncan [1, 3]. Definition 7.2 and Theorem 7.1 are from van der Corput [7]. In this paper, one finds also many other results on distribution functions mod 1, such as a characterization of the set of distribution functions mod 1 of a sequence of distinct elements in I and the result of Exercise 7.10. Some further results are presented in Chapter 2, Section 4. The term "almost u.d. mod 1" was coined by Pjateckiĭ-Šapiro [2]. See also Ammann [1] and Chauvineau [6]. Lower and upper distribution functions mod 1 were introduced by Koksma [2; 4, Kap. 8], where Theorem 7.7 may also be found. For further results and related notions, see Chauvineau [6]. Theorem 7.5 is due to Schoenberg [1] and, in a slightly different form, to Wiener [1]. For a strengthening of the result, see Keogh and Petersen [1]. Theorem 7.6 is again from Schoenberg [1]. Kuipers [5] proves the theorem using methods of Haviland [1] and proves also some related results. For a quantitative version and a multidimensional generalization, see Hlawka [10, 16]. The results concerning ($\log n$) and related sequences can already be found in Pólya and Szegö [1, II. Abschn., Aufg. 179–182]. See also Thorp and Whitley [1] and the notes in Section 2. Theorems 7.8 and 7.9 are from Schoenberg [2]. For other types of metric results, we refer to Cigler and Volkmann [1].

For the construction of sequences having a prescribed a.d.f. (mod 1), see von Mises [1] and Chapter 2, Section 4. Various other results are shown in Chauvineau [1, 2, 6], Lacaze [1], Niederreiter [3], and Scoville [1]. The generalization of the theory to an abstract setting is to be found in Hlawka [3] and Chapter 3. As to the investigation of special sequences, the classical case to be studied was $(\phi(n)/n)$, $n = 1, 2, \ldots$, where ϕ denotes the Euler phi function. Schoenberg [1] established that this sequence has a continuous a.d.f. (mod 1) (see also M. Kac [4, Chapter 4]), and Erdös [1] showed that this a.d.f. (mod 1) is singular. A quantitative refinement is due to Faĭnleĭb [2], who improved earlier quantitative results of Tyan [1], Faĭnleĭb [1], and Ilyasov [1]. A related problem was studied by Diamond [1]. For other sequences arising from number theory, see Roos [1], Elliott [3], and van de Lune [1]. Zame [1] characterizes the a.d.f. (mod 1) of certain lacunary and exponentially increasing sequences, and Pjateckiĭ-Šapiro [1] does the same for the sequences $(a^n x)$, $n = 1, 2, \ldots$, where $a > 1$ is an integer (for the latter case, see also the notes in Section 8). Many papers have been devoted to the study of the distribution of additive number-theoretic functions. Such functions possessing an asymptotic d.f. are characterized by the celebrated theorem of Erdös and Wintner [1]. Later developments of this aspect are surveyed in M. Kac [3], Kubilius [1], and Galambos [1]. Various sufficient conditions for u.d. mod 1 of such functions can be found in Corrádi and Kátai [1], Delange [3, 5, 6, 11], Erdös [3], Kubilius [1, Chapter 4], and Levin and Faĭnleĭb [1–3]. Additive functions u.d. mod 1 were characterized by Delange [9, 10], and those having an a.d.f. (mod 1) were characterized by Elliott [1].

A detailed treatment of the general theory of summation methods is given in the books of Hardy [2], Cooke [1], Knopp [1], G. M. Petersen [2], Zeller and Beekmann [1] (with extensive bibliography), and Peyerimhoff [1]. For a proof of Theorem 7.12, see Hardy [2, Sections 3.2–3.3]. Some isolated results on u.d. mod 1 with respect to summation methods can already be found in Weyl [4] and Pólya and Szegö [1, II. Abschn., Aufg. 173]. The first systematic study is due to Tsuji [2], who considered weighted means $(\mathbf{R}, p_n)$. Some of his results were improved by Kano [1] and Kemperman [4]. Further results on weighted means are shown in Cigler [1, 10], Gerl [2], and Schnabl [1]. Our discussion of asymptotic distribution mod 1 with respect to positive Toeplitz matrices follows Cigler [1,

10]. A more general approach is presented in Chapter 3, Section 4. Blum and Mizel [1] extend the Weyl criterion for summation methods to two-sided sequences. Chauvineau [6] discusses (**H**, 2)-u.d. mod 1 under the name "équirépartition en moyenne (mod 1)," apparently without realizing that this is equivalent to u.d. mod 1 by Theorem 7.13 and Example 7.1. An interesting unresolved problem is the characterization of the matrix methods **A** for which all sequences $(n\theta)$, θ irrational, are **A**-u.d. mod 1 (see Cigler [10] and Dowidar [1] for partial results). Kemperman [4] studies the **A**-a.d.f. (mod 1) of slowly growing sequences.

Exercises

7.1. A sequence (x_n) has the continuous a.d.f. (mod 1) $g(x)$ if and only if (7.3) holds for every Riemann-integrable function f on [0, 1].

7.2. Explain why Theorem 7.2 breaks down for discontinuous $g(x)$.

7.3. If (x_n) has a continuous a.d.f. (mod 1) and m is a nonzero integer, then (mx_n) has a continuous a.d.f. (mod 1).

7.4. The sequence (x_n) is constructed as follows. For $n \geq 1$, let $k(n) \geq -1$ be the unique integer with $2^{k(n)} < n \leq 2^{k(n)+1}$. Then define $x_n = 1/n$ if $k(n)$ is even, $x_n = 0$ if $k(n)$ is odd. Prove that (x_n) has an a.d.f. (mod 1) but that $(-x_n)$ does not have one.

7.5. If (x_n) has the continuous a.d.f. (mod 1) $g(x)$, then $\lim_{N\to\infty} A([0, x]; N)/N = g(x)$ for $0 \leq x \leq 1$.

7.6. Construct an example to show that $\lim_{N\to\infty} A([0, x]; N)/N$ need not exist if (x_n) has a discontinuous a.d.f. (mod 1).

7.7. Let $F_1, F_2, \ldots$ be a sequence of uniformly bounded functions on [0, 1]. Prove that there exists a subsequence $F_{n_1}, F_{n_2}, \ldots$ such that $\lim_{k\to\infty} F_{n_k}(x)$ exists for every x from a given countable subset C of [0, 1]. *Hint:* Let $C = \{x_1, x_2, \ldots\}$; then $(F_n(x_1))$ contains a convergent subsequence, say $(F_{n1}(x_1))$; furthermore, $(F_{n1}(x_2))$ contains a convergent subsequence, and so on; show that a certain "diagonal" sequence satisfies the desired property.

7.8. If $N_1, N_2, \ldots$ is an increasing sequence of positive integers with $\lim_{k\to\infty} A([0, t_0); N_k)/N_k = \alpha$ for some $t_0 \in [0, 1]$, then (x_n) has a d.f. (mod 1) $z(x)$ with $z(t_0) = \alpha$. *Hint:* Use the method of Theorem 7.1.

7.9. The sequence (x_n) has the a.d.f. (mod 1) $g(x)$ if and only if $g(x)$ is the only d.f. (mod 1) of (x_n). *Hint:* Use Exercise 7.8 for the sufficiency part.

7.10. If $z_1(x), z_2(x), \ldots$ are d.f.'s (mod 1) of (x_n) with $\lim_{n\to\infty} z_n(x) = z(x)$ for $0 \leq x \leq 1$, then $z(x)$ is a d.f. (mod 1) of (x_n).

7.11. Let (x_n) be the sequence constructed as follows. We first have a block of 1! terms $\frac{1}{2}$, then a block of length 2! consisting of alternating terms $\frac{1}{4}$ and $\frac{3}{4}$, then a block of 3! terms $\frac{1}{2}$, and so on. Thus, the first terms of the sequence are $\frac{1}{2}, \frac{1}{4}, \frac{3}{4}, \frac{1}{2}, \frac{1}{2}, \frac{1}{2}, \frac{1}{2}, \frac{1}{2}, \frac{1}{2}, \frac{1}{4}, \frac{3}{4}, \ldots$. Prove that the lower d.f. (mod 1) and the upper d.f. (mod 1) of (x_n) are not d.f.'s (mod 1) of (x_n).

7.12. Prove that the functions $z(x, \xi)$ appearing in (7.10) represent all d.f.'s (mod 1) of the sequence $(\log n)$.

7.13. For a fixed $a \in \mathbb{R}$, the sequence $(a \log n)$ is not almost u.d. mod 1.

7.14. Construct a sequence that is almost u.d. mod 1 but not u.d. mod 1.

7.15. Prove that $(\sin n)$, $n = 1, 2, \ldots$, has an a.d.f. (mod 1) and determine it.

7.16. Let $f(t)$, $t \geq 1$, have a positive derivative such that $f(t) \to \infty$ and $tf'(t) \to 0$ as $t \to \infty$. Then the sequence $(f(n))$, $n = 1, 2, \ldots$, has the lower d.f. (mod 1) $\varphi(x)$ given by $\varphi(x) = 0$ for $0 \leq x < 1$ and $\varphi(1) = 1$, and the upper d.f. (mod 1) $\Phi(x)$ given by $\Phi(0) = 0$ and $\Phi(x) = 1$ for $0 < x \leq 1$.

7.17. Show that (7.7) is equivalent to

$$\lim_{H\to\infty} \frac{1}{H} \sum_{h=1}^{H} |\omega_h| = 0.$$

Hint: Use the Cauchy-Schwarz inequality.

7.18. Let (x_n) be a sequence in I having the a.d.f. (mod 1) $g(x)$, and let $\psi(t)$ be a continuous increasing function on $[0, 1]$ with $\psi(0) = 0$ and $\psi(1) = 1$. Then the sequence $(\psi(x_n))$ has the a.d.f. (mod 1) $g(\eta(x))$, where $\eta(x)$ is the inverse function of $\psi(t)$.

7.19. If $g(x)$ is the continuous a.d.f. (mod 1) of the sequence (x_n) in I, then the sequence $(g(x_n))$ is u.d. mod 1. Does this also hold for discontinuous $g(x)$?

7.20. If the sequence (x_n) in $(0, 1)$ is u.d. mod 1, then $(1/x_n)$ is not u.d. mod 1.

7.21. If the sequence (x_n) has the continuous asymptotic d.f. $f(\xi)$, then the sequence $(|x_n|)$ has the asymptotic d.f. $g(\xi)$ given by $g(\xi) = 0$ for $\xi < 0$ and $g(\xi) = f(\xi) - f(-\xi)$ for $\xi \geq 0$.

7.22. Prove Theorems 7.10 and 7.11 in detail.

7.23. Let (x_n) be a sequence for which $A([0, x); N) = Nx + o(\sqrt{N})$ for $0 \leq x \leq 1$. Then (x_n) is $(\mathbf{E}, s)$-u.d. mod 1 for all $s > 0$. *Hint:* Use Hardy [2, Theorem 149].

7.24. Prove that $(\mathbf{R}, 1/n \log(n + 1))$ includes $(\mathbf{R}, 1/n)$.

7.25. Prove that $(\mathbf{R}, 1/n)$ includes $(\mathbf{C}, 1)$ but not conversely.

7.26. Let $g(x)$, $x > 0$, be a positive, increasing, twice continuously differentiable function with $g(x) \to \infty$ as $x \to \infty$ and $g'(x)$ nonincreasing. Let $f(x)$, $x > 0$, be an increasing continuously differentiable function with $f'(x) \to 0$ as $x \to \infty$. Then the sequence $(f(g(n)))$, $n = 1, 2, \ldots$, has the continuous $(\mathbf{R}, g'(n))$-a.d.f. (mod 1) $h(x)$ if and only if $(f(n))$ has the a.d.f. (mod 1) $h(x)$. *Hint:* Use Theorem 7.11 and the Euler summation formula.

7.27. If $f(x)$ satisfies the conditions in Exercise 7.26 and if the sequence $(f(n))$, $n = 1, 2, \ldots$, is u.d. mod 1, then the sequence $(f(n^\sigma))$, $n = 1, 2, \ldots$, with $0 < \sigma \leq 1$ is u.d. mod 1.

7.28. Let $y_n = \frac{1}{2} - x_n$ for $n \geq 1$, where (x_n) is the sequence from Exercise 7.4. Prove that for some equivalence class $\tilde{g}$ we have

$$\lim_{N\to\infty} \frac{1}{N} \sum_{n=1}^{N} f(y_n) = \int_0^1 f(x)\, d\tilde{g}(x)$$

for all real-valued continuous functions f on $[0, 1]$, but that (y_n) has no a.d.f. (mod 1).

8. NORMAL NUMBERS

Definition and Relation to Uniform Distribution Mod 1

Let α be a real number, and let $b \geq 2$ be an integer. Then α has a unique b-adic expansion, that is, an expansion of the form

$$\alpha = [\alpha] + \sum_{n=1}^{\infty} \frac{a_n}{b^n} = [\alpha] \,.\, a_1 a_2 \cdots a_n \cdots, \tag{8.1}$$

where the "digits" a_n are integers with $0 \leq a_n < b$ for $n \geq 1$, and also $a_n < b - 1$ for infinitely many n. If a is a digit with respect to the base b (i.e., an integer with $0 \leq a < b$) and if N is a positive integer, we let $A_b(a; N; \alpha)$ denote the number of a_n, $1 \leq n \leq N$, in (8.1) for which $a_n = a$. If α is fixed, we write $A_b(a; N)$ instead of $A_b(a; N; \alpha)$. More generally, let $B_k = b_1 b_2 \cdots b_k$, $k \geq 1$, be a given block of digits of length k. Denote by $A_b(B_k; N) = A_b(B_k; N; \alpha)$ the number of occurrences of the block B_k in the block of digits $a_1 a_2 \cdots a_N$. In other words, $A_b(B_k; N)$ is the number of n, $1 \leq n \leq N - k + 1$, such that $a_{n+j-1} = b_j$ for $1 \leq j \leq k$.

DEFINITION 8.1. The number α is called *simply normal to the base b* if

$$\lim_{N\to\infty} \frac{A_b(a; N)}{N} = \frac{1}{b} \qquad \text{for } a = 0, 1, \ldots, b-1. \tag{8.2}$$

The number α is called *normal to the base b* if

$$\lim_{N\to\infty} \frac{A_b(B_k; N)}{N} = \frac{1}{b^k} \qquad \text{for all } k \geq 1 \text{ and all } B_k. \tag{8.3}$$

Obviously, a number normal to the base b is also simply normal to the same base. The converse is not true in general. For instance, the number given by the 2-adic expansion $0.010101 \cdots$ is certainly simply normal to the base 2 but not normal to the base 2 since the block of digits 11 of length 2 does not occur in the expansion.

THEOREM 8.1. The number α is normal to the base b if and only if the sequence $(b^n\alpha)$, $n = 0, 1, \ldots$, is u.d. mod 1.

PROOF. Let (8.1) be the b-adic expansion of α, and let ω be the sequence $(b^n\alpha)$, $n = 0, 1, \ldots$. Consider a block $B_k = b_1b_2\cdots b_k$ of k digits. Then, for $m \geq 1$, the block $a_ma_{m+1}\cdots a_{m+k-1}$ is identical with B_k if and only if

$$\alpha = [\alpha] + \sum_{n=1}^{m-1}\frac{a_n}{b^n} + \frac{b_1}{b^m} + \cdots + \frac{b_k}{b^{m+k-1}} + \sum_{n=m+k}^{\infty}\frac{a_n}{b^n},$$

or

$$\{b^{m-1}\alpha\} = \frac{b_1b^{k-1} + \cdots + b_k}{b^k} + \sum_{n=k+1}^{\infty}\frac{a_{n+m-1}}{b^n},$$

or

$$\{b^{m-1}\alpha\} \in \left[\frac{b_1b^{k-1} + \cdots + b_k}{b^k}, \frac{b_1b^{k-1} + \cdots + b_k + 1}{b^k}\right) = J(B_k).$$

It follows that $A_b(B_k; N) = A(J(B_k); N - k + 1; \omega)$. Now suppose that ω is u.d. mod 1. Then

$$\lim_{N\to\infty}\frac{A_b(B_k; N)}{N} = \lim_{N\to\infty}\frac{A(J(B_k); N - k + 1; \omega)}{N - k + 1}\cdot\frac{N - k + 1}{N} = \frac{1}{b^k},$$

and so α is normal to the base b. Conversely, if α is normal to the base b, then

$$\lim_{N\to\infty}\frac{A(J(B_k); N; \omega)}{N} = \lim_{N\to\infty}\frac{A_b(B_k; N + k - 1)}{N + k - 1}\cdot\frac{N + k - 1}{N} = \frac{1}{b^k}.$$

This holds for all B_k. Therefore, $\lim_{N\to\infty} A(J; N; \omega)/N = \lambda(J)$, the length of J, for all half-open subintervals J of I satisfying the following property: the end points of J are rationals whose denominators are powers of b. Now let E be an arbitrary half-open subinterval of I, and let $\varepsilon > 0$ be given. Choose intervals J_1 and J_2 of the above type with $J_1 \subseteq E \subseteq J_2$ and $\lambda(E) - \varepsilon < \lambda(J_1) \leq \lambda(J_2) < \lambda(E) + \varepsilon$. Then, for sufficiently large N,

$$\frac{A(E; N; \omega)}{N} \geq \frac{A(J_1; N; \omega)}{N} > \lambda(J_1) - \varepsilon > \lambda(E) - 2\varepsilon,$$

and similarly $A(E; N; \omega)/N < \lambda(E) + 2\varepsilon$. Hence, $\lim_{N\to\infty} A(E; N; \omega)/N = \lambda(E)$. ■

COROLLARY 8.1. Almost all real numbers α (in the sense of Lebesgue measure) are normal to the base b.

PROOF. This follows immediately from Theorems 4.1 and 8.1. ■

DEFINITION 8.2. The number α is called *absolutely normal* if it is normal to all bases $b \geq 2$.

COROLLARY 8.2. Almost all real numbers α (in the sense of Lebesgue measure) are absolutely normal.

PROOF. This follows from Corollary 8.1, since the countable union of null sets is still a null set. ■

Further Results

LEMMA 8.1. Let α be a real number and $b \geq 2$. Suppose that there exists a constant C such that for any nonnegative continuous function f on $[0, 1]$,

$$\overline{\lim_{N\to\infty}} \frac{1}{N} \sum_{n=0}^{N-1} f(\{b^n\alpha\}) \leq C \int_0^1 f(x)\, dx. \tag{8.4}$$

Then α is normal to the base b.

PROOF. By Theorem 7.1 there exists a d.f. (mod 1) $z(x)$ of the sequence $(b^n\alpha)$, $n = 0, 1, \ldots$. Thus, for some increasing sequence $N_1, N_2, \ldots$ of positive integers, we have

$$\lim_{i\to\infty} \frac{1}{N_i} \sum_{n=0}^{N_i-1} f(\{b^n\alpha\}) = \int_0^1 f(x)\, dz(x) \tag{8.5}$$

for all continuous functions f on $[0, 1]$. It follows from (8.4) that

$$\int_0^1 f(x)\, dz(x) \leq C \int_0^1 f(x)\, dx \tag{8.6}$$

for any nonnegative continuous function f on $[0, 1]$, and so $z(x)$ must be continuous on $[0, 1]$. In particular, (8.5) will even hold for all f on $[0, 1]$ that are continuous except for finitely many jumps (see the proof of Theorem 7.4). If f is such a function on $[0, 1]$, then $g(x) = f(\{bx\})$, $0 \leq x \leq 1$, is of the same type. Applying (8.5) to $g(x)$, we find

$$\lim_{i\to\infty} \frac{1}{N_i} \sum_{n=0}^{N_i-1} f(\{b^{n+1}\alpha\}) = \int_0^1 f(\{bx\})\, dz(x). \tag{8.7}$$

On the other hand, we have

$$\left| \frac{1}{N_i} \sum_{n=0}^{N_i-1} f(\{b^{n+1}\alpha\}) - \frac{1}{N_i} \sum_{n=0}^{N_i-1} f(\{b^n\alpha\}) \right| \leq \frac{2M}{N_i},$$

where $M = \sup_{0\leq x\leq 1} |f(x)|$, and so the two limits in (8.5) and (8.7) are identical. Consequently, we get

$$\int_0^1 f(\{bx\})\, dz(x) = \int_0^1 f(x)\, dz(x).$$

By induction, we arrive at

$$\int_0^1 f(\{b^n x\})\, dz(x) = \int_0^1 f(x)\, dz(x) \qquad \text{for all integers } n \geq 0. \tag{8.8}$$

For a continuous f on $[0, 1]$, we have, according to Corollary 8.1,

$$\lim_{N\to\infty} \frac{1}{N} \sum_{n=0}^{N-1} f(\{b^n t\}) = \int_0^1 f(x)\, dx \qquad \text{for all } t \in [0, 1]\backslash E, \tag{8.9}$$

with a Lebesgue null set $E \subseteq [0, 1]$. Then, by (8.6),

$$\int_E dz(x) = 0, \tag{8.10}$$

where the integral is a Lebesgue-Stieltjes integral. Since

$$\left| \frac{1}{N} \sum_{n=0}^{N-1} f(\{b^n x\}) \right| \leq M \qquad \text{for } 0 \leq x \leq 1,$$

where M has the same meaning as above, we may use the dominated convergence theorem, together with (8.8), (8.9), and (8.10), to obtain

$$\begin{aligned}\int_0^1 f(x)\, dz(x) &= \lim_{N\to\infty} \int_0^1 \left(\frac{1}{N} \sum_{n=0}^{N-1} f(\{b^n x\}) \right) dz(x) \\ &= \int_0^1 \left(\lim_{N\to\infty} \frac{1}{N} \sum_{n=0}^{N-1} f(\{b^n t\}) \right) dz(t) \\ &= \int_0^1 \left(\int_0^1 f(x)\, dx \right) dz(t) = \int_0^1 f(x)\, dx.\end{aligned}$$

Since f was an arbitrary continuous function on $[0, 1]$, it follows that $z(x) = x$ for $0 \leq x \leq 1$. Therefore, $z(x) = x$ is the only d.f. (mod 1) of the sequence $(b^n\alpha)$, $n = 0, 1, \ldots$, and so this sequence is u.d. mod 1 by Exercise 7.9. An application of Theorem 8.1 completes the argument. ■

The condition (8.4) is of course also necessary. Lemma 8.1 will be instrumental in the proof of the subsequent theorem, which leads to a new characterization of normal numbers (Theorem 8.3).

THEOREM 8.2. Let $k \geq 2$ be an integer. A real number α is normal to the base b if and only if α is normal to the base b^k.

PROOF. Let α be normal to the base b. For every nonnegative continuous function f on $[0, 1]$ we have the inequality

$$\frac{1}{N} \sum_{n=0}^{N-1} f(\{b^{kn}\alpha\}) \leq k \frac{1}{kN} \sum_{n=0}^{kN-1} f(\{b^n\alpha\}). \tag{8.11}$$

Since the right-hand side of (8.11) converges to $k\int_0^1 f(x)\,dx$ by assumption, we obtain

$$\overline{\lim_{N\to\infty}}\ \frac{1}{N}\sum_{n=0}^{N-1} f(\{b^{kn}\alpha\}) \leq k\int_0^1 f(x)\,dx,$$

and so α is normal to the base b^k by Lemma 8.1. Conversely, if α is normal to the base b^k, then $(b^{kn}\alpha)$, $n = 0, 1, \ldots$, is u.d. mod 1 and so $(b^{kn+j}\alpha)$, $n = 0, 1, \ldots$, is u.d. mod 1 for each $j = 0, 1, \ldots, k-1$. Thus, $(b^n\alpha)$, $n = 0, 1, \ldots$, is u.d. mod 1 as the superposition of k sequences that are u.d. mod 1 (see Exercise 2.12). Hence, α is normal to the base b. ∎

THEOREM 8.3. The real number α is normal to the base b if and only if α is simply normal to all of the bases $b, b^2, b^3, \ldots$.

PROOF. One implication is obvious: if α is normal to the base b, then α is normal to all of the bases $b, b^2, b^3, \ldots$ by Theorem 8.2, and so *a fortiori* simply normal to all of these bases. Now suppose that α is simply normal to all of the bases $b, b^2, b^3, \ldots$, and let (8.1) be the b-adic expansion of α. Then, for $r \geq 1$,

$$\alpha = [\alpha] + \sum_{n=1}^{\infty}\frac{a_n^{(r)}}{b^{rn}} \quad \text{with} \quad a_n^{(r)} = \sum_{j=0}^{r-1} a_{nr-j}b^j$$

is the b^r-adic expansion of α. By the simple normality of α to the base b^r, every integer t with $0 \leq t < b^r$ occurs among the $a_n^{(r)}$ with asymptotic frequency b^{-r}. Now $t = b_1b^{r-1} + \cdots + b_{r-1}b + b_r$ with certain digits $b_1, \ldots, b_r$ (with respect to the base b), and so $t = a_n^{(r)}$ precisely if $a_{(n-1)r+1} = b_1, \ldots, a_{nr} = b_r$. Therefore, the hypothesis may also be expressed as follows:

Every block $b_1b_2\cdots b_r$ occurs among the blocks $a_na_{n+1}\cdots a_{n+r-1}$, $n \equiv 1$ (mod r), with asymptotic frequency b^{-r}. (*)

Let now $B_k = b_1b_2\cdots b_k$ be an arbitrary block of digits with respect to the base b. To estimate $A_b(B_k; N)$, we choose an integer $r > k$ and write $N = ur + v$ with $0 \leq v < r$. For $j \geq 0$, we denote by $A(B_k; jr+1, (j+1)r)$ the number of n, $jr + 1 \leq n \leq (j+1)r - k + 1$, such that $B_k = a_na_{n+1}\cdots a_{n+k-1}$. If we replace $A_b(B_k; (u+1)r)$ by $\sum_{j=0}^{u} A(B_k; jr+1, (j+1)r)$, we have neglected the blocks $a_na_{n+1}\cdots a_{n+k-1}$ that contain both a_{jr} and a_{jr+1} for some j, $1 \leq j \leq u$. For every j, $1 \leq j \leq u$, these are exactly $k-1$ blocks. Thus,

$$\begin{aligned} A_b(B_k; N) &\leq A_b(B_k; (u+1)r) \\ &\leq \sum_{j=0}^{u} A(B_k; jr+1, (j+1)r) + u(k-1). \end{aligned} \tag{8.12}$$

We also have

$$A_b(B_k; N) \geq A_b(B_k; ur) \geq \sum_{j=0}^{u-1} A(B_k; jr+1, (j+1)r). \tag{8.13}$$

We consider $\sum_{j=0}^{s} A(B_k; jr + 1, (j + 1)r)$ for an integer $s \geq 0$. Let C denote an extension of B_k to a block of digits of length r, say

$$C = c_1 \cdots c_p b_1 \cdots b_k d_1 \cdots d_{r-k-p}.$$

Two such C are considered to be distinct if they are either distinct as blocks of length r or if b_1 has a different place. Denote by $A^*(C; s)$ the number of j, $0 \leq j \leq s$, such that $C = a_{jr+1}a_{jr+2} \cdots a_{(j+1)r}$. Since B_k occurs in

$$a_{jr+1}a_{jr+2} \cdots a_{(j+1)r}$$

exactly as often as there are distinct C such that $C = a_{jr+1}a_{jr+2} \cdots a_{(j+1)r}$, we have

$$\sum_{j=0}^{s} A(B_k; jr + 1, (j + 1)r) = \sum_{C} A^*(C; s).$$

Therefore, from (8.12) and (8.13),

$$\frac{u}{ur + v} \sum_{C} \frac{A^*(C; u - 1)}{u} \leq \frac{A_b(B_k; N)}{N} \leq \frac{u + 1}{ur + v} \sum_{C} \frac{A^*(C; u)}{u + 1} + \frac{u(k - 1)}{ur + v}.$$

By (*) each C has the asymptotic frequency b^{-r}. Letting N tend to infinity (or equivalently $u \to \infty$), we obtain

$$\frac{1}{r} \sum_{C} b^{-r} \leq \varliminf_{N\to\infty} \frac{A_b(B_k; N)}{N} \leq \varlimsup_{N\to\infty} \frac{A_b(B_k; N)}{N} \leq \frac{1}{r} \sum_{C} b^{-r} + \frac{k - 1}{r}.$$

Since there are $r - k + 1$ possibilities to choose the place of b_1 and, the place of b_1 being fixed, there are b^{r-k} possibilities to assign values to the $r - k$ new digits, we have altogether $(r - k + 1)b^{r-k}$ possibilities for C. Therefore,

$$\frac{r - k + 1}{rb^k} \leq \varliminf_{N\to\infty} \frac{A_b(B_k; N)}{N} \leq \varlimsup_{N\to\infty} \frac{A_b(B_k; N)}{N} \leq \frac{r - k + 1}{rb^k} + \frac{k - 1}{r}.$$

Letting $r \to \infty$, we arrive at (8.3). ■

Notes

Normal and absolutely normal numbers were introduced by Borel [1; 2, pp. 194–201], who showed the metric results in Corollaries 8.1 and 8.2. Borel's original characterization of normality is that of Exercise 8.4. The equivalence of this characterization and our Definition 8.1 was first shown by Niven and Zuckerman [1] (see also Cassels [7], Knuth [2, Chapter 3], and Niven [1, Chapter 8]). Theorem 8.1 is from Wall [1] (see also Niven [1, Chapter 8] and Postnikov [8, Chapter 3]). The characterization of normality enunciated in Theorem 8.3 is that of Pillai [2]. For other proofs, see Maxfield [1] and Niven [1, Chapter 8]. Long [1] shows that α is normal to the base b if and only if there exist positive integers $m_1 < m_2 < \cdots$ such that α is simply normal to all of the bases b^{m_i}, $i \geq 1$; no finite set of m_i suffices. Jager [1] characterizes normality in terms of certain digit shift transformations. The characterization in Exercise 8.6 is due to Mendès France [1, 3, 4]. For another criterion, see Chapter 3, Exercise 3.10.

Sierpiński [3] and Lebesgue [1] gave partially explicit constructions of normal numbers. In fact, they even constructed absolutely normal numbers. A simple example is due to Champernowne [1], who showed that the number θ in Example 2.2 is normal to the base 10. Proofs of this result can also be found in Niven [1, Chapter 8], Pillai [1, 2], and Postnikov [5]. If (p_n) is an increasing sequence of positive integers that is sufficiently dense, such as the sequence of primes, then the decimal $0.p_1p_2\cdots$ is normal to the base 10 (Copeland and Erdös [1]). If f is a polynomial all of whose values for $n = 1, 2, \ldots$ are positive integers, then the decimal $0.f(1)f(2)\cdots$ is normal to the base 10 (Davenport and Erdös [1]). Further constructions of normal numbers can be found in Postnikov [8, Chapter 3], W. M. Schmidt [3] (absolutely normal numbers), Spears and Maxfield [1], Stoneham [4, 5, 6], and Ville [1, Chapter 1]. Another constructive approach is via so-called "normal periodic systems of digits" (Good [1], Rees [1], Korobov [4, 5, 6, 9]). There are also intimate relations between the construction of sequences completely u.d. mod 1 (see Chapter 3, Section 3) and the construction of normal numbers (Knuth [2, Chapter 3], Korobov [1, 2, 5, 10]). Some constructive results may also be found in the papers on the sequences $(b^n\alpha)$ mentioned in the notes in Section 3 of Chapter 2. All the known normal numbers have been constructed *ad hoc*. It is not known whether irrationals of number-theoretic interest, such as e, π, $\sqrt{2}$, $\log 2, \ldots$, are normal. However, statistical studies of the digits of such numbers have been carried out. We refer to Beyer, Metropolis, and Neergaard [1] (good bibliography), [2], Dutka [1], and Stoneham [2]. Numbers having other prescribed digit (or block) frequencies have also been constructed. See Copeland [1], Postnikov [5; 8, Chapter 3], Postnikov and Pjateckiĭ-Šapiro [1, 2], Postnikova [3], Šahov [3, 4], Ville [1, Chapter 1], and von Mises [1].

Lemma 8.1 is essentially due to Pjateckiĭ-Šapiro [1]. It was improved by Postnikov [1], and a best-possible version was established by Pjateckiĭ-Šapiro [4]. See also Pjateckiĭ-Šapiro [3], Postnikov [5], and Kemperman [2]. For some generalizations, see Postnikov and Pjateckiĭ-Šapiro [1] and Moskvin [1]. In essence, Lemma 8.1 is a result from ergodic theory. The connection between normal numbers and ergodic theory was first pointed out by Riesz [1]. Since the transformation $\alpha \to \{b\alpha\}$, $\alpha \in I$, is ergodic with respect to Lebesgue measure, the individual ergodic theorem implies that for every Lebesgue-integrable function f on [0, 1] we have $\lim_{N\to\infty}(1/N)\sum_{n=1}^{N} f(\{b^n\alpha\}) = \int_0^1 f(x)\,dx$ for almost all $\alpha \in I$. Without using ergodic theory, this was proved earlier by Raikov [1] and Fortet [1]. The relation between ergodic theory and normal numbers is exploited further in Cigler [2, 9] (see also Blum and Hanson [1]), Franklin [1], Furstenberg [3], Hartman, Marczewski, and Ryll-Nardzewski [1], and Postnikov [7; 8, Chapter 3]. For various other proofs of Borel's metric theorem (Corollary 8.1) not using ergodic theory, see Hardy and Littlewood [1], Sierpiński [3], M. Kac [4, Chapter 2], Ducray [1], and the literature in Koksma [4, pp. 116–118], where also quantitative refinements are mentioned (see also the notes in Chapter 2, Section 3). Further metric results concerning normal numbers (or certain generalizations thereof) were shown by Franklin [1, 2], Postnikov and Pjateckiĭ-Šapiro [1], Sanders [1], and Schweiger [1]. Maxfield [2] notes that the set of simply normal numbers to the base b is of the first category (see also Šalát [1] and Schweiger [3]). There is a vast literature concerning the Hausdorff dimension of sets of nonnormal numbers, or, more generally, of sets defined by digit properties. We mention Besicovitch [1], Best [1], Beyer [1], Cigler [5], Cigler and Volkmann [1], Colebrook [1], Eggleston [1, 2], Erdös and Taylor [1], Helson and Kahane [1], Knſchal [1, 2], Mendès France [2, 3], Nagasaka [1], and Volkmann [1, 2, 4, 6, 7].

An interesting problem is that of the relation between normality with respect to different bases. A rather easy result is that in Exercise 8.5. W. M. Schmidt [3] proves the following general theorem: Partition the possible bases 2, 3, . . . into two classes such that numbers that are rational powers of each other lie in the same class; then the set of numbers that are normal to the bases in the first class and not normal to the bases in the second class has the power of the continuum. For earlier results, see W. M. Schmidt [2] and Cassels [10]. From more general standpoints, the problem has been discussed by W. M. Schmidt [4], Colebrook and Kemperman [1], and Schweiger [2].

Some variants of the definition of normality occur in the literature. Normality of order k (see Exercise 8.7) is discussed in Ville [1, Chapter 1], Good [1], Rees [1], Knuth [2, Chapter 3], and Long [2]. See also Mahler [6]. Besicovitch [2] introduced "(j, ε)-normality," which was further studied by Davenport and Erdös [1], Hanson [1], and Stoneham [1, 3, 5, 6]. The papers of Korobov [23, 24] treat a related question. Pjateckiĭ-Šapiro [1] characterizes the a.d.f. (mod 1) of a sequence $(b^n\alpha)$. See also Postnikov [8, Chapter 3]. Further results concerning these sequences and a.d.f.'s (mod 1) can be found in Cigler [2, 9], Colebrook [1], Colebrook and Kemperman [1], Helson and Kahane [1], and Kemperman [2]. The u.d. mod 1 of these sequences with respect to summation methods is briefly discussed in Schnabl [1]. Closely related sequences have been studied by Korobov [7], Franklin [1, 2], and Mendès France [5]. If $q(n)$ denotes the b-adic sum of digits of the positive integer n, then $(q(n)\theta)$, $n = 1, 2, \ldots$, is u.d. mod 1 for every irrational θ (Mendès France [4]). This is still true if n runs through the prime numbers (Olivier [1]). See also Mendès France [9]. The function $q(n)$ is also connected with the theory of P.V. numbers (see Example 4.2). For this and related matters, see Mendès France [5, 7], Senge and Straus [1], and Bésineau [1, 2]. Some investigations have also been carried out for nonintegral bases $\beta > 1$, mostly from the metric point of view (Rényi [1], Gel'fond [1], Parry [1], Cigler [4], Roos [1], Galambos [2, 3, 4]). For the sequences $(\beta^n\alpha)$, see Helson and Kahane [1], Kulikova [1], and Mendès France [4].

Maxfield [2] calls $(\alpha_1, \ldots, \alpha_k)$ a *normal k-tuple to the base b* if the sequence

$$((b^n\alpha_1, \ldots, b^n\alpha_k)),$$

$n = 1, 2, \ldots$, is u.d. mod 1 in $\mathbb{R}^k$. The analogue of Corollary 8.1 for this case was already shown by Hardy and Littlewood [1]. Maxfield [2] proves a result for normal k-tuples that implies, in particular, that every nonzero rational multiple of a normal number is normal. The k-tuple $(\alpha_1, \ldots, \alpha_k)$ is called *jointly normal to the bases* $b_1, \ldots, b_k$ if the sequence $((b_1{}^n\alpha_1, \ldots, b_k{}^n\alpha_k))$, $n = 1, 2, \ldots$, is u.d. mod 1 in $\mathbb{R}^k$. For results on the above two concepts (e.g., construction of such k-tuples), see Korobov [8, 13, 14], Postnikova [1], Starčenko [1], and Volkmann [4]. More generally, one may consider the u.d. mod 1 in $\mathbb{R}^k$ of a sequence $(\mathbf{A}^n\mathbf{x})$, $n = 1, 2, \ldots$, where $\mathbf{A}$ is a given $k \times k$ matrix with integer entries and $\mathbf{x} \in \mathbb{R}^k$. The transformation $\mathbf{x} \to \{\mathbf{A}\mathbf{x}\}$, $\mathbf{x} \in I^k$, is ergodic with respect to Lebesgue measure if and only if $\mathbf{A}$ is nonsingular and none of its eigenvalues is a root of unity (Rohlin [1], Auslander [1]). In this case, the generalization of Borel's metric theorem follows from the individual ergodic theorem, and an analogue of Lemma 8.1 can also be shown (Postnikov [3; 8, Chapter 3], Polosuev [3, 5], Starčenko [1], Cigler [2, 9], Muhutdinov [1]). Further work on $(\mathbf{A}^n\mathbf{x})$ was done by Cigler [2, 9], Franklin [3], Polosuev [3, 5], Postnikov [8, Chapter 3], W. M. Schmidt [4], Schweiger [2], and Uchiyama [2]. For an abstract approach to normality, we refer to Cigler [3] (see also the notes in Chapter 4, Section 1).

A set M of real numbers is called a *normal set* ("*ensemble normal*") if there exists a sequence (λ_n) in $\mathbb{R}$ such that $(\lambda_n x)$ is u.d. mod 1 if and only if $x \in M$. The set of numbers normal to the base b is a normal set by Theorem 8.1. Other nontrivial normal sets are $\mathbb{Q}\backslash\{0\}$ (Rauzy [2], Zame [5]), the complement in $\mathbb{R}$ of a real algebraic number field, and the set of transcendental numbers (Meyer [1–4], Mendès France [6], Zame [5]). A countable intersection of normal sets is a normal set (Dress [2], Rauzy [1]). A countable union of normal sets need not be a normal set (Mendès France [7], Rauzy [1]). A characterization of normal sets was given by Rauzy [1]. For further results, see Colombeau [1] (to a large part superseded by Dress [2]), Dress and Mendès France [1], Lesca and Mendès France [1], Méla [1], Mendès France [5, 7], and Zame [5]. For the related question of so-called "anormal" numbers with respect to a sequence (λ_n), see Kahane and Salem [1], Kahane [1], and Amice [2]. Mendès France [9] studies sequences with "empty spectrum," that is, sequences (x_n) for which $(x_n + n\alpha)$, $n = 1, 2, \ldots$, is u.d. mod 1 for all real α.

An interesting application of normal numbers occurs in the foundation of probability theory, namely, in von Mises's theory of collectives, which is an attempt to develop probability theory from the notion of relative frequency rather than from measure theory. We refer to Copeland [1], von Mises [1; 2, Chapter 1], Ville [1], and Postnikova [2], and the literature given there. Other applications of normal numbers can be found in Couot [3], Knuth [2, Section 3.5], Mendès France [4], Postnikov [5], and Veech [3].

Exercises

8.1. Prove that the expansion (8.1) is unique.

8.2. There are rational numbers that are simply normal to the base b, but a normal number to the base b is necessarily irrational.

8.3. Give an example of an irrational number that is not normal (to the base 2, say).

8.4. The number α is normal to the base b if and only if each of the numbers $\alpha, b\alpha, b^2\alpha, \ldots$ is simply normal to all of the bases $b, b^2, b^3, \ldots$.

8.5. If b_1 and b_2 are integers ≥ 2 such that one is a rational power of the other, then α is normal to the base b_1 if and only if α is normal to the base b_2.

8.6. Let $\phi_n(x)$, $n = 0, 1, \ldots$, be the Rademacher functions as defined in Exercise 2.1 in Chapter 2. Suppose that α is not a dyadic rational. Then α is normal to the base 2 if and only if

$$\lim_{N\to\infty} \frac{1}{N} \sum_{n=0}^{N-1} \phi_{n+k_1}(\alpha) \cdots \phi_{n+k_s}(\alpha) = 0$$

holds for all $s \geq 1$ and all distinct nonnegative integers $k_1, \ldots, k_s$. *Hint:* Use some of the results of Exercises 2.1–2.4 in Chapter 2.

8.7. For a positive integer k, the number α is called *normal of order k to the base b* if (8.3) holds for all blocks B_k of fixed length k. Prove that the following is an equivalent condition: the sequence $(b^n\alpha)$ satisfies $\lim_{N\to\infty} A([0, x); N)/N = x$ for all rationals x, $0 < x < 1$, with denominator b^k.

8.8. Prove that α is normal to the base b if there exists a constant C such that the sequence $(b^n\alpha)$ satisfies $\overline{\lim}_{N\to\infty} A([\beta, \gamma); N)/N \leq C(\gamma - \beta)$ for all $[\beta, \gamma) \subseteq I$. *Hint:* Compare with the proof of Lemma 8.1.

8.9. Whenever α is normal to the base b, then so is $r\alpha$ for every nonzero rational r. *Hint:* Use Lemma 8.1 or Exercise 8.8.

8.10. Give a proof of Theorem 5.3 based on the theory of normal numbers (suppose $p \geq 2$).

8.11. The set of irrationals is a normal set.

8.12. The set of nonzero reals is a normal set.

9. CONTINUOUS DISTRIBUTION MOD 1

Basic Results

Let $f(t)$ be a real-valued Lebesgue-measurable function defined for $0 \leq t < \infty$. Let $0 \leq a < b \leq 1$ and let $c_{[a,b)}(x)$ be the characteristic function of the interval $[a, b)$. For $T > 0$, let $T(a, b)$ be the set of t, $0 \leq t \leq T$, for which $\{f(t)\} \in [a, b)$. Evidently $T(a, b)$ has a Lebesgue measure $\lambda(T(a, b))$, and we have

$$\lambda(T(a, b)) = \int_0^T c_{[a,b)}(\{f(t)\})\, dt.$$

Definition 9.1. If for all $[a, b) \subseteq I$ we have

$$\lim_{T\to\infty} \frac{\lambda(T(a, b))}{T} = b - a,$$

then the function $f(t)$ is said to be *continuously uniformly distributed mod 1* (abbreviated c.u.d. mod 1).

THEOREM 9.1. The Lebesgue-measurable function $f(t)$ defined on $[0, \infty)$ is c.u.d. mod 1 if and only if for each real-valued continuous function w on $[0, 1]$,

$$\lim_{T\to\infty} \frac{1}{T} \int_0^T w(\{f(t)\})\, dt = \int_0^1 w(x)\, dx. \tag{9.1}$$

THEOREM 9.2. The Lebesgue-measurable function $f(t)$ defined on $[0, \infty)$ is c.u.d. mod 1 if and only if

$$\lim_{T\to\infty} \frac{1}{T} \int_0^T e^{2\pi i h f(t)}\, dt = 0 \qquad \text{for all integers } h \neq 0. \tag{9.2}$$

The proofs of these theorems can be given in a way analogous to the proofs of Theorems 1.1 and 2.1 and are left to the reader.

EXAMPLE 9.1. Obvious examples of c.u.d. mod 1 functions are the linear functions $f(t) = at + b$, $t \geq 0$, with $a \neq 0$. On the other hand, the function $f(t) = \log(t + 1)$, $t \geq 0$, is not c.u.d. mod 1. This function does not satisfy (9.2) as can be derived from the proof given in Example 2.4. ■

EXAMPLE 9.2. The function e^t, $t \geq 0$, is c.u.d. mod 1. Consider the expression

$$\frac{1}{T}\int_0^T e^{2\pi i h e^t}\,dt \qquad \text{with an integer } h \neq 0, \tag{9.3}$$

and put $e^t = x$; then (9.3) becomes

$$\frac{1}{T}\int_1^{e^T} \frac{e^{2\pi i h x}}{x}\,dx. \tag{9.4}$$

Consider now the real and imaginary parts of (9.4). Then

$$\frac{1}{T}\int_1^{e^T} \frac{\cos 2\pi h x}{x}\,dx = \frac{1}{T}\int_1^{e^\tau} \cos 2\pi h x\,dx$$

for some τ, $0 \leq \tau \leq T$, according to the second mean value theorem of integral calculus, and thus the real part of (9.4) goes to 0 as $T \to \infty$. The same holds with respect to the imaginary part of (9.4). ■

As we have already remarked in Section 4, one does not know whether or not the sequence (e^n), $n = 1, 2, \ldots$, is u.d. mod 1.

The second part of Example 9.1 is a special case of Theorem 9.3, below. We first establish a simple auxiliary result.

LEMMA 9.1. If $F(u)$, $u \geq 0$, is twice differentiable, if $uF''(u)$ is bounded, and if $F(u)/u \to 0$ as $u \to \infty$, then $F'(u) \to 0$ as $u \to \infty$.

PROOF. For $u > 0$ and $\eta > 0$, we have

$$F(u + \eta) - F(u) = \eta F'(u) + \tfrac{1}{2}\eta^2 F''(u + \theta\eta) \qquad \text{with } 0 < \theta < 1,$$

or, solving for $F'(u)$,

$$F'(u) = \frac{F(u + \eta) - F(u)}{\eta} - \tfrac{1}{2}\eta F''(u + \theta\eta). \tag{9.5}$$

Choose $\varepsilon > 0$, let M be an upper bound for $|uF''(u)|$, and set $\delta = \varepsilon/M$. With $\eta = \delta u$ in (9.5), we get

$$F'(u) = \left(\frac{F(u + \delta u)}{u + \delta u}\cdot\frac{1 + \delta}{\delta} - \frac{F(u)}{u}\cdot\frac{1}{\delta}\right)$$

$$- \frac{\delta}{2(1 + \theta\delta)}(u + \theta\,\delta u)F''(u + \theta\,\delta u). \tag{9.6}$$

Since the difference in parentheses tends to 0 as $u \to \infty$, it is in absolute value $<\varepsilon/2$ for $u > u_0(\varepsilon)$. But the remaining term on the right of (9.6) is in absolute value $<\varepsilon/2$, and this completes the argument. ■

THEOREM 9.3. Let $f(t)$, $t \geq 0$, be differentiable and let $tf'(t)$ be bounded. Then $f(t)$ is not c.u.d. mod 1.

PROOF. Suppose (9.2) would hold for $h = 1$. Then, setting $F(u) = \int_0^u \cos 2\pi f(t)\,dt$, we have $F(u)/u \to 0$ as $u \to \infty$. However, since $F'(u) = \cos 2\pi f(u)$ and $F''(u) = -2\pi f'(u) \sin 2\pi f(u)$, it follows that $uF''(u)$ is bounded, and so, according to Lemma 9.1, we would have $\cos 2\pi f(u) \to 0$ as $u \to \infty$. In the same way, it would follow that $\sin 2\pi f(u) \to 0$ as $u \to \infty$, and so, $e^{2\pi i f(u)} \to 0$ as $u \to \infty$, an obvious contradiction. ∎

THEOREM 9.4. Let $f(t)$ be a Lebesgue-measurable function defined on $[0, \infty)$. Let $\Delta f(t) = f(t+1) - f(t)$ be monotone, and furthermore let $\Delta f(t) \to 0$ and $t\,|\Delta f(t)| \to \infty$ as $t \to \infty$. Then $f(t)$ is c.u.d. mod 1.

PROOF. The result can be shown in a manner similar to that in which Theorem 2.5 was proved. ∎

THEOREM 9.5. For $t \geq 0$, let $f(t) = g(\log(t+1))$, where g is an increasing convex function on $[0, \infty)$ such that $\lim_{t\to\infty} g(t)/t = \infty$. Then $f(t)$ is c.u.d. mod 1.

PROOF. It suffices to prove (9.2) for integers $h > 0$. Set $u = 2\pi h f(t) = \varphi(t)$ and $t = \psi(u)$, the inverse function of $\varphi(t)$. Since g is convex, it follows that $t\varphi'(t)$ is an increasing function [$\varphi'(t)$ may cease to exist on a countable set of points where we define $\varphi'(t)$ suitably]. If $t\varphi'(t)$ were bounded above, then $\varphi(t) = O(\log t)$, a contradiction. Hence, $t\varphi'(t) \to \infty$ as $t \to \infty$. Now $t\varphi'(t) = \psi(u)/\psi'(u)$, and so, $\psi'(u)/\psi(u)$ is a decreasing function of u that tends to 0 as $u \to \infty$. For $T > 0$, consider the integral

$$J = \int_0^T \sin \varphi(t)\,dt.$$

Applying the substitution $u = \varphi(t)$ and setting $\varphi(0) = \rho_0$ and $\varphi(T) = \rho$, we obtain for $\rho_0 < \tau < \rho$ by repeated application of the second mean value theorem,

$$\begin{aligned} J &= \int_{\rho_0}^{\rho} \psi'(u) \sin u\,du = \int_{\rho_0}^{\tau} \psi'(u) \sin u\,du + \int_{\tau}^{\rho} \psi(u) \frac{\psi'(u)}{\psi(u)} \sin u\,du \\ &= \int_{\rho_0}^{\tau} \psi'(u) \sin u\,du + \frac{\psi'(\tau)}{\psi(\tau)} \int_{\tau}^{\tau_2} \psi(u) \sin u\,du \qquad (\tau < \tau_2 < \rho) \\ &= \int_{\rho_0}^{\tau} \psi'(u) \sin u\,du + \frac{\psi'(\tau)}{\psi(\tau)} \psi(\tau_2) \int_{\tau_1}^{\tau_2} \sin u\,du \qquad (\tau < \tau_1 < \tau_2) \\ &= \int_{\rho_0}^{\tau} \psi'(u) \sin u\,du + \frac{\psi'(\tau)}{\psi(\tau)} A, \end{aligned}$$

where $|A| \leq 2\psi(\rho) = 2T$. Let $\varepsilon > 0$ be arbitrary. Take τ so large that $\psi'(\tau)/\psi(\tau) \leq \varepsilon$, and then ρ so large that

$$\left| \int_{\rho_0}^{\tau} \psi'(u) \sin u \, du \right| \leq \varepsilon \psi(\rho).$$

Then $|J| \leq 3\varepsilon T$ for sufficiently large T, and so, $J = o(T)$. The integral $\int_0^T \cos \varphi(t) \, dt$ can be dealt with similarly. ■

Relations Between U.D. Mod 1 and C.U.D. Mod 1

THEOREM 9.6. Let $f(t)$, $t \geq 0$, be a real-valued Lebesgue-measurable function. Then under each of the following conditions $f(t)$ is c.u.d. mod 1:

(a) The sequence $(f(n + t))$, $n = 0, 1, \ldots$, is u.d. mod 1 for almost all t in $[0, 1]$.

(b) The sequence $(f(nt))$, $n = 1, 2, \ldots$, is u.d. mod 1 for almost all t in $[0, 1]$.

PROOF. (a) For an integer $h \neq 0$, we have for almost all t in $[0, 1]$,

$$\lim_{N \to \infty} \frac{1}{N} \sum_{n=0}^{N-1} e^{2\pi i h f(n+t)} = 0.$$

But then by the dominated convergence theorem,

$$\begin{aligned} 0 &= \int_0^1 \left(\lim_{N \to \infty} \frac{1}{N} \sum_{n=0}^{N-1} e^{2\pi i h f(n+t)} \right) dt \\ &= \lim_{N \to \infty} \int_0^1 \left(\frac{1}{N} \sum_{n=0}^{N-1} e^{2\pi i h f(n+t)} \right) dt \\ &= \lim_{N \to \infty} \frac{1}{N} \sum_{n=0}^{N-1} \int_n^{n+1} e^{2\pi i h f(u)} \, du \\ &= \lim_{N \to \infty} \frac{1}{N} \int_0^N e^{2\pi i h f(u)} \, du. \end{aligned}$$

Hence,

$$\frac{1}{T} \int_0^T e^{2\pi i h f(u)} \, du$$

tends to 0 if $T \to \infty$ in an arbitrary way.

(b) For an integer $h \neq 0$, we have for almost all t in $[0, 1]$,

$$\lim_{N \to \infty} \frac{1}{N} \sum_{n=1}^{N} e^{2\pi i h f(nt)} = 0.$$

Again, by integration and interchange of orders of operations we find

$$0 = \lim_{N\to\infty} \frac{1}{N} \sum_{n=1}^{N} \int_0^1 e^{2\pi i h f(nt)}\,dt$$

$$= \lim_{N\to\infty} \frac{1}{N} \sum_{n=1}^{N} \frac{1}{n} \int_0^n e^{2\pi i h f(u)}\,du = \lim_{N\to\infty} \frac{1}{N} \sum_{n=1}^{N} \frac{1}{n} \sum_{j=1}^{n} c_j,$$

where

$$c_j = \int_{j-1}^{j} e^{2\pi i h f(u)}\,du \qquad \text{for } j \geq 1.$$

In other words, the sequence (c_j), $j = 1, 2, \ldots$, is summable by the Hölder means (**H**, 2) to the value 0 (see Example 7.1). It follows that (c_j) is summable by the Cesàro means (**C**, 2) to the value 0 (see Example 7.1). Since $|c_j| \leq 1$ for $j \geq 1$, the Tauberian theorem mentioned in the proof of Theorem 7.13 yields

$$\lim_{N\to\infty} \frac{1}{N} \sum_{j=1}^{N} c_j = \lim_{N\to\infty} \frac{1}{N} \int_0^N e^{2\pi i h f(u)}\,du = 0. \quad \blacksquare$$

THEOREM 9.7. If $f(t)$, $t \geq 0$, has a continuous derivative of constant sign, if $f(t)$ is c.u.d. mod 1, and if $f(t)/t \to 0$ as $t \to \infty$, then the sequence $(f(n))$, $n = 1, 2, \ldots$, is u.d. mod 1.

PROOF. We use the Euler summation formula (see Example 2.4). For an integer $h \neq 0$, we get

$$\sum_{n=1}^{N} e^{2\pi i h f(n)} = \int_1^N e^{2\pi i h f(t)}\,dt + \tfrac{1}{2}(e^{2\pi i h f(1)} + e^{2\pi i h f(N)})$$
$$+ 2\pi i h \int_1^N (\{t\} - \tfrac{1}{2}) f'(t) e^{2\pi i h f(t)}\,dt.$$

Denote the third term on the right by $A(N)$. Then

$$\frac{|A(N)|}{N} \leq \frac{\pi\,|h|}{N} \int_1^N |f'(t)|\,dt \leq \frac{\pi\,|h|}{N} (|f(N)| + |f(1)|),$$

and by the assumption regarding $f(t)/t$ we have that $|A(N)|/N \to 0$ as $N \to \infty$. The proof can now be completed easily. $\blacksquare$

THEOREM 9.8. Let $f(t)$, $t \geq 0$, have a continuous derivative and suppose $f'(t) \log t \to C$, a nonzero constant, as $t \to \infty$. Then the sequence $(f(n))$, $n = 1, 2, \ldots$, is u.d. mod 1.

PROOF. First we show that $f(t)$ is c.u.d. mod 1. For $t \geq t_0 \geq 1$, f' has a constant sign, say $f'(t) > 0$ for $t \geq t_0$. If g is the inverse function of f, we have for $T > t_0$,

$$\frac{1}{T} \int_{t_0}^{T} e^{2\pi i h f(t)}\,dt = \frac{1}{T} \int_{f(t_0)}^{f(T)} \frac{e^{2\pi i h u}}{f'(g(u))}\,du,$$

the right member of which goes to 0 as $T \to \infty$ if the expression

$$\frac{1}{T} \int_{f(t_0)}^{f(T)} e^{2\pi i h u} \log g(u)\, du \tag{9.7}$$

goes to 0 as $T \to \infty$. Consider the real and the imaginary part of (9.7) and use the second mean value theorem. One obtains

$$\frac{1}{T} \int_{f(t_0)}^{f(T)} (\cos 2\pi h u) \log g(u)\, du = \frac{\log T}{T} \int_{f(\xi)}^{f(T)} \cos 2\pi h u\, du$$

for some ξ with $t_0 \le \xi \le T$, and a similar expression for the imaginary part of (9.7). Hence, (9.7) goes to 0 as $T \to \infty$, and thus, $f(t)$ is c.u.d. mod 1. Furthermore, the condition $f'(t) \log t \to C$ as $t \to \infty$ implies that $f(t)/t \to 0$ as $t \to \infty$. Now apply Theorem 9.7, and the proof is complete. ∎

Multidimensional Case

The notion of c.u.d. mod 1 can be extended to systems of measurable functions. We use the same notation as in Section 6.

DEFINITION 9.2. Let $\mathbf{f}(t) = (f_1(t), \ldots, f_s(t))$ be a system of s measurable functions defined on $[0, \infty)$. For $T > 0$ and $[\mathbf{a}, \mathbf{b}) \subseteq I^s$, let $T(\mathbf{a}, \mathbf{b})$ be the set of t, $0 \le t \le T$, for which $\{\mathbf{f}(t)\} \in [\mathbf{a}, \mathbf{b})$. The system $\mathbf{f}(t)$ is said to be *c.u.d. mod 1 in* $\mathbb{R}^s$ if

$$\lim_{T \to \infty} \frac{\lambda(T(\mathbf{a}, \mathbf{b}))}{T} = \prod_{j=1}^{s} (b_j - a_j)$$

holds for all $[\mathbf{a}, \mathbf{b}) \subseteq I^s$.

THEOREM 9.9. The system $\mathbf{f}(t)$ of measurable functions is c.u.d. mod 1 in $\mathbb{R}^s$ if and only if for each lattice point $\mathbf{h} \in \mathbb{Z}^s$, $\mathbf{h} \ne \mathbf{0}$,

$$\lim_{T \to \infty} \frac{1}{T} \int_0^T e^{2\pi i \langle \mathbf{h}, \mathbf{f}(t) \rangle}\, dt = 0.$$

For the proof we refer to the remarks following Theorem 6.2.

COROLLARY 9.1. The system $\mathbf{f}(t)$ of measurable functions is c.u.d. mod 1 in $\mathbb{R}^s$ if and only if, for each lattice point $\mathbf{h} \in \mathbb{Z}^s$, $\mathbf{h} \ne \mathbf{0}$, the function $\langle \mathbf{h}, \mathbf{f}(t) \rangle$ is c.u.d. mod 1.

PROOF. This follows readily from Theorems 9.2 and 9.9. ∎

EXAMPLE 9.3. In the euclidean plane, consider a rectilinear uniform motion defined by $x = x(t) = a + \theta_1 t$, $y = y(t) = b + \theta_2 t$, $0 \le t < \infty$, a, b, θ_1, and θ_2 are real, and where θ_1 and θ_2 are linearly independent over the rationals. Consider the time interval $0 \le t \le T$. The sum of the time subintervals during which the moving point is mod 1 in the rectangular interval $[\alpha_1, \alpha_2) \times [\beta_1, \beta_2) \subseteq I^2$ is given by the measure of the set $E \cap [0, T]$, where

$$E = \{0 \leq t < \infty : \alpha_1 \leq a_1 + \theta_1 t < \alpha_2 (\text{mod } 1), \beta_1 \leq b + \theta_2 t < \beta_2 (\text{mod } 1)\}.$$

Now the system of functions $(a + \theta_1 t, b + \theta_2 t)$ is c.u.d. mod 1 in $\mathbb{R}^2$ by Corollary 9.1 and Example 9.1. Hence, in the limit, the fraction of the time that the moving point is mod 1 in the considered rectangle is equal to $(\alpha_2 - \alpha_1)(\beta_2 - \beta_1)$, the area of the rectangle. ∎

Notes

Functions c.u.d. mod 1 were already discussed by Weyl [4]. He established the Weyl criterion and showed that any nonconstant polynomial is c.u.d. mod 1. For other early work on c.u.d. mod 1, see Steinhaus [1] and Tsuji [1]. Theorem 9.2 is contained as a special case in the general Weyl criterion of Kuipers and Stam [1]. Theorem 9.3 is from Kuipers and Meulenbeld [3] and Kuiper [1], and Theorem 9.5 is a result of Tsuji [1]. Related theorems can be found in Kuipers [1], Kuipers and Meulenbeld [1, 2], and Kuiper [1]. We refer also to Exercises 9.8, 9.9, and 9.13. Theorem 9.6 was shown by Ryll-Nardzewski [1], and Theorems 9.7 and 9.8 are from Kuipers [3], who also showed other results concerning the relation between c.u.d. mod 1 and u.d. mod 1 of sequences. The multidimensional case was already considered by Weyl [4], who proved the result in Example 9.3 (see also Weyl [3] and Pólya and Szegö [1, II. Abschn., Aufg. 185, 186]). This result has important applications in areas such as statistical mechanics, kinetic gas theory, and stellar dynamics. A closely related fact is enunciated in Exercise 9.29 and is again due to Weyl [3, 4]. It may also be found in Pólya and Szegö [1, II. Abschn., Aufg. 187]. That the trajectory must be dense in the unit square (resp. cube) was shown earlier by König and Szücs [1] (see also Sudan [1]).

Kuipers [2, 7] proves that $f(t) = \alpha t + \beta \sin t$, $\beta \neq 0$, is c.u.d. mod 1 iff $\pi\alpha$ is irrational, and other results concerning functions with periodic derivatives. Difference theorems for c.u.d. mod 1 were established by Kuipers [1] and Hlawka [9]. C.u.d. mod 1 in sequences of intervals was studied by Kuipers and Meulenbeld [4] and Kuipers [4]. Interesting connections between c.u.d. mod 1 and the notion of independence for functions were pointed out by Steinhaus [1] and Kuipers [7] (for a discrete analogue, see Chapter 5, Section 1). The quantitative study of functions c.u.d. mod 1 was initiated by Hlawka [9]. Many of the results in Chapter 2 on discrepancy of sequences have continuous analogues. See also Holewijn [1], Mück [1], Stackelberg [1], and Fleischer [3]. Hlawka [24] shows that "almost all" continuous functions (in the sense of Wiener measure) defined on $[0, \infty)$ are c.u.d. mod 1. Refinements are due to Fleischer [3] and Stackelberg [1], the stronger result being that of the latter author (compare with the notes in Chapter 2, Section 1). Fleischer [2] generalizes Hlawka's theorem to the multidimensional case. Kuipers and van der Steen [1] prove continuous analogues of Theorems 4.2 and 4.3 and improve some metric results of Holewijn [1]. See also Mück [1].

Various extensions and refinements of the notion of c.u.d. mod 1 have been considered. Kuipers [6] discusses well distributed functions mod 1. Hlawka [9] and Holewijn [1] study c.u.d. mod 1 with respect to summation methods and weight functions. Loynes [1] discusses c.u.d. mod 1 of stochastic processes. Kuipers [1] introduces distribution functions (mod 1) of measurable functions and proves an analogue of Theorem 7.6 (see also Haviland [1]). If (9.1) is required to hold for the characteristic functions of all Lebesgue-measurable subsets of I (or, equivalently, for all Lebesgue-integrable functions w on $[0, 1]$), one arrives at the notion of c^{III}-*u.d. mod 1* (Kuipers [1]). This was studied further by Kuiper [1], Kuipers [2], and Kuipers and Meulenbeld [1, 2]. A notion of c^{II}-u.d. mod 1 was defined by Kuipers [1] but was then shown to be equivalent to c^{III}-u.d. mod 1 (Kuipers and Meulenbeld [1]).

A further extension is of course to the multidimensional case, and Definition 9.2 may in fact be generalized to systems of functions of several variables. See Tsuji [1], Kuipers [1, 4], Meulenbeld [1], Kuipers and Meulenbeld [3, 5], and Couot [4]. As to c.u.d. on surfaces, there is an isolated result of I. S. Kac [1] and a more systematic study by Gerl [3]. A theory of c.u.d. in p-adic number fields was developed by Chauvineau [5, 6]. For functions on groups, see the notes in Chapter 4, Section 1. Other general viewpoints are discussed in Helmberg [6] and Kemperman [3]. Another notion of c.u.d. is presented in Chapter 5, Section 1.

Exercises

9.1. Prove Theorem 9.1.

9.2. Prove Theorem 9.2.

9.3. Prove Theorem 9.1 with "continuous" replaced by "Riemann-integrable."

9.4. Let $f(t)$ be c.u.d. mod 1, and let $g(t)$, $t \geq 0$, be a Lebesgue-measurable function such that $\lim_{t\to\infty} (f(t) - g(t))$ exists. Prove that $g(t)$ is c.u.d. mod 1.

9.5. Let $(f_n(t))$, $n = 1, 2, \ldots$, be a sequence of functions c.u.d. mod 1 that converges uniformly on $[0, \infty)$ to the limit function $f(t)$. Then $f(t)$ is c.u.d. mod 1.

9.6. If $f(t)$ is c.u.d. mod 1, then $mf(t)$ is c.u.d. mod 1 for all nonzero integers m.

9.7. If $f(t)$ is c.u.d. mod 1, then $f(ct)$, where c is a positive constant, has the same property.

9.8. Generalize the preceding exercise as follows: If $f(t)$ is c.u.d. mod 1 and if $\varphi(t)$, $t \geq 0$, is a nonnegative differentiable function with $\varphi'(t) \to c$, a positive constant, as $t \to \infty$, then $f(\varphi(t))$ is c.u.d. mod 1.

9.9. A Lebesgue-measurable periodic function $f(t)$, $t \geq 0$, of period $p > 0$ is c.u.d. mod 1 if and only if

$$\int_0^p e^{2\pi i h f(t)}\, dt = 0 \qquad \text{for } h = 1, 2, \ldots.$$

9.10. Prove that $f(t) = \sin t$, $t \geq 0$, is not c.u.d. mod 1.

9.11. Show that Lemma 9.1 holds as well if in both limit relations zero is replaced by the same constant c.

9.12. Prove Theorem 9.4 in detail.

9.13. If $f(t)$ has a positive nondecreasing derivative on $[0, \infty)$, then $f(t)$ is c.u.d. mod 1. *Hint:* Use the method in Example 9.2.

9.14. A nonconstant function on $[0, \infty)$ of the form $f(t) = \sum_{i=1}^{k} a_i t^{s_i}$ with $a_i, s_i \in \mathbb{R}$ and $s_k > s_{k-1} > \cdots > s_1 \geq 0$ is c.u.d. mod 1.

9.15. Let $f(t)$, $t \geq 0$, be a Lebesgue-measurable function. If for every positive integer h the function $f(t+h) - f(t)$ is c.u.d. mod 1, then $f(t)$ is c.u.d. mod 1.

9.16. Let $f(t)$ be a Lebesgue-measurable function defined for $t \geq 0$, and let k be a positive integer. Define $\Delta^k f(t)$ recursively by $\Delta^1 f(t) = \Delta f(t)$ and $\Delta^j f(t) = \Delta(\Delta^{j-1} f(t))$ for $j \geq 2$. Suppose $\Delta^k f(t)$ is monotone in t with $\Delta^k f(t) \to 0$ and $t\,|\Delta^k f(t)| \to \infty$ as $t \to \infty$. Then $f(t)$ is c.u.d. mod 1.

9.17. For a positive integer k, let $f(t)$ be k times differentiable on $[0, \infty)$ with $f^{(k)}(t) \to c$, a nonzero constant, as $t \to \infty$. Prove that $f(t)$ is c.u.d. mod 1.

9.18. For each $\sigma > 0$ the function $f(t) = t^\sigma \sin 2\pi t$ is c.u.d. mod 1. *Hint:* Use Theorem 9.6.

9.19. The function $f(t) = \sqrt{t} + \sin 2\pi(t + 1/(t+1))$, $t \geq 0$, is c.u.d. mod 1. *Hint:* Use Theorem 9.6.

9.20. Let $g(t)$ be a periodic Lebesgue-measurable function on $[0, \infty)$ with period $p > 0$, and suppose the Lebesgue-measurable function $f(t)$, $t \geq 0$, is such that $(f(np + u))$, $n = 0, 1, \ldots,$ is u.d. mod 1 for almost all u in $[0, p]$. Then $f(t) + g(t)$ is c.u.d. mod 1.

9.21. The function $f(t) = \sqrt[3]{\log(t+1)} + \sqrt{\log(t+1)}$, $t \geq 0$, is not c.u.d. mod 1.

9.22. Let (x_n), $n = 0, 1, \ldots,$ be a sequence of real numbers. Prove that (x_n) is u.d. mod 1 if and only if the function $f(t) = x_{[t]}$, $t \geq 0$, is c.u.d. mod 1.

9.23. Let (x_n) be u.d. mod 1. Prove that the function $f(t)$ from Exercise 9.22 is c.u.d. mod 1 but not c^{III}-u.d. mod 1.

9.24. Use Theorem 9.3 to prove: If $f(t)$ satisfies the conditions of this theorem, then $(f(n))$, $n = 1, 2, \ldots,$ is not u.d. mod 1 (compare with Theorem 2.6).

9.25. The sequence $(\log n + \sin(\log n))$, $n = 1, 2, \ldots,$ is not u.d. mod 1. More generally, if $\varphi(u)$, $u \geq 0$, is a differentiable function with bounded derivative and if $f(t)$, $t \geq 0$, is a nonnegative differentiable function with $tf'(t)$ bounded, then the sequence $(\varphi(f(n)))$, $n = 1, 2, \ldots,$ is not u.d. mod 1.

9.26. Let the system of functions $\mathbf{f}(t) = (f_1(t), \ldots, f_s(t))$ be c.u.d. mod 1 in $\mathbb{R}^s$. Then for any real-valued continuous function $w(u_1, \ldots, u_s)$ on $\bar{I}^s$ we have

$$\lim_{T\to\infty} \frac{1}{T} \int_0^T w(\{f_1(t)\}, \ldots, \{f_s(t)\})\, dt = \int_0^1 \cdots \int_0^1 w(u_1, \ldots, u_s)\, du_1 \cdots du_s.$$

The converse is also true.

9.27. If the real numbers $\theta_1, \theta_2, \ldots, \theta_s$ are linearly independent over the rationals, then the system of functions $(\theta_1 t, \theta_2 t, \ldots, \theta_s t)$ is c.u.d. mod 1 in $\mathbb{R}^s$.

9.28. If $\theta_1, \theta_2, \ldots, \theta_s$ are nonzero real numbers, then the system of functions $(\theta_1 t^s, \theta_2 t^{s-1}, \ldots, \theta_s t)$ is c.u.d. mod 1 in $\mathbb{R}^s$.

9.29. A point moves inside a square of unit area and its motion is rectilinear and with constant speed. At the sides of the square it is reflected symmetrically. The tangent of the angle between a side of the square and the initial direction of the motion is an irrational number. Consider a subinterval S of the square and let $S(T)$ be the time that the point is in S after T time units. Then $\lim_{T\to\infty} S(T)/T$ is equal to the area of S.

2

DISCREPANCY

In the previous chapter, we studied uniform distribution modulo 1 from a purely qualitative point of view. We were mainly interested in deciding whether a given sequence is uniformly distributed at all. But looking at various uniformly distributed sequences, one will realize that there exist sequences with a very good distribution behavior, whereas other sequences might just barely be uniformly distributed. It is the aim of this chapter to introduce a quantity (the so-called *discrepancy* of the sequence) that measures the deviation of the sequence from an ideal distribution. This will enable us to distinguish between sequences with "good" uniform distribution and sequences with "bad" uniform distribution. We investigate the quantitative aspects of several important theorems from the first chapter, and thereby obtain results that complement or go beyond those statements. Interesting applications of the concept of discrepancy to problems in numerical analysis can be found in Section 5. We remark that throughout this chapter the counting functions $A(E; N)$ introduced in Sections 1 and 6 of the previous chapter are understood to be defined also for finite sequences containing at least N terms.

1. DEFINITION AND BASIC PROPERTIES

One-Dimensional Case

DEFINITION 1.1. Let $x_1, \ldots, x_N$ be a finite sequence of real numbers. The number

$$D_N = D_N(x_1, \ldots, x_N) = \sup_{0 \leq \alpha < \beta \leq 1} \left| \frac{A([\alpha, \beta); N)}{N} - (\beta - \alpha) \right| \quad (1.1)$$

is called the *discrepancy* of the given sequence. For an infinite sequence ω of real numbers (or for a finite sequence containing at least N terms), the discrepancy $D_N(\omega)$ is meant to be the discrepancy of the initial segment formed by the first N terms of ω.

It is clear from the above definition that when proving an assertion about the discrepancy of a given sequence, no loss of generality will result if we assume the sequence to be contained in I.

The pertinence of the concept of discrepancy in the theory of u.d. mod 1 is revealed by the following criterion.

THEOREM 1.1. The sequence ω is u.d. mod 1 if and only if $\lim_{N\to\infty} D_N(\omega) = 0$.

PROOF. The sufficiency of the condition is obvious. To show the necessity, we choose an integer $m \geq 2$. For $0 \leq k \leq m - 1$, let I_k denote the interval $I_k = [k/m, (k + 1)/m)$. Since ω is u.d. mod 1, there exists a positive integer $N_0 = N_0(m)$ such that for $N \geq N_0$ and for every $k = 0, 1, \ldots, m - 1$ we have

$$\frac{1}{m}\left(1 - \frac{1}{m}\right) \leq \frac{A(I_k; N)}{N} \leq \frac{1}{m}\left(1 + \frac{1}{m}\right). \tag{1.2}$$

Now consider an arbitrary subinterval $J = [\alpha, \beta)$ of I. There clearly exist intervals J_1 and J_2, finite unions of intervals I_k, such that $J_1 \subseteq J \subseteq J_2$, $\lambda(J) - \lambda(J_1) < 2/m$, and $\lambda(J_2) - \lambda(J) < 2/m$. From (1.2) we get for all $N \geq N_0$:

$$\lambda(J_1)\left(1 - \frac{1}{m}\right) \leq \frac{A(J_1; N)}{N} \leq \frac{A(J; N)}{N} \leq \frac{A(J_2; N)}{N} \leq \lambda(J_2)\left(1 + \frac{1}{m}\right).$$

Consequently, we obtain

$$\left(\lambda(J) - \frac{2}{m}\right)\left(1 - \frac{1}{m}\right) < \frac{A(J; N)}{N} < \left(\lambda(J) + \frac{2}{m}\right)\left(1 + \frac{1}{m}\right),$$

and, using $\lambda(J) \leq 1$,

$$-\frac{3}{m} - \frac{2}{m^2} < \frac{A(J; N)}{N} - \lambda(J) < \frac{3}{m} + \frac{2}{m^2} \qquad \text{for all } N \geq N_0. \tag{1.3}$$

Since the bounds in (1.3) are independent of J, we arrive at $D_N(\omega) \leq (3/m) + (2/m^2)$ for all $N \geq N_0$. But $(3/m) + (2/m^2)$ can be made arbitrarily small, and so the proof is complete. ∎

In particular, the above theorem enunciates the following interesting fact: Whenever ω is u.d. mod 1, then $\lim_{N\to\infty} A(J; N)/N = \lambda(J)$ uniformly in all subintervals $J = [\alpha, \beta)$ of I.

A meaningful concept of discrepancy (i.e., one for which a criterion similar to Theorem 1.1 holds) may also be defined with respect to a continuous distribution function. More explicitly, let f be a nondecreasing function on [0, 1] with $f(0) = 0$ and $f(1) = 1$, and let ω be a sequence of real numbers. Then the expression

$$D_N(\omega; f) = \sup_{0 \le \alpha < \beta \le 1} \left| \frac{A([\alpha, \beta); N; \omega)}{N} - (f(\beta) - f(\alpha)) \right| \tag{1.4}$$

may be called the *discrepancy of* ω *with respect to* f. By using the same ideas as in the proof of Theorem 1.1, it can be shown that if f is continuous, then the sequence ω has f as its a.d.f. (mod 1) precisely if $\lim_{N\to\infty} D_N(\omega; f) = 0$ (see Exercise 1.3). An analogous criterion fails to hold for discontinuous f (see the counterexample in Exercise 1.4).

For the time being, we will be satisfied with a simple estimate for D_N that already shows that D_N cannot converge too rapidly to zero. Refinements will be shown in Section 2.

THEOREM 1.2. For any sequence of N numbers, we have

$$\frac{1}{N} \le D_N \le 1. \tag{1.5}$$

PROOF. The right-hand side inequality is evident from the definition. Now choose $\varepsilon > 0$, and let $x \in I$ be one of the elements of the sequence. Consider the interval $J = [x, x + \varepsilon) \cap I$. Since $x \in J$, we have $A(J; N)/N - \lambda(J) \ge (1/N) - \lambda(J) \ge (1/N) - \varepsilon$. This implies $D_N \ge (1/N) - \varepsilon$, and the desired inequality is established. ∎

EXAMPLE 1.1. We may have $D_N = 1/N$, for instance when the elements of the sequence are the numbers $0, 1/N, 2/N, \ldots, (N-1)/N$ in some order. Consider an arbitrary half-open subinterval J of I. There is a unique integer k with $0 \le k \le N - 1$ and $k/N < \lambda(J) \le (k+1)/N$. Then J contains at least k and at most $k + 1$ elements from the above sequence. Consequently, we obtain $|A(J; N)/N - \lambda(J)| \le 1/N$ for all J. ∎

It is sometimes useful to slightly restrict the family of intervals over which the supremum is formed in the definition of discrepancy. The most important type of restriction is to consider only the intervals $[0, \alpha)$ with $0 < \alpha \le 1$.

DEFINITION 1.2. For a finite sequence $x_1, \ldots, x_N$ of real numbers, we define

$$D_N^* = D_N^*(x_1, \ldots, x_N) = \sup_{0 < \alpha \le 1} \left| \frac{A([0, \alpha); N)}{N} - \alpha \right|. \tag{1.6}$$

The definition of D_N^* is extended in the same way as in Definition 1.1.

THEOREM 1.3. The discrepancies D_N and D_N^* are related by the inequality

$$D_N^* \leq D_N \leq 2D_N^*. \tag{1.7}$$

PROOF. The first inequality follows immediately from the definitions. To obtain the second one, we note that $A([\alpha, \beta); N) = A([0, \beta); N) - A([0, \alpha); N)$, where $0 \leq \alpha < \beta \leq 1$, and therefore,

$$\left|\frac{A([\alpha, \beta); N)}{N} - (\beta - \alpha)\right| \leq \left|\frac{A([0, \beta); N)}{N} - \beta\right| + \left|\frac{A([0, \alpha); N)}{N} - \alpha\right|.$$

Taking suprema yields the desired result. ■

COROLLARY 1.1. The sequence ω is u.d. mod 1 if and only if

$$\lim_{N\to\infty} D_N^*(\omega) = 0.$$

PROOF. This is an immediate consequence of Theorems 1.1 and 1.3. ■

There is a simple alternative representation for the discrepancy D_N^* as a maximum of finitely many numbers. We note that the discrepancy D_N^* (and D_N, for that matter) of the numbers $x_1, x_2, \ldots, x_N$ in I does not depend on the order of those elements. Therefore, we may assume without loss of generality that the x_i are ordered according to their magnitude.

THEOREM 1.4. Let $x_1 \leq x_2 \leq \cdots \leq x_N$ be N numbers in I. Then their discrepancy D_N^* is given by

$$\begin{aligned} D_N^* &= \max_{i=1,\ldots,N} \max\left(\left|x_i - \frac{i}{N}\right|, \left|x_i - \frac{i-1}{N}\right|\right) \\ &= \frac{1}{2N} + \max_{i=1,\ldots,N}\left|x_i - \frac{2i-1}{2N}\right|. \end{aligned} \tag{1.8}$$

PROOF. For notational convenience, we set $x_0 = 0$ and $x_{N+1} = 1$. The distinct values of the numbers x_i, $0 \leq i \leq N + 1$, define a subdivision of $[0, 1]$. Therefore,

$$\begin{aligned} D_N^* &= \max_{\substack{i=0,\ldots,N\\ x_i<x_{i+1}}} \sup_{x_i<\alpha\leq x_{i+1}} \left|\frac{A([0, \alpha); N)}{N} - \alpha\right| \\ &= \max_{\substack{i=0,\ldots,N\\ x_i<x_{i+1}}} \sup_{x_i<\alpha\leq x_{i+1}} \left|\frac{i}{N} - \alpha\right|. \end{aligned}$$

Whenever $x_i < x_{i+1}$, the function $g_i(\alpha) = |(i/N) - \alpha|$ attains its maximum in $[x_i, x_{i+1}]$ at one of the end points of the interval. Consequently, we have

$$D_N^* = \max_{\substack{i=0,\ldots,N\\ x_i<x_{i+1}}} \max\left(\left|\frac{i}{N} - x_i\right|, \left|\frac{i}{N} - x_{i+1}\right|\right). \tag{1.9}$$

We show now that we may drop the restriction $x_i < x_{i+1}$ in the first maximum. So suppose we have $x_i < x_{i+1} = x_{i+2} = \cdots = x_{i+r} < x_{i+r+1}$ with some $r \geq 2$. The indices not admitted in the first maximum in (1.9) are the integers $i + j$ with $1 \leq j \leq r - 1$. We shall prove that the numbers

$$\left|\frac{i+j}{N} - x_{i+j}\right| \quad \text{and} \quad \left|\frac{i+j}{N} - x_{i+j+1}\right|$$

with $1 \leq j \leq r - 1$, which are excluded in (1.9), are in fact dominated by numbers already occurring in (1.9). For $1 \leq j \leq r - 1$, we get by the same reasoning as above (consider the function $h_{i+1}(y) = |y - x_{i+1}|$):

$$\left|\frac{i+j}{N} - x_{i+j}\right| = \left|\frac{i+j}{N} - x_{i+1}\right| < \max\left(\left|\frac{i}{N} - x_{i+1}\right|, \left|\frac{i+r}{N} - x_{i+1}\right|\right)$$
$$= \max\left(\left|\frac{i}{N} - x_{i+1}\right|, \left|\frac{i+r}{N} - x_{i+r}\right|\right),$$

and both numbers in the last maximum occur in (1.9). Exactly the same argument holds for $|(i + j)/N - x_{i+j+1}|$, $1 \leq j \leq r - 1$. Thus, we arrive at

$$D_N^* = \max_{i=0,\ldots,N} \max\left(\left|\frac{i}{N} - x_i\right|, \left|\frac{i}{N} - x_{i+1}\right|\right)$$
$$= \max_{i=1,\ldots,N} \max\left(\left|\frac{i}{N} - x_i\right|, \left|\frac{i-1}{N} - x_i\right|\right).$$

The last step is valid because the only terms we dropped are $|(0/N) - x_0|$ and $|(N/N) - x_{N+1}|$, both of which are zero. The second identity in (1.8) is clear. ■

COROLLARY 1.2. For any sequence of N numbers in I we have $D_N^* \geq 1/2N$, with equality only for the sequence $1/2N, 3/2N, \ldots, (2N - 1)/2N$ or its rearrangements.

We arrive at the following interpretation for D_N^*: The discrepancy D_N^* of a sequence of N terms in I in natural order is obtained by adding to the smallest possible value $1/2N$ of D_N^* the maximal deviation of the given sequence from the increasing sequence of N terms with least discrepancy. The finite sequence $1/2N$, $3/2N, \ldots,$ $(2N - 1)/2N$ shows also that the constant 2 in Theorem 1.3 is best possible.

Multidimensional Discrepancy

The definitions of D_N and D_N^* may be extended in a rather obvious fashion to sequences in $\mathbb{R}^k$ (for results on u.d. mod 1 in $\mathbb{R}^k$, see Chapter 1, Section 6).

DEFINITION 1.3. Let $\mathbf{x}_1, \ldots, \mathbf{x}_N$ be a finite sequence in $\mathbb{R}^k$. The discrepancies D_N and D_N^* are defined by

$$D_N = D_N(\mathbf{x}_1, \ldots, \mathbf{x}_N) = \sup_J \left| \frac{A(J; N)}{N} - \lambda(J) \right|, \tag{1.10}$$

$$D_N^* = D_N^*(\mathbf{x}_1, \ldots, \mathbf{x}_N) = \sup_{J^*} \left| \frac{A(J^*; N)}{N} - \lambda(J^*) \right|, \tag{1.11}$$

where J runs through all subintervals of I^k of the form $J = \{(x_1, \ldots, x_k) \in I^k : \alpha_i \leq x_i < \beta_i \text{ for } 1 \leq i \leq k\}$, and J^* runs through all subintervals of I^k of the form $J^* = \{(x_1, \ldots, x_k) \in I^k : 0 \leq x_i < \beta_i \text{ for } 1 \leq i \leq k\}$. Moreover, λ denotes the k-dimensional Lebesgue measure. For infinite sequences, the same convention as in Definition 1.1 applies.

EXAMPLE 1.2. The relation between D_N and D_N^* in several dimensions is very much similar to what was shown for $k = 1$ in Theorem 1.3. The idea is to start from an arbitrary subinterval J of I^k and represent it in terms of intervals of type J^*. Let us describe the details for $k = 2$, the higher-dimensional cases being completely similar. Let $J = \{(x_1, x_2) \in I^2 : \alpha_1 \leq x_1 < \beta_1 \text{ and } \alpha_2 \leq x_2 < \beta_2\} = [\alpha_1, \beta_1) \times [\alpha_2, \beta_2)$ with $0 \leq \alpha_i < \beta_i \leq 1$, $i = 1, 2$, be a subinterval of I^2. Then $J = \{([0, \beta_1) \times [0, \beta_2)) \backslash ([0, \alpha_1) \times [0, \beta_2))\} \backslash \{([0, \beta_1) \times [0, \alpha_2)) \backslash ([0, \alpha_1) \times [0, \alpha_2))\} = (J_1^* \backslash J_2^*) \backslash (J_3^* \backslash J_4^*)$. Furthermore, $\lambda(J) = \lambda(J_1^*) - \lambda(J_2^*) - \lambda(J_3^*) + \lambda(J_4^*)$ and $A(J; N) = A(J_1^*; N) - A(J_2^*; N) - A(J_3^*; N) + A(J_4^*; N)$. By the same reasoning as in the proof of Theorem 1.3, we arrive at $D_N^* \leq D_N \leq 4D_N^*$. In the general k-dimensional case, we will have $D_N^* \leq D_N \leq 2^k D_N^*$. ■

It can be shown as in Theorem 1.1 that a sequence ω is u.d. mod 1 in $\mathbb{R}^k$ if and only if $\lim_{N\to\infty} D_N(\omega) = 0$, or, equivalently, $\lim_{N\to\infty} D_N^*(\omega) = 0$. Moreover, the proof of Theorem 1.2 carries over to yield $D_N \geq 1/N$ for the multidimensional case as well (see Section 2 for a stronger result).

Isotropic Discrepancy

We are led to another interesting notion of discrepancy in $\mathbb{R}^k$ if we extend the supremum in Definition 1.3 over a much larger class of subsets of I^k, namely, over all convex subsets. We note that every bounded convex set C in $\mathbb{R}^k$ has a finite Lebesgue measure $\lambda(C)$ and is even measurable in the sense of Jordan.

DEFINITION 1.4. Let $\mathbf{x}_1, \ldots, \mathbf{x}_N$ be a finite sequence in $\mathbb{R}^k$. The *isotropic discrepancy* $J_N = J_N(\mathbf{x}_1, \ldots, \mathbf{x}_N)$ is defined to be

$$J_N = \sup_{C \in \mathscr{C}} \left| \frac{A(C; N)}{N} - \lambda(C) \right|, \tag{1.12}$$

where $\mathscr{C}$ is the family of all convex subsets of I^k. The definition of J_N is extended as in Definition 1.3.

Actually, we will show that one need not take all convex subsets of I^k to define J_N. It suffices to consider a special class of convex sets, namely, closed or open convex polytopes. We recall that a *closed convex polytope* is defined to be the convex hull of a finite number of points in $\mathbb{R}^k$, that is, the smallest convex set containing these points. By an *open convex polytope*, we mean the interior (in the usual topology of $\mathbb{R}^k$) of a closed convex polytope. We shall use the following elementary theorem on convex sets. For every convex set C in $\mathbb{R}^k$ and every point $\mathbf{x} \in \mathbb{R}^k$ that does not lie in the interior of C, there exists a hyperplane H through $\mathbf{x}$ such that C is entirely contained in one of the two closed half-spaces defined by H. The hyperplane H is called a *supporting hyperplane of C*.

THEOREM 1.5. The isotropic discrepancy J_N is also given by

$$J_N = \sup_{P \in \mathscr{P}} \left| \frac{A(P; N)}{N} - \lambda(P) \right|, \tag{1.13}$$

where $\mathscr{P}$ denotes the family of all closed convex polytopes and of all open convex polytopes contained in I^k.

PROOF. We shall show that $J_N = \sup_{Q \in \mathscr{Q}} |A(Q; N)/N - \lambda(Q)|$, where $\mathscr{Q}$ denotes the family of all closed convex polytopes and of all open convex polytopes contained in $\bar{I}^k$. Then we are done, because, first of all, every open convex polytope contained in $\bar{I}^k$ is also contained in I^k. Moreover, since w.l.o.g. the given finite sequence is contained in I^k, there will exist, for every $\varepsilon > 0$ and for every closed convex polytope Q contained in $\bar{I}^k$, a closed convex polytope $P = P(\varepsilon)$ contained in I^k with $A(Q; N) = A(P; N)$ and $\lambda(Q) \geq \lambda(P) > \lambda(Q) - \varepsilon$. Thus, the supremum extended over $\mathscr{Q}$ is the same as the supremum extended over $\mathscr{P}$.

The symbols int M and $\bar{M}$ shall denote the interior and the closure of a set M, respectively. For $C \in \mathscr{C}$, we have $A(\text{int } C; N) \leq A(C; N) \leq A(\bar{C}; N)$, and also $\lambda(\text{int } C) = \lambda(C) = \lambda(\bar{C})$, since C is Jordan-measurable. Therefore,

$$\left| \frac{A(C; N)}{N} - \lambda(C) \right| \leq \max \left(\left| \frac{A(\text{int } C; N)}{N} - \lambda(\text{int } C) \right|, \left| \frac{A(\bar{C}; N)}{N} - \lambda(\bar{C}) \right| \right),$$

and so, it suffices to consider closed or open convex sets in $\bar{I}^k$. The same argument yields that we need only prove

$$J_N = \sup_{R \in \mathscr{R}} \left| \frac{A(R; N)}{N} - \lambda(R) \right|$$

where $\mathscr{R}$ is the class of all sets R of the form $R = R_1 \cap \bar{I}^k$ for some closed or open convex polytope R_1 in $\mathbb{R}^k$ (note that both int R and $\bar{R}$ are in $\mathscr{Q}$).

It suffices to show that for each closed or open convex set C in $\bar{I}^k$, there exist P and Q from $\mathscr{R}$ with $A(P; N) = A(Q; N) = A(C; N)$ and $\lambda(P) \leq \lambda(C) \leq \lambda(Q)$. For then we get

$$\left| \frac{A(C; N)}{N} - \lambda(C) \right| \leq \max \left(\left| \frac{A(P; N)}{N} - \lambda(P) \right|, \left| \frac{A(Q; N)}{N} - \lambda(Q) \right| \right),$$

and we are done.

Let $\mathbf{x}_1, \ldots, \mathbf{x}_N$ be the given finite sequence in I^k. Let a closed or open convex set $C \subseteq \bar{I}^k$ be given, and suppose that C contains exactly the elements $\mathbf{x}_{i_1}, \ldots, \mathbf{x}_{i_r}$ of the given sequence. Let P be the convex hull of those points (or the empty set if C contains no elements of the sequence). Then $P \in \mathscr{R}$ and $P \subseteq C$; moreover, $A(P; N) = A(C; N)$ and $\lambda(P) \leq \lambda(C)$.

Now to the construction of Q. Let $\mathbf{x}_{j_1}, \ldots, \mathbf{x}_{j_s}$ be the elements of the given sequence that are not contained in C. If $s = 0$, then we simply put $Q = \bar{I}^k$. Thus, $s > 0$ in the sequel. If C is not open (hence, compact), then it is clear that we can enlarge C to a convex set C' (not necessarily contained in $\bar{I}^k$) that still does not contain $\mathbf{x}_{j_1}, \ldots, \mathbf{x}_{j_s}$ but has $\mathbf{x}_{i_1}, \ldots, \mathbf{x}_{i_r}$ as interior points. If C is open, we set $C' = C$. Through each point $\mathbf{x}_{j_m}$ with $1 \leq m \leq s$ there is a supporting hyperplane H_m of C' such that C' lies entirely in a closed half-space T_m defined by H_m. We note that the set $Q_1 = \bigcap_{m=1}^{s} T_m$ contains C'. Let $T_m{}^0$ be the open half-space corresponding to T_m. We claim that the set $Q = T_1{}^0 \cap \cdots \cap T_s{}^0 \cap \bar{I}^k$ satisfies all our requirements. We clearly have $Q \in \mathscr{R}$. Since $\mathbf{x}_{j_m} \notin T_m{}^0$ for all $1 \leq m \leq s$, the set Q contains none of the points $\mathbf{x}_{j_1}, \ldots, \mathbf{x}_{j_s}$. On the other hand, the points $\mathbf{x}_{i_1}, \ldots, \mathbf{x}_{i_r}$ are interior points of C', and so, they are all contained in Q. Therefore, $A(Q; N) = A(C; N)$. Furthermore, $\lambda(Q) = \lambda(Q_1 \cap \bar{I}^k) \geq \lambda(C' \cap \bar{I}^k) \geq \lambda(C)$, and the proof is complete. ■

THEOREM 1.6. For every sequence of N points in $\mathbb{R}^k$ we have

$$D_N \leq J_N \leq (4k\sqrt{k} + 1)D_N^{1/k}. \tag{1.14}$$

PROOF. The first inequality is evident, since every interval is a convex set. The basic idea in the proof of the second inequality is the following one. We start out from a convex subset of I^k, and we may assume for simplicity that it is a polytope $P \in \mathscr{P}$ (by using the previous theorem). Then we enclose P between two sets P_1 and P_2, both of them finite unions of intervals, such that $P_1 \subseteq P \subseteq P_2$. The exact form of P_1 and P_2 will be given later on. We note that

$$\left(\frac{A(P_1; N)}{N} - \lambda(P_1)\right) + (\lambda(P_1) - \lambda(P)) \leq \frac{A(P; N)}{N} - \lambda(P)$$

$$\leq \left(\frac{A(P_2; N)}{N} - \lambda(P_2)\right) + (\lambda(P_2) - \lambda(P)),$$

and so,

$$\left|\frac{A(P;N)}{N} - \lambda(P)\right| \le \max_{i=1,2}\left|\frac{A(P_i;N)}{N} - \lambda(P_i)\right| + \max_{i=1,2}|\lambda(P_i) - \lambda(P)|. \quad (1.15)$$

The sets P_1 and P_2 are constructed as follows. Let r be an arbitrary positive integer. For every lattice point $(h_1, h_2, \ldots, h_k)$ with $0 \le h_j < r$ for all $1 \le j \le k$, define an interval $J^{(r)}_{h_1h_2\cdots h_k} = \{(x_1, x_2, \ldots, x_k) \in \mathbb{R}^k\colon h_j/r \le x_j < (h_j + 1)/r$ for $1 \le j \le k\}$. The collection $\mathscr{L}^{(r)}$ of all those intervals forms a partition of I^k. We define $P_1 = P_1^{(r)}$ to be the union of all those intervals from $\mathscr{L}^{(r)}$ that are entirely contained in P. Moreover, let $P_2 = P_2^{(r)}$ be the union of all those intervals from $\mathscr{L}^{(r)}$ that have a nonvoid intersection with P. Then we clearly have $P_1 \subseteq P \subseteq P_2$. If we fix $k - 1$ integers $h_1, \ldots, h_{k-1}$ satisfying the above restriction, then the integers $h, 0 \le h < r$, with $J^{(r)}_{h_1\cdots h_{k-1}h} \subseteq P$ are consecutive integers (there can be no "gaps" because of the convexity of P). Therefore, the union of these intervals $J^{(r)}_{h_1\cdots h_{k-1}h}$ is again an interval. It follows that P_1 can be written as the union of at most r^{k-1} pairwise disjoint intervals. In exactly the same way, it can be shown that P_2 can be written as the union of at most r^{k-1} pairwise disjoint intervals. Consequently, we have

$$\max_{i=1,2}\left|\frac{A(P_i;N)}{N} - \lambda(P_i)\right| \le r^{k-1}D_N. \quad (1.16)$$

We shall now estimate $\lambda(P_2) - \lambda(P)$. We observe that the diameter of each $J^{(r)}_{h_1h_2\cdots h_k}$ (i.e., the supremum of the distances between two points of the set) is $\delta = (1/r)\sqrt{k}$. Therefore, each point of P_2 has a distance at most δ from some point of P. Hence, P_2 is contained in the set Q constructed as follows. For each one of the hyperplanes H forming the boundary of P, consider a parallel hyperplane H' at orthogonal distance δ in the open half-space determined by H that does not contain P (thus, in both half-spaces in the degenerate case where P is entirely contained in a hyperplane). Intersecting the closed half-spaces that are determined by those H' and that contain P, and intersecting the resulting set with $\bar{I}^k$, we obtain the closed convex polytope Q. We have then $\lambda(P_2) - \lambda(P) \le \lambda(Q\backslash P)$.

From the construction of Q, it should be clear that $\lambda(Q\backslash P)$ is at most δ times the surface of Q. To give a formal proof, we choose a point $\mathbf{y}$ in the interior of Q not belonging to P. Since the boundary of Q is compact, there is a point $\mathbf{x}$ in the boundary of Q that has the least distance from $\mathbf{y}$ among all points from the boundary of Q. Then the line joining $\mathbf{y}$ and $\mathbf{x}$ is orthogonal to the hyperplane (or one of the hyperplanes) defining Q in which $\mathbf{x}$ lies. For otherwise, the line through $\mathbf{y}$ orthogonal to a hyperplane in which $\mathbf{x}$ lies will intersect that hyperplane in a point $\mathbf{x}_1 \notin Q$ (because of the minimality property of $\mathbf{x}$). The line segment from $\mathbf{y}$ (in the interior of Q) to $\mathbf{x}_1$ (outside of Q) will meet the boundary of Q in a point $\mathbf{x}_2$, say. But $\mathbf{x}_2$ lies then closer to $\mathbf{y}$ than $\mathbf{x}$ does, a contradiction. Moreover, since $\mathbf{y} \notin P$, the distance from $\mathbf{y}$ to $\mathbf{x}$ will be at most δ. We have thus shown that every point in $Q\backslash P$ lies in one of the "prisms" of height δ erected over each $(k-1)$-dimensional face of Q and directed toward the interior of Q. Our estimate for $\lambda(Q\backslash P)$ is then immediate.

We now apply the following classical theorem (see Bonnesen and Fenchel [1, p. 47], Eggleston [3, Chapter 5]): If K_1, K_2 are two bounded closed convex sets with K_1 contained in K_2, then the surface of K_1 cannot exceed the surface of K_2. The closed convex polytope Q is contained in $\bar{I}^k$, and so the surface of Q is at most $2k$, since the surface of $\bar{I}^k$ consists of $2k$ "unit squares." We have therefore shown $\lambda(P_2) - \lambda(P) \le \lambda(Q\backslash P) \le 2k\sqrt{k}/r$. In exactly the same way, one proves $\lambda(P) - \lambda(P_1) \le 2k\sqrt{k}/r$. Combining (1.15), (1.16), and the last two inequalities, we arrive at

$$\left|\frac{A(P; N)}{N} - \lambda(P)\right| \le r^{k-1} D_N + \frac{2k\sqrt{k}}{r},$$

and since the upper bound is independent of P, we infer that

$$J_N \le r^{k-1} D_N + \frac{2k\sqrt{k}}{r}.$$

This estimate holds for all positive integers r. We choose $r = [D_N^{-1/k}]$, and we obtain $J_N \le (4k\sqrt{k} + 1)D_N^{1/k}$. ∎

COROLLARY 1.3. The sequence ω in $\mathbb{R}^k$ is u.d. mod 1 in $\mathbb{R}^k$ if and only if $\lim_{N\to\infty} J_N(\omega) = 0$.

Notes

The first instance where the notion of discrepancy is studied in its own right is in a paper of Bergström [2], who used the rather cumbersome term *Intensitätsdispersion*. Quantitative investigations for various u.d. mod 1 sequences had been carried out earlier (see, e.g., Section 3). The term "discrepancy" was probably coined by van der Corput. The first extensive study of discrepancy was undertaken by van der Corput and Pisot [1], who showed Theorem 1.2 but also some deeper results. For Theorem 1.1, see Weyl [4].

The discrepancy D_N^* may be interpreted as the supremum norm of the function

$$g(x) = \frac{A([0, x); N)}{N} - x$$

on the interval $\bar{I}$. We arrive at other notions of discrepancy by taking the L^p norm of that function, that is,

$$D_N^{(p)} = \left(\int_0^1 \left|\frac{A([0, x); N)}{N} - x\right|^p dx\right)^{1/p}$$

for $1 \leq p < \infty$. The number $D_N^{(p)}$ may be called the L^p *discrepancy* of the given sequence. In this context, the discrepancies D_N and D_N^* are sometimes referred to as the *extreme discrepancies*. The L^2 discrepancy has been studied in some detail by Halton [4], Halton and Zaremba [1], Niederreiter [13], Sobol' [7], Warnock [1], and Zaremba [2, 3]. The L^p discrepancies may of course also be defined in the multidimensional case. For various other notions of discrepancy, see Hlawka [25], Mück and Philipp [1], and Niederreiter [13, 14].

The term *isotropic discrepancy* was coined by Zaremba [7]. The study of J_N goes back to Hlawka [12]. Theorem 1.5 was shown by Niederreiter [10] and even in a stronger form. Some open problems on isotropic discrepancy and the proof of Theorem 1.6 can be found in Niederreiter [5]. See also Mück and Philipp [1]. Earlier results in this direction are in Hlawka [12, 25]. An example of Zaremba [4] shows that the exponent $1/k$ in (1.14) cannot be improved. For further results on isotropic discrepancy, see the last paragraph of these notes and the notes in Sections 2 and 5. We refer to Bonnesen and Fenchel [1] and Eggleston [3] for an exposition of the theory of convex sets.

Formula (1.8) is from Niederreiter [10], who also gave multidimensional analogues. A generalization is also contained in Niederreiter [14]. Bateman [1] makes use of the discrepancy in a geometric problem. A discrete counterpart of discrepancy was studied by Niederreiter [6, 9], Meijer and Niederreiter [1], Tijdeman [2], and Meijer [5]. For results on the sequences in Exercise 1.8, see Erdös and Rényi [1] and Hlawka [7].

Statements such as Theorem 1.1 and Corollary 1.3 are special results of the theory of uniformity classes (see Billingsley and Topsøe [1] and the notes in Chapter 3, Section 1). Theorem 1.1 follows also from the Pólya-Cantelli theorem on the pointwise convergence of monotone functions (Fréchet [1, pp. 319–321]).

The notion of discrepancy can be viewed as a special case of a notion arising in the theory of empirical distribution functions. Let $\xi_1, \xi_2, \ldots$ be independent random variables on a probability space $(X, \mathscr{B}, \mu)$ with a common continuous distribution function $F(t)$. For $x \in X$ and a positive integer N, the *empirical distribution function* $F_N(t, x)$ is defined as N^{-1} times the number of $\xi_n(x)$, $1 \leq n \leq N$, with $\xi_n(x) \leq t$. The so-called *two-sided Kolmogorov test* is then given by $G_N(x) = \sup_{t \in \mathbb{R}} |F_N(t, x) - F(t)|$. If $\xi_1, \xi_2, \ldots$ are the coordinate projections of the infinite-dimensional unit cube, then the definition of $G_N(x)$ just reduces to the definition of the discrepancy D_N^* (see Exercise 1.6). By the Glivenko-Cantelli theorem, one has $\lim_{N\to\infty} G_N(x) = 0$ μ-a.e. Many quantitative refinements are known. We state them for D_N^*, although they hold in the general case as well. The Kolmogorov-Smirnov limit theorem (Kolmogorov [2], Smirnov [1]) yields

$$\lim_{N\to\infty} \lambda_\infty(\{\omega \in I^\infty : \sqrt{N} D_N^*(\omega) \leq \alpha\}) = 1 - 2 \sum_{j=1}^{\infty} (-1)^{j+1} e^{-2j^2\alpha^2} \quad \text{for} \quad \alpha > 0.$$

Chung [1] shows a law of the iterated logarithm:

$$\overline{\lim_{N\to\infty}} \frac{\sqrt{2N} D_N^*(\omega)}{\sqrt{\log \log N}} = 1 \ \lambda_\infty\text{-a.e.}$$

See also Cassels [6]. For further results and references, see Feller [1], Doob [1], Donsker [1], Dvoretzky, Kiefer, and Wolfowitz [1], Darling [1], von Mises [2, Chapter 9], Billingsley [2, Chapter 2], and Niederreiter [14, 15]. The theory may be extended to the multidimensional case by starting with a sequence of vector-valued random variables and defining empirical distribution functions in the corresponding $\mathbb{R}^k$ in the obvious manner. For this case, results of the Kolmogorov-Smirnov type were established by Kiefer and Wolfowitz [1] and Kiefer [1]. Again one has a law of the iterated logarithm:

$$\overline{\lim_{N\to\infty}} \frac{\sqrt{2N}D_N^*(\omega)}{\sqrt{\log\log N}} = 1 \ (\lambda_k)_\infty\text{-a.e.},$$

where D_N^* and λ_k are the discrepancy resp. Lebesgue measure in I^k (Kiefer [1]). See also Philipp [6, 8] and Zaremba [8]. For $k = 2$, Philipp [9] proves a law of the iterated logarithm for the isotropic discrepancy J_N. In this context, one may also note that Stackelberg [1] has shown a law of the iterated logarithm for c.u.d. mod 1.

Exercises

1.1. Prove $\lim_{N\to\infty} D_N^*(\omega) = 0$ for a u.d. mod 1 sequence ω in $\mathbb{R}$ by using an indirect argument.

1.2. Prove the analogue of Theorem 1.1 for sequences in $\mathbb{R}^k$, $k \geq 2$.

1.3. Let f be a continuous nondecreasing function on $\bar{I}$ with $f(0) = 0$ and $f(1) = 1$. Using the definition of $D_N(\omega; f)$ given in (1.4), prove that ω has f as its a.d.f. (mod 1) if and only if $\lim_{N\to\infty} D_N(\omega; f) = 0$.

1.4. Let $\omega = (x_n)$ be given by $x_n = 1/(n + 1)$ for $n \geq 1$. Show that ω has an a.d.f. (mod 1) f but that $D_N(\omega; f) = 1$ for all $N \geq 1$.

1.5. Prove that

$$D_N(x_1, \ldots, x_N) = \sup_{0 \leq \alpha \leq \beta \leq 1} \left| \frac{A([\alpha, \beta]; N)}{N} - (\beta - \alpha) \right|.$$

1.6. Prove that

$$D_N^*(x_1, \ldots, x_N) = \sup_{0 \leq \alpha \leq 1} \left| \frac{A([0, \alpha]; N)}{N} - \alpha \right|.$$

1.7. For a sequence of N elements in I with discrepancy D_N, show that a fixed value $c \in I$ can be attained by the sequence at most $[ND_N]$ times.

1.8. Let $\omega = (x_n)$ be a sequence in $\mathbb{R}$, and let m be a nonzero integer. Prove that the discrepancy $D_N(\sigma)$ of the sequence $\sigma = (mx_n)$ satisfies $D_N(\sigma) \leq |m|\, D_N(\omega)$.

1.9. Provide a justification for the assertion that the result in Exercise 1.8 does not hold for nonintegral rational m.

1.10. Provide a justification for the assertion that the result in Exercise 1.8 does not hold for irrational m.

1.11. For $n \geq 1$, let ω_n be the finite sequence $0, 1/n^2, 4/n^2, \ldots, (n - 1)^2/n^2$. Prove that $\lim_{n\to\infty} D_n^*(\omega_n) = \frac{1}{4}$.

1.12. For positive integers k and n, let $\omega_n^{(k)}$ be the finite sequence $0, 1/n^k, 2^k/n^k, \ldots, (n - 1)^k/n^k$. Prove that $\lim_{k\to\infty}\lim_{n\to\infty} D_n^*(\omega_n^{(k)}) = 1$.

1.13. If a sequence of N elements in I^k has discrepancy D_N, then every subinterval of I^k of volume greater than D_N contains at least one point of the sequence.

1.14. Let ω be a sequence of N elements in $\mathbb{R}^k$. For $1 \leq i \leq k$, let $\omega^{(i)}$ be the ith coordinate sequence of ω. Then $D_N(\omega) \geq D_N(\omega^{(i)})$ for $1 \leq i \leq k$. A similar result holds for D_N^*.

2. ESTIMATION OF DISCREPANCY

Lower Bounds: Roth's Method

We shall first establish results concerning lower bounds for D_N^* that improve the trivial estimate in the preceding section. We need a sequence of lemmas, culminating in an inequality given in Lemma 2.5.

We consider the distribution of N points $\mathbf{p}_1, \ldots, \mathbf{p}_N$ in I^k with $k \geq 2$. Let $\mathbf{p}_i$ be given by $\mathbf{p}_i = (\alpha_{i1}, \ldots, \alpha_{ik})$ for $1 \leq i \leq N$. The function ψ is defined by $\psi(x) = -1$ for $0 \leq x < \frac{1}{2}$ and $\psi(x) = 1$ for $\frac{1}{2} \leq x < 1$ and then extended to $\mathbb{R}$ with period 1. We choose an integer n with $2^{(k-1)(n-1)} > N$ that will be specified later on. For each $(k-1)$-tuple $(r_1, \ldots, r_{k-1})$ of integers with $0 \leq r_j \leq n-1$ for $1 \leq j \leq k-1$, we define a function $G_{r_1 \cdots r_{k-1}}$ on $\mathbb{R}^k$ as follows: If there is an i, $1 \leq i \leq N$, such that

$$([2^{r_1}\alpha_{i1}], \ldots, [2^{r_{k-1}}\alpha_{i,k-1}], [2^{(k-1)(n-1)-r_1-\cdots-r_{k-1}}\alpha_{ik}]) = ([x_1], \ldots, [x_k]), \tag{2.1}$$

then we put $G_{r_1 \cdots r_{k-1}}(x_1, \ldots, x_k) = 0$. Otherwise, we set

$$G_{r_1 \cdots r_{k-1}}(x_1, \ldots, x_k) = \psi(x_1) \cdots \psi(x_k).$$

Furthermore, we define

$$F_{r_1 \cdots r_{k-1}}(x_1, \ldots, x_k) = G_{r_1 \cdots r_{k-1}}(2^{r_1}x_1, \ldots, 2^{r_{k-1}}x_{k-1}, 2^{(k-1)(n-1)-r_1-\cdots-r_{k-1}}x_k) \tag{2.2}$$

and

$$F(x_1, \ldots, x_k) = \sum_{r_1, \ldots, r_{k-1}=0}^{n-1} F_{r_1 \cdots r_{k-1}}(x_1, \ldots, x_k). \tag{2.3}$$

LEMMA 2.1. For given $r_1, \ldots, r_{k-1}$ from above and j with $1 \leq j \leq k-1$, let a and b be two integral multiples of 2^{-r_j}, say $a = h2^{-r_j}$ and $b = m2^{-r_j}$ with $h < m$. Then, for any fixed $x_1, \ldots, x_{j-1}, x_{j+1}, \ldots, x_k$, we have

$$\int_a^b F_{r_1 \cdots r_{k-1}}(x_1, \ldots, x_k)\, dx_j = 0. \tag{2.4}$$

Similarly, if $e = p2^{r_1+\cdots+r_{k-1}-(k-1)(n-1)}$ and $f = q2^{r_1+\cdots+r_{k-1}-(k-1)(n-1)}$ with integers $p < q$, then for any fixed $x_1, \ldots, x_{k-1}$ we have

$$\int_e^f F_{r_1 \cdots r_{k-1}}(x_1, \ldots, x_k)\, dx_k = 0. \tag{2.5}$$

PROOF. Using the substitution $t = 2^{r_j}x_j$, we obtain

$$\int_a^b F_{r_1\cdots r_{k-1}}(x_1, \ldots, x_k)\, dx_j$$

$$= \int_a^b G_{r_1\cdots r_{k-1}}(2^{r_1}x_1, \ldots, 2^{r_{k-1}}x_{k-1}, 2^{(k-1)(n-1)-r_1-\cdots-r_{k-1}}x_k)\, dx_j$$

$$= 2^{-r_j}\int_h^m G_{r_1\cdots r_{k-1}}(2^{r_1}x_1, \ldots, t, \ldots, 2^{r_{k-1}}x_{k-1}, 2^{(k-1)(n-1)-r_1-\cdots-r_{k-1}}x_k)\, dt.$$

Split up the interval $[h, m)$ into subintervals of the form $[c, c+1)$ with integers c. It follows from the definition of $G_{r_1\cdots r_{k-1}}$ that the integrand will be identical to zero on certain of these intervals. On the remaining intervals, the integrand will be equal to $\psi(2^{r_1}x_1)\cdots\psi(t)\cdots\psi(2^{r_{k-1}}x_{k-1})\psi(2^{(k-1)(n-1)-r_1-\cdots-r_{k-1}}x_k)$. But $\int_s^{s+1}\psi(t)\,dt = 0$ for any real number s, and (2.4) follows. The proof of (2.5) is analogous. ∎

LEMMA 2.2.

$$\int_0^1\cdots\int_0^1 x_1\cdots x_k F(x_1, \ldots, x_k)\, dx_1\cdots dx_k$$
$$\geq n^{k-1}2^{-2(k-1)(n-1)-2k}(2^{(k-1)(n-1)} - N). \quad (2.6)$$

PROOF. It suffices to prove that

$$\int_0^1\cdots\int_0^1 x_1\cdots x_k F_{r_1\cdots r_{k-1}}(x_1, \ldots, x_k)\, dx_1\cdots dx_k$$
$$\geq 2^{-2(k-1)(n-1)-2k}(2^{(k-1)(n-1)} - N) \quad (2.7)$$

holds for all $r_1, \ldots, r_{k-1}$ under consideration. Using the substitution $t_j = 2^{r_j}x_j$ for $1 \leq j \leq k-1$ and $t_k = 2^{(k-1)(n-1)-r_1-\cdots-r_{k-1}}x_k$, we get

$$\int_0^1\cdots\int_0^1 x_1\cdots x_k F_{r_1\cdots r_{k-1}}(x_1, \ldots, x_k)\, dx_1\cdots dx_k$$

$$= \int_0^1\cdots\int_0^1 x_1\cdots x_k$$
$$\cdot\, G_{r_1\cdots r_{k-1}}(2^{r_1}x_1\ \ldots, 2^{r_{k-1}}x_{k-1}, 2^{(k-1)(n-1)-r_1-\cdots-r_{k-1}}x_k)\, dx_1\cdots dx_k$$

$$= 2^{-2(k-1)(n-1)}\int_0^{2^{(k-1)(n-1)-r_1-\cdots-r_{k-1}}}\int_0^{2^{r_{k-1}}}\cdots\int_0^{2^{r_1}} t_1\cdots t_k$$
$$\cdot\, G_{r_1\cdots r_{k-1}}(t_1, \ldots, t_k)\, dt_1\cdots dt_k.$$

By the definition of $G_{r_1\cdots r_{k-1}}$, we have

$$\int_{h_k}^{h_k+1}\cdots\int_{h_1}^{h_1+1} t_1\cdots t_k G_{r_1\cdots r_{k-1}}(t_1, \ldots, t_k)\, dt_1\cdots dt_k = 0$$

whenever $(h_1, \ldots, h_k)$ is a lattice point with

$$(h_1, \ldots, h_k) = ([2^{r_1}\alpha_{i1}], \ldots, [2^{r_{k-1}}\alpha_{i,k-1}], [2^{(k-1)(n-1)-r_1-\cdots-r_{k-1}}\alpha_{ik}])$$

for some i, $1 \leq i \leq N$. Therefore,

$$\int_0^1 \cdots \int_0^1 x_1 \cdots x_k F_{r_1 \cdots r_{k-1}}(x_1, \ldots, x_k)\, dx_1 \cdots dx_k$$

$$= 2^{-2(k-1)(n-1)} \sum_{(h_1,\ldots,h_k)}^{*} \int_{h_k}^{h_k+1} \cdots \int_{h_1}^{h_1+1} t_1 \cdots t_k \psi(t_1) \cdots \psi(t_k)\, dt_1 \cdots dt_k,$$

where $\sum^*_{(h_1,\ldots,h_k)}$ stands for the sum over all lattice points $(h_1, \ldots, h_k)$ with $0 \leq h_j < 2^{r_j}$ for $1 \leq j \leq k-1$ and $0 \leq h_k < 2^{(k-1)(n-1)-r_1-\cdots-r_{k-1}}$, and $(h_1, \ldots, h_k)$ not identical with one of the lattice points $([2^{r_1}\alpha_{i1}], \ldots, [2^{r_{k-1}}\alpha_{i,k-1}], [2^{(k-1)(n-1)-r_1-\cdots-r_{k-1}}\alpha_{ik}])$, $1 \leq i \leq N$. We note that $\sum^*_{(h_1,\ldots,h_k)}$ is a sum over at least $2^{(k-1)(n-1)} - N$ lattice points. For any integer h, we have

$$\int_h^{h+1} t\psi(t)\, dt = -\int_h^{h+(1/2)} t\, dt + \int_{h+(1/2)}^{h+1} t\, dt = \tfrac{1}{4}.$$

Therefore,

$$\sum_{(h_1,\ldots,h_k)}^{*} \int_{h_k}^{h_k+1} \cdots \int_{h_1}^{h_1+1} t_1 \cdots t_k \psi(t_1) \cdots \psi(t_k)\, dt_1 \cdots dt_k$$

$$\geq 2^{-2k}(2^{(k-1)(n-1)} - N),$$

and (2.7) is established. ■

LEMMA 2.3.

$$\int_0^1 \cdots \int_0^1 F^2(x_1, \ldots, x_k)\, dx_1 \cdots dx_k \leq n^{k-1}. \tag{2.8}$$

PROOF.

$$\int_0^1 \cdots \int_0^1 F^2(x_1, \ldots, x_k)\, dx_1 \cdots dx_k$$

$$= \sum_{r_1,\ldots,r_{k-1}=0}^{n-1} \int_0^1 \cdots \int_0^1 F^2_{r_1 \cdots r_{k-1}}(x_1, \ldots, x_k)\, dx_1 \cdots dx_k$$

$$+ \sum_{\substack{(r_1,\ldots,r_{k-1})\\(s_1,\ldots,s_{k-1})\\(r_1,\ldots,r_{k-1}) \neq (s_1,\ldots,s_{k-1})}} \int_0^1 \cdots \int_0^1 F_{r_1 \cdots r_{k-1}}(x_1, \ldots, x_k)$$

$$\cdot F_{s_1 \cdots s_{k-1}}(x_1, \ldots, x_k)\, dx_1 \cdots dx_k.$$

Since $|F_{r_1 \cdots r_{k-1}}| \leq 1$ for all $r_1, \ldots, r_{k-1}$, the first term in the above expression is at most n^{k-1}. We prove now that each term in the second sum is zero. So choose $(r_1, \ldots, r_{k-1}) \neq (s_1, \ldots, s_{k-1})$; then there exists j, $1 \leq j \leq k-1$, such that $r_j \neq s_j$. Without loss of generality, we may assume $r_j < s_j$. We will be done once we show that for all fixed $x_1, \ldots, x_{j-1}, x_{j+1}, \ldots, x_k$ we have

$$\int_0^1 F_{r_1\cdots r_{k-1}}(x_1, \ldots, x_k)F_{s_1\cdots s_{k-1}}(x_1, \ldots, x_k)\, dx_j = 0. \tag{2.9}$$

The substitution $t = 2^{s_j}x_j$ transforms the integral in (2.9) into

$$2^{-s_j}\int_0^{2^{s_j}} G_{r_1\cdots r_{k-1}}(2^{r_1}x_1, \ldots, 2^{r_j-s_j}t, \ldots)G_{s_1\cdots s_{k-1}}(2^{s_1}x_1, \ldots, t, \ldots)\, dt. \tag{2.10}$$

We split up the interval $[0, 2^{s_j})$ into subintervals $[c, c+1)$ with integers c. In an interval of the latter type, the integrand in (2.10) is either identical to zero or equal to $\psi(2^{r_1}x_1)\cdots\psi(2^{r_j-s_j}t)\cdots\psi(2^{s_1}x_1)\cdots\psi(t)\cdots$. It suffices therefore to show that $\int_c^{c+1}\psi(2^{r_j-s_j}t)\psi(t)\,dt = 0$ for any integer c. But this is almost trivial, since $r_j < s_j$ implies that $\psi(2^{r_j-s_j}t)$ is constant on $[c, c+1)$ and since $\int_c^{c+1}\psi(t)\,dt = 0$. ■

LEMMA 2.4. For $1 \leq i \leq N$, we have

$$\int_{\alpha_{ik}}^1 \cdots \int_{\alpha_{i1}}^1 F(x_1, \ldots, x_k)\, dx_1 \cdots dx_k = 0. \tag{2.11}$$

PROOF. It suffices to prove that

$$\int_{\alpha_{ik}}^1 \cdots \int_{\alpha_{i1}}^1 F_{r_1\cdots r_{k-1}}(x_1, \ldots, x_k)\, dx_1 \cdots dx_k = 0 \tag{2.12}$$

for all i and all $r_1, \ldots, r_{k-1}$. For fixed i and for $1 \leq j \leq k-1$, let a_j be the least integral multiple of 2^{-r_j} that is $\geq \alpha_{ij}$, and let a_k be the least integral multiple of $2^{r_1+\cdots+r_{k-1}-(k-1)(n-1)}$ which is $\geq \alpha_{ik}$. Then $\int_{\alpha_{ik}}^1 \cdots \int_{\alpha_{i1}}^1 = \int_{\alpha_{ik}}^{a_k} \cdots \int_{\alpha_{i1}}^{a_1} +$ (sum of integrals in which, for at least one variable x_j, we integrate over an interval $[a_j, 1]$). The first integral on the right-hand side is zero, since for all $(x_1, \ldots, x_k)$ in the interval $[\alpha_{i1}, a_1) \times \cdots \times [\alpha_{ik}, a_k)$, we have $([2^{r_1}x_1], \ldots, [2^{r_{k-1}}x_{k-1}], [2^{(k-1)(n-1)-r_1-\cdots-r_{k-1}}x_k]) = ([2^{r_1}\alpha_{i1}], \ldots, [2^{r_{k-1}}\alpha_{i,k-1}], [2^{(k-1)(n-1)-r_1-\cdots-r_{k-1}}\alpha_{ik}])$ and therefore, $F_{r_1\cdots r_{k-1}}(x_1, \ldots, x_k) = 0$ by definition. In each of the remaining integrals, interchange the order of integration so that the inner integral is taken over the interval $[a_j, 1]$ with respect to the variable x_j. By Lemma 2.1, the inner integral is then zero, and we are done. ■

LEMMA 2.5. For $(x_1, \ldots, x_k) \in \bar{I}^k$, let $A(x_1, \ldots, x_k)$ denote the number of points $\mathbf{p}_i$, $1 \leq i \leq N$, in the interval $[0, x_1) \times \cdots \times [0, x_k)$. Then,

$$\int_0^1 \cdots \int_0^1 (A(x_1, \ldots, x_k) - Nx_1\cdots x_k)^2\, dx_1 \cdots dx_k > c_k(\log N)^{k-1} \tag{2.13}$$

with an absolute constant $c_k > 0$ only depending on k.

PROOF. For $1 \leq i \leq N$, let $f_i(x_1, \ldots, x_k)$ be the characteristic function of the interval $(\alpha_{i1}, 1] \times \cdots \times (\alpha_{ik}, 1]$. Then $A(x_1, \ldots, x_k) = \sum_{i=1}^N f_i(x_1, \ldots, x_k)$. Therefore,

$$\int_0^1 \cdots \int_0^1 A(x_1, \ldots, x_k) F(x_1, \ldots, x_k)\, dx_1 \cdots dx_k$$
$$= \sum_{i=1}^N \int_0^1 \cdots \int_0^1 f_i(x_1, \ldots, x_k) F(x_1, \ldots, x_k)\, dx_1 \cdots dx_k$$
$$= \sum_{i=1}^N \int_{\alpha_{ik}}^1 \cdots \int_{\alpha_{i1}}^1 F(x_1, \ldots, x_k)\, dx_1 \cdots dx_k = 0$$

by Lemma 2.4. Hence, using Lemma 2.2 we obtain

$$\int_0^1 \cdots \int_0^1 (Nx_1 \cdots x_k - A(x_1, \ldots, x_k)) F(x_1, \ldots, x_k)\, dx_1 \cdots dx_k$$
$$= N \int_0^1 \cdots \int_0^1 x_1 \cdots x_k F(x_1, \ldots, x_k)\, dx_1 \cdots dx_k$$
$$\geq N n^{k-1} 2^{-2(k-1)(n-1)-2k} (2^{(k-1)(n-1)} - N).$$

Then

$$N^2 n^{2k-2} 2^{-4(k-1)(n-1)-4k} (2^{(k-1)(n-1)} - N)^2$$
$$\leq \left(\int_0^1 \cdots \int_0^1 (Nx_1 \cdots x_k - A(x_1, \ldots, x_k)) F(x_1, \ldots, x_k)\, dx_1 \cdots dx_k \right)^2$$
$$\leq \left(\int_0^1 \cdots \int_0^1 (Nx_1 \cdots x_k - A(x_1, \ldots, x_k))^2\, dx_1 \cdots dx_k \right)$$
$$\cdot \left(\int_0^1 \cdots \int_0^1 F^2(x_1, \ldots, x_k)\, dx_1 \cdots dx_k \right)$$

by the Cauchy-Schwarz inequality. It follows from Lemma 2.3 that

$$\int_0^1 \cdots \int_0^1 (Nx_1 \cdots x_k - A(x_1, \ldots, x_k))^2\, dx_1 \cdots dx_k$$
$$\geq N^2 n^{k-1} 2^{-4(k-1)(n-1)-4k} (2^{(k-1)(n-1)} - N)^2.$$

Let n be the unique integer for which $2N \leq 2^{(k-1)(n-1)} < 2^k N$. Then the integral in (2.13) is $\geq N^4 n^{k-1} 2^{-4(k-1)(n-1)} 2^{-4k} > 2^{-8k} n^{k-1}$. But $(k-1)(n-1) \geq (\log N)/(\log 2) + 1$, and so,

$$n \geq \frac{\log N}{(k-1)\log 2} + \frac{k}{k-1}.$$

Therefore, (2.13) holds with $c_k = 2^{-8k}((k-1)\log 2)^{1-k}$. ■

THEOREM 2.1. For any sequence of N points in $\mathbb{R}^k$ with $k \geq 2$ we have

$$ND_N^* > C_k(\log N)^{(k-1)/2} \tag{2.14}$$

with an absolute constant $C_k > 0$ only depending on k.

PROOF. This follows immediately from Lemma 2.5. ■

If f and g are two functions with $g > 0$, then we write $f = \Omega(g)$ in case $f \neq o(g)$. In order to verify $f(N) = \Omega(g(N))$, it suffices to show that there is a constant $b > 0$ such that $|f(N)| \geq bg(N)$ holds for infinitely many positive integers N.

THEOREM 2.2. For any infinite sequence ω in $\mathbb{R}^k$ with $k \geq 1$, we have

$$ND_N^*(\omega) > C_k'(\log N)^{k/2} \tag{2.15}$$

for infinitely many positive integers N, where $C_k' > 0$ is an absolute constant only depending on k. In particular, we get $ND_N^*(\omega) = \Omega((\log N)^{k/2})$.

PROOF. Let $\omega = (\mathbf{x}_n)$ be a given sequence in I^k, and suppose that $\mathbf{x}_n = (\alpha_{n1}, \ldots, \alpha_{nk})$ for $n \geq 1$. For fixed $N \geq 1$, consider the finite sequence of points $\mathbf{p}_1, \ldots, \mathbf{p}_N$ in I^{k+1} given by $\mathbf{p}_i = (\alpha_{i1}, \ldots, \alpha_{ik}, (i-1)/N)$ for $1 \leq i \leq N$. It follows from Theorem 2.1 that there exist $x_1, \ldots, x_{k+1}$ with $0 < x_j \leq 1$ for $1 \leq j \leq k+1$ such that

$$|A(x_1, \ldots, x_{k+1}) - Nx_1 \cdots x_{k+1}| > C_{k+1}(\log N)^{k/2}.$$

Now let m be the positive integer with $(m-1)/N < x_{k+1} \leq m/N$. We note that $A(x_1, \ldots, x_{k+1})$ is the number of i, $1 \leq i \leq N$, for which $0 \leq \alpha_{ij} < x_j$ for all $1 \leq j \leq k$ and $0 \leq (i-1)/N < x_{k+1}$. But since the last condition is equivalent to $1 \leq i \leq m$, we arrive at $A(x_1, \ldots, x_{k+1}) = A([0, x_1) \times \cdots \times [0, x_k); m; \omega)$. It follows that

$$\begin{aligned}
&|A([0, x_1) \times \cdots \times [0, x_k); m; \omega) - mx_1 \cdots x_k| \\
&\quad \geq |A(x_1, \ldots, x_{k+1}) - Nx_1 \cdots x_{k+1}| - |Nx_1 \cdots x_{k+1} - mx_1 \cdots x_k| \\
&\quad > C_{k+1}(\log N)^{k/2} - x_1 \cdots x_k\, |Nx_{k+1} - m| \\
&\quad > C_{k+1}(\log N)^{k/2} - 1 > C_k'(\log N)^{k/2}
\end{aligned}$$

for sufficiently large N. Thus, we have shown that for every sufficiently large N there exists m with $1 \leq m \leq N$ such that $mD_m^*(\omega) > C_k'(\log N)^{k/2} \geq C_k'(\log m)^{k/2}$. The desired result follows then immediately. ■

COROLLARY 2.1: Van Aardenne-Ehrenfest Theorem. For any infinite sequence ω in $\mathbb{R}^k$ with $k \geq 1$, we have $\overline{\lim}_{N\to\infty} ND_N^*(\omega) = \infty$.

EXAMPLE 2.1. For $k = 1$, we cannot replace "infinitely many positive integers N" by "almost all positive integers N" in the statement of Theorem 2.2. Consider the infinite sequence

$$0, \frac{1}{2}, \frac{1}{4}, \frac{3}{4}, \frac{1}{8}, \frac{3}{8}, \frac{5}{8}, \frac{7}{8}, \ldots, \frac{1}{2^n}, \frac{3}{2^n}, \ldots, \frac{2^n - 1}{2^n}, \ldots.$$

In an initial segment of length $N = 2^n$, we find exactly all rationals $k/2^n$ with $0 \le k < 2^n$. For such an N, we therefore get $D_N^* = 1/N$ by Theorem 1.4. It follows that, for this sequence, we have $D_N^* = 1/N$ for infinitely many N. Many other examples of this type can be constructed. ■

EXAMPLE 2.2. We have seen in the proof of Theorem 2.2 that as soon as $ND_N^* > f(N)$ for any N points in I^{k+1}, then a result of the following type is implied: For any N points in I^k there exists m with $1 \le m \le N$ such that $mD_m^* > f(N) - 1$. We shall show that essentially the converse is also valid.

So suppose it is true that for any N points in I^k, $k \ge 1$, there exists m with $1 \le m \le N$ such that $mD_m^* > f(N)$. We claim that then $ND_N^* > \frac{1}{2}f(N)$ holds for any N points in I^{k+1}. Let $\mathbf{p}_1, \ldots, \mathbf{p}_N$ be arbitrary N points in I^{k+1} with discrepancy D_N^* and $\mathbf{p}_i = (\alpha_{i1}, \ldots, \alpha_{i,k+1})$ for $1 \le i \le N$. Without loss of generality, we may assume that they are ordered in such a way that $\alpha_{1,k+1} \le \alpha_{2,k+1} \le \cdots \le \alpha_{N,k+1}$. Define $\mathbf{q}_i = (\alpha_{i1}, \ldots, \alpha_{ik}, (i-1)/N)$ for $1 \le i \le N$, and let $\tilde{D}_N^*$ be the discrepancy of the points $\mathbf{q}_1, \ldots, \mathbf{q}_N$ in I^{k+1}. By Theorem 1.4 and Exercise 1.14, we have $|\alpha_{i,k+1} - (i-1)/N| \le D_N^*$ for $1 \le i \le N$.

We need the following auxiliary result: If $\mathbf{y}_1, \ldots, \mathbf{y}_N$ and $\mathbf{z}_1, \ldots, \mathbf{z}_N$ are points in I^{k+1} with discrepancies $D_N^{*(1)}$ and $D_N^{*(2)}$, if $\mathbf{y}_i = (\beta_{i1}, \ldots, \beta_{ik}, \gamma_i)$ for $1 \le i \le N$ and $\mathbf{z}_i = (\beta_{i1}, \ldots, \beta_{ik}, \delta_i)$ for $1 \le i \le N$, and if $|\gamma_i - \delta_i| \le \varepsilon$ for $1 \le i \le N$, then $|D_N^{*(1)} - D_N^{*(2)}| \le \varepsilon$. We denote by $A^{(1)}(x_1, \ldots, x_{k+1})$ the number of $\mathbf{y}_i$ in $[0, x_1) \times \cdots \times [0, x_{k+1})$, and $A^{(2)}(x_1, \ldots, x_{k+1})$ is defined similarly with respect to the $\mathbf{z}_i$. Since $A^{(1)}(x_1, \ldots, x_k, x_{k+1} - \varepsilon) \le A^{(2)}(x_1, \ldots, x_{k+1}) \le A^{(1)}(x_1, \ldots, x_k, x_{k+1} + \varepsilon)$, it follows that

$$\left| \frac{A^{(2)}(x_1, \ldots, x_{k+1})}{N} - x_1 \cdots x_{k+1} \right| \le D_N^{*(1)} + \varepsilon x_1 \cdots x_k \le D_N^{*(1)} + \varepsilon$$

in case $0 \le x_i \le 1$ for $1 \le i \le k+1$, and so $D_N^{*(2)} \le D_N^{*(1)} + \varepsilon$. Interchanging the roles of the $\mathbf{y}_i$ and $\mathbf{z}_i$, we obtain $D_N^{*(1)} \le D_N^{*(2)} + \varepsilon$, and the assertion follows.

The above auxiliary result implies that $|D_N^* - \tilde{D}_N^*| \le D_N^*$; hence, $\tilde{D}_N^* \le 2D_N^*$. Let τ be the finite sequence of points $\mathbf{q}_1', \ldots, \mathbf{q}_N'$ in I^k defined by $\mathbf{q}_i' = (\alpha_{i1}, \ldots, \alpha_{ik})$ for $1 \le i \le N$. By hypothesis, there exists m with $1 \le m \le N$ and $x_1, \ldots, x_k$ with $0 < x_i \le 1$ for $1 \le i \le k$ such that $|A([0, x_1) \times \cdots \times [0, x_k); m; \tau) - mx_1 \cdots x_k| > f(N)$. By the same reasoning as in the proof of Theorem 2.2, it follows that $A([0, x_1) \times \cdots \times [0, x_k); m; \tau) = A(x_1, \ldots,$

$x_k, m/N)$ = the number of $\mathbf{q}_i$ in $[0, x_1) \times \cdots \times [0, x_k) \times [0, m/N)$. Therefore, $|A(x_1, \ldots, x_k, m/N) - N((m/N)x_1 \cdots x_k)| > f(N)$, and so, $N\tilde{D}_N^* > f(N)$. Consequently, we get $ND_N^* \geq \frac{1}{2}N\tilde{D}_N^* > \frac{1}{2}f(N)$. ■

Lower Bounds: Schmidt's Method

For $k = 1$, an improvement of Theorem 2.2 can be shown by using a different method. We need two auxiliary results and some notation.

Let (x_n) be a given infinite sequence in I. For $N \geq 1$ and $0 \leq x \leq 1$, we set $R_N(x) = A([0, x); N) - Nx$. Since $R_N(0) = R_N(1)$, we may extend $R_N(x)$ with period 1 to $\mathbb{R}$. Moreover, we write $R_N(x, y) = R_N(y) - R_N(x)$. By K, L, and L' we shall denote intervals of integers of the type $(a, b]$ where a, b are integers with $0 \leq a < b$. For intervals K and arbitrary x, y, we put $g^+(K, x, y) = \max_{n \in K} R_n(x, y)$ and $g^-(K, x, y) = \min_{n \in K} R_n(x, y)$. For a pair of intervals L, L' put

$$h(L, L', x, y) = \max(g^-(L, x, y) - g^+(L', x, y), g^-(L', x, y) - g^+(L, x, y)).$$

Furthermore, we set $g^+(K, y) = g^+(K, 0, y)$, $g^-(K, y) = g^-(K, 0, y)$, and $h(K, y) = g^+(K, y) - g^-(K, y)$.

LEMMA 2.6. Suppose L, L' are subintervals of some interval K. Then, for any x and y, we have

$$h(K, x) + h(K, y) \geq h(L, L', x, y) + \tfrac{1}{2}(h(L, x) + h(L, y) + h(L', x) + h(L', y)). \quad (2.16)$$

PROOF. Without loss of generality, we may assume $h(L, L', x, y) = g^-(L, x, y) - g^+(L', x, y)$, for otherwise we just interchange the roles of L and L'. For every $n \in L$ and every $n' \in L'$, we have $R_n(x, y) - R_{n'}(x, y) \geq h(L, L', x, y)$, and so,

$$R_n(y) - R_n(x) - R_{n'}(y) + R_{n'}(x) \geq h(L, L', x, y). \quad (2.17)$$

There are integers $m(x)$, $n(x)$, $m(y)$, $n(y)$ in L with $R_{m(x)}(x) = g^+(L, x)$, $R_{n(x)}(x) = g^-(L, x)$, $R_{m(y)}(y) = g^+(L, y)$, and $R_{n(y)}(y) = g^-(L, y)$. Then,

$$R_{m(x)}(x) - R_{n(x)}(x) = h(L, x), \quad (2.18)$$

$$R_{m(y)}(y) - R_{n(y)}(y) = h(L, y). \quad (2.19)$$

Similarly, there are integers $m'(x)$, $n'(x)$, $m'(y)$, $n'(y)$ in L' with

$$R_{m'(x)}(x) - R_{n'(x)}(x) = h(L', x), \quad (2.20)$$

$$R_{m'(y)}(y) - R_{n'(y)}(y) = h(L', y). \quad (2.21)$$

Now add the four equations (2.18), (2.19), (2.20), and (2.21) and the two inequalities resulting from (2.17) by applying it first with $n = m(x)$, $n' = m'(y)$, and then with $n = n(y)$, $n' = n'(x)$, and we obtain

$$c_1 + c_2 + c_3 + c_4 \geq 2h(L, L', x, y) + h(L, x) + h(L, y) + h(L', x) + h(L', y ,$$

where $c_1 = R_{m'(x)}(x) - R_{n(x)}(x)$, $c_2 = R_{m'(y)}(x) - R_{n(y)}(x)$, $c_3 = R_{m(y)}(y) - R_{n'(y)}(y)$, $c_4 = R_{m(x)}(y) - R_{n'(x)}(y)$. Since $h(K, x) \geq c_1$, $h(K, x) \geq c_2$, $h(K, y) \geq c_3$, $h(K, y) \geq c_4$, the lemma follows. ■

LEMMA 2.7. Let t be a positive integer. Then, for any interval K of length at least 4^t and any y, we have

$$4^{-t}\sum_{j=1}^{4^t} h(K, y + j4^{-t}) \geq 2^{-5}t. \tag{2.22}$$

PROOF. We shall proceed by induction on t. First, we note that

$$R_{n+1}(y + \tfrac{1}{2}) - R_n(y + \tfrac{1}{2}) - (R_{n+1}(y) - R_n(y))$$

is an integer minus $(n + 1)(y + \frac{1}{2}) - n(y + \frac{1}{2}) - (n + 1)y + ny$ and hence is an integer minus $\frac{1}{2}$. Let K be an interval containing at least the integers n and $n + 1$. Then $h(K, y + \frac{1}{2}) + h(K, y) \geq |R_{n+1}(y + \frac{1}{2}) - R_n(y + \frac{1}{2})| + |R_{n+1}(y) - R_n(y)| \geq |R_{n+1}(y + \frac{1}{2}) - R_n(y + \frac{1}{2}) - (R_{n+1}(y) - R_n(y))| \geq \frac{1}{2}$, and so (2.22) holds for $t = 1$.

Now let $K = (a, b]$ be an interval of length at least 4^{t+1}. We set $L = (a, a + 4^t]$ and $L' = (a + 2 \cdot 4^t, a + 3 \cdot 4^t]$. Because of the periodicity of $h(K, y)$, we may assume $0 \leq y < 4^{-t-1}$. For $0 \leq j \leq 4^{t+1}$, put $z_j = y + j4^{-t-1}$. For $1 \leq j \leq 4^{t+1} - 1$, let d_j be the number of integers $m \in (a + 4^t, a + 2 \cdot 4^t]$ with $x_m \in [z_{j-1}, z_j)$; for $j = 4^{t+1}$, let d_j be the number of integers $m \in (a + 4^t, a + 2 \cdot 4^t]$ with $x_m \in [z_{j-1}, 1) \cup [0, z_0)$. Now choose integers $n \in L$ and $n' \in L'$. Then $R_{n'}(z_{j-1}, z_j) - R_n(z_{j-1}, z_j) = e_j - (n' - n)(z_j - z_{j-1})$ for $1 \leq j \leq 4^{t+1}$, where e_j is the number of integers $m \in (n, n']$ with $x_m \in [z_{j-1}, z_j)$ (resp. $x_m \in [z_{j-1}, 1) \cup [0, z_0)$ for $j = 4^{t+1}$). It follows that

$$R_{n'}(z_{j-1}, z_j) - R_n(z_{j-1}, z_j) \geq d_j - 3 \cdot 4^t 4^{-t-1} = d_j - \tfrac{3}{4}.$$

This yields $h(L, L', z_{j-1}, z_j) \geq d_j - \frac{3}{4}$, and if d_j is positive, then

$$h(L, L', z_{j-1}, z_j) \geq \tfrac{1}{4}d_j.$$

Together with Lemma 2.6 we obtain

$$h(K, z_{j-1}) + h(K, z_j) \geq \tfrac{1}{4}d_j + \tfrac{1}{2}(h(L, z_{j-1}) + h(L, z_j) + h(L', z_{j-1}) + h(L', z_j)). \tag{2.23}$$

Since $h(K, z_j) \geq h(L, z_j)$ and $h(K, z_j) \geq h(L', z_j)$ for $0 \leq j \leq 4^{t+1}$, (2.23) also holds in the case $d_j = 0$. We divide the sum

$$\sum_{j=1}^{4^{t+1}} h(L, z_j) \tag{2.24}$$

into four parts, according to the residue class of j modulo 4. Each of these four parts is a sum like the one on the left-hand side of (2.22). By induction hypothesis, each of these sums has the lower bound $4^t 2^{-5} t$, and so the sum (2.24) is at least $4 \cdot 4^t 2^{-5} t = 2^{-3} 4^t t$. The same lower bound holds if $h(L, z_j)$ in (2.24) is replaced by $h(L, z_{j-1})$, $h(L', z_{j-1})$, or $h(L', z_j)$. We now take the sum of (2.23) over $j = 1, 2, \ldots, 4^{t+1}$, and using $h(K, z_0) = h(K, z_{4^{t+1}})$, we obtain

$$2 \sum_{j=1}^{4^{t+1}} h(K, z_j) \geq \tfrac{1}{4} \sum_{j=1}^{4^{t+1}} d_j + 2(2^{-3} 4^t t) = 4^{t-1}(1 + t).$$

Dividing by $2 \cdot 4^{t+1}$, we obtain (2.22) with t replaced by $t + 1$. ■

THEOREM 2.3. For any infinite sequence ω in $\mathbb{R}$ we have

$$N D_N^*(\omega) > c \log N \tag{2.25}$$

for infinitely many positive integers N, where $c > 0$ is an absolute constant. In particular, we get $N D_N^*(\omega) = \Omega(\log N)$.

PROOF. We shall show that for every N there exists m with $1 \leq m \leq N$ and

$$m D_m^*(\omega) > c \log N. \tag{2.26}$$

Suppose first that $N \geq 4^{32}$. There is an integer $t \geq 32$ with $4^t \leq N < 4^{t+1}$. Apply Lemma 2.7 with $K = (0, 4^t]$ and any particular y, and it follows that $h(K, x) \geq 2^{-5} t$ for some $x \in I$. Thus, there are integers $p, q \in K$ with $R_p(x) - R_q(x) \geq 2^{-5} t$. Since either $R_p(x)$ or $R_q(x)$ must be $\geq 2^{-6} t$ in absolute value, it follows that there is an integer m with $1 \leq m \leq 4^t \leq N$ and $|R_m(x)| \geq 2^{-6} t$ and, hence, with $m D_m^*(\omega) \geq 2^{-6} t$. Since $t \geq 32$, we have $t \geq \frac{32}{33}(t + 1)$, whence

$$m D_m^*(\omega) \geq 2^{-6} \tfrac{32}{33}(t + 1) > \frac{\log N}{66 \log 4}.$$

In the case $1 \leq N < 4^{32}$, we use Corollary 1.2 to obtain $D_1^*(\omega) \geq \frac{1}{2} > (\log N)/66 \log 4$. Thus, in any case, we get (2.26) with $c = (66 \log 4)^{-1}$. ■

COROLLARY 2.2. For any sequence of N points in $\mathbb{R}^2$ we have $N D_N^* > c' \log N$ with an absolute constant $c' > 0$.

PROOF. This follows from (2.26) and Example 2.2. An admissible value for the constant is $c' = (132 \log 4)^{-1}$. ■

Upper Bounds

One important technique to get upper bounds is to estimate the discrepancy in terms of the exponential sums occurring in Weyl's criterion and then to use well-established methods from analytic number theory to estimate these exponential sums. In fact, we shall present two results of this type that are known as LeVeque's inequality and the theorem of Erdös-Turán, respectively. We consider only the one-dimensional case.

Suppose that $x_1, \ldots, x_N$ are N points in I. For $0 \leq x \leq 1$, we again set $R_N(x) = A([0, x); N) - Nx$.

LEMMA 2.8.

$$\int_0^1 R_N^2(x)\,dx = \left(\sum_{n=1}^{N}(x_n - \tfrac{1}{2})\right)^2 + \frac{1}{2\pi^2}\sum_{h=1}^{\infty}\frac{1}{h^2}\left|\sum_{n=1}^{N} e^{2\pi i h x_n}\right|^2. \tag{2.27}$$

PROOF. We observe that $R_N(x)$ is a piecewise linear function in $[0, 1]$ with only finitely many discontinuities at $x = x_1, x_2, \ldots, x_N$; in addition, we have $R_N(0) = R_N(1)$. The function $R_N(x)$ can therefore be expanded into a Fourier series $\sum_{h=-\infty}^{\infty} a_h e^{2\pi i h x}$ which will represent $R_N(x)$ apart from finitely many points. The Fourier coefficients are given by

$$a_h = \int_0^1 R_N(x) e^{-2\pi i h x}\,dx.$$

For $1 \leq n \leq N$, let $c_n(x)$ be the characteristic function of the interval $(x_n, 1]$. Then $A([0, x); N) = \sum_{n=1}^{N} c_n(x)$, and so,

$$\begin{aligned} a_0 &= \int_0^1 R_N(x)\,dx \\ &= \sum_{n=1}^{N}\int_0^1 c_n(x)\,dx - N\int_0^1 x\,dx \\ &= -\sum_{n=1}^{N}(x_n - \tfrac{1}{2}). \end{aligned} \tag{2.28}$$

For $h \neq 0$, we obtain

$$\begin{aligned} a_h &= \sum_{n=1}^{N}\int_0^1 c_n(x) e^{-2\pi i h x}\,dx - N\int_0^1 x e^{-2\pi i h x}\,dx \\ &= \sum_{n=1}^{N}\int_{x_n}^1 e^{-2\pi i h x}\,dx + \frac{N}{2\pi i h} \\ &= \frac{1}{2\pi i h}\sum_{n=1}^{N}(e^{-2\pi i h x_n} - 1) + \frac{N}{2\pi i h} \\ &= \frac{1}{2\pi i h}\sum_{n=1}^{N} e^{-2\pi i h x_n}. \end{aligned} \tag{2.29}$$

By Parseval's identity, we have

$$\int_0^1 R_N^2(x)\,dx = a_0^2 + 2\sum_{h=1}^{\infty} |a_h|^2,$$

and the desired result follows immediately. ■

THEOREM 2.4: LeVeque's Inequality. The discrepancy D_N of the finite sequence $x_1, \ldots, x_N$ in I satisfies

$$D_N \leq \left(\frac{6}{\pi^2} \sum_{h=1}^{\infty} \frac{1}{h^2} \left| \frac{1}{N} \sum_{n=1}^{N} e^{2\pi i h x_n} \right|^2 \right)^{1/3}. \tag{2.30}$$

PROOF. We put $S_N = \sum_{n=1}^{N} (x_n - \frac{1}{2})$, and $T_N(x) = (1/N)(R_N(x) + S_N)$ for $0 \leq x \leq 1$. The function T_N is piecewise linear (the linear pieces having a slope of -1) and left continuous and has only finitely many discontinuities, each of which is a jump by a positive number. Since $T_N(0) = T_N(1)$, we can extend T_N to $\mathbb{R}$ with a period of 1.

Let α and β be numbers from $[0, 1]$ with $T_N(\alpha) > 0$ and $T_N(\beta) < 0$. Such numbers exist because of $\int_0^1 T_N(x)\,dx = 0$, which follows in turn from (2.28). In the interval $[\alpha, \alpha + T_N(\alpha)]$, the graph of T_N will not lie below the line segment joining the points $(\alpha, T_N(\alpha))$ and $(\alpha + T_N(\alpha), 0)$ in the coordinate plane. By the periodicity of T_N, there exists $\beta_1 \in [\alpha, \alpha + 1]$ with $T_N(\beta_1) = T_N(\beta)$. In the interval $[\beta_1 + T_N(\beta_1), \beta_1]$, the graph of T_N will not lie above the line segment joining $(\beta_1 + T_N(\beta_1), 0)$ and $(\beta_1, T_N(\beta_1))$; therefore, the graph of $|T_N|$ will not lie below the line segment joining $(\beta_1 + T_N(\beta_1), 0)$ and $(\beta_1, -T_N(\beta_1))$. Moreover, the intervals $[\alpha, \alpha + T_N(\alpha)]$ and $[\beta_1 + T_N(\beta_1), \beta_1]$ can have at most one point in common, because of the properties that the graph of T_N satisfies there. It follows that

$$\begin{aligned}\int_0^1 T_N^2(x)\,dx &= \int_\alpha^{\alpha+1} T_N^2(x)\,dx \\ &\geq \int_\alpha^{\alpha+T_N(\alpha)} T_N^2(x)\,dx + \int_{\beta_1+T_N(\beta_1)}^{\beta_1} T_N^2(x)\,dx \\ &\geq \tfrac{1}{3}T_N^3(\alpha) + \tfrac{1}{3}(-T_N(\beta_1))^3 = \tfrac{1}{3}T_N^3(\alpha) + \tfrac{1}{3}(-T_N(\beta))^3.\end{aligned}$$

For nonnegative real numbers r and s, put $t = \frac{1}{2}(r + s)$ and $u = \frac{1}{2}(r - s)$. Then $r^3 + s^3 = (t + u)^3 + (t - u)^3 = 2t^3 + 6tu^2 \geq 2t^3 = \frac{1}{4}(r + s)^3$. We apply this inequality with $r = T_N(\alpha)$ and $s = -T_N(\beta)$ and obtain

$$\frac{1}{12}(T_N(\alpha) - T_N(\beta))^3 \leq \int_0^1 T_N^2(x)\,dx. \tag{2.31}$$

It is then evident that (2.31) even holds for all α and β in $[0, 1]$. Using the definition of T_N, it follows that

$$\frac{1}{12}\left(\frac{R_N(\alpha) - R_N(\beta)}{N}\right)^3 \leq \int_0^1 T_N^2(x)\,dx$$

holds for all $\alpha, \beta \in [0, 1]$, and so

$$\frac{1}{12} D_N^{\,3} \leq \int_0^1 T_N^{\,2}(x)\, dx. \tag{2.32}$$

Lemma 2.8 and (2.28) enable us to compute the integral on the right-hand side of (2.32):

$$\begin{aligned}\int_0^1 T_N^{\,2}(x)\, dx &= \frac{1}{N^2}\int_0^1 R_N^{\,2}(x)\, dx + \frac{2}{N^2} S_N \int_0^1 R_N(x)\, dx + \frac{1}{N^2} S_N^{\,2} \\ &= \frac{1}{N^2}\int_0^1 R_N^{\,2}(x)\, dx - \frac{1}{N^2} S_N^{\,2} \\ &= \frac{1}{2\pi^2}\sum_{h=1}^{\infty}\frac{1}{h^2}\left|\frac{1}{N}\sum_{n=1}^{N} e^{2\pi i h x_n}\right|^2.\end{aligned}$$

We combine this result with (2.32), and the proof is complete. ■

We remark that LeVeque's inequality holds for any finite real sequence $x_1, \ldots, x_N$ not necessarily contained in I, since both sides of the inequality only depend on the fractional parts of the numbers involved.

EXAMPLE 2.3. The constant $6/\pi^2$ in LeVeque's inequality is best possible. Choose $x_1 = x_2 = \cdots = x_N = 0$. Then $D_N = 1$, and the right-hand side is equal to

$$\left(\frac{6}{\pi^2}\sum_{h=1}^{\infty}\frac{1}{h^2}\right)^{1/3} = 1. \quad ■$$

It can also be shown that the exponent $\frac{1}{3}$ in LeVeque's inequality is best possible (see notes). We turn our attention to the second important theorem dealing with the relation between discrepancies and exponential sums.

THEOREM 2.5: Theorem of Erdös-Turán. For any finite sequence $x_1, \ldots,$ x_N of real numbers and any positive integer m, we have

$$D_N \leq \frac{6}{m+1} + \frac{4}{\pi}\sum_{h=1}^{m}\left(\frac{1}{h} - \frac{1}{m+1}\right)\left|\frac{1}{N}\sum_{n=1}^{N} e^{2\pi i h x_n}\right|. \tag{2.33}$$

PROOF. We set $\Delta_N(x) = (1/N)R_N(x)$ for $0 \leq x \leq 1$ and extend this function with period 1 to $\mathbb{R}$. We consider first a sequence $x_1, \ldots, x_N$ in I for which

$$\int_0^1 \Delta_N(x)\, dx = 0. \tag{2.34}$$

We set $S_h = (1/N)\sum_{n=1}^{N} e^{2\pi i h x_n}$ for $h \in \mathbb{Z}$. Then from (2.29),

$$\frac{S_h}{-2\pi i h} = \int_0^1 \Delta_N(x) e^{2\pi i h x}\, dx \qquad \text{for } h \neq 0. \tag{2.35}$$

Choose a positive integer m, and let a be a real number to be determined later. From (2.34) and (2.35), it follows that

$$\begin{aligned}\sum_{h=-m}^{m}{}^{*} (m+1-|h|)e^{-2\pi iha}\frac{S_h}{-2\pi ih} &= \int_0^1 \Delta_N(x)\left(\sum_{h=-m}^{m}(m+1-|h|)e^{2\pi ih(x-a)}\right)dx\\ &= \int_{-a}^{1-a}\Delta_N(x+a)\left(\sum_{h=-m}^{m}(m+1-|h|)e^{2\pi ihx}\right)dx, \qquad (2.36)\end{aligned}$$

where the asterisk indicates that $h = 0$ is deleted from the range of summation. Because of the periodicity of the integrand, the last integral may also be taken over $[-\frac{1}{2}, \frac{1}{2}]$. We note that

$$\sum_{h=-m}^{m}(m+1-|h|)e^{2\pi ihx} = \frac{\sin^2(m+1)\pi x}{\sin^2 \pi x}, \qquad (2.37)$$

where the right-hand side is interpreted as $(m + 1)^2$ in case x is an integer. We infer then from (2.36) that

$$\begin{aligned}\left|\int_{-1/2}^{1/2}\Delta_N(x+a)\frac{\sin^2(m+1)\pi x}{\sin^2\pi x}\,dx\right| &\le \frac{1}{2\pi}\sum_{h=-m}^{m}{}^{*}(m+1-|h|)\frac{|S_h|}{|h|}\\ &= \frac{1}{\pi}\sum_{h=1}^{m}(m+1-h)\frac{|S_h|}{h}. \qquad (2.38)\end{aligned}$$

We either have $\Delta_N(b) = -D_N^*$ or $\Delta_N(b+0) = D_N^*$ for some $b \in I$. We treat only the second alternative, the first one being completely similar. For $b < t \le b + D_N^*$, we have $\Delta_N(t) = D_N^* + \Delta_N(t) - \Delta_N(b+0) \ge D_N^* + b - t$. Now choose $a = b + \frac{1}{2}D_N^*$. Then $\Delta_N(x+a) \ge \frac{1}{2}D_N^* - x$ for $|x| < \frac{1}{2}D_N^*$. Consequently, we get

$$\begin{aligned}&\int_{-1/2}^{1/2}\Delta_N(x+a)\frac{\sin^2(m+1)\pi x}{\sin^2\pi x}\,dx\\ &= \left(\int_{(-1/2)D_N^*}^{(1/2)D_N^*} + \int_{-1/2}^{(-1/2)D_N^*} + \int_{(1/2)D_N^*}^{1/2}\right)\Delta_N(x+a)\frac{\sin^2(m+1)\pi x}{\sin^2\pi x}\,dx\\ &\ge \int_{(-1/2)D_N^*}^{(1/2)D_N^*}(\tfrac{1}{2}D_N^* - x)\frac{\sin^2(m+1)\pi x}{\sin^2\pi x}\,dx\\ &\quad - D_N^*\int_{-1/2}^{(-1/2)D_N^*}\frac{\sin^2(m+1)\pi x}{\sin^2\pi x}\,dx - D_N^*\int_{(1/2)D_N^*}^{1/2}\frac{\sin^2(m+1)\pi x}{\sin^2\pi x}\,dx\\ &= D_N^*\int_0^{(1/2)D_N^*}\frac{\sin^2(m+1)\pi x}{\sin^2\pi x}\,dx - 2D_N^*\int_{(1/2)D_N^*}^{1/2}\frac{\sin^2(m+1)\pi x}{\sin^2\pi x}\,dx,\end{aligned}$$

(2.39)

using the evenness of $(\sin^2 (m+1)\pi x)/\sin^2 \pi x$. The integral of this function over $[0, \frac{1}{2}]$ is $(m+1)/2$ by (2.37). Therefore, from (2.39),

$$\int_{-1/2}^{1/2} \Delta_N(x+a) \frac{\sin^2 (m+1)\pi x}{\sin^2 \pi x}\, dx$$

$$\geq D_N^* \int_0^{1/2} \frac{\sin^2 (m+1)\pi x}{\sin^2 \pi x}\, dx - 3D_N^* \int_{(1/2)D_N^*}^{1/2} \frac{\sin^2 (m+1)\pi x}{\sin^2 \pi x}\, dx$$

$$\geq \frac{m+1}{2} D_N^* - 3D_N^* \int_{(1/2)D_N^*}^{1/2} \frac{dx}{4x^2} \geq \frac{m+1}{2} D_N^* - \frac{3}{2},$$

where we used $\sin \pi x \geq 2x$ for $0 \leq x \leq \frac{1}{2}$. Combining this with (2.38), we arrive at

$$D_N^* \leq \frac{3}{m+1} + \frac{2}{\pi} \sum_{h=1}^{m} \left(\frac{1}{h} - \frac{1}{m+1}\right) |S_h|.$$

If $\tilde{D}_N$ denotes the discrepancy extended over all half-open intervals mod 1, then

$$\tilde{D}_N \leq \frac{6}{m+1} + \frac{4}{\pi} \sum_{h=1}^{m} \left(\frac{1}{h} - \frac{1}{m+1}\right) |S_h|. \tag{2.40}$$

In particular, this proves (2.33) in case (2.34) is satisfied.

We shall show that for any finite sequence $x_1, \ldots, x_N$ in I, there exists $c \in I$ such that the shifted sequence $\{x_1 + c\}, \ldots, \{x_N + c\}$ satisfies (2.34). This will prove the theorem, since both the left-hand and the right-hand side of (2.40) are invariant under the transition from $x_1, \ldots, x_N$ to the shifted sequence. By (2.28), we have to prove the existence of a $c \in I$ for which $(1/N) \sum_{n=1}^{N} \{x_n + c\} = \frac{1}{2}$. For any $c \in I$, we have

$$\frac{1}{N} \sum_{n=1}^{N} (\{x_n + c\} - x_n) = \frac{1}{N} \sum_{x_n < 1-c} c + \frac{1}{N} \sum_{x_n \geq 1-c} (c-1) = \Delta_N(1-c). \tag{2.41}$$

Therefore, it remains to show that

$$\Delta_N(1-c) = \frac{1}{2} - \frac{1}{N} \sum_{n=1}^{N} x_n = s,$$

say, for some $c \in I$. We consider only the case $s > 0$, the case $s < 0$ being completely analogous. Since $\int_0^1 \Delta_N(t)\, dt = s$, we have $\Delta_N(x) \geq s$ for some $x \in (0, 1)$. But since $\Delta_N(1) = 0$ and since Δ_N is piecewise linear with positive jumps only, the function Δ_N must attain the value s in the interval $[x, 1)$. ∎

When applying this theorem, we shall usually work with the following version thereof: There exists an absolute constant C such that

$$D_N \leq C\left(\frac{1}{m} + \sum_{h=1}^{m} \frac{1}{h} \left| \frac{1}{N} \sum_{n=1}^{N} e^{2\pi i h x_n} \right|\right) \tag{2.42}$$

for any real numbers $x_1, \ldots, x_N$ and any positive integer m.

It will sometimes happen that the sequence for which we have to estimate the discrepancy can be decomposed into a number of subsequences with small discrepancies. We prove a simple theorem pertaining to this situation.

THEOREM 2.6. For $1 \leq i \leq k$, let ω_i be a sequence of N_i elements from $\mathbb{R}$ with discrepancy $D_{N_i}(\omega_i)$. Let ω be a superposition of $\omega_1, \ldots, \omega_k$, that is, a sequence obtained by listing in some order the terms of the ω_i. We set $N = N_1 + \cdots + N_k$, which will be the number of elements of ω. Then,

$$D_N(\omega) \leq \sum_{i=1}^{k} \frac{N_i}{N} D_{N_i}(\omega_i). \tag{2.43}$$

We have also

$$D_N^*(\omega) \leq \sum_{i=1}^{k} \frac{N_i}{N} D_{N_i}^*(\omega_i). \tag{2.44}$$

PROOF. Let $J = [\alpha, \beta)$ be a subinterval of I. Then, by the construction of ω, we have $A(J; N; \omega) = \sum_{i=1}^{k} A(J; N_i; \omega_i)$. Therefore,

$$\left| \frac{A(J; N; \omega)}{N} - \lambda(J) \right| = \left| \sum_{i=1}^{k} \frac{N_i}{N} \left(\frac{A(J; N_i; \omega_i)}{N_i} - \lambda(J) \right) \right| \leq \sum_{i=1}^{k} \frac{N_i}{N} D_{N_i}(\omega_i),$$

and taking the supremum on the left-hand side completes the proof of the first inequality. The second inequality follows similarly. ∎

Notes

The important property of the discrepancy enunciated in Corollary 2.1 was already touched upon by van der Corput [7] when he stated that he did not know of any infinite sequence ω in I having $ND_N^*(\omega)$ uniformly bounded. This conjecture was first confirmed by van Aardenne-Ehrenfest [1, 2]. Theorems 2.1 and 2.2 are from Roth [1], who also shows a weaker form of the result given in Example 2.2. Theorem 2.3 and Corollary 2.2 are results of W. M. Schmidt [14]. We have improved the constants. It will follow from Section 3 that these two lower bounds are best possible, apart from the values of the constants. We note also that the lower bound in Lemma 2.5 was found to be best possible (at least for $k = 2$) by Davenport [1] and Halton and Zaremba [1]. In connection with Lemma 2.5, the following result of Sobol' [5] is of interest: For any N points in I^k, $k \geq 2$, we have

$$\int_0^1 \cdots \int_0^1 |A(x_1, \ldots, x_k) - Nx_1 \cdots x_k| \, dx_1 \cdots dx_k > \tfrac{1}{4} - \varepsilon_k(N),$$

where $0 < \varepsilon_k(N) < \frac{1}{4}$ and $\varepsilon_k(N) = O(N^{-1} \log^{k-2} N)$. This is best possible in the sense that one can always find sequences for which the integral is less than $\frac{1}{4}$. For $k = 1$, the integral is at least $\frac{1}{4}$, and this value is attained for the sequences occurring in Corollary 1.2 (Sobol' [3]). For small N, points in I^2 with minimal discrepancy were tabulated by White [1].

A collection of interesting problems on irregularities of distribution can be found in Erdös [6, 7]. Some of these problems have been settled by W. M. Schmidt [6, 13, 15]. For instance, he shows that for any infinite sequence in $\mathbb{R}$ there are at most countably many $x \in I$ for which $R_n(x)$ can remain bounded as a function of n. A weaker result in this direction was shown by Lesca [3]. An interesting variant of irregularities of distribution was studied by Berlekamp and Graham [1] (see also Steinhaus [2, Problems 6 and 7]). Their result shows, in particular, that for any c satisfying (2.26) we must have $c < (\log 17)^{-1}$. For related investigations, see de Bruijn and Erdös [1].

A nontrivial lower bound is known for the isotropic discrepancy J_N in $\mathbb{R}^k$. Namely, we have always $J_N \geq d_k N^{-(k+1)/2k}$ with an absolute constant $d_k > 0$ depending only on k, a result of Zaremba [7]. W. M. Schmidt has announced an improvement. We also refer to an important sequence of papers by W. M. Schmidt [7, 8, 9, 11] in which irregularities of distribution occurring in special classes of convex sets, such as rectangles, balls, and spherical caps, are studied. See also Niederreiter [17].

Lemma 2.8 is due to Koksma (unpublished). Our proof follows Kuipers [9]. Theorem 2.4 with D_N^* instead of D_N was shown by LeVeque [9]. In the same paper, the exponent $\frac{1}{3}$ is verified to be best possible. A more general inequality is due to Elliott [4]. A multi-dimensional version of Theorem 2.4 is not known.

Theorem 2.5, with unspecified constants, can be found in a paper of Erdös and Turán [3]. For another proof, see Yudin [1]. Our proof follows Niederreiter and Philipp [2] where one can also find somewhat improved values for the constants. See Elliott [4], Faĭnleĭb [2], and Niederreiter and Philipp [1, 2] for generalizations. A weaker version of this theorem was already known to van der Corput and Koksma (see Koksma [4, p. 101]). Van der Corput and Pisot [1] proved that

$$D_N \leq 2\delta + \sum_{h=1}^{\infty} \min\,(1/h,\ 1/\delta h^2)\ \left|(1/N) \sum_{n=1}^{N} e^{2\pi i h x_n}\right|$$

for every $\delta > 0$. See also Jagerman [3]. A generalization of Theorem 2.5 to several dimensions can also be given, and we shall refer to it as the theorem of Erdös-Turán-Koksma. For a lattice point $\mathbf{h} = (h_1, \ldots, h_s)$ in $\mathbb{Z}^s$, define $\|\mathbf{h}\| = \max_{1 \leq j \leq s} |h_j|$ and

$$r(\mathbf{h}) = \prod_{j=1}^{s} \max\,(|h_j|, 1).$$

For $\mathbf{x}, \mathbf{y} \in \mathbb{R}^s$, let $\langle \mathbf{x}, \mathbf{y} \rangle$ denote the standard inner product. Now let $\mathbf{x}_1, \ldots, \mathbf{x}_N$ be a finite sequence of points in $\mathbb{R}^s$. Then, for any positive integer m, we have

$$D_N \leq C_s\left(\frac{1}{m} + \sum_{0<\|\mathbf{h}\|\leq m} \frac{1}{r(\mathbf{h})}\left|\frac{1}{N}\sum_{n=1}^{N} e^{2\pi i \langle \mathbf{h}, \mathbf{x}_n \rangle}\right|\right),$$

where the constant C_s only depends on the dimension s. This result was proved at about the same time, but independently, by Koksma [11] and Szüsz [1]. An explicit value for C_s is $C_s = 2s^2 3^{s+1}$. For a generalization of the inequality, see Niederreiter and Philipp [1, 2].

The inequality in Theorem 2.6 was used in disguised form by many authors. It was first stated explicitly as a lemma in Bergström [2]. See also Cassels [4] and Niederreiter [2].

Starting from Theorem 1.4, the problem of finding upper bounds for D_N^* can be linked with convex programming techniques (Niederreiter [10]). This approach was also used in Niederreiter [2]. A survey of methods for estimating discrepancy is given in Niederreiter [5].

Exercises

2.1. The auxiliary functions $\psi(2^n x)$ used in the lemmas preceding Theorem 2.1 are variants of the so-called *Rademacher functions* $\phi_n(x)$, $n = 0, 1, 2, \ldots$. Let $\phi_0(x)$ be the function of period 1 that is given by $\phi_0(x) = 1$

for $0 < x < \frac{1}{2}$, $\phi_0(x) = -1$ for $\frac{1}{2} < x < 1$, $\phi_0(0) = \phi_0(\frac{1}{2}) = 0$. For $n \geq 1$, define $\phi_n(x) = \phi_0(2^n x)$. Prove that these functions may also be given by $\phi_n(x) = \operatorname{sign} \sin 2^{n+1}\pi x$ for $n \geq 0$.

2.2. Suppose x is not a dyadic rational, and let

$$x = d_0(x) + \sum_{n=1}^{\infty} \frac{d_n(x)}{2^n}$$

be its dyadic expansion. Prove that $d_n(x) = \frac{1}{2}(1 - \phi_{n-1}(x))$ for $n \geq 1$.

2.3. Prove that the Rademacher functions $\phi_0, \phi_1, \ldots$ form an orthonormal system over $[0, 1]$. (Note: The Rademacher functions are even independent.)

2.4. The so-called *Walsh functions* are defined as follows: Let $N = 2^{n_1} + 2^{n_2} + \cdots + 2^{n_k}$ with $n_1 > n_2 > \cdots > n_k \geq 0$ be the dyadic expansion of the positive integer N; then set $\chi_N(x) = \phi_{n_1}(x) \cdots \phi_{n_k}(x)$. Define also $\chi_0(x) = 1$. Prove that the functions $\chi_0, \chi_1, \ldots$ form an orthonormal system over $[0, 1]$. (Note: The system of Walsh functions is even complete.)

2.5. Show that for every sequence of N points in I^2 with discrepancy D_N^* there exists a sequence of the form $(0, s_0/N), (1/N, s_1/N), \ldots, ((N-1)/N, s_{N-1}/N)$, where the integers $s_0, \ldots, s_{N-1}$ are a permutation of $0, \ldots, N-1$, and with discrepancy $\bar{D}_N^*$ satisfying $\bar{D}_N^* \leq 4D_N^*$. *Hint:* Compare with Example 2.2.

2.6. Prove that LeVeque's inequality implies the sufficiency part of the Weyl criterion.

2.7. Prove that the theorem of Erdös-Turán implies the sufficiency part of the Weyl criterion.

2.8. Give a detailed proof of (2.37).

2.9. If $\tilde{D}_N$ denotes the discrepancy extended over all half-open intervals mod 1, prove in detail that $\tilde{D}_N \leq 2D_N^*$.

2.10. Prove in detail that the discrepancy $\tilde{D}_N$ in Exercise 2.9 is invariant under shifts of the sequence mod 1.

2.11. Suppose C_1 and C_2 are positive constants such that the inequality

$$D_N \leq \frac{C_1}{m+1} + C_2 \sum_{h=1}^{m} \left(\frac{1}{h} - \frac{1}{m+1}\right) \left| \frac{1}{N} \sum_{n=1}^{N} e^{2\pi i h x_n} \right|$$

holds for any real numbers $x_1, \ldots, x_N$ and any positive integer m. Prove that necessarily $C_1 + C_2 \geq 2$.

2.12. Prove that the constant C_1 in Exercise 2.11 must satisfy $C_1 \geq 1$.

2.13. Suppose $x_1, \ldots, x_N$ is a finite sequence such that, for some real number $\lambda > 0$, we have $|\sum_{n=1}^{N} e^{2\pi i h x_n}| \leq h^\lambda$ for all integers h with $1 \leq h \leq N^{1/(\lambda+1)}$. Prove that $D_N \leq CN^{-1/(\lambda+1)}$ with a constant C only depending on λ.

2.14. Consider the infinite sequence ω given by

$$\frac{1}{2}, \frac{1}{4}, \frac{3}{4}, \ldots, \frac{1}{2k}, \frac{3}{2k}, \ldots, \frac{2k-1}{2k}, \ldots.$$

Using Theorem 2.6, prove that $D_N(\omega) = O(1/\sqrt{N})$. Show also $D_N(\omega) = \Omega(1/\sqrt{N})$.

3. SPECIAL SEQUENCES

Almost-Arithmetic Progressions

We shall first discuss a general class of sequences that is of great theoretical interest (see notes). For a reason that will be apparent from the subsequent definition, these sequences are called *almost-arithmetic progressions.*

DEFINITION 3.1. For $0 \leq \delta < 1$ and $\varepsilon > 0$, a finite sequence $x_1 < x_2 < \cdots < x_N$ in I is called an almost-arithmetic progression-(δ, ε) if there exists an η, $0 < \eta \leq \varepsilon$, such that the following conditions are satisfied: (i) $0 \leq x_1 \leq \eta + \delta\eta$; (ii) $\eta - \delta\eta \leq x_{n+1} - x_n \leq \eta + \delta\eta$ for $1 \leq n \leq N-1$; (iii) $1 - \eta - \delta\eta \leq x_N < 1$.

If $\delta = 0$, then we have a true arithmetic progression with difference η. It is clear that an almost-arithmetic progression-(δ, ε) is also an almost-arithmetic progression-(δ', ε') whenever $\delta \leq \delta'$ and $\varepsilon \leq \varepsilon'$.

THEOREM 3.1. Let $x_1 < x_2 < \cdots < x_N$ be an almost-arithmetic progression-(δ, ε) and let η be the positive real number corresponding to the sequence according to Definition 3.1. Then

$$D_N^* \leq \frac{1}{N} + \frac{\delta}{1 + \sqrt{1 - \delta^2}} \qquad \text{for } \delta > 0, \tag{3.1}$$

$$D_N^* \leq \min\left(\eta, \frac{1}{N}\right) \qquad \text{for } \delta = 0. \tag{3.2}$$

PROOF. Let us first consider the simple case $\delta = 0$. Here we have $x_i = x_1 + (i-1)\eta$ for $1 \leq i \leq N$. We estimate D_N^* by using Theorem 1.4. We note that $x_i - (i-1)/N = x_1 + (i-1)(\eta - (1/N))$. We distinguish two cases, depending on whether $\eta \geq 1/N$ or $\eta < 1/N$. If $\eta \geq 1/N$, then

$$0 \leq x_i - \frac{i-1}{N} \leq x_1 + (N-1)\left(\eta - \frac{1}{N}\right) = x_N - \frac{N-1}{N} \leq \frac{1}{N},$$

and subtracting $1/N$, we get $-1/N \leq x_i - (i/N) \leq 0$. If $\eta < 1/N$, then

$$\eta \geq x_i - \frac{i-1}{N} \geq x_1 + (N-1)\left(\eta - \frac{1}{N}\right) = x_N - \frac{N-1}{N} \geq -\eta + \frac{1}{N},$$

and again subtracting $1/N$, we get $\eta - (1/N) \geq x_i - (i/N) \geq -\eta$. In any case, we have

$$\max\left(\left|x_i - \frac{i}{N}\right|, \left|x_i - \frac{i-1}{N}\right|\right) \leq \min\left(\eta, \frac{1}{N}\right)$$

for all i with $1 \leq i \leq N$, and so,

$$D_N^* \leq \min\left(\eta, \frac{1}{N}\right).$$

Now we take the case $\delta > 0$. It follows immediately from Definition 3.1 that

$$(i-1)(\eta - \delta\eta) \leq x_i \leq i(\eta + \delta\eta) \qquad \text{for } 1 \leq i \leq N. \tag{3.3}$$

Similarly, the inequality

$$1 - (N - i + 1)(\eta + \delta\eta) \leq x_i \leq 1 - (N - i)(\eta - \delta\eta) \tag{3.4}$$

holds for $1 \leq i \leq N$. We shall again use the representation for D_N^* given in Theorem 1.4. Let us first estimate $x_i - (i/N)$ from above. We use the one of (3.3) and (3.4) that gives the better upper bound for a given i, and therefore, we distinguish two cases.

Case 1: $i(\eta + \delta\eta) \leq 1 - (N - i)(\eta - \delta\eta)$. We note that this is equivalent to

$$i \leq \frac{1 - N\eta(1-\delta)}{2\delta\eta}.$$

It clearly suffices to consider $N\eta \leq 1/(1-\delta)$, for otherwise, the present case could not occur at all. We have

$$x_i - \frac{i}{N} \leq i\left(\eta + \delta\eta - \frac{1}{N}\right)$$

from (3.3). Let us assume for the moment that $\eta + \delta\eta - (1/N) \geq 0$. Then $1/(1+\delta) \leq N\eta \leq 1/(1-\delta)$. Furthermore,

$$x_i - \frac{i}{N} \leq i\left(\eta + \delta\eta - \frac{1}{N}\right) \leq \frac{1 - N\eta(1-\delta)}{2\delta\eta}\left(\eta + \delta\eta - \frac{1}{N}\right)$$

$$= \frac{2N\eta - 1 - N^2\eta^2(1-\delta^2)}{2\,\delta N\eta}.$$

Set

$$N\eta = t \qquad \text{and} \qquad h(t) = \frac{2t - 1 - t^2(1-\delta^2)}{2\,\delta t}.$$

Since $h(t)$ has an absolute maximum in the interval $[1/(1+\delta), 1/(1-\delta)]$ at $t_0 = 1/\sqrt{1-\delta^2}$, we get

$$x_i - \frac{i}{N} \le h(t_0) = \frac{\delta}{1+\sqrt{1-\delta^2}}. \tag{3.5}$$

If $\eta + \delta\eta - (1/N) < 0$, then (3.5) is trivially true since $x_i - i/N$ is then negative.

Case 2: $i(\eta + \delta\eta) > 1 - (N-i)(\eta - \delta\eta)$. This is equivalent to $i > (1 - N\eta(1-\delta))/2\delta\eta$. The present case can only occur if

$$(1 - N\eta(1-\delta))/2\delta\eta \le N, \text{ or } N\eta \ge 1/(1+\delta).$$

We have

$$x_i - \frac{i}{N} \le 1 - (N-i)(\eta - \delta\eta) - \frac{i}{N} = (N-i)\left(\frac{1}{N} + \delta\eta - \eta\right).$$

As above, it suffices to consider $(1/N) + \delta\eta - \eta \ge 0$, or $N\eta \le 1/(1-\delta)$. Then

$$\begin{aligned} x_i - \frac{i}{N} &\le (N-i)\left(\frac{1}{N} + \delta\eta - \eta\right) \\ &\le \left(N - \frac{1 - N\eta(1-\delta)}{2\,\delta\eta}\right)\left(\frac{1}{N} + \delta\eta - \eta\right) \\ &= \frac{2N\eta - 1 - N^2\eta^2(1-\delta^2)}{2\,\delta N\eta} \le \frac{\delta}{1+\sqrt{1-\delta^2}}. \end{aligned} \tag{3.6}$$

Working in the same way with the lower bounds in (3.3) and (3.4), we obtain

$$\frac{i-1}{N} - x_i \le \frac{\delta}{1+\sqrt{1-\delta^2}} \qquad \text{for } 1 \le i \le N. \tag{3.7}$$

Combining (3.5), (3.6), and (3.7) results in the following inequalities for $1 \le i \le N$:

$$-\frac{1}{N} - \frac{\delta}{1+\sqrt{1-\delta^2}} \le x_i - \frac{i}{N} \le \frac{\delta}{1+\sqrt{1-\delta^2}},$$

$$-\frac{\delta}{1+\sqrt{1-\delta^2}} \le x_i - \frac{i-1}{N} \le \frac{1}{N} + \frac{\delta}{1+\sqrt{1-\delta^2}}.$$

Therefore,

$$\max\left(\left|x_i - \frac{i}{N}\right|, \left|x_i - \frac{i-1}{N}\right|\right) \le \frac{1}{N} + \frac{\delta}{1+\sqrt{1-\delta^2}}$$

for $1 \le i \le N$, and an application of Theorem 1.4 completes the proof. ■

Diophantine Approximation

A most important class of u.d. mod 1 sequences is given by the sequences $(n\alpha)$, $n = 1, 2, \ldots$, with α irrational. The discrepancy of $(n\alpha)$ will depend on the finer arithmetical properties of α. Therefore, we shall start with some brief remarks on diophantine approximation and the classification of irrational numbers.

DEFINITION 3.2. For a real number t, let $\langle t\rangle$ denote the distance from t to the nearest integer, namely,

$$\langle t\rangle = \min_{n\in\mathbb{Z}} |t - n| = \min(\{t\}, 1 - \{t\}). \tag{3.8}$$

DEFINITION 3.3. Let ψ be a nondecreasing positive function that is defined at least for all positive integers. The irrational number α is said to be *of type* $<\psi$ if $q\langle q\alpha\rangle \geq 1/\psi(q)$ holds for all positive integers q. If ψ is a constant function, then an irrational α of type $<\psi$ is also called *of constant type*.

DEFINITION 3.4. Let η be a positive real number or infinity. The irrational number α is said to be *of type* η if η is the supremum of all γ for which $\underline{\lim}_{q\to\infty} q^\gamma\langle q\alpha\rangle = 0$, where q runs through the positive integers.

By Dirichlet's theorem (see also (3.9)), we have $\underline{\lim}_{q\to\infty} q^\gamma\langle q\alpha\rangle = 0$ for any $\gamma < 1$ and for any irrational α. Therefore the type η of an irrational number will always satisfy $\eta \geq 1$. There is, of course, a close connection between the above definitions.

LEMMA 3.1. The irrational number α is of type η if and only if η is the infimum of all real numbers τ for which there exists a positive constant $c = c(\tau, \alpha)$ such that α is of type $<\psi$, where $\psi(q) = cq^{\tau-1}$.

PROOF. Let η be finite. Then for any $\varepsilon > 0$ we have $\underline{\lim}_{q\to\infty} q^{\eta-\varepsilon}\langle q\alpha\rangle = 0$ and $\underline{\lim}_{q\to\infty} q^{\eta+\varepsilon}\langle q\alpha\rangle > 0$ whenever α is of type η. From the first statement, it follows that for any positive c there is a positive integer q with $q\langle q\alpha\rangle < 1/cq^{\eta-1-\varepsilon}$. Therefore, α is not of type $<\psi$ for any ψ of the form $\psi(q) = cq^{\eta-1-\varepsilon}$. But from the second statement above, we conclude that for any $\varepsilon > 0$ there is a positive constant $d(\varepsilon, \alpha)$ such that $q^{\eta+\varepsilon}\langle q\alpha\rangle \geq d(\varepsilon, \alpha)$ holds for all q. Therefore, $q\langle q\alpha\rangle \geq d(\varepsilon, \alpha)/q^{\eta-1+\varepsilon}$ for all q, and α is of type $<\psi$ with $\psi(q) = (1/d(\varepsilon, \alpha))q^{\eta-1+\varepsilon}$. These arguments are easily seen to be reversible. If η is infinity, then the same ideas can be carried out with obvious modifications (the statement of the lemma is, of course, interpreted to mean that no such numbers τ with the indicated property exist). ■

We review very briefly some facts about continued fractions. Let $\alpha = [a_0, a_1, a_2, \ldots]$ be the continued fraction expansion of an irrational α, where a_0 is an integer and $a_1, a_2, \ldots$ are positive integers, the so-called *partial quotients*. For $n \geq 0$, the *nth convergent to* α is defined as

$$r_n = [a_0, a_1, \ldots, a_n].$$

The rationals r_n may be obtained by the following algorithm. Define $p_{-2} = 0$, $p_{-1} = 1$, $p_i = a_i p_{i-1} + p_{i-2}$ for $i \geq 0$; define $q_{-2} = 1$, $q_{-1} = 0$, $q_i = a_i q_{i-1} + q_{i-2}$ for $i \geq 0$. An easy induction argument shows that $r_n = p_n/q_n$ for $n \geq 0$. Moreover, the fractions p_n/q_n, $n \geq 0$, are in reduced form. For later reference, we note that $1 = q_0 \leq q_1 < q_2 < \cdots < q_i < \cdots$. For all $n \geq 0$, we have

$$\left|\alpha - \frac{p_n}{q_n}\right| < \frac{1}{q_n q_{n+1}} \leq \frac{1}{q_n^{\,2}}. \tag{3.9}$$

In particular, we get $\lim_{n\to\infty} r_n = \alpha$, which explains the term *convergent*. The irrational α is of constant type (see Definition 3.3) if and only if there exists a constant K such that $a_i \leq K$ for all $i \geq 1$. In the latter case, α is said to have *bounded partial quotients*. An interesting class of such α is formed by the quadratic irrationals.

Discrepancy of $(n\alpha)$

The following rule of thumb holds: The smaller the type of α, the smaller the discrepancy of $(n\alpha)$. In other words, the irrationals α that are badly approximable by rationals are those for which $(n\alpha)$ shows a good distribution behavior. Let us turn to the details. The initial step in the estimation process is to apply the theorem of Erdös-Turán. In the sequel, α will always denote a fixed irrational number.

LEMMA 3.2. The discrepancy $D_N(\omega)$ of $\omega = (n\alpha)$ satisfies

$$D_N(\omega) \leq C\left(\frac{1}{m} + \frac{1}{N}\sum_{h=1}^{m}\frac{1}{h\langle h\alpha\rangle}\right) \tag{3.10}$$

for any positive integer m, where C is an absolute constant.

PROOF. By (2.42) we have

$$D_N(\omega) \leq C\left(\frac{1}{m} + \frac{1}{N}\sum_{h=1}^{m}\frac{1}{h}\left|\sum_{n=1}^{N} e^{2\pi i h n\alpha}\right|\right)$$

for any positive integer m. Now

$$\left|\sum_{n=1}^{N} e^{2\pi i h n\alpha}\right| \leq \frac{2}{|e^{2\pi i h\alpha} - 1|} = \frac{1}{|\sin \pi h\alpha|}.$$

We note that $|\sin \pi h\alpha| = \sin \pi\langle h\alpha\rangle$, since $\langle h\alpha\rangle$ is either of the form $h\alpha - p$ or $p - h\alpha$ for some integer p. Then $\sin \pi x \geq 2x$ for $0 \leq x \leq \frac{1}{2}$ implies $1/|\sin \pi h\alpha| \leq 1/(2\langle h\alpha\rangle)$ for all $h \geq 1$. The desired inequality follows immediately. ■

LEMMA 3.3. Let α be of type $<\psi$. Then,

$$\sum_{h=1}^{m} \frac{1}{h\langle h\alpha\rangle} = O\left(\psi(2m)\log m + \sum_{h=1}^{m}\frac{\psi(2h)\log h}{h}\right). \tag{3.11}$$

PROOF. By Abel's summation formula, we have

$$\sum_{h=1}^{m} \frac{1}{h\langle h\alpha\rangle} = \sum_{h=1}^{m} \frac{s_h}{h(h+1)} + \frac{s_m}{m+1}, \tag{3.12}$$

where $s_h = \sum_{j=1}^{h} 1/\langle j\alpha\rangle$. For $0 \leq p < q \leq h$, we obtain

$$\langle q\alpha \pm p\alpha\rangle = \langle (q \pm p)\alpha\rangle \geq \frac{1}{(q \pm p)\psi(q \pm p)} \geq \frac{1}{2h\psi(2h)}.$$

It follows that

$$|\langle q\alpha\rangle - \langle p\alpha\rangle| \geq \frac{1}{2h\psi(2h)} \qquad \text{for } 0 \leq p < q \leq h. \tag{3.13}$$

But (3.13) implies that in each of the intervals

$$\left[0, \frac{1}{2h\psi(2h)}\right), \left[\frac{1}{2h\psi(2h)}, \frac{2}{2h\psi(2h)}\right), \ldots, \left[\frac{h}{2h\psi(2h)}, \frac{h+1}{2h\psi(2h)}\right)$$

there is at most one number of the form $\langle j\alpha\rangle$, $1 \leq j \leq h$, with no such number lying in the first interval. Therefore,

$$s_h = \sum_{j=1}^{h} \frac{1}{\langle j\alpha\rangle} \leq \sum_{j=1}^{h} \frac{2h\psi(2h)}{j} = O(h\psi(2h)\log h).$$

Using this with (3.12), we arrive at (3.11). ■

THEOREM 3.2. Let α be of finite type η. Then, for every $\varepsilon > 0$, the discrepancy $D_N(\omega)$ of $\omega = (n\alpha)$ satisfies

$$D_N(\omega) = O(N^{(-1/\eta)+\varepsilon}). \tag{3.14}$$

PROOF. Let $\varepsilon > 0$ be fixed. By Lemma 3.1, there exists $c > 0$ such that α is of type $<\psi$ with $\psi(q) = cq^{\eta-1+(\varepsilon/2)}$. Then Lemma 3.3 implies

$$\sum_{h=1}^{m} \frac{1}{h\langle h\alpha\rangle} = O(m^{\eta-1+\varepsilon}) + O\left(\sum_{h=1}^{m} h^{\eta-2+\varepsilon}\right) = O(m^{\eta-1+\varepsilon}).$$

Combining this with Lemma 3.2, we get

$$D_N(\omega) \le C_1\left(\frac{1}{m} + \frac{1}{N} m^{\eta-1+\varepsilon}\right) \qquad \text{for all } m \ge 1.$$

Now choose $m = [N^{1/\eta}]$, and the desired result follows. ■

EXAMPLE 3.1. In particular, if α is of type $\eta = 1$, then $ND_N(\omega) = O(N^\varepsilon)$ for every $\varepsilon > 0$. There is an important class of irrationals that have type $\eta = 1$, namely, the algebraic irrationals. This follows from the famous theorem of Thue-Siegel-Roth: For every irrational algebraic number α and for every $\varepsilon > 0$, there exists a positive constant $c = c(\alpha, \varepsilon)$ such that

$$\left|\alpha - \frac{p}{q}\right| \ge \frac{c}{q^{2+\varepsilon}}$$

holds for all integers $q > 0$ and p. For the proof we refer to the literature. ■

EXAMPLE 3.2. Let α be of constant type. Then Lemma 3.3 implies

$$\sum_{h=1}^{m} \frac{1}{h\langle h\alpha\rangle} = O(\log^2 m).$$

Therefore,

$$D_N(\omega) \le C_2\left(\frac{1}{m} + \frac{1}{N}\log^2 m\right)$$

for all $m \ge 1$. Choosing $m = N$, we obtain $ND_N(\omega) = O(\log^2 N)$ in this case. We will improve this result in the sequel. ■

Theorem 3.2 is best possible in the following sense.

THEOREM 3.3. Let α be of finite type η. Then, for every $\varepsilon > 0$, the discrepancy $D_N(\omega)$ of $\omega = (n\alpha)$ satisfies

$$D_N(\omega) = \Omega(N^{(-1/\eta)-\varepsilon}). \tag{3.15}$$

PROOF. For given $\varepsilon > 0$, there exists $0 < \delta < \eta$ with $1/(\eta - \delta) = (1/\eta) + \varepsilon$. By Definition 3.4, we have $\underline{\lim}_{q\to\infty} q^{\eta-(\delta/2)}\langle q\alpha\rangle = 0$. In particular, we get $\langle q\alpha\rangle < q^{-\eta+(\delta/2)}$ for infinitely many positive integers q. Hence, there are infinitely many positive integers q and corresponding integers p such that $|\alpha - (p/q)| < q^{-1-\eta+(\delta/2)}$. Take one such q, and set $N = [q^{\eta-\delta}]$. We have $\alpha = (p/q) + \theta q^{-1-\eta+(\delta/2)}$ with $|\theta| < 1$. For $1 \le n \le N$, we get then $n\alpha = np/q + \theta_n$ with $|\theta_n| = |n\theta q^{-1-\eta+(\delta/2)}| < q^{-1-(\delta/2)}$. Now let us look at the fractional parts $\{\alpha\}, \{2\alpha\}, \ldots, \{N\alpha\}$. It follows from the above that none of these numbers lies in the interval $J = [q^{-1-(\delta/2)}, q^{-1} - q^{-1-(\delta/2)})$. Therefore,

$$D_N(\omega) \ge \left|\frac{A(J; N)}{N} - \lambda(J)\right| = \lambda(J).$$

For sufficiently large q, we have $\lambda(J) \geq 1/2q$. On the other hand, the definition of N implies $q^{\eta-\delta} \leq 2N$. Combining these inequalities, we arrive at $D_N(\omega) \geq cN^{-1/(\eta-\delta)} = cN^{(-1/\eta)-\varepsilon}$, with a constant $c > 0$ just depending on η and ε. But since we have infinitely many q to choose from, this lower bound for $D_N(\omega)$ holds for infinitely many values of N. ■

There is one case where an improvement of Theorem 3.2 seems to be worthwhile, namely, when α is of constant type (see also Example 3.2). For such α, the sequence $(n\alpha)$ has a very small discrepancy. In fact, the discrepancy has the least order of magnitude possible in the light of Theorem 2.3.

THEOREM 3.4. Suppose the irrational $\alpha = [a_0, a_1, a_2, \ldots]$ has bounded partial quotients, say $a_i \leq K$ for $i \geq 1$. Then the discrepancy $D_N(\omega)$ of $\omega = (n\alpha)$ satisfies $ND_N(\omega) = O(\log N)$. More exactly, we have

$$ND_N(\omega) \leq 3 + \left(\frac{1}{\log \xi} + \frac{K}{\log (K+1)}\right) \log N, \tag{3.16}$$

where $\xi = (1 + \sqrt{5})/2$.

PROOF. Let $1 = q_0 \leq q_1 < q_2 < \cdots$ be the denominators of the convergents to α. For given $N \geq 1$, there exists $r \geq 0$ such that $q_r \leq N < q_{r+1}$. By the division algorithm, we have $N = b_r q_r + N_{r-1}$ with $0 \leq N_{r-1} < q_r$. We note that $(a_{r+1} + 1)q_r \geq q_{r+1} > N$, and so, $b_r \leq a_{r+1}$. If $r > 0$, we may write $N_{r-1} = b_{r-1}q_{r-1} + N_{r-2}$ with $0 \leq N_{r-2} < q_{r-1}$. Again we find $b_{r-1} \leq a_r$. Continuing in this manner, we arrive at a representation for N of the form $N = \sum_{i=0}^{r} b_i q_i$ with $0 \leq b_i \leq a_{i+1}$ for $0 \leq i \leq r$, and $b_r \geq 1$.

We decompose the given sequence $\{\alpha\}, \{2\alpha\}, \ldots, \{N\alpha\}$ into b_r sequences $(\{n\alpha\})$ where n runs through q_r consecutive integers, into b_{r-1} sequences $(\{n\alpha\})$ where n runs through q_{r-1} consecutive integers, and so on. We estimate first the discrepancy of such a finite sequence $(\{n\alpha\})$ where n runs through q_i consecutive integers, say $n = n_0 + j$ with $1 \leq j \leq q_i$. By (3.9) we have

$$\alpha = \frac{p_i}{q_i} + \frac{\theta}{q_i q_{i+1}} \qquad \text{with } |\theta| < 1.$$

Therefore,

$$\{n\alpha\} = \left\{n_0\alpha + \frac{jp_i}{q_i} + \frac{j\theta}{q_i q_{i+1}}\right\}.$$

Since $(p_i, q_i) = 1$, the numbers $n_0\alpha + jp_i/q_i$, $1 \leq j \leq q_i$, considered mod 1, form a sequence of q_i equidistant points with distance $1/q_i$. Since $|j\theta/q_i q_{i+1}| < 1/q_{i+1}$ for $1 \leq j \leq q_i$, the sequence $(\{n\alpha\})$, $n_0 + 1 \leq n \leq n_0 + q_i$, is obtained by shifting mod 1 the elements $\{n_0\alpha + (jp_i/q_i)\}$, $1 \leq j \leq q_i$, either all to the right or all to the left by a distance less than $1/q_{i+1}$ (the direction of the shift will depend on the sign of θ). It is then easily seen that the discrepancy D_{q_i} of the finite sequence $(\{n\alpha\})$, $n_0 + 1 \leq n \leq n_0 + q_i$, satisfies

$$D_{q_i} \leq \frac{1}{q_i} + \frac{1}{q_{i+1}}. \tag{3.17}$$

By Theorem 2.6 and the way in which we decomposed the original sequence $\{\alpha\}, \{2\alpha\}, \ldots, \{N\alpha\}$, we infer from (3.17) that

$$ND_N(\omega) \leq \sum_{i=0}^{r} b_i\left(\frac{q_i}{q_{i+1}} + 1\right) \leq r + 1 + \sum_{i=0}^{r} b_i. \tag{3.18}$$

We claim that

$$\sum_{i=0}^{r} b_i \leq 1 + \frac{K}{\log (K+1)} \log N. \tag{3.19}$$

We prove (3.19) by induction on r. For $N \geq 1$, we put $\sigma(N) = \sum_{i=0}^{r} b_i$. If $q_0 < q_1$, then the smallest possible r is $r = 0$, and a corresponding N satisfies $1 \leq \sigma(N) = N < q_1 \leq K$. If $q_0 = q_1 = 1$, then the smallest possible r is $r = 1$, and a corresponding N satisfies $1 \leq \sigma(N) = N < q_2 \leq K + 1$. For the first step in the induction, it therefore suffices to show

$$N \leq 1 + \frac{K}{\log (K+1)} \log N \qquad \text{for } 1 \leq N < K + 1.$$

This follows by considering the function $f(x) = x - (K/\log (K+1)) \log x$ on the interval $1 \leq x \leq K + 1$ and noting that $f(1) = f(K+1) = 1$ and that f is concave upward on the entire interval.

Now take an arbitrary N with $1 < q_r \leq N < q_{r+1}$, and write $N = b_r q_r + N_{r-1}$ with $0 \leq N_{r-1} < q_r$. Suppose for the moment that $N_{r-1} > 0$. Then $\sigma(N) = b_r + \sigma(N_{r-1})$, and the induction hypothesis yields $\sigma(N) \leq 1 + b_r + (K/\log (K+1)) \log N_{r-1}$. Now $N > b_r N_{r-1} + N_{r-1} = (b_r + 1)N_{r-1}$, and so,

$$\sigma(N) \leq 1 + b_r + \frac{K}{\log (K+1)} \log \frac{N}{b_r + 1}.$$

But this inequality holds for $N_{r-1} = 0$ as well, since then $\sigma(N) = b_r$ and $N/(b_r + 1) = b_r q_r/(b_r + 1) \geq 1$. To complete the argument, we have to prove $b_r \leq (K/\log (K+1)) \log (b_r + 1)$. But this follows immediately from $1 \leq b_r \leq a_{r+1} \leq K$ and the fact that $g(x) = x/\log (x+1)$ is increasing for $x > 0$.

To estimate r, one first proves by induction that $q_i \geq \xi^{i-1}$ for $i \geq 0$. Therefore, $N \geq q_r \geq \xi^{r-1}$, or $r \leq (\log N)/\log \xi + 1$. Then (3.16) follows from (3.18) and (3.19). ■

The Van der Corput Sequence

We shall now exhibit a sequence that has an extremely small discrepancy. In fact, no infinite sequence has yet been found that has a uniformly smaller discrepancy than the one that we are going to construct. For optimal estimates of the discrepancy of this sequence, see the notes.

We define the so-called *van der Corput sequence* (x_n) as follows: For $n \geq 1$, let $n - 1 = \sum_{j=0}^{s} a_j 2^j$ be the dyadic expansion of $n - 1$. Then we set $x_n = \sum_{j=0}^{s} a_j 2^{-j-1}$. The sequence (x_n) is then clearly contained in the unit interval.

THEOREM 3.5. The discrepancy $D_N(\omega)$ of the van der Corput sequence $\omega = (x_n)$ satisfies

$$ND_N(\omega) \leq \frac{\log(N+1)}{\log 2}. \tag{3.20}$$

PROOF. We represent a given $N \geq 1$ by its dyadic expansion $N = 2^{h_1} + 2^{h_2} + \cdots + 2^{h_s}$ with $h_1 > h_2 > \cdots > h_s \geq 0$. Partition the interval $[1, N]$ of integers into subintervals $M_1, M_2, \ldots, M_s$ as follows: For $1 \leq j \leq s$, we put $M_j = [2^{h_1} + 2^{h_2} + \cdots + 2^{h_{j-1}} + 1,\ 2^{h_1} + 2^{h_2} + \cdots + 2^{h_j}]$; for $j = 1$, the expression $2^{h_1} + \cdots + 2^{h_{j-1}}$ is an empty sum, and thus meant to be 0.

An integer $n \in M_j$ can be written in the form $n = 1 + 2^{h_1} + 2^{h_2} + \cdots + 2^{h_{j-1}} + \sum_{i=0}^{h_j-1} a_i 2^i$ with $a_i = 0$ or 1. In fact, we get all 2^{h_j} integers in M_j if we let the a_i run through all possible combinations of 0 and 1. It follows that

$$x_n = 2^{-h_1-1} + \cdots + 2^{-h_{j-1}-1} + \sum_{i=0}^{h_j-1} a_i 2^{-i-1} = y_j + \sum_{i=0}^{h_j-1} a_i 2^{-i-1},$$

where y_j only depends on j, and not on n. If n runs through M_j, then $\sum_{i=0}^{h_j-1} a_i 2^{-i-1}$ runs through all fractions $0, 2^{-h_j}, \ldots, (2^{h_j} - 1)2^{-h_j}$ in some order. Moreover, we note that $0 \leq y_j < 2^{-h_j}$. We conclude that if the elements x_n with $n \in M_j$ are ordered according to their magnitude, then we obtain a sequence ω_j of 2^{h_j} elements that is an almost-arithmetic progression with parameters $\delta = 0$ and $\eta = 2^{-h_j}$. It is then easily seen that the discrepancy of each ω_j, multiplied by the number of elements in ω_j, is at most 1. Combining this with (2.43) and the fact that $x_1, \ldots, x_N$ is decomposed into s sequences ω_j, we obtain $ND_N(\omega) \leq s$.

It remains to estimate s in terms of N. But $N \geq 2^{s-1} + 2^{s-2} + \cdots + 2^0 = 2^s - 1$, and so, $s \leq (\log(N+1))/\log 2$. ∎

Notes

Almost-arithmetic progressions were introduced by O'Neil [1]. Their theoretical importance stems from the following criterion. The sequence (x_n) in I is u.d. mod 1 if and only if the following condition holds: for any three positive numbers δ, ε, and ε', there exists $\bar{N} = \bar{N}(\delta, \varepsilon, \varepsilon')$ such that for all $N > \bar{N}$, the initial segment $x_1, \ldots, x_N$ can be decomposed into almost-arithmetic progressions-(δ, ε) with at most N_0 elements left over, where $N_0 < \varepsilon' N$. The upper bound for the discrepancy of almost-arithmetic progressions was established by Niederreiter [2]. As an application, it is shown that the sequences $\omega = (f(n))$ appearing in Chapter 1, Exercise 2.22, satisfy $D_N(\omega) = O(f(N)/N + 1/Nf'(N))$. See also Drewes [1].

The classical treatises on continued fractions are Perron [1] and Khintchine [7]. A very interesting geometric approach is carried out in the book of Stark [1]. Concerning the theory of diophantine approximations, we mention above all the extensive report of Koksma [4] and the monograph of Cassels [9]. Very readable accounts of the subject are given in the books of Niven [1, 3], as well as in LeVeque [5], Rademacher [1], Hardy and Wright [1], and Lang [1]. Because of the intimate connection, most of the books on diophantine approximation will also cover continued fractions to some extent.

Results of the type of Lemma 3.3 were already known to Hardy and Littlewood [4, 5]. They also gave lower bounds for sums of the form $\Sigma_{h=1}^{m} \langle h\alpha \rangle^{-n}$ with $n \geq 1$. Detailed proofs for these lower bounds can be found in Haber and Osgood [2]. Using continued fractions in the same way as in the proof of Theorem 3.4, the estimate for s_h and with it Lemma 3.3 can be slightly improved (see Lang [1] and Exercises 3.11 and 3.12), but this does not yield any improvement on Theorem 3.2. A survey of the literature on this subject prior to 1936 is given in Koksma [4, Kap. 9]. See also Hardy and Littlewood [6], Muromskiĭ [1], and Kruse [1].

Theorem 3.2 was first shown by Hecke [1] and Ostrowski [1]. Related investigations were carried out almost simultaneously by Hardy and Littlewood [3, 4] and by Behnke [1]. A very detailed analysis was undertaken by Behnke [2], who also showed the Ω-result in Theorem 3.3.

Theorem 3.4 is from Niederreiter [13] and improves earlier results of Ostrowski [1], Behnke [2], and Zaremba [1]. In the last paper, the case $K = 1$ is investigated in more detail, and the inequality $ND_N(\omega) \leq \frac{7}{6} \log (6N)$ is established in this special case. Sós (unpublished) has shown for the case $K = 1$ that $ND_N(\omega) \leq c_N \log N$ for $N \geq 2$, where $\lim_{N \to \infty} c_N = 1$. See also Gillet [1, 2]. In this context, we note that Behnke [2] already showed $ND_N(\omega) = \Omega(\log N)$ for any sequence $\omega = (n\alpha)$, thus establishing a special case of Theorem 2.3. Improving theorems of Hecke [1] and Ostrowski [2], Kesten [4] shows the following remarkable result: Let $R_N(t)$ be the remainder function with respect to the sequence $(n\alpha)$, α irrational; then $R_N(b) - R_N(a)$ with $0 \leq a \leq b \leq 1$ and $b - a < 1$ is bounded in N if and only if $b - a = \{j\alpha\}$ for some integer j. For simpler proofs, see Furstenberg, Keynes, and Shapiro [1], K. Petersen [1], Petersen and Shapiro [1], and L. Shapiro [1]. See also Lesca [5]. A similar problem in two dimensions was investigated by Szüsz [2, 3]. For probabilistic quantitative results on $(n\alpha)$, see Kesten [1, 2, 3]. It follows from (3.18) and metric theorems of Khintchine [2] that for every positive nondecreasing function g such that $\Sigma_{n=1}^{\infty} (g(n))^{-1}$ converges, the discrepancy $D_N(\omega)$ of $\omega = (n\alpha)$ satisfies

$$ND_N(\omega) = O((\log N)g(\log \log N))$$

for almost all α.

In Example 3.1 we mentioned the celebrated theorem of Thue-Siegel-Roth. The result was a long-standing conjecture of Siegel and was verified in the fundamental paper of Roth [2]. In a slightly simplified form, Roth's proof has since appeared in several books, for example, Cassels [9], LeVeque [5], and in the revised edition of Landau [1]. A multi-dimensional analogue was recently established by W. M. Schmidt [12].

Various subsequences of $(n\alpha)$ have received great attention, notably lacunary sequences. Most of the papers on sequences with gap conditions in which a quantitative viewpoint is pursued are of a probabilistic nature. See Khintchine [3, 5], M. Kac [1, 2] (and Leonov

[1, 2] for a generalization), Fortet [1], Erdös and Koksma [1], Cassels [1], Erdös and Gál [1], Mineev [1], Ciesielskí and Kesten [1], I.S. and L. Gál [1], Kátai [1], Postnikov [8, Section 15], Ibragimov [1, 2], Muhutdinov [2, 3], Philipp [4, 5, 6, 7; 8, Chapter 4; 10], and R. C. Baker [3]. A general theorem of Gál and Koksma [1, 2] is often used in these investigations. Surveys and further literature are given in Koksma [4, Kap. 9], M. Kac [3], and Gaposhkin [1]. A metric theorem on a wider class of subsequences of $(n\alpha)$ was shown by R. C. Baker [4], thereby improving a result of Salem [2]. For slowly growing sequences, see R. C. Baker [6]. Deterministic results on the sequences $(q^n\alpha)$ with an integer $q > 1$ were given by Koksma [8], Korobov [7, 11, 13, 21], Korobov and Postnikov [1], Kulikova [2], Postnikov [5, 6; 8, Section 14], Postnikova [5], and Usol'cev [1, 3]. For a related result, see Postnikov [2]. Korobov [13] and Polosuev [1, 6] studied analogous problems in several dimensions. Quantitative results on the sequence $(n^2\alpha)$ and related sequences can be found in Behnke [2], Bergström [2], and Jagerman [1, 2]. For $(p_n\alpha)$, where (p_n) is the sequence of primes, see Vinogradov [4, 5] and Hua [1]. Concerning "irregularities of distribution" of subsequences of $(n\alpha)$, see Cohen [1], Davenport [2], and Hewitt and Zuckerman [1].

The distribution of small fractional parts of $n\alpha$ was studied by LeVeque [6, 7, 8], Erdös [5], and Ennola [1]. See also W. M. Schmidt [1, 5], Szüsz [4], Philipp [4, 7], Gallagher [1, 2], and Ennola [2]. The related notion of *suites eutaxiques* has been studied by Lesca [2] and de Mathan [3, 4, 5].

For $((n\alpha_1, \ldots, n\alpha_k))$ and related multidimensional sequences, see Hartman [1], Hlawka [16, 28], Karimov [2], Verbickiĭ [1], and Zinterhof [1]. Niederreiter [5] has shown that if $\alpha_1, \ldots, \alpha_k$ are real algebraic numbers with $1, \alpha_1, \ldots, \alpha_k$ linearly independent over the rationals, then the discrepancy of $\omega = ((n\alpha_1, \ldots, n\alpha_k))$ satisfies $D_N(\omega) = O(N^{-1+\varepsilon})$ for every $\varepsilon > 0$. See also Exercise 3.17 and Niederreiter [13].

Many authors considered sequences $(f(n))$ with a polynomial f and estimated the discrepancy or exponential sums in terms of these sequences. A survey of the early literature on this subject is given in Koksma [4, Kap. 9]. Van der Corput and Pisot [1] used the Vinogradov-van der Corput method. Their results were superseded by the refined method of Vinogradov, an exposition of which may be found in the monographs of Vinogradov [5], Hua [1, 2], and Walfisz [1]. For low-degree polynomials, further improvements were obtained by Rodosskiĭ [1]. See also Vinogradov [2, 4, 6, 7, 8] for related results. Karacuba [1] treats the case where f grows somewhat faster than a polynomial. Kovalevskaja [1] considers multidimensional polynomial sequences.

The sequence in Theorem 3.5 was introduced by van der Corput [7]. Haber [1] improved (3.20) to $ND_N^*(\omega) \leq (\log N)/(3 \log 2) + O(1)$, and showed that the constant $1/(3 \log 2)$ is best possible. The strongest result is that of Tijdeman (unpublished), who proved $ND_N^*(\omega) \leq (\log N)/(3 \log 2) + 1$ and $\overline{\lim}_{N\to\infty} (ND_N^*(\omega) - (\log N)/3 \log 2) \geq \frac{4}{9} + (\log 3)/3 \log 2$. For an application of the van der Corput sequence, see Knuth [2, Section 3.5]. A two-dimensional version of the van der Corput sequence was constructed by Roth [1]. Improving results of Gabai [1, 2], the discrepancy of Roth's sequence was computed by Halton and Zaremba [1], who also proposed a modified version of Roth's sequence with a smaller discrepancy. See also White [2]. There are generalizations to arbitrary dimensions. For integers $m \geq 2$ and $n \geq 0$, let $\phi_m(n)$ be the m-adic fraction obtained by "reflecting" the m-adic representation of n in the "decimal point" (ϕ_m is called a *radical-inverse function*). Then the *van der Corput-Halton sequence* in I^k, $k \geq 1$, is defined by $((\phi_{m_1}(n), \phi_{m_2}(n), \ldots, \phi_{m_k}(n)))$, $n = 0, 1, \ldots,$ where $m_1, \ldots, m_k$ are pairwise relatively prime (Halton [1]). For $k = 1$ and $m_1 = 2$, we just get the van der Corput sequence. The *Hammersley sequence of order N* in I^k, $k \geq 2$, is defined by $((n/N, \phi_{p_1}(n), \ldots, \phi_{p_{k-1}}(n)))$, $n = 0, 1, \ldots, N - 1$, where $p_1, \ldots, p_{k-1}$ are the first $k - 1$ primes (Hammersley [1], Halton [1]). The following discrepancy estimates hold (Halton [1]): $ND_N^*(\omega) \leq C_k \log^k N$ for all $N \geq 2$ for the van der Corput-Halton sequence, and $ND_N^* \leq C_k' \log^{k-1} N$ for the Hammersley sequence of order $N \geq 2$, where C_k and C_k' are certain numerical constants independent of N (see also Meijer [3]). It is a widely held belief that no infinite (resp. finite) sequence can have a discrepancy of smaller order of magnitude than the van der Corput-Halton (resp. Hammersley) sequence. Compare also with Section 2. A survey of these sequences may also be found in Halton [3]. A van der Corput-Halton type sequence in the infinite-dimensional unit cube was studied by Sobol' [4, 7]. A very detailed study of sequences in I^k based on dyadic rationals was carried out by Sobol' [6, 7], and Sobol' [1, 2, 3, 7] also investigated the van der Corput-Halton and Hammersley sequences with respect to numerical integration (see also Section 5). For sequences that are of great importance for numerical analysis, namely, so-called pseudorandom numbers generated by congruential methods, discrepancy estimates were established by Jagerman [3] and Niederreiter [4, 18].

For further quantitative results on special classes of sequences, we refer to Cassels [2, 3, 4], Drewes [1], Erdös [2], Erdös and Koksma [2], Erdös and Turán [1, 2, 3, 4], Korobov [14, 18], LeVeque [1, 2, 3], Mineev [2], Sanders [1], and Usol'cev [2].

Exercises

3.1. Prove that the discrepancy of the sequence $\omega = (an^\sigma)$, $a > 0$, $0 < \sigma < 1$, satisfies $D_N(\omega) = O(N^{\tau-1})$ with $\tau = \max(\sigma, 1 - \sigma)$.

3.2. Give an example of a sequence $\omega = (an^{1/2})$, $a > 0$, for which $D_N(\omega) = \Omega(N^{-1/2})$.

3.3. Prove that the discrepancy of the sequence $\omega = (a \log^\sigma n)$, $a > 0$, $\sigma > 1$, satisfies $D_N(\omega) = O(\log^{1-\sigma} N)$.

3.4. Let ψ be a positive function such that $\sum_{q=1}^{\infty} \psi(q)$ converges. Then for almost all numbers α (in the sense of Lebesgue measure) there are only finitely many integers $q > 0$ and p such that $|q\alpha - p| < \psi(q)$. *Hint:* It suffices to consider $\alpha \in \bar{I}$; choose $\varepsilon > 0$ and Q with $\sum_{q=Q}^{\infty} \psi(q) < \varepsilon$; estimate the measure of the set $M = \{\alpha \in \bar{I}$: there are $q \geq Q$ and p with $|\alpha - (p/q)| < \psi(q)/q\}$.

3.5. Let $\varepsilon > 0$ be given. Prove that almost all α are of type $<c(\alpha) \log^{1+\varepsilon} 2q$, where $c(\alpha)$ is a positive constant that may depend on α.

3.6. Find the continued fraction expansion of $\sqrt{2}$, $\sqrt{3}$, and $(1 + \sqrt{2})/3$.

3.7. Prove that $p_n q_{n-1} - p_{n-1} q_n = (-1)^{n-1}$ for $n \geq -1$ and $p_n q_{n-2} - p_{n-2} q_n = (-1)^n a_n$ for $n \geq 0$.

3.8. Let α be of type $<\psi$, and let $q_{n+1} > 1$ be the denominator of a convergent to α. Prove that $\langle h\alpha \rangle \geq 1/q_n \psi(q_n)$ for $1 \leq h < q_{n+1}$.

3.9. Let α be of type $<\psi$, and let $q_n < q_{n+1}$ be the denominators of two consecutive convergents to α. Using ideas from the proof of Theorem 3.4 and the result of Exercise 3.8, prove that

$$\sum_{\substack{j=1 \\ h_0 + j < q_{n+1}}}^{q_n} \frac{1}{\langle (h_0 + j)\alpha \rangle} \leq c q_n (\psi(q_n) + \log q_n),$$

where $q_n - 1 \leq h_0 < q_{n+1}$ and $c > 0$ is an absolute constant.

3.10. Deduce from Exercise 3.9 that for α of type $<\psi$ we have

$$\sum_{h=q_n}^{q_{n+1}-1} \frac{1}{\langle h\alpha\rangle} \leq c q_{n+1}(\psi(q_n) + \log q_n) \qquad \text{for all } n \geq 0.$$

3.11. Let α be of type $<\psi$. Use Exercise 3.10 to show that

$$\sum_{h=1}^{m} \frac{1}{\langle h\alpha\rangle} = \mathrm{O}(m \log m + m\psi(m)).$$

3.12. Show the following improvement of Lemma 3.3: If α is of type $<\psi$, then

$$\sum_{h=1}^{m} \frac{1}{h\langle h\alpha\rangle} = \mathrm{O}\left(\log^2 m + \psi(m) + \sum_{h=1}^{m} \frac{\psi(h)}{h}\right).$$

Hint: Use Abel's summation formula and Exercise 3.11.

3.13. Prove that for given $\varepsilon > 0$, the discrepancy $D_N(\omega)$ of $\omega = (n\alpha)$ satisfies $ND_N(\omega) = \mathrm{O}(\log^{2+\varepsilon} N)$ for almost all α. *Hint:* Use Exercise 3.5.

3.14. Suppose that for all irrationals α of type $<\psi$, we have $\sum_{h=1}^{m} 1/\{h\alpha\} = \mathrm{O}(f(m))$. Then prove that also $\sum_{h=1}^{m} 1/\langle h\alpha\rangle = \mathrm{O}(f(m))$ for all such α. (Note: The converse is trivial.)

3.15. For a fixed positive integer m, define $g(n) = 1/n(n+1)$ for $1 \leq n < m$ and $g(m) = 1/m$. For a lattice point $(n_1, \ldots, n_k) \in \mathbb{Z}^k$ with $1 \leq n_i \leq m$ for $1 \leq i \leq k$, define $f(n_1, \ldots, n_k) = \prod_{i=1}^{k} g(n_i)$. Suppose s is an arbitrary function defined at least for the lattice points $\mathbf{h} \in \mathbb{Z}^k$ with $0 < \|\mathbf{h}\| \leq m$. Then the identity

$$\sum_{0<\|\mathbf{h}\|\leq m} r^{-1}(\mathbf{h})s(\mathbf{h}) = \sum_{n_1,\ldots,n_k=1}^{m} f(n_1, \ldots, n_k) \sum_{\substack{\mathbf{h}=(h_1,\ldots,h_k)\\ |h_j|\leq n_j}}^{*} s(\mathbf{h})$$

holds, where the asterisk signalizes deletion of the origin from the range of summation. For the definition of $\|\mathbf{h}\|$ and $r(\mathbf{h})$, see the notes in Section 2.

3.16. Let $\alpha_1, \ldots, \alpha_k$ be irrationals such that $1, \alpha_1, \ldots, \alpha_k$ are linearly independent over the rationals. Suppose there exists $\eta \geq 1$ and $c > 0$ such that $r^\eta((h_1, \ldots, h_k)) \langle h_1\alpha_1 + \cdots + h_k\alpha_k\rangle \geq c$ for all lattice points $(h_1, \ldots, h_k) \neq (0, \ldots, 0)$. Verify that

$$\sum_{\substack{\mathbf{h}=(h_1,\ldots,h_k)\\ |h_j|\leq n_j}}^{*} \langle h_1\alpha_1 + \cdots + h_k\alpha_k\rangle^{-1} = \mathrm{O}(r^\eta(\mathbf{n}) \log r(\mathbf{n})),$$

where $\mathbf{n} = (n_1, \ldots, n_k)$ is a lattice point with $n_i \geq 1$ for $1 \leq i \leq k$. *Hint:* Compare with the proof of Lemma 3.3.

3.17. Let $\alpha_1, \ldots, \alpha_k$ satisfy the same conditions as in Exercise 3.16. Show that the discrepancy $D_N(\omega)$ of the sequence $\omega = ((n\alpha_1, \ldots, n\alpha_k))$, $n = 1, 2, \ldots,$ satisfies $D_N(\omega) = O(N^{-1}\log^{k+1} N)$ for $\eta = 1$ and $D_N(\omega) = O(N^{-1/((\eta-1)k+1)} \log N)$ for $\eta > 1$. *Hint:* Use the theorem of Erdös-Turán-Koksma and the results of Exercises 3.15 and 3.16.

3.18. Let (x_n) be the van der Corput sequence. For an integer $h \geq 1$, put $N = \sum_{i=0}^{h-1} 2^{2i}$ and $\alpha = \sum_{i=0}^{h-1} 2^{-2i-1}$. For $1 \leq j \leq h$, define y_j and the finite sequence ω_j as in the proof of Theorem 3.5. Now show that $A([0, \alpha); 2^{2(h-j)}; \omega_j) = (\alpha - y_j)2^{2(h-j)} + \frac{1}{2}$ for $1 \leq j \leq h$. *Hint:* Use that ω_j is equidistant with difference $2^{-2(h-j)}$.

3.19. Deduce from Exercise 3.18 that for the discrepancy $D_N^*(\omega)$ of the van der Corput sequence ω we have $ND_N^*(\omega) > (\log N)/6 \log 2$ for infinitely many N.

3.20. Prove that the discrepancy $D_N(\omega)$ of the sequence $\omega = (n \log n)$ satisfies $D_N(\omega) = O(N^{-1/5} \log^{2/5} N)$. *Hint:* Use Theorem 2.7 in Chapter 1.

3.21. Prove that the discrepancy $D_N(\omega)$ of the sequence $\omega = (n \log \log en)$ satisfies $D_N(\omega) = O(N^{-1/5}(\log^{1/5} N)(\log \log N)^{2/5})$. *Hint:* Use Theorem 2.7 in Chapter 1.

4. REARRANGEMENT OF SEQUENCES

Dense Sequences and Uniform Distribution

In this section, we shed some new light on the property of u.d. mod 1 by showing that it depends more on the order in which the terms of the sequence are given than on the specific nature of the terms themselves. The central result will be that an everywhere dense sequence in the unit interval can fit any prescribed distribution behavior if one just rearranges the terms in a suitable manner. In particular, any everywhere dense sequence in the unit interval can be rearranged so as to yield a u.d. sequence mod 1.

Before we set out to prove these results, we study the relationship between the discrepancies of sequences for which corresponding terms are close.

THEOREM 4.1. Let $x_1, x_2, \ldots, x_N$ and $y_1, y_2, \ldots, y_N$ be two finite sequences in I. Suppose $\varepsilon_1, \varepsilon_2, \ldots, \varepsilon_N$ are nonnegative numbers such that $|x_n - y_n| \leq \varepsilon_n$ for $1 \leq n \leq N$. Then, for any $\varepsilon \geq 0$, we have

$$|D_N(x_1, \ldots, x_N) - D_N(y_1, \ldots, y_N)| \leq 2\varepsilon + \frac{\bar{N}(\varepsilon)}{N}, \tag{4.1}$$

where $\bar{N}(\varepsilon)$ denotes the number of n, $1 \leq n \leq N$, such that $\varepsilon_n > \varepsilon$.

PROOF. Let ω and τ stand for the sequences $x_1, \ldots, x_N$ and $y_1, \ldots, y_N$, respectively. Let $J = [\alpha, \beta)$ be a subinterval of I. Consider first the interval $J_1 = [\alpha - \varepsilon, \beta + \varepsilon) \cap I$. Whenever $y_n \in J$, then either $x_n \in J_1$ or $\varepsilon_n > \varepsilon$. Therefore, $A(J; N; \tau) \leq A(J_1; N; \omega) + \bar{N}(\varepsilon)$. Furthermore, we have

$$A(J_1; N; \omega) = N\lambda(J_1) + \delta_1 N D_N(\omega)$$

with $|\delta_1| \leq 1$. It follows that

$$\begin{aligned} A(J; N; \tau) - N\lambda(J) &\leq N(\lambda(J_1) - \lambda(J)) + \delta_1 N D_N(\omega) + \bar{N}(\varepsilon) \\ &\leq 2\varepsilon N + N D_N(\omega) + \bar{N}(\varepsilon). \end{aligned} \tag{4.2}$$

On the other hand, consider the interval $J_2 = [\alpha + \varepsilon, \beta - \varepsilon)$. This interval might be empty, in which case the subsequent estimates will hold trivially. Whenever $x_n \in J_2$, then either $y_n \in J$ or $\varepsilon_n > \varepsilon$. Thus, $A(J_2; N; \omega) \leq A(J; N; \tau) + \bar{N}(\varepsilon)$. We have $A(J_2; N; \omega) = N\lambda(J_2) + \delta_2 N D_N(\omega)$ with $|\delta_2| \leq 1$. It follows that

$$\begin{aligned} A(J; N; \tau) - N\lambda(J) &\geq N(\lambda(J_2) - \lambda(J)) + \delta_2 N D_N(\omega) - \bar{N}(\varepsilon) \\ &\geq -2\varepsilon N - N D_N(\omega) - \bar{N}(\varepsilon). \end{aligned} \tag{4.3}$$

Combining (4.2) and (4.3), we conclude that

$$\left| \frac{A(J; N; \tau)}{N} - \lambda(J) \right| \leq D_N(\omega) + 2\varepsilon + \frac{\bar{N}(\varepsilon)}{N}.$$

This upper bound is independent of the chosen interval J, and so,

$$D_N(\tau) \leq D_N(\omega) + 2\varepsilon + \frac{\bar{N}(\varepsilon)}{N}.$$

Interchanging the roles of ω and τ, we arrive at

$$D_N(\omega) \leq D_N(\tau) + 2\varepsilon + \frac{\bar{N}(\varepsilon)}{N},$$

and the proof is complete. ■

COROLLARY 4.1. Let $x_1, \ldots, x_N$ and $y_1, \ldots, y_N$ be two finite sequences in I, and suppose there is an $\varepsilon \geq 0$ with $|x_n - y_n| \leq \varepsilon$ for $1 \leq n \leq N$. Then,

$$|D_N(x_1, \ldots, x_N) - D_N(y_1, \ldots, y_N)| \leq 2\varepsilon. \tag{4.4}$$

PROOF. Choose $\varepsilon_n = \varepsilon$ for $1 \leq n \leq N$ in Theorem 4.1. ■

THEOREM 4.2. Let $\omega = (x_n)$ be an infinite sequence in I. Furthermore, we are given a sequence (y_n) that is everywhere dense in I. Let $h(x)$ be an increasing nonnegative function defined for $x \geq 1$, and satisfying

$$\lim_{x \to \infty} h(x) = \infty.$$

Then one can rearrange the sequence (y_n) so as to yield a sequence σ satisfying

$$|D_N(\omega) - D_N(\sigma)| \leq \frac{h(N)}{N} \tag{4.5}$$

for all sufficiently large N.

PROOF. Without loss of generality, we may assume $h(1) = 0$, for otherwise, we replace $h(x)$ by $h(x) - h(1)$. Let $g(t)$ be the inverse function of $\frac{1}{4}h(x)$. We note that $g(0) = 1$, and so, $g(n) > 0$ for every positive integer n. Since the sequence (y_n) is everywhere dense, we can find for every x_n an element y_{k_n} such that $|x_n - y_{k_n}| \leq 1/g(n)$, and also $k_n \neq k_m$ whenever $n \neq m$. Let ε be a number with $0 < \varepsilon \leq 1$ that will be specified later on. By Theorem 4.1, the sequence $\tau = (y_{k_n})$ satisfies

$$|D_N(\omega) - D_N(\tau)| \leq 2\varepsilon + \frac{\bar{N}(\varepsilon)}{N},$$

where $\bar{N}(\varepsilon)$ is the number of elements $1/g(n)$, $1 \leq n \leq N$, such that $1/g(n) > \varepsilon$. Now $1/g(n) > \varepsilon$ is equivalent to $g(n) < 1/\varepsilon$, or $n < \frac{1}{4}h(1/\varepsilon)$, and thus $\bar{N}(\varepsilon) \leq [\frac{1}{4}h(1/\varepsilon)]$. We take $\varepsilon = 1/N$, and obtain

$$|D_N(\omega) - D_N(\tau)| \leq \frac{2}{N} + \frac{h(N)}{4N}. \tag{4.6}$$

We now define a rearrangement $\sigma = (u_n)$ of (y_n) in the following way. We put $u_n = y_{k_n}$ if n is not of the form $[g(m)] + 1$ for some positive integer m. The remaining elements of (y_n) are those of the form y_{k_n} with $n = [g(m)] + 1$ and those that are no y_{k_n} at all. We enumerate those elements in an arbitrary fashion, say $t_1, t_2, \ldots$. We still have to define u_n for the n of the form $[g(m)] + 1$. We order the distinct n of this form according to their magnitude: $n_1 < n_2 < n_3 < \cdots$. We define then $u_{n_i} = t_i$. The sequence $\sigma = (u_n)$ is then a rearrangement of (y_n).

We will estimate $|D_N(\tau) - D_N(\sigma)|$. Let c_J denote the characteristic function of a subinterval J of I. We have

$$\left|\left(\frac{A(J;N;\sigma)}{N} - \lambda(J)\right) - \left(\frac{A(J;N;\tau)}{N} - \lambda(J)\right)\right| = \left|\frac{1}{N}\sum_{n=1}^{N}(c_J(u_n) - c_J(y_{k_n}))\right|.$$

But $c_J(u_n) - c_J(y_{k_n}) = 0$ whenever n is not of the form $[g(m)] + 1$. Therefore,

$$\left|\frac{1}{N}\sum_{n=1}^{N}(c_J(u_n) - c_J(y_{k_n}))\right| \leq \frac{1}{N}\sum_{\substack{n=1 \\ n=[g(m)]+1}}^{N} 1 \leq \frac{1}{N}\sum_{\substack{m \\ [g(m)]+1\leq N}} 1 \leq \frac{1}{N}\sum_{\substack{m \\ g(m)\leq N}} 1$$

$$= \frac{1}{N}\sum_{\substack{m \\ m\leq(1/4)h(N)}} 1 = \frac{1}{N}\,[\tfrac{1}{4}h(N)] \leq \frac{h(N)}{4N}.$$

This implies

$$\left|\frac{A(J;N;\sigma)}{N} - \lambda(J)\right| \leq \left|\frac{A(J;N;\tau)}{N} - \lambda(J)\right| + \frac{h(N)}{4N},$$

and so, $D_N(\sigma) \leq D_N(\tau) + h(N)/4N$. Similarly, one shows $D_N(\tau) \leq D_N(\sigma) + h(N)/4N$, and consequently,

$$|D_N(\tau) - D_N(\sigma)| \leq \frac{h(N)}{4N}. \tag{4.7}$$

Thus, we get from (4.6) and (4.7) that

$$|D_N(\omega) - D_N(\sigma)| \leq |D_N(\omega) - D_N(\tau)| + |D_N(\tau) - D_N(\sigma)|$$

$$\leq \frac{2}{N} + \frac{h(N)}{2N},$$

and this for all positive integers N. But for sufficiently large N we have $h(N) \geq 4$, and therefore,

$$|D_N(\omega) - D_N(\sigma)| \leq \frac{h(N)}{2N} + \frac{h(N)}{2N} = \frac{h(N)}{N}. \quad \blacksquare$$

COROLLARY 4.2. Any sequence that is everywhere dense in I can be rearranged to a u.d. sequence mod 1.

PROOF. For the sequence $\omega = (x_n)$, we choose a u.d. sequence mod 1; and for $h(x)$ we choose a function that tends sufficiently slowly to infinity, say $h(x) = \log x$. By the previous theorem, there is a rearrangement σ of the given everywhere dense sequence (y_n) with $D_N(\sigma) \leq D_N(\omega) + (\log N)/N$ for sufficiently large N. It follows that $\lim_{N\to\infty} D_N(\sigma) = 0$, and so, σ is u.d. mod 1. $\blacksquare$

Farey Points

For the subsequent example, and also for later use, the following lemma will be helpful.

LEMMA 4.1. Let (c_n) be a bounded sequence of complex numbers. Divide the sequence into nonempty blocks, the first block consisting of the first A_1 elements of (c_n), the second block consisting of the next A_2 elements, and so on. Label the A_k elements in the kth block $d_1^{(k)}, \ldots, d_{A_k}^{(k)}$. Suppose that $\lim_{k\to\infty} A_{k+1}/(A_1 + \cdots + A_k) = 0$ and $\lim_{k\to\infty} (1/A_k)\sum_{\nu=1}^{A_k} d_\nu^{(k)} = c$. Then,

$$\lim_{N\to\infty} \frac{1}{N} \sum_{n=1}^{N} c_n = c.$$

PROOF. For $k \geq 1$, put $B_k = A_1 + \cdots + A_k$. Then,

$$\frac{1}{B_k}\sum_{n=1}^{B_k} c_n = \frac{1}{B_k}\left(A_1 \cdot \frac{1}{A_1}\sum_{\nu=1}^{A_1} d_\nu^{(1)} + \cdots + A_k \cdot \frac{1}{A_k}\sum_{\nu=1}^{A_k} d_\nu^{(k)}\right),$$

and so, by Cauchy's theorem, we have

$$\lim_{k\to\infty} \frac{1}{B_k}\sum_{n=1}^{B_k} c_n = c. \tag{4.8}$$

Now let $N \geq A_1$ be given. Then N can be written in the form $N = B_k + r$ with $0 \leq r < A_{k+1}$. We get

$$\frac{1}{N}\sum_{n=1}^{N} c_n = \frac{B_k}{N}\left(\frac{1}{B_k}\sum_{n=1}^{B_k} c_n\right) + \frac{1}{N}\sum_{n=B_k+1}^{N} c_n.$$

The first term on the right tends to c because of $1 \leq N/B_k < 1 + (A_{k+1}/B_k)$ and (4.8). The second term is dominated in absolute value by $(A_{k+1}/B_k)M$, where M is an upper bound for $|c_n|$, and so tends to zero. ∎

EXAMPLE 4.1. Here is an arrangement of the rationals in I into a u.d. sequence mod 1. We choose the lexicographic ordering $\frac{0}{1}, \frac{1}{2}, \frac{1}{3}, \frac{2}{3}, \frac{1}{4}, \frac{3}{4}, \ldots$, where we write down, for successive values of $n = 1, 2, \ldots$, all rationals r/n with $0 \leq r < n$ and $(r, n) = 1$. For technical reasons, we consider a sequence that is identical mod 1 with the above sequence, namely $\frac{1}{1}, \frac{1}{2}, \frac{1}{3}, \frac{2}{3}, \frac{1}{4}, \frac{3}{4}, \ldots$. This sequence consists of blocks S_n, $n = 1, 2, \ldots$, of all rationals r/n with $1 \leq r \leq n$ and $(r, n) = 1$. For $0 \leq \alpha \leq 1$, let $A_\alpha(n)$ be the number of terms from S_n in $[0, \alpha]$. The number of fractions (reduced or not) of the form r/n, $1 \leq r \leq n$, in $[0, \alpha]$ is exactly $[n\alpha]$. Now we look at this number from a different angle. If we reduce those fractions r/n, group them according to the new denominators (which have to be positive divisors d of n), and count them anew, then we arrive at the basic identity $\sum_{d|n} A_\alpha(d) = [n\alpha]$. By the Moebius inversion formula, we obtain $A_\alpha(n) = \sum_{d|n} \mu(n/d)[d\alpha]$. Therefore, $A_\alpha(n) = \alpha\sum_{d|n} \mu(n/d)d - \sum_{d|n} \mu(n/d)\{d\alpha\} = \alpha\phi(n) - \sum_{d|n} \mu(n/d)\{d\alpha\}$, and so

$$\left|\frac{A_\alpha(n)}{\phi(n)} - \alpha\right| \leq \frac{1}{\phi(n)}\sum_{d|n} |\mu(d)|. \tag{4.9}$$

We claim that $\lim_{n\to\infty} f(n) = 0$, where $f(n) = (1/\phi(n))\sum_{d|n} |\mu(d)|$. We note that $f(n)$ is a multiplicative number-theoretic function, and for such functions $\lim_{n\to\infty} f(n) = 0$ is equivalent to $f(n)$ tending to 0 as n runs through the prime powers.

Thus, we need only consider $f(n)$ for prime powers $n = p^a$. But then $f(p^a) = 2/(p^{a-1}(p-1))$, which clearly tends to 0 if $p^a \to \infty$. From (4.9) we infer that $\lim_{n\to\infty} A_\alpha(n)/\phi(n) = \alpha$, and we now invoke Lemma 4.1. It remains then to show that $\lim_{n\to\infty} \phi(n+1)/\Phi(n) = 0$, where $\Phi(n) = \phi(1) + \cdots + \phi(n)$ for $n \geq 1$.

We work with an elementary estimate for $\Phi(n)$. Much sharper results are known in analytic number theory. We observe that $\Phi(n)$ can be interpreted as the number of lattice points (x, y) with $1 \leq x \leq y \leq n$ and x and y relatively prime. Then $B(n) = 2\Phi(n) - 1$ is the total number of lattice points (x, y) with $1 \leq x \leq n$, $1 \leq y \leq n$, and x, y relatively prime. But

$$B(n) = n^2 - \sum_{d=2}^{n} \sum_{\substack{1\leq x\leq n\\ 1\leq y\leq n\\ (x,y)=d}} 1 \geq n^2 - \sum_{d=2}^{n}\left(\frac{n}{d}\right)^2 \geq n^2 - n^2\sum_{d=2}^{\infty}\frac{1}{d^2} = n^2\left(2 - \frac{\pi^2}{6}\right).$$

Consequently, $\Phi(n) \geq cn^2$ for some positive constant c, and so $\phi(n+1)/\Phi(n) \leq (n+1)/cn^2$ settles the proof. ■

Rearrangements and Distribution Functions

We shall now prove analogues of Corollary 4.2 with respect to a.d.f.'s (mod 1) in the sense of Definition 7.1 in Chapter 1. We will use results of the present chapter to show that for every nondecreasing function f on $\bar{I}$ with $f(0) = 0$ and $f(1) = 1$ there really exists a sequence having this function f as its a.d.f. (mod 1), a problem that was left open in Chapter 1, Section 7. We start with establishing an even stronger result for continuous f.

LEMMA 4.2. Let f be a continuous, nondecreasing function on $\bar{I}$ with $f(0) = 0$ and $f(1) = 1$. Then there exists a sequence ω in I such that

$$|A([0, \alpha); N; \omega) - Nf(\alpha)| \leq \frac{\log (N+1)}{\log 2} \tag{4.10}$$

holds for all $N \geq 1$ and $0 \leq \alpha \leq 1$.

PROOF. Let $\tau = (y_n)$ be the van der Corput sequence constructed in Section 3. By the continuity and monotonicity of f, the set $I_n = \{\beta \in I: f(\beta) \leq y_n\}$ is a closed subinterval of I (or a singleton) for each $n \geq 1$. Let x_n be the largest element of I_n. We claim that $\omega = (x_n)$ satisfies (4.10).

Let us first show that for any $n \geq 1$ and $0 \leq \alpha \leq 1$, the inequalities

$x_n \geq \alpha$ and $y_n \geq f(\alpha)$ imply each other. If $x_n \geq \alpha$, then $x_n \in I_n$ implies $y_n \geq f(x_n) \geq f(\alpha)$; conversely, if $y_n \geq f(\alpha)$, then $\alpha \in I_n$, and so, $\alpha \leq x_n$.

It follows that $x_n < \alpha$ precisely if $y_n < f(\alpha)$, and therefore,

$$A([0, \alpha); N; \omega) = A([0, f(\alpha)); N; \tau)$$

for all $N \geq 1$ and $0 \leq \alpha \leq 1$. An application of Theorem 3.5 completes the proof. ■

To achieve the transition from the continuous case to the general case, we need a lemma from real analysis.

LEMMA 4.3. Let f be a nondecreasing function on the compact interval $[a, b]$. Then there exists a sequence $g_1, g_2, \ldots, g_k, \ldots$ of continuous nondecreasing functions on $[a, b]$, satisfying $g_k(a) = f(a)$ and $g_k(b) = f(b)$ for $k \geq 1$, which converges pointwise to f, that is, $\lim_{k\to\infty} g_k(\beta) = f(\beta)$ for all $\beta \in [a, b]$.

PROOF. For each $k \geq 1$, we choose a finite sequence $a = \alpha_0^{(k)} < \alpha_1^{(k)} < \cdots < \alpha_{m_k}^{(k)} = b$, with $\alpha_{i+1}^{(k)} - \alpha_i^{(k)} < 1/k$ for $0 \leq i < m_k$, that contains all points $\alpha \in (a, b)$ with $f(\alpha + 0) - f(\alpha - 0) > 1/k$ (note that there can be only finitely many such α). Let g_k be the function with $g_k(\alpha_i^{(k)}) = f(\alpha_i^{(k)})$ for $0 \leq i \leq m_k$ that is linear on the intervals $[\alpha_i^{(k)}, \alpha_{i+1}^{(k)}]$, $0 \leq i < m_k$. Then g_k is clearly continuous and nondecreasing on $[a, b]$, and $g_k(a) = f(a)$ and $g_k(b) = f(b)$. It remains to verify that the g_k converge pointwise to f (this is, of course, trivial for the end points a and b).

Consider first the case where $\beta \in (a, b)$ is a point of discontinuity of f. Then $f(\beta + 0) - f(\beta - 0) > 0$, and so, from some k on, we will have $\beta = \alpha_{i_k}^{(k)}$ for some i_k, $0 < i_k < m_k$. Thus, $g_k(\beta) = f(\beta)$ for sufficiently large k, and everything is clear.

Now let f be continuous at $\beta \in (a, b)$ and let $\varepsilon > 0$ be given. Then for all sufficiently large k (say $k \geq k_0$) we will have $f(\beta) - \varepsilon < f(\gamma) < f(\beta) + \varepsilon$ for $\gamma \in (\beta - (1/k), \beta + (1/k))$. But for each k we have $\alpha_i^{(k)} \leq \beta \leq \alpha_{i+1}^{(k)}$ for some $i = i(k)$, $0 \leq i < m_k$. Since $\alpha_{i+1}^{(k)} - \alpha_i^{(k)} < 1/k$, both $\alpha_i^{(k)}$ and $\alpha_{i+1}^{(k)}$ lie in $(\beta - (1/k), \beta + (1/k))$. Hence, for $k \geq k_0$ we obtain $f(\alpha_i^{(k)}) > f(\beta) - \varepsilon$ and $f(\alpha_{i+1}^{(k)}) < f(\beta) + \varepsilon$; thus, $g_k(\alpha_i^{(k)}) > f(\beta) - \varepsilon$ and $g_k(\alpha_{i+1}^{(k)}) < f(\beta) + \varepsilon$. Now $g_k(\alpha_i^{(k)}) \leq g_k(\beta) \leq g_k(\alpha_{i+1}^{(k)})$, and so, $f(\beta) - \varepsilon < g_k(\beta) < f(\beta) + \varepsilon$ for all $k \geq k_0$. ■

THEOREM 4.3. Let f be a nondecreasing function on $\bar{I}$ with $f(0) = 0$ and $f(1) = 1$. Then there exists a sequence in I having f as its a.d.f. (mod 1). We can even find such a sequence with all elements distinct.

PROOF. By Lemma 4.3 there exists a sequence $g_1, g_2, \ldots, g_k, \ldots$ of nondecreasing continuous functions on $\bar{I}$, with $g_k(0) = 0$ and $g_k(1) = 1$ for

$k \geq 1$, that converges pointwise to f. For each g_k, we can find a sequence τ_k of points $x_1^{(k)}, x_2^{(k)}, \ldots, x_n^{(k)}, \ldots$ in I satisfying (4.10). We note that g_k is uniformly continuous; therefore, there exists $\varepsilon_k > 0$ such that $g_k(\alpha) - g_k(\alpha - \delta) < 1/k$ holds for all $0 \leq \delta \leq \varepsilon_k$ and all $\alpha \in [\delta, 1]$. Let ω_k denote a sequence in I of the form $y_1^{(k)}, y_2^{(k)}, \ldots, y_n^{(k)}, \ldots$ such that $x_n^{(k)} \leq y_n^{(k)} \leq x_n^{(k)} + \varepsilon_k$ for all $n \geq 1$. Now let ω be the sequence $y_1^{(1)}, y_1^{(2)}, y_2^{(2)}, y_1^{(3)}, y_2^{(3)}, y_3^{(3)}, \ldots, y_1^{(k)}, \ldots, y_k^{(k)}, \ldots$. Thus, ω is constructed by listing successively the first term of ω_1, the first two terms of $\omega_2, \ldots,$ the first k terms of ω_k, and so on. If we choose elements $y_n^{(k)}$, $1 \leq n \leq k$, that are distinct from the preceding elements in ω (this is possible since the $y_n^{(k)}$ come from an interval of positive length), then ω even consists of distinct elements. We shall show that $\lim_{N\to\infty} A([0, \alpha); N; \omega)/N = f(\alpha)$ for $0 \leq \alpha \leq 1$.

By Lemma 4.1, it suffices to prove $\lim_{k\to\infty} A([0, \alpha); k; \omega_k)/k = f(\alpha)$ for $0 \leq \alpha \leq 1$. We have

$$\left| \frac{A([0, \alpha); k; \omega_k)}{k} - f(\alpha) \right| \leq \frac{1}{k} |A([0, \alpha); k; \omega_k) - A([0, \alpha); k; \tau_k)| + \left| \frac{A([0, \alpha); k; \tau_k)}{k} - g_k(\alpha) \right| + |g_k(\alpha) - f(\alpha)|. \quad (4.11)$$

On the right-hand side of (4.11), the second term is at most $(\log(k + 1))/k \log 2$, and so, it tends to 0, as does the third term.

To estimate the first term, we note that $y_n^{(k)} < \alpha$ implies $x_n^{(k)} < \alpha$, and so, $A([0, \alpha); k; \omega_k) \leq A([0, \alpha); k; \tau_k)$. Now $A([0, \alpha); k; \tau_k) - A([0, \alpha); k; \omega_k)$ is equal to the number of subscripts n, $1 \leq n \leq k$, for which $x_n^{(k)} < \alpha$, but $y_n^{(k)} \geq \alpha$. But for such n it follows that $\alpha - \varepsilon_k \leq x_n^{(k)} < \alpha$. Put $[\alpha - \varepsilon_k, \alpha) \cap [0, \alpha) = [\alpha - \delta_k, \alpha)$, where $0 \leq \delta_k \leq \varepsilon_k$. Then,

$$\begin{aligned} &|A([0, \alpha); k; \omega_k) - A([0, \alpha); k; \tau_k)| \\ &\quad \leq A([\alpha - \delta_k, \alpha); k; \tau_k) \\ &\quad = (A([0, \alpha); k; \tau_k) - kg_k(\alpha)) - (A([0, \alpha - \delta_k); k; \tau_k) - kg_k(\alpha - \delta_k)) \\ &\qquad + k(g_k(\alpha) - g_k(\alpha - \delta_k)) \\ &\quad \leq \frac{2 \log(k + 1)}{\log 2} + 1. \end{aligned}$$

It follows that the first term on the right-hand side of (4.11) also tends to 0 as $k \to \infty$. ∎

We are now in the position to prove a generalization of Corollary 4.2 on the rearrangement of everywhere dense sequences in the unit interval.

THEOREM 4.4. Let f be a nondecreasing function on $\bar{I}$ with $f(0) = 0$ and $f(1) = 1$. Then any everywhere dense sequence in I can be rearranged so as to yield a sequence having f as its a.d.f. (mod 1).

PROOF. Let $\omega = (x_n)$ be a sequence of distinct elements in I having f as its a.d.f. (mod 1) (such an ω exists by Theorem 4.3), and let (y_n) be a given everywhere dense sequence in I. We decompose ω into blocks, the kth block consisting of the x_n with $(k-1)k/2 < n \leq k(k+1)/2$. For each k, take the k elements of the kth block and order them according to their magnitude, say $\alpha_1^{(k)} < \alpha_2^{(k)} < \cdots < \alpha_k^{(k)}$. Now choose k numbers $\beta_1^{(k)}, \ldots, \beta_k^{(k)}$ with $\alpha_1^{(k)} < \beta_1^{(k)} < \alpha_2^{(k)} < \beta_2^{(k)} < \cdots < \alpha_{k-1}^{(k)} < \beta_{k-1}^{(k)} < \alpha_k^{(k)} < \beta_k^{(k)} < 1$ and such that each $\beta_i^{(k)}$ is a y_n. Let $\xi = (z_n)$ be the sequence $\beta_1^{(1)}, \beta_1^{(2)}, \beta_2^{(2)}, \ldots, \beta_1^{(k)}, \ldots, \beta_k^{(k)}, \ldots$. Clearly, $z_n = y_{k_n}$ for some k_n. Moreover, the construction of ξ shows that we can suppose $k_n \neq k_m$ whenever $n \neq m$.

We claim that ξ has the a.d.f. (mod 1) f. It suffices to show

$$\lim_{k\to\infty} \frac{2}{k(k+1)} A\left([0, \alpha); \frac{k(k+1)}{2}; \xi\right) = f(\alpha)$$

for $0 \leq \alpha \leq 1$, since this implies $\lim_{N\to\infty} A([0, \alpha); N; \xi)/N = f(\alpha)$. Comparing the ith block of ω and ξ, we see that in $[0, \alpha)$ there are at least as many $\alpha_r^{(i)}$, $1 \leq r \leq i$, as there are $\beta_s^{(i)}$, $1 \leq s \leq i$. On the other hand, the excess of the number of $\alpha_r^{(i)}$ in $[0, \alpha)$ over the number of $\beta_s^{(i)}$ in $[0, \alpha)$ can be at most 1. Therefore, $0 \leq A([0, \alpha); k(k+1)/2; \omega) - A([0, \alpha); k(k+1)/2; \xi) \leq k$, and so,

$$\lim_{k\to\infty} \frac{2}{k(k+1)} A\left([0, \alpha); \frac{k(k+1)}{2}; \xi\right) = \lim_{k\to\infty} \frac{2}{k(k+1)} A\left([0, \alpha); \frac{k(k+1)}{2}; \omega\right) = f(\alpha).$$

Finally, we take the sequence ξ and fill in the remaining y_n at large gaps so that we do not disturb the distribution behavior of ξ (compare with the proof of Theorem 4.2). We enumerate those y_n that are not contained in ξ, together with those z_n for which n is a perfect square, in some order: $u_1, u_2, \ldots$. Then we define a sequence $\tau = (t_n)$ by $t_n = z_n$ if n is not a perfect square, and $t_n = u_m$ if $n = m^2$. Evidently, this sequence τ is a rearrangement of (y_n). Furthermore, if we compare the first N terms of τ with the first N terms of ξ, then we observe that they differ in at most $[\sqrt{N}]$ terms. Hence,

$$|A([0, \alpha); N; \tau) - A([0, \alpha); N; \xi)| \leq \sqrt{N}$$

for $0 \leq \alpha \leq 1$, and so,

$$\lim_{N\to\infty} \frac{A([0, \alpha); N; \tau)}{N} = \lim_{N\to\infty} \frac{A([0, \alpha); N; \xi)}{N} = f(\alpha). \quad \blacksquare$$

Notes

The fact that an everywhere dense sequence can be rearranged to a sequence with small discrepancy was already known to van der Corput [7]. Theorem 4.2 is essentially from the same source. Our proof is taken from Hlawka [22], where Theorem 4.1 can be found as well; see also Niederreiter [5]. An analogous result in a general setting is given in Niederreiter [1]. Corollary 4.1 was first shown by van der Corput and Pisot [1].

The results enunciated in Corollary 4.2 and Theorem 4.4 were first shown by von Neumann [1] and van der Corput [7], respectively. It was van der Corput [7] who established the following strong result along the same lines: Let g and G be two nondecreasing functions in $\bar{I}$ with $g(x) \leq G(x)$ for $0 \leq x \leq 1$, $g(0) = G(0) = 0$, and $g(1) = G(1) = 1$. Then any everywhere dense sequence in I can be rearranged to a sequence such that the set of d.f.'s (mod 1) of the new sequence consists exactly of all nondecreasing functions f in $\bar{I}$ satisfying $g(x) \leq f(x) \leq G(x)$ for $0 \leq x \leq 1$. In a more general context, we shall return to this subject in Section 2 of Chapter 3.

A detailed study of the relation between order properties of a sequence and its distribution behavior was carried out by Niederreiter [3].

The result given in Example 4.1 may also be verified by using the Weyl criterion. For the blocks S_n, this is done in Pólya and Szegö [1, II. Abschn., nos. 188–189], whereas the infinite sequence is treated in Erdös, Kac, van Kampen, and Wintner [1] and Kac, van Kampen, and Wintner [1]. Further results on the distribution of this sequence can be found in Franel [4], Neville [1], Bateman [1], Delange [7], Huxley [1], and Niederreiter [16]. Another application of the method in Example 4.1 is given in Niederreiter [4].

Exercises

4.1. Prove an analogue of Corollary 4.1 for D_N^*.

4.2. Prove an analogue of Theorem 4.1 for D_N^*.

4.3. Let $x_1, \ldots, x_N$ and $y_1, \ldots, y_N$ be two finite sequences in I with $|x_n - y_n| \leq \varepsilon$ for $1 \leq n \leq N$. Suppose $a_1 \leq a_2 \leq \cdots \leq a_N$ is the sequence of x_n ordered according to their magnitude and that $b_1 \leq b_2 \leq \cdots \leq b_N$ is the sequence of y_n ordered according to their magnitude. Prove that $|a_n - b_n| \leq \varepsilon$ for $1 \leq n \leq N$.

4.4. Deduce from the result of the preceding exercise an alternative proof for the analogue of Corollary 4.1 for D_N^*.

4.5. Let (x_n) be an arbitrary sequence in I, and let f be a positive nonincreasing function defined at least for all positive integers and for which $\lim_{N\to\infty} Nf(N) = \infty$. Prove that there is a rearrangement σ of (x_n) for which $D_N(\sigma) \geq 1 - f(N)$ holds for all $N \geq 1$.

4.6. Consider the set of all rationals of the form $k/2^m$ with $1 \leq k \leq 2^m - 1$, k odd, and $m \geq 1$. What is a u.d. mod 1 arrangement of these numbers without repetitions?

4.7. Consider the set of rationals in the preceding exercise in lexicographic order, that is, $\frac{1}{2}, \frac{1}{4}, \frac{3}{4}, \frac{1}{8}, \frac{3}{8}, \frac{5}{8}, \frac{7}{8}, \ldots$. Show that this sequence is not u. d. mod 1.

4.8. Prove that the sequence in Exercise 4.7 has more than one continuous d.f. (mod 1).

4.9. Strengthen the assertion in Exercise 4.7 by showing that for this sequence ω there are infinitely many N such that $D_N^*(\omega) \geq \frac{1}{6} + 1/4N$.

4.10. Consider the sequence ω given by $\frac{0}{1}, \frac{0}{2}, \frac{1}{2}, \frac{0}{3}, \frac{1}{3}, \frac{2}{3}, \frac{0}{4}, \frac{1}{4}, \frac{2}{4}, \frac{3}{4}, \ldots$. Prove that there is a positive constant c such that $ND_N(\omega) \geq c\sqrt{N}$ holds for all N.

4.11. Complementary to Exercise 4.10, prove that for the sequence ω considered there, we have $ND_N(\omega) \leq C\sqrt{N}$ for some constant C and all N.

4.12. Prove that $\Phi(n) = (3/\pi^2)n^2 + O(n \log n)$, where $\Phi(n) = \sum_{j=1}^{n} \phi(j)$.

4.13. Analogous to Exercise 4.10 but with the sequence from Example 4.1. *Hint:* Use the result of Exercise 4.12.

4.14. Show that some rearrangement of the sequence $(\log n)$ is u.d. mod 1.

4.15. Same as Exercise 4.14 but with $(\phi(n)/n)$.

4.16. Same as Exercise 4.14 but with $(\sqrt{\log n})$.

4.17. Same as Exercise 4.14 but with $(\log \log n)$, $n = 2, 3, \ldots$.

4.18. Let f be a continuous increasing function on $\bar{I}$ with $f(0) = 0$ and $f(1) = 1$, and let g be its inverse function. For a sequence $\omega = (x_n)$ in I, let $\tau = (y_n)$ be defined by $y_n = g(x_n)$. Prove that $D_N(\omega) = D_N(\tau; f)$ for all N; see (1.4) for the definition of the latter discrepancy.

4.19. Show that Theorem 4.2 need not be true if $h(x)$ is a bounded function.

5. NUMERICAL INTEGRATION

Koksma's Inequality

We have seen in Section 1 of Chapter 1 that whenever (x_n) is a u.d. sequence mod 1 in I and f is Riemann-integrable in $\bar{I}$, then $\lim_{N\to\infty} (1/N)\sum_{n=1}^{N} f(x_n) = \int_0^1 f(x)\,dx$. In other words, a Riemann integral over [0, 1] may be approximated to any degree of accuracy by arithmetic means of values of the integrand at points in I forming a u.d. sequence mod 1. It is this very fact on which efficient numerical methods of computing integrals can be based. A very remarkable application of number theory indeed! Of course, any numerical method claiming practicality has to be accompanied by some *a priori* estimate of the error that we commit. And this is exactly the point where the notion of discrepancy comes in. The quality of the approximation of the integral by arithmetic means of the said type is linked directly to the discrepancy of the sequence (x_n) of "nodes." The better the sequence (x_n) is distributed, the faster an approximation we can expect. It is one of the aims of this section to provide a justification for this intuitive statement.

LEMMA 5.1. Let $x_1 \le x_2 \le \cdots \le x_N$ be N given points in I, and let f be a function on $\bar{I}$ of bounded variation. Then, with $x_0 = 0$ and $x_{N+1} = 1$, we have the identity

$$\frac{1}{N}\sum_{n=1}^{N} f(x_n) - \int_0^1 f(t)\,dt = \sum_{n=0}^{N}\int_{x_n}^{x_{n+1}}\left(t - \frac{n}{N}\right) df(t). \tag{5.1}$$

PROOF. Using integration by parts and Abel's summation formula, we obtain

$$\begin{aligned}\sum_{n=0}^{N}\int_{x_n}^{x_{n+1}}\left(t - \frac{n}{N}\right) df(t) &= \int_0^1 t\,df(t) - \sum_{n=0}^{N}\frac{n}{N}(f(x_{n+1}) - f(x_n))\\ &= [tf(t)]_{t=0}^{1} - \int_0^1 f(t)\,dt + \frac{1}{N}\sum_{n=0}^{N-1} f(x_{n+1}) - f(1)\\ &= \frac{1}{N}\sum_{n=1}^{N} f(x_n) - \int_0^1 f(t)\,dt. \quad \blacksquare\end{aligned}$$

THEOREM 5.1: Koksma's Inequality. Let f be a function on $\bar{I}$ of bounded variation $V(f)$, and suppose we are given N points $x_1, \ldots, x_N$ in I with discrepancy D_N^*. Then,

$$\left|\frac{1}{N}\sum_{n=1}^{N} f(x_n) - \int_0^1 f(t)\,dt\right| \le V(f)D_N^*. \tag{5.2}$$

PROOF. Without loss of generality, we may assume $x_1 \le x_2 \le \cdots \le x_N$. Thus, we can apply Lemma 5.1. For fixed n with $0 \le n \le N$, we have

$$\left|t - \frac{n}{N}\right| \le \max\left(\left|x_n - \frac{n}{N}\right|, \left|x_{n+1} - \frac{n}{N}\right|\right) \le D_N^* \qquad \text{for } x_n \le t \le x_{n+1}$$

by Theorem 1.4, and the desired inequality follows immediately. $\blacksquare$

It is clear that Theorem 5.1 remains valid for any N real numbers $x_1, \ldots, x_N$ if we only suppose in addition that f be periodic with period 1. Before passing on, we note an application of Koksma's inequality to the estimation of exponential sums that is, in a sense, a simple counterpart to Theorem 2.4.

COROLLARY 5.1. Let $x_1, \ldots, x_N$ be N real numbers with discrepancy D_N^*. Then,

$$\left|\frac{1}{N}\sum_{n=1}^{N} e^{2\pi i x_n}\right| \le 4D_N^*. \tag{5.3}$$

PROOF. Let L denote the left-hand side of (5.3). Then

$$\frac{1}{N}\sum_{n=1}^{N} e^{2\pi i x_n} = Le^{2\pi i\theta}$$

for some $\theta \in I$. Since L is real, we obtain

$$L = \frac{1}{N} \sum_{n=1}^{N} e^{2\pi i(x_n - \theta)} = \frac{1}{N} \sum_{n=1}^{N} \cos 2\pi(x_n - \theta).$$

Using the remark following Theorem 5.1, we apply Koksma's inequality to the points $x_1, \ldots, x_N$ and to the function $f(t) = \cos 2\pi(t - \theta)$. Since $V(f) = 4$ on $\bar{I}$, we arrive at the desired inequality. ■

EXAMPLE 5.1. If the function f has a continuous derivative in $\bar{I}$, then the Riemann-Stieltjes integrals in the above proofs may be replaced by Riemann integrals. We arrive then at the inequality $|(1/N)\sum_{n=1}^{N} f(x_n) - \int_0^1 f(t)\,dt| \leq D_N^* \int_0^1 |f'(t)|\,dt$, where f may also be complex-valued. For real-valued f, we have $\int_0^1 |f'(t)|\,dt = V(f)$ (see Exercise 5.10). ■

Some Remarkable Identities

In case the function f in Theorem 5.1 is of a simple type (for instance linear), then sharper results can be shown. For $f(t) = t$, we obtain the following identity.

THEOREM 5.2. For any points $x_1, \ldots, x_N$ in I we have

$$\left(\sum_{n=1}^{N} x_n - \frac{N}{2}\right)^2 = \int_0^1 R_N^2(t)\,dt - \int_0^1 \left(\sum_{n=1}^{N} \{x_n + t\} - \frac{N}{2}\right)^2 dt, \tag{5.4}$$

where R_N is the function introduced in Section 2.

PROOF. By (2.28) we have $\int_0^1 R_N(t)\,dt = -S_N$, where $S_N = \sum_{n=1}^{N} x_n - (N/2)$. Putting $S_N(t) = \sum_{n=1}^{N} \{x_n + t\} - (N/2)$ for $0 \leq t \leq 1$, we get from (2.41) that

$$S_N(t) - S_N = R_N(1 - t). \tag{5.5}$$

A simple change of variable shows that $\int_0^1 R_N(1 - t)\,dt = \int_0^1 R_N(t)\,dt = -S_N$, and so, from (5.5),

$$\int_0^1 S_N(t)\,dt = S_N + \int_0^1 R_N(1 - t)\,dt = 0. \tag{5.6}$$

Combining (5.5) and (5.6) we find

$$\int_0^1 R_N^2(t)\,dt = \int_0^1 R_N^2(1 - t)\,dt = \int_0^1 S_N^2(t)\,dt - 2S_N \int_0^1 S_N(t)\,dt + S_N^2$$

$$= \int_0^1 S_N^2(t)\,dt + S_N^2. \quad ■$$

In particular, we have then

$$\left|\frac{1}{N}\sum_{n=1}^{N} x_n - \frac{1}{2}\right|^2 \leq D_N^{*2} - \int_0^1 \left(\frac{1}{N}\sum_{n=1}^{N}\{x_n + t\} - \frac{1}{2}\right)^2 dt,$$

an improvement on Koksma's inequality. Let us prove a not so well-known identity that is in the same vein as Theorem 5.2.

THEOREM 5.3. For any points $x_1, \ldots, x_N$ in I, we have

$$\int_0^1 R_N^2(t)\,dt = \tfrac{1}{3}N^2 + N\sum_{n=1}^{N} x_n^2 + \sum_{n=1}^{N} x_n - 2\sum_{n=1}^{N}\sum_{m=1}^{n} \max(x_n, x_m). \quad (5.7)$$

PROOF. We have $R_N(t) = \sum_{n=1}^{N} c_t(x_n) - Nt$, where c_t denotes the characteristic function of $[0, t)$. Therefore,

$$\int_0^1 R_N^2(t)\,dt = N^2\int_0^1 t^2\,dt - 2N\sum_{n=1}^{N}\int_0^1 tc_t(x_n)\,dt + \sum_{n=1}^{N}\sum_{m=1}^{N}\int_0^1 c_t(x_n)c_t(x_m)\,dt$$

$$= \tfrac{1}{3}N^2 - 2N\sum_{n=1}^{N}\int_{x_n}^1 t\,dt + \sum_{n=1}^{N}\sum_{m=1}^{N}\int_{\max(x_n,x_m)}^1 dt$$

$$= \tfrac{1}{3}N^2 - N\sum_{n=1}^{N}(1 - x_n^2) + \sum_{n=1}^{N}\sum_{m=1}^{N}(1 - \max(x_n, x_m))$$

$$= \tfrac{1}{3}N^2 + N\sum_{n=1}^{N} x_n^2 + \sum_{n=1}^{N} x_n - 2\sum_{n=1}^{N}\sum_{m=1}^{n}\max(x_n, x_m). \quad ■$$

EXAMPLE 5.2. If $x_1, \ldots, x_N$ are ordered according to their magnitude, that is, $x_1 \leq x_2 \leq \cdots \leq x_N$, then from (5.7) we obtain $\int_0^1 R_N^2(t)\,dt = \frac{1}{3}N^2 + N\sum_{n=1}^{N} x_n^2 + \sum_{n=1}^{N} x_n - 2\sum_{n=1}^{N} nx_n$. It follows that

$$\int_0^1 R_N^2(t)\,dt = \sum_{n=1}^{N}(x_n - \tfrac{1}{2}) + N\sum_{n=1}^{N}(x_n - (n/N))^2 - \tfrac{1}{6}, \quad (5.8)$$

using $\sum_{n=1}^{N} n^2 = \frac{1}{6}N(N+1)(2N+1)$. We note that the second sum in (5.8) is closely connected with D_N^* because of Theorem 1.4. ■

An Error Estimate for Continuous Functions

We shall need the following auxiliary notion. Let f be a continuous function on $\bar{I}$. Then its *modulus of continuity* M is given by

$$M(h) = \sup_{\substack{x,y\in\bar{I} \\ |x-y|\leq h}} |f(x) - f(y)| \qquad \text{for } 0 \leq h \leq 1.$$

Since f is uniformly continuous on $\bar{I}$, we have $\lim_{h\to 0+0} M(h) = 0$.

THEOREM 5.4. Suppose the continuous function f on $\bar{I}$ has the modulus of continuity M. Let $x_1, x_2, \ldots, x_N$ be any N points in I with discrepancy D_N^*. Then

$$\left|\frac{1}{N}\sum_{n=1}^{N} f(x_n) - \int_0^1 f(t)\,dt\right| \leq M(D_N^*). \tag{5.9}$$

PROOF. Without loss of generality, we may assume that $x_1 \leq x_2 \leq \cdots \leq x_N$. We note that $\int_0^1 f(t)\,dt = \sum_{n=1}^{N} \int_{(n-1)/N}^{n/N} f(t)\,dt$. By the mean value theorem for integrals, we have $\int_{(n-1)/N}^{n/N} f(t)\,dt = (1/N)f(\xi_n)$ for some ξ_n with $(n-1)/N < \xi_n < n/N$. Therefore,

$$\frac{1}{N}\sum_{n=1}^{N} f(x_n) - \int_0^1 f(t)\,dt = \frac{1}{N}\sum_{n=1}^{N}(f(x_n) - f(\xi_n)).$$

We will be done if we show that $|x_n - \xi_n| \leq D_N^*$ for $1 \leq n \leq N$. Now, whenever $x_n \geq \xi_n$, then $|x_n - \xi_n| < |x_n - (n-1)/N| \leq D_N^*$ by Theorem 1.4. Likewise, if $x_n < \xi_n$, then $|x_n - \xi_n| < |x_n - (n/N)| \leq D_N^*$ by the same theorem. ■

A weaker version of this theorem that more closely resembles Koksma's inequality may be given.

COROLLARY 5.2. Under the hypotheses of Theorem 5.4, we have

$$\left|\frac{1}{N}\sum_{n=1}^{N} f(x_n) - \int_0^1 f(t)\,dt\right| \leq (ND_N^* + 1)M\left(\frac{1}{N}\right) \leq 3ND_N^*M\left(\frac{1}{N}\right). \tag{5.10}$$

PROOF. To prove the first inequality, it suffices to show that $M(D_N^*) \leq (ND_N^* + 1)M(1/N)$. Choose $x, y \in \bar{I}$ with $y \leq x \leq y + D_N^*$. We insert points between y and x in the following manner:

$$y,\ y + \frac{1}{N},\ y + \frac{2}{N},\ \ldots,\ y + \frac{k-1}{N},\ x,$$

where

$$\frac{k-1}{N} \leq x - y < \frac{k}{N}.$$

We note that $k = [N(x-y)] + 1 \leq ND_N^* + 1$. Now

$$\begin{aligned} &|f(x) - f(y)| \\ &\quad\leq \sum_{i=1}^{k-1}\left|f\left(y + \frac{i}{N}\right) - f\left(y + \frac{i-1}{N}\right)\right| + \left|f(x) - f\left(y + \frac{k-1}{N}\right)\right| \\ &\quad\leq kM\left(\frac{1}{N}\right) \leq (ND_N^* + 1)M\left(\frac{1}{N}\right), \end{aligned}$$

and forming the requisite supremum on the left-hand side completes the argument. To prove the second inequality, we simply observe that $2ND_N^* \geq 1$ by Corollary 1.2. ■

The Koksma-Hlawka Inequality

We want to generalize Koksma's inequality to several dimensions. To this end, we first have to describe under what circumstances a function of several variables is said to be of bounded variation. Suppose we are given a function $f(\mathbf{x}) = f(x^{(1)}, \ldots, x^{(k)})$ on $\bar{I}^k$ with $k \geq 2$. By a *partition P* of $\bar{I}^k$, we mean a set of k finite sequences $\eta_0^{(j)}, \eta_1^{(j)}, \ldots, \eta_{m_j}^{(j)}$ $(j = 1, \ldots, k)$ with $0 = \eta_0^{(j)} \leq \eta_1^{(j)} \leq \cdots \leq \eta_{m_j}^{(j)} = 1$ for $j = 1, \ldots, k$. In connection with such a partition, we define, for each $j = 1, \ldots, k$, an operator Δ_j by

$$\begin{aligned}\Delta_j f(x^{(1)}, \ldots, x^{(j-1)}, \eta_i^{(j)}, x^{(j+1)}, \ldots, x^{(k)}) \\ = f(x^{(1)}, \ldots, x^{(j-1)}, \eta_{i+1}^{(j)}, x^{(j+1)}, \ldots, x^{(k)}) \\ - f(x^{(1)}, \ldots, x^{(j-1)}, \eta_i^{(j)}, x^{(j+1)}, \ldots, x^{(k)}) \quad (5.11)\end{aligned}$$

for $0 \leq i < m_j$. Operators with different subscripts obviously commute, and $\Delta_{j_1, \ldots, j_p}$ will stand for $\Delta_{j_1} \cdots \Delta_{j_p}$. Such an operator commutes with summation over variables on which it does not act.

DEFINITION 5.1. For a function f on $\bar{I}^k$, we set

$$V^{(k)}(f) = \sup_P \sum_{i_1=0}^{m_1-1} \cdots \sum_{i_k=0}^{m_k-1} |\Delta_{1,\ldots,k} f(\eta_{i_1}^{(1)}, \ldots, \eta_{i_k}^{(k)})|, \quad (5.12)$$

where the supremum is extended over all partitions P of $\bar{I}^k$. If $V^{(k)}(f)$ is finite, then f is said to be *of bounded variation on $\bar{I}^k$ in the sense of Vitali.*

It follows immediately from the definition of the Δ-operators that whenever the function f on $\bar{I}^k$ actually depends on less than k variables, then $V^{(k)}(f) = 0$. This is not a very healthy state of affairs, since such a function f might still be extremely irregular. Thus, to arrive at a more suitable notion of variation, we also have to take into account the behavior of f on the various faces of $\bar{I}^k$. This leads to the following definition.

DEFINITION 5.2. Let f be a function of bounded variation on $\bar{I}^k$ in the sense of Vitali. Suppose the restriction of f to each face F of $\bar{I}^k$ of dimension $1, 2, \ldots, k-1$ is of bounded variation on F in the sense of Vitali. Then f is said to be *of bounded variation on $\bar{I}^k$ in the sense of Hardy and Krause*.

The functions of bounded variation in the sense of Hardy and Krause are the ones for which a multidimensional version of Koksma's inequality can be proved. To have an idea of how to proceed, let us first analyze the proof of the one-dimensional case. The essential step in the pivotal Lemma 5.1 was integration by parts of a Riemann-Stieltjes integral. But it is well known that the applicability of integration by parts to such integrals is a consequence of the Abel summation formula. It will therefore be an important task to generalize this formula to several dimensions.

First, we introduce another operator; namely, we define for $1 \leq j \leq k$:

$$\Delta_j^* f(x^{(1)}, \ldots, x^{(k)}) = f(x^{(1)}, \ldots, x^{(j-1)}, 1, x^{(j+1)}, \ldots, x^{(k)}) - f(x^{(1)}, \ldots, x^{(j-1)}, 0, x^{(j+1)}, \ldots, x^{(k)}). \quad (5.13)$$

The remarks about the Δ-operators also hold for this class of operators. In particular, we again let $\Delta^*_{j_1,\ldots,j_p}$ stand for $\Delta^*_{j_1} \cdots \Delta^*_{j_p}$.

Given any expression $F(r, \ldots, r+p-1; r+p, \ldots, s)$ depending only on the partition of variables $i_r, \ldots, i_s$ into the sets $\{i_r, \ldots, i_{r+p-1}\}$ and $\{i_{r+p}, \ldots, i_s\}$, the summation symbol

$$\sum_{r,\ldots,s;p}^* F(r, \ldots, r+p-1; r+p, \ldots, s)$$

will denote the sum of all the expressions derived from $F(r, \ldots, r+p-1; r+p, \ldots, s)$ by replacing the given partition of $\{i_r, \ldots, i_s\}$ successively by all other partitions of this set into a set of p and a set of $s-r-p+1$ variables, each partition being taken exactly once. If either $p = 0$ or $p = s - r + 1$, one of the sets becomes empty; in order to avoid troublesome exceptions, the sum will be interpreted, in such cases, as being reduced to one term.

LEMMA 5.2. Let P be a partition of $\bar{I}^k$, consisting of the k sequences $\eta_0^{(j)}, \eta_1^{(j)}, \ldots, \eta_{m_j}^{(j)}$ $(j = 1, \ldots, k)$, and let Q be a second partition of $\bar{I}^k$ consisting of the k sequences $\xi_0^{(j)}, \xi_1^{(j)}, \ldots, \xi_{m_j+1}^{(j)}$ $(j = 1, \ldots, k)$. Furthermore, let $f(\mathbf{x})$ and $g(\mathbf{x})$ be two given functions on $\bar{I}^k$. Then,

$$\sum_{i_1=0}^{m_1-1} \cdots \sum_{i_k=0}^{m_k-1} f(\xi_{i_1+1}^{(1)}, \ldots, \xi_{i_k+1}^{(k)}) \Delta_{1,\ldots,k} g(\eta_{i_1}^{(1)}, \ldots, \eta_{i_k}^{(k)})$$
$$= \sum_{p=0}^{k} (-1)^p \sum_{1,\ldots,k;p}^* \Delta^*_{p+1,\ldots,k} \sum_{i_1=0}^{m_1} \cdots \sum_{i_p=0}^{m_p} g(\eta_{i_1}^{(1)}, \ldots, \eta_{i_p}^{(p)}, x^{(p+1)}, \ldots, x^{(k)})$$
$$\cdot \Delta_{1,\ldots,p} f(\xi_{i_1}^{(1)}, \ldots, \xi_{i_p}^{(p)}, x^{(p+1)}, \ldots, x^{(k)}). \quad (5.14)$$

On the right-hand side, when $p = 0$, the summation symbols referring to $i_1, \ldots, i_p$, as well as $\Delta_{1,\ldots,p}$, are understood to disappear, and similarly, when $p = k$, then $\Delta^*_{p+1,\ldots,k}$ should be disregarded, the variables $x^{(p+1)}, \ldots, x^{(k)}$ disappearing altogether.

PROOF. We proceed by induction on the dimension k. When $k = 1$, then (5.14) reduces to

$$\sum_{i_1=0}^{m_1-1} f(\xi_{i_1+1}^{(1)}) \Delta_1 g(\eta_{i_1}^{(1)}) = \Delta_1^*(g(x^{(1)}) f(x^{(1)})) - \sum_{i_1=0}^{m_1} g(\eta_{i_1}^{(1)}) \Delta_1 f(\xi_{i_1}^{(1)}), \quad (5.15)$$

which, in a simplified notation, reads

$$\sum_{i=0}^{m-1} f(\xi_{i+1})(g(\eta_{i+1}) - g(\eta_i))$$
$$= g(1)f(1) - g(0)f(0) - \sum_{i=0}^{m} g(\eta_i)(f(\xi_{i+1}) - f(\xi_i)).$$

But this is just the Abel summation formula (note that $\eta_0 = \xi_0 = 0$ and $\eta_m = \xi_{m+1} = 1$).

Assume that the proposition holds for variables with superscripts $2, \ldots, k$. In the corresponding version of (5.14) we substitute $\Delta_1 g$ for g, and summing with respect to i_1 from 0 to $m_1 - 1$, we find

$$\sum_{i_1=0}^{m_1-1} \cdots \sum_{i_k=0}^{m_k-1} f(\xi_{i_1+1}^{(1)}, \ldots, \xi_{i_k+1}^{(k)})\, \Delta_{1,\ldots,k} g(\eta_{i_1}^{(1)}, \ldots, \eta_{i_k}^{(k)})$$
$$= \sum_{p=0}^{k-1} (-1)^p \sum_{2,\ldots,k;p}^{*} \Delta_{p+2,\ldots,k}^{*} \sum_{i_1=0}^{m_1-1} \sum_{i_2=0}^{m_2} \cdots \sum_{i_{p+1}=0}^{m_{p+1}}$$
$$\Delta_1 g(\eta_{i_1}^{(1)}, \ldots, \eta_{i_{p+1}}^{(p+1)}, x^{(p+2)}, \ldots, x^{(k)})$$
$$\cdot \Delta_{2,\ldots,p+1} f(\xi_{i_1+1}^{(1)}, \xi_{i_2}^{(2)}, \ldots, \xi_{i_{p+1}}^{(p+1)}, x^{(p+2)}, \ldots, x^{(k)}). \quad (5.16)$$

We have to show that the right-hand sides of (5.16) and (5.14) are identical. We consider (5.15) and replace f by $\Delta_{2,\ldots,p+1} f(x^{(1)}, \xi_{i_2}^{(2)}, \ldots, \xi_{i_{p+1}}^{(p+1)}, x^{(p+2)}, \ldots, x^{(k)})$ and g by $g(x^{(1)}, \eta_{i_2}^{(2)}, \ldots, \eta_{i_{p+1}}^{(p+1)}, x^{(p+2)}, \ldots, x^{(k)})$, both expressions being considered as functions of $x^{(1)}$. Then we obtain

$$\sum_{i_1=0}^{m_1-1} \Delta_1 g(\eta_{i_1}^{(1)}, \ldots, \eta_{i_{p+1}}^{(p+1)}, x^{(p+2)}, \ldots, x^{(k)})$$
$$\cdot \Delta_{2,\ldots,p+1} f(\xi_{i_1+1}^{(1)}, \xi_{i_2}^{(2)}, \ldots, \xi_{i_{p+1}}^{(p+1)}, x^{(p+2)}, \ldots, x^{(k)})$$
$$= \Delta_1^{*}(g(x^{(1)}, \eta_{i_2}^{(2)}, \ldots, \eta_{i_{p+1}}^{(p+1)}, x^{(p+2)}, \ldots, x^{(k)})$$
$$\cdot \Delta_{2,\ldots,p+1} f(x^{(1)}, \xi_{i_2}^{(2)}, \ldots, \xi_{i_{p+1}}^{(p+1)}, x^{(p+2)}, \ldots, x^{(k)}))$$
$$- \sum_{i_1=0}^{m_1} g(\eta_{i_1}^{(1)}, \ldots, \eta_{i_{p+1}}^{(p+1)}, x^{(p+2)}, \ldots, x^{(k)})$$
$$\cdot \Delta_{1,\ldots,p+1} f(\xi_{i_1}^{(1)}, \ldots, \xi_{i_{p+1}}^{(p+1)}, x^{(p+2)}, \ldots, x^{(k)}).$$

This identity holds for $p = 0, \ldots, k-1$ if we understand it in the same way as (5.14). We apply to both sides of it the operator

$$\sum_{p=0}^{k-1} (-1)^p \sum_{2,\ldots,k;p}^{*} \Delta_{p+2,\ldots,k}^{*} \sum_{i_2=0}^{m_2} \cdots \sum_{i_{p+1}=0}^{m_{p+1}} .$$

In the new identity, the left-hand side coincides with the right-hand side of (5.16), while the right-hand side becomes

$$\sum_{p=0}^{k-1}(-1)^p \sum^*_{2,\dots,k;p} \Delta^*_{1,p+2,\dots,k} \sum_{i_2=0}^{m_2}\cdots\sum_{i_{p+1}=0}^{m_{p+1}} g(x^{(1)}, \eta_{i_2}^{(2)}, \dots, \eta_{i_{p+1}}^{(p+1)}, x^{(p+2)}, \dots, x^{(k)})$$
$$\cdot \Delta_{2,\dots,p+1} f(x^{(1)}, \xi_{i_2}^{(2)}, \dots, \xi_{i_{p+1}}^{(p+1)}, x^{(p+2)}, \dots, x^{(k)})$$
$$+\sum_{p=0}^{k-1}(-1)^{p+1} \sum^*_{2,\dots,k;p} \Delta^*_{p+2,\dots,k} \sum_{i_1=0}^{m_1}\cdots\sum_{i_{p+1}=0}^{m_{p+1}} g(\eta_{i_1}^{(1)}, \dots, \eta_{i_{p+1}}^{(p+1)}, x^{(p+2)}, \dots, x^{(k)})$$
$$\cdot \Delta_{1,\dots,p+1} f(\xi_{i_1}^{(1)}, \dots, \xi_{i_{p+1}}^{(p+1)}, x^{(p+2)}, \dots, x^{(k)}). \tag{5.17}$$

It remains to show that this expression is equal to the right-hand side of (5.14). Indeed, the terms corresponding to $p = 0$ on the right-hand side of (5.14) and in the first part of (5.17) are the same. Similarly, the term corresponding to $p = k$ on the right-hand side of (5.14) is equal to the term corresponding to $p = k - 1$ in the second part of (5.17). Finally, if $0 < p < k$, the corresponding group of terms in (5.14) can be split into two parts according to the effect of the operator $\sum^*$, namely, the sum of all the terms in which 1 appears as a subscript of Δ^* and the sum of all the other terms; now the first part is identical with the term corresponding to the same value of p in the first part of (5.17), whereas the second part is identical with the term corresponding to $p - 1$ in the second part of (5.17). ■

EXAMPLE 5.3. To elucidate the complex formula (5.14), let us write down the case $k = 2$ in detail. We get

$$\sum_{i=0}^{m-1}\sum_{j=0}^{n-1} f(\xi_{i+1}^{(1)}, \xi_{j+1}^{(2)})(g(\eta_{i+1}^{(1)}, \eta_{j+1}^{(2)}) - g(\eta_{i+1}^{(1)}, \eta_j^{(2)}) - g(\eta_i^{(1)}, \eta_{j+1}^{(2)}) + g(\eta_i^{(1)}, \eta_j^{(2)}))$$
$$= g(1,1)f(1,1) - g(1,0)f(1,0) - g(0,1)f(0,1) + g(0,0)f(0,0)$$
$$- \sum_{i=0}^{m} g(\eta_i^{(1)}, 1)(f(\xi_{i+1}^{(1)}, 1) - f(\xi_i^{(1)}, 1))$$
$$+ \sum_{i=0}^{m} g(\eta_i^{(1)}, 0)(f(\xi_{i+1}^{(1)}, 0) - f(\xi_i^{(1)}, 0))$$
$$- \sum_{j=0}^{n} g(1, \eta_j^{(2)})(f(1, \xi_{j+1}^{(2)}) - f(1, \xi_j^{(2)}))$$
$$+ \sum_{j=0}^{n} g(0, \eta_j^{(2)})(f(0, \xi_{j+1}^{(2)}) - f(0, \xi_j^{(2)}))$$
$$+ \sum_{i=0}^{m}\sum_{j=0}^{n} g(\eta_i^{(1)}, \eta_j^{(2)})(f(\xi_{i+1}^{(1)}, \xi_{j+1}^{(2)}) - f(\xi_{i+1}^{(1)}, \xi_j^{(2)})$$
$$- f(\xi_i^{(1)}, \xi_{j+1}^{(2)}) + f(\xi_i^{(1)}, \xi_j^{(2)})). \quad ■$$

THEOREM 5.5: Koksma-Hlawka Inequality. Let $f(\mathbf{x})$ be of bounded variation on $\bar{I}^k$ in the sense of Hardy and Krause. Let ω be the finite sequence of points $\mathbf{x}_1, \ldots, \mathbf{x}_N$ in I^k, and let $\omega_{j_1,\ldots,j_p}$ denote the projection of the sequence ω on the $(k-p)$-dimensional face of $\bar{I}^k$ defined by $x^{(j_1)} = \cdots = x^{(j_p)} = 1$. Then we have

$$\left| \frac{1}{N} \sum_{n=1}^{N} f(\mathbf{x}_n) - \int_{\bar{I}^k} f(\mathbf{x})\, d\mathbf{x} \right| \leq \sum_{p=1}^{k} \sum_{1,\ldots,k;p}^{*} D_N^*(\omega_{p+1,\ldots,k}) V^{(p)}(f(\ldots, 1, \ldots, 1)), \tag{5.18}$$

where $V^{(p)}(f(\ldots, 1, \ldots, 1))$ denotes the p-dimensional variation of $f(x^{(1)}, \ldots, x^{(p)}, 1, \ldots, 1)$ on $\bar{I}^p$ in the sense of Vitali and where the term of the sum corresponding to $p = k$ is understood to be $D_N^*(\omega)V^{(k)}(f)$. The discrepancy $D_N^*(\omega_{p+1,\ldots,k})$ is computed in the face of $\bar{I}^k$ in which $\omega_{p+1,\ldots,k}$ is contained.

PROOF. For a subset M of I^k, we abbreviate the counting function $A(M; N; \omega)$ by $A(M)$. We define a function g on $\bar{I}^k$ by

$$g(\mathbf{x}) = g(x^{(1)}, \ldots, x^{(k)}) = \frac{1}{N} A([0, x^{(1)}) \times \cdots \times [0, x^{(k)})) - x^{(1)} \cdots x^{(k)}. \tag{5.19}$$

We note that

$$D_N^*(\omega) = \sup_{\mathbf{x} \in \bar{I}^k} |g(\mathbf{x})|,$$

and

$$D_N^*(\omega_{p+1,\ldots,k}) = \sup_{(x^{(1)},\ldots,x^{(p)}) \in \bar{I}^p} |g(x^{(1)}, \ldots, x^{(p)}, 1, \ldots, 1)|.$$

For $1 \leq n \leq N$, let us put $\mathbf{x}_n = (x_n^{(1)}, \ldots, x_n^{(k)})$.

By an *admissible double partition* of $\bar{I}^k$, we shall mean a pair (P, Q) of partitions P and Q of $\bar{I}^k$ satisfying the following conditions. First of all, P consists of the k sequences $\eta_0^{(j)}, \eta_1^{(j)}, \ldots, \eta_{m_j}^{(j)}$ $(j = 1, \ldots, k)$, and Q consists of the k sequences $\xi_0^{(j)}, \xi_1^{(j)}, \ldots, \xi_{m_j+1}^{(j)}$ $(j = 1, \ldots, k)$, and these are related by

$$0 = \xi_0^{(j)} = \eta_0^{(j)} \leq \xi_1^{(j)} < \eta_1^{(j)} \leq \xi_2^{(j)} < \eta_2^{(j)} \leq \cdots < \eta_{m_j}^{(j)} = \xi_{m_j+1}^{(j)} = 1 \quad \text{for } j = 1, \ldots, k. \tag{5.20}$$

Moreover, for each $j = 1, \ldots, k$, the sequence $\xi_1^{(j)}, \ldots, \xi_{m_j}^{(j)}$ should at least contain the numbers $x_1^{(j)}, \ldots, x_N^{(j)}$.

With such an admissible double partition, we can apply Lemma 5.2 with the given function f and the function g from (5.19). First, studying the left-hand side of (5.14), we obtain

$$\sum_{i_1=0}^{m_1-1} \cdots \sum_{i_k=0}^{m_k-1} f(\xi_{i_1+1}^{(1)}, \ldots, \xi_{i_k+1}^{(k)}) \Delta_{1,\ldots,k} g(\eta_{i_1}^{(1)}, \ldots, \eta_{i_k}^{(k)})$$

$$= \frac{1}{N} \sum_{i_1=0}^{m_1-1} \cdots \sum_{i_k=0}^{m_k-1} f(\xi_{i_1+1}^{(1)}, \ldots, \xi_{i_k+1}^{(k)}) \Delta_{1,\ldots,k} A([0, \eta_{i_1}^{(1)}) \times \cdots \times [0, \eta_{i_k}^{(k)}))$$

$$- \sum_{i_1=0}^{m_1-1} \cdots \sum_{i_k=0}^{m_k-1} f(\xi_{i_1+1}^{(1)}, \ldots, \xi_{i_k+1}^{(k)}) \Delta_{1,\ldots,k} \eta_{i_1}^{(1)} \cdots \eta_{i_k}^{(k)}. \quad (5.21)$$

Now

$$\Delta_{1,\ldots,k} A([0, \eta_{i_1}^{(1)}) \times \cdots \times [0, \eta_{i_k}^{(k)}))$$

$$= \Delta_{1,\ldots,k-1}(A([0, \eta_{i_1}^{(1)}) \times \cdots \times [0, \eta_{i_k+1}^{(k)})) - A([0, \eta_{i_1}^{(1)}) \times \cdots \times [0, \eta_{i_k}^{(k)})))$$

$$= \Delta_{1,\ldots,k-1} A([0, \eta_{i_1}^{(1)}) \times \cdots \times [0, \eta_{i_{k-1}}^{(k-1)}) \times [\eta_{i_k}^{(k)}, \eta_{i_k+1}^{(k)}))$$

$$= \cdots = A([\eta_{i_1}^{(1)}, \eta_{i_1+1}^{(1)}) \times \cdots \times [\eta_{i_k}^{(k)}, \eta_{i_k+1}^{(k)})).$$

Thus, the first term on the right-hand side of (5.21) reduces to

$$\frac{1}{N} \sum_{i_1=0}^{m_1-1} \cdots \sum_{i_k=0}^{m_k-1} f(\xi_{i_1+1}^{(1)}, \ldots, \xi_{i_k+1}^{(k)}) A([\eta_{i_1}^{(1)}, \eta_{i_1+1}^{(1)}) \times \cdots \times [\eta_{i_k}^{(k)}, \eta_{i_k+1}^{(k)})). \quad (5.22)$$

Hence, only those k-tuples $(i_1, \ldots, i_k)$ have to be taken into account for which there is an $\mathbf{x}_n$, $1 \le n \le N$, in the interval $[\eta_{i_1}^{(1)}, \eta_{i_1+1}^{(1)}) \times \cdots \times [\eta_{i_k}^{(k)}, \eta_{i_k+1}^{(k)})$. But whenever this happens, the condition (5.20) and the additional condition on Q imply that $\mathbf{x}_n = (\xi_{i_1+1}^{(1)}, \ldots, \xi_{i_k+1}^{(k)})$. Therefore, (5.22) is nothing else but $(1/N)\sum_{n=1}^{N} f(\mathbf{x}_n)$. Consequently, the left-hand side of (5.14) reads

$$\frac{1}{N} \sum_{n=1}^{N} f(x_n) - \sum_{i_1=0}^{m_1-1} \cdots \sum_{i_k=0}^{m_k-1} f(\xi_{i_1+1}^{(1)}, \ldots, \xi_{i_k+1}^{(k)}) \Delta_{1,\ldots,k} \eta_{i_1}^{(1)} \cdots \eta_{i_k}^{(k)}. \quad (5.23)$$

Now let us take a look at the right-hand side of (5.14). The important fact we need is that $g(\mathbf{x}) = 0$ whenever at least one coordinate of $\mathbf{x}$ vanishes, and moreover, $g(1, \ldots, 1) = 0$. The term corresponding to $p = 0$ on the right-hand side of (5.14), namely, $\Delta_{1,\ldots,k}^{*} g(x^{(1)}, \ldots, x^{(k)}) f(x^{(1)}, \ldots, x^{(k)})$, is therefore zero. Furthermore, for $1 \le p \le k$, only those terms are left where all the variables $x^{(p+1)}, \ldots, x^{(k)}$ are replaced by 1. It follows that the right-hand side of (5.14) reduces to

$$\sum_{p=1}^{k} (-1)^p \sum_{1,\ldots,k;p}^{*} \sum_{i_1=0}^{m_1} \cdots \sum_{i_p=0}^{m_p} g(\eta_{i_1}^{(1)}, \ldots, \eta_{i_p}^{(p)}, 1, \ldots, 1)$$

$$\cdot \Delta_{1,\ldots,p} f(\xi_{i_1}^{(1)}, \ldots, \xi_{i_p}^{(p)}, 1, \ldots, 1). \quad (5.24)$$

We have thus established the identity between the expressions in (5.23) and in (5.24). Now we estimate the absolute value of (5.24). An upper bound is certainly given by

$$\sum_{p=1}^{k} \sum_{1,\ldots,k;p}^{*} \sum_{i_1=0}^{m_1} \cdots \sum_{i_p=0}^{m_p} |g(\eta_{i_1}^{(1)}, \ldots, \eta_{i_p}^{(p)}, 1, \ldots, 1)|$$

$$\cdot |\Delta_{1,\ldots,p} f(\xi_{i_1}^{(1)}, \ldots, \xi_{i_p}^{(p)}, 1, \ldots, 1)|.$$

The absolute value involving g can be bounded uniformly by $D_N^*(\omega_{p+1,\ldots,k})$, the remaining sum over $i_1, \ldots, i_p$ is dominated by $V^{(p)}(f(\ldots, 1, \ldots, 1))$. This entails

$$\left| \frac{1}{N} \sum_{n=1}^{N} f(\mathbf{x}_n) - \sum_{i_1=0}^{m_1-1} \cdots \sum_{i_k=0}^{m_k-1} f(\xi_{i_1+1}^{(1)}, \ldots, \xi_{i_k+1}^{(k)}) \Delta_{1,\ldots,k} \eta_{i_1}^{(1)} \cdots \eta_{i_k}^{(k)} \right|$$

$$\leq \sum_{p=1}^{k} \sum_{1,\ldots,k;p}^{*} D_N^*(\omega_{p+1,\ldots,k}) V^{(p)}(f(\ldots, 1, \ldots, 1)). \quad (5.25)$$

We remark that $\Delta_{1,\ldots,k} \eta_{i_1}^{(1)} \cdots \eta_{i_k}^{(k)} = (\eta_{i_1+1}^{(1)} - \eta_{i_1}^{(1)}) \cdots (\eta_{i_k+1}^{(k)} - \eta_{i_k}^{(k)})$, and therefore, (5.20) implies that the sum over $i_1, \ldots, i_k$ on the left of (5.25) is nothing else but a Riemann sum for $\int_{\bar{I}^k} f(\mathbf{x})\, d\mathbf{x}$. The other terms in (5.25) are independent of the chosen admissible double partition (P, Q). The proof is therefore completed by letting (P, Q) run through a sequence of admissible double partitions with

$$\max_{1 \leq j \leq k} \max_{0 \leq i < m_j} (\eta_{i+1}^{(j)} - \eta_i^{(j)}) \to 0. \quad \blacksquare \quad (5.26)$$

Along the same lines, a multidimensional companion to Example 5.1 can be shown to exist.

THEOREM 5.6. Let $f(\mathbf{x})$ be a function on $\bar{I}^k$ for which the partial derivative $\partial^k f / \partial x^{(1)} \cdots \partial x^{(k)}$ is continuous on $\bar{I}^k$. For a finite sequence ω of points $\mathbf{x}_1, \ldots, \mathbf{x}_N$ in I^k, we have then

$$\left| \frac{1}{N} \sum_{n=1}^{N} f(\mathbf{x}_n) - \int_{\bar{I}^k} f(\mathbf{x})\, d\mathbf{x} \right| \leq \sum_{p=1}^{k} \sum_{1,\ldots,k;p}^{*} D_N^*(\omega_{p+1,\ldots,k})$$

$$\cdot \int_{\bar{I}^p} \left| \frac{\partial^p f(x^{(1)}, \ldots, x^{(p)}, 1, \ldots, 1)}{\partial x^{(1)} \cdots \partial x^{(p)}} \right| dx^{(1)} \cdots dx^{(p)}, \quad (5.27)$$

where $\omega_{p+1,\ldots,k}$ and $D_N^*(\omega_{p+1,\ldots,k})$ have the same meaning as in Theorem 5.5. Moreover, for $p = k$, the symbol $\omega_{p+1,\ldots,k}$ is understood to be ω.

PROOF. Let $g(\mathbf{x})$ be defined by (5.19). We have shown that the expressions in (5.23) and (5.24) are identical for a given admissible double partition. We observed also that the second term in (5.23) is a Riemann sum for the integral $\int_{\bar{I}^k} f(\mathbf{x})\, d\mathbf{x}$. Now the sum

$$\sum_{i_1=0}^{m_1} \cdots \sum_{i_p=0}^{m_p} g(\eta_{i_1}^{(1)}, \ldots, \eta_{i_p}^{(1)}, 1, \ldots, 1)\, \Delta_{1,\ldots,p} f(\xi_{i_1}^{(1)}, \ldots, \xi_{i_p}^{(p)}, 1, \ldots, 1)$$

occurring in (5.24) is a Riemann-Stieltjes sum for the Riemann-Stieltjes integral $\int_{\bar{I}^p} g(x^{(1)}, \ldots, x^{(p)}, 1, \ldots, 1)\, df(x^{(1)}, \ldots, x^{(p)}, 1, \ldots, 1)$, the latter being equal to the Riemann integral

$$\int_{\bar{I}^p} g(x^{(1)}, \ldots, x^{(p)}, 1, \ldots, 1) \frac{\partial^p f(x^{(1)}, \ldots, x^{(p)}, 1, \ldots, 1)}{\partial x^{(1)} \cdots \partial x^{(p)}}\, dx^{(1)} \cdots dx^{(p)}.$$

Therefore, if we choose a sequence of admissible double partitions satisfying (5.26), then we arrive at the identity

$$\frac{1}{N}\sum_{n=1}^{N} f(\mathbf{x}_n) - \int_{\bar{I}^k} f(\mathbf{x})\, d\mathbf{x} = \sum_{p=1}^{k} (-1)^p \sum_{1,\ldots,k;p}^{*} \int_{\bar{I}^p} g(x^{(1)}, \ldots, x^{(p)}, 1, \ldots, 1)$$
$$\cdot \frac{\partial^p f(x^{(1)}, \ldots, x^{(p)}, 1, \ldots, 1)}{\partial x^{(1)} \cdots \partial x^{(p)}}\, dx^{(1)} \cdots dx^{(p)}. \quad (5.28)$$

Taking absolute values and estimating $|g(x^{(1)}, \ldots, x^{(p)}, 1, \ldots, 1)|$ by $D_N^*(\omega_{p+1,\ldots,k})$, we obtain the desired inequality. ■

Good Lattice Points

We indicate how a certain kind of low-discrepancy sequences used in numerical integration can be found.

THEOREM 5.7. Let p be a prime number and $k \geq 2$. Then there exists a lattice point $\mathbf{g} = (g_1, \ldots, g_k)$, with $1 \leq g_j \leq p - 1$ for $1 \leq j \leq k$, such that the discrepancy D_p of the sequence $((n/p)\mathbf{g})$, $n = 1, 2, \ldots, p$, satisfies

$$D_p \leq c_k \frac{\log^k p}{p}, \quad (5.29)$$

where c_k is an effectively computable constant only depending on k.

PROOF. We use the theorem of Erdös-Turán-Koksma (see notes in Section 2) with the same notation introduced there. Let $\mathbf{g} = (g_1, \ldots, g_k)$ be a lattice point with $1 \leq g_j \leq p - 1$ for $1 \leq j \leq k$. Applying the said theorem with $m = p - 1$, we obtain

$$D_p \leq C\left(\frac{1}{p-1} + \sum_{0 < \|\mathbf{h}\| < p} \frac{1}{r(\mathbf{h})} \left| \frac{1}{p} \sum_{n=1}^{p} e^{2\pi i (n/p)\langle \mathbf{h}, \mathbf{g}\rangle} \right|\right).$$

Now the exponential sum occurring here is either 0 or p, depending on whether $\langle \mathbf{h},\mathbf{g}\rangle \not\equiv 0 \pmod p$ or $\langle \mathbf{h},\mathbf{g}\rangle \equiv 0 \pmod p$. Therefore,

$$D_p \le C\left(\frac{1}{p-1} + \sum_{\substack{0<\|\mathbf{h}\|<p \\ \langle \mathbf{h},\mathbf{g}\rangle\equiv 0 \pmod p}} r^{-1}(\mathbf{h})\right).$$

To prove our result, it thus suffices to show that there exists $\mathbf{g}$ for which

$$\sum_{\substack{0<\|\mathbf{h}\|<p \\ \langle \mathbf{h},\mathbf{g}\rangle\equiv 0 \pmod p}} r^{-1}(\mathbf{h}) \le c_k' \frac{\log^k p}{p}.$$

This is accomplished by an averaging procedure. We consider the sum

$$S = \frac{1}{(p-1)^k} \sum_{\mathbf{g}}{}^{*} \sum_{\substack{0<\|\mathbf{h}\|<p \\ \langle \mathbf{h},\mathbf{g}\rangle\equiv 0 \pmod p}} r^{-1}(\mathbf{h}), \tag{5.30}$$

where $\sum_{\mathbf{g}}^{*}$ denotes the sum over all $(p-1)^k$ lattice points $\mathbf{g}$ running in the competition. We count how often a fixed lattice point $\mathbf{h} = (h_1, \ldots, h_k)$ with $0 < \|\mathbf{h}\| < p$ occurs in the inner sum in (5.30). This will happen as often as there are lattice points $\mathbf{g} = (g_1, \ldots, g_k)$ of the type under consideration for which

$$h_1 g_1 + \cdots + h_k g_k \equiv 0 \pmod p. \tag{5.31}$$

We note that $h_j \not\equiv 0 \pmod p$ for some coordinate h_j of $\mathbf{h}$. To satisfy the congruence (5.31), we may therefore pick arbitrary $g_1, \ldots, g_{j-1}, g_{j+1}, \ldots, g_k$, the remaining number g_j being uniquely determined in the least residue system mod p. Of course, g_j may turn out to be zero, thus producing no acceptable lattice point $\mathbf{g}$; but, at any rate, the number of $\mathbf{g}$ satisfying (5.31) can be at most $(p-1)^{k-1}$. Consequently, we get

$$S \le \frac{1}{p-1} \sum_{0<\|\mathbf{h}\|<p} r^{-1}(\mathbf{h}) < \frac{2}{p} \sum_{0\le\|\mathbf{h}\|<p} r^{-1}(\mathbf{h}).$$

Now

$$\sum_{0\le\|\mathbf{h}\|<p} r^{-1}(\mathbf{h}) = \sum_{h_1=-p+1}^{p-1} \cdots \sum_{h_k=-p+1}^{p-1} \frac{1}{\max(1,|h_1|)\cdots\max(1,|h_k|)}$$
$$= \left(\sum_{h=-p+1}^{p-1} \frac{1}{\max(1,|h|)}\right)^k = \left(3 + 2\sum_{h=2}^{p-1} \frac{1}{h}\right)^k.$$

For $p \ge 3$, we have $3 + 2\sum_{h=2}^{p-1} 1/h < 3 + 2\log p < 5\log p$. It is verified by inspection that this upper bound also holds for $p = 2$. Altogether, we have shown

$$S < \frac{2}{p}(5\log p)^k.$$

It follows from the definition of S that there is a lattice point $\mathbf{g}$ for which

$$\sum_{\substack{0<\|\mathbf{h}\|<p \\ \langle\mathbf{h},\mathbf{g}\rangle\equiv 0 \,(\mathrm{mod}\, p)}} r^{-1}(\mathbf{h}) < \frac{2}{p}(5\log p)^k. \quad \blacksquare \tag{5.32}$$

EXAMPLE 5.4. Lattice points $\mathbf{g}$ that satisfy condition (5.32) might be called *good lattice points* (*modulo* p). Let us exhibit how these lattice points may be used in numerical integration. Suppose f is a function on $\mathbb{R}^k$ represented by the multiple Fourier series $f(\mathbf{x}) = \sum_{\mathbf{h}} c_{\mathbf{h}} e^{2\pi i\langle\mathbf{h},\mathbf{x}\rangle}$, where the sum is over all lattice points $\mathbf{h}$ in $\mathbb{Z}^k$. Assume that the Fourier coefficients $c_{\mathbf{h}}$ with $\mathbf{h} \neq \mathbf{0}$ satisfy $|c_{\mathbf{h}}| \leq Mr^{-q}(\mathbf{h})$ for some constants $M > 0$ and $q > 1$. The Fourier series is then absolutely and uniformly convergent. Choose a prime p and a good lattice point $\mathbf{g}$ mod p. We note that the Fourier coefficient $c_{\mathbf{h}}$ corresponding to $\mathbf{h} = \mathbf{0}$ is just $\int_{\bar{I}^k} f(\mathbf{x})\, d\mathbf{x}$. Therefore,

$$\left|\frac{1}{p}\sum_{n=1}^{p} f\left(\frac{n}{p}\mathbf{g}\right) - \int_{\bar{I}^k} f(\mathbf{x})\, d\mathbf{x}\right| = \left|\sum_{\|\mathbf{h}\|>0} c_{\mathbf{h}}\left(\frac{1}{p}\sum_{n=1}^{p} e^{2\pi i(n/p)\langle\mathbf{h},\mathbf{g}\rangle}\right)\right|$$
$$= \left|\sum_{\substack{\|\mathbf{h}\|>0 \\ \langle\mathbf{h},\mathbf{g}\rangle\equiv 0\,(\mathrm{mod}\, p)}} c_{\mathbf{h}}\right| \leq M \sum_{\substack{\|\mathbf{h}\|>0 \\ \langle\mathbf{h},\mathbf{g}\rangle\equiv 0\,(\mathrm{mod}\, p)}} r^{-q}(\mathbf{h}).$$

We split up this last sum into two parts. We first consider the sum over those $\mathbf{h}$ for which all coordinates are multiples of p; that is, $\mathbf{h} = p\mathbf{a}$ with a lattice point $\mathbf{a} \neq \mathbf{0}$. We observe that for those $\mathbf{h}$, the condition $\langle\mathbf{h}, \mathbf{g}\rangle \equiv 0 \pmod p$ holds automatically. We have

$$\sum_{\substack{\mathbf{h}=p\mathbf{a} \\ \|\mathbf{a}\|>0}} r^{-q}(\mathbf{h}) < p^{-q}\sum_{\mathbf{a}} r^{-q}(\mathbf{a})$$
$$= p^{-q}\sum_{a_1=-\infty}^{\infty}\cdots\sum_{a_k=-\infty}^{\infty} (\max(1,|a_1|))^{-q}\cdots(\max(1,|a_k|))^{-q}$$
$$= p^{-q}\left(\sum_{a=-\infty}^{\infty}(\max(1,|a|))^{-q}\right)^k = p^{-q}\left(1 + 2\sum_{a=1}^{\infty} a^{-q}\right)^k$$
$$= p^{-q}(1 + 2\zeta(q))^k, \tag{5.33}$$

where ζ denotes the Riemann zeta-function.

Let the sum over the remaining lattice points $\mathbf{h}$ be denoted by $\sum$. These lattice points are uniquely represented in the form $\mathbf{h} = \mathbf{h}^* + p\mathbf{a}$ with a lattice point $\mathbf{a}$ and a lattice point $\mathbf{h}^* = (h_1^*, \ldots, h_k^*)$ satisfying

$$\langle\mathbf{h}^*, \mathbf{g}\rangle \equiv 0 \,(\mathrm{mod}\, p), \quad \|\mathbf{h}^*\| > 0, \quad \text{and} \quad -\frac{p}{2} < h_j^* \leq \frac{p}{2} \text{ for } 1 \leq j \leq k. \tag{5.34}$$

Therefore,

$$\sum = \sum_{\mathbf{a}}\sum_{\mathbf{h}^*}{}^{*} r^{-q}(\mathbf{h}^* + p\mathbf{a}), \tag{5.35}$$

where $\sum_{\mathbf{h}^*}^*$ stands for the sum over all $\mathbf{h}^*$ satisfying (5.34). We claim that

$$r(\mathbf{h}^* + p\mathbf{a}) \geq r(\mathbf{h}^*)r(\mathbf{a}). \tag{5.36}$$

It suffices to show that

$$\max(1, |h_j^* + pa_j|) \geq \max(1, |h_j^*|) \max(1, |a_j|) \tag{5.37}$$

holds for $1 \leq j \leq k$, where $\mathbf{a} = (a_1, \ldots, a_k)$. But (5.37) is trivial whenever $h_j^* = 0$ or $a_j = 0$. If both h_j^* and a_j are nonzero, then

$$|h_j^* + pa_j| \geq p\,|a_j| - |h_j^*| \geq p\,|a_j| - \frac{p}{2} = \frac{p}{2}(2\,|a_j| - 1) \geq |h_j^*|\,|a_j|,$$

and so (5.37) is shown in all cases. From (5.35) and (5.36) we infer

$$\sum \leq \sum_{\mathbf{a}} \sum_{\mathbf{h}^*}^* r^{-q}(\mathbf{h}^*) r^{-q}(\mathbf{a}) = \left(\sum_{\mathbf{a}} r^{-q}(\mathbf{a})\right)\left(\sum_{\mathbf{h}^*}^* r^{-q}(\mathbf{h}^*)\right).$$

The first sum was already evaluated explicitly in (5.33), namely,

$$\sum_{\mathbf{a}} r^{-q}(\mathbf{a}) = (1 + 2\zeta(q))^k.$$

Recalling that $\mathbf{g}$ is a good lattice point, we obtain for the second sum

$$\sum_{\mathbf{h}^*}^* r^{-q}(\mathbf{h}^*) \leq \sum_{\substack{0 < \|\mathbf{h}\| < p \\ \langle \mathbf{h}, \mathbf{g}\rangle \equiv 0 \pmod p}} r^{-q}(\mathbf{h}) \leq \left(\sum_{\substack{0 < \|\mathbf{h}\| < p \\ \langle \mathbf{h}, \mathbf{g}\rangle \equiv 0 \pmod p}} r^{-1}(\mathbf{h})\right)^q$$
$$< \left(\frac{2}{p}\right)^q (5 \log p)^{kq}.$$

Thus, we finally arrive at the inequality

$$\left|\frac{1}{p}\sum_{n=1}^{p} f\left(\frac{n}{p}\mathbf{g}\right) - \int_{I^k} f(\mathbf{x})\, d\mathbf{x}\right| < M(1 + 2\zeta(q))^k \frac{1 + 2^q(5 \log p)^{kq}}{p^q},$$

and so at an error term of the order $(\log p)^{kq}/p^q$. ∎

Notes

Theorem 5.1 is from Koksma [6]. For a special case, see Pólya and Szegö [1, II. Abschn., Aufg. 9]. Earlier investigations in this direction mostly concentrated on sums of the type $(1/N)\sum_{n=1}^{N} f(\{n\alpha\})$ where α is irrational and f is a Bernoulli polynomial (note that in this case $\int_0^1 f(t)\, dt = 0$). The first result seems to be due to Lerch [1], who showed

$$\sum_{n=1}^{N} (\{n\alpha\} - \tfrac{1}{2}) = O(\log N)$$

for α with bounded partial quotients, thereby answering problems posed by Franel [1, 2]. The subject was taken up again by Hecke [1], Ostrowski [1], Hardy and Littlewood

[3, 4], and Behnke [1, 2]. A detailed account of these results can be found in Koksma [4, Kap. 9]. More recent references are Hartman [2], Sós [1], Mikolás [1], and Lesca [5]. For results on $(1/N)\Sigma_{n=1}^{N}(\{q^n\alpha\} - 1/2)$ with irrational α and an integer $q > 1$, see Korobov [3, 4, 6].

An interesting unsolved problem occurs in connection with Corollary 5.1 (see Exercises 5.7, 5.8, and 5.9). Van der Corput and Pisot [1] proved the weaker inequality

$$\left|\frac{1}{N}\sum_{n=1}^{N} e^{2\pi i x_n}\right| \leq 2\pi D_N.$$

The first result of this type is in Behnke [2]. Postnikova [4] estimates these exponential sums, given the distribution of the x_n in intervals of small length (see Judin [1] for a multidimensional version). A number of integral identities such as Theorems 5.2 and 5.3 were given by Koksma [7]. See also Koksma [9, 13] and Exercises 5.12, 5.13, 5.32, and 5.33 for further results along these lines.

Theorem 5.4 was found by Niederreiter [5]. The upper bound $3ND_N^*M(1/N)$ in Corollary 5.2 is an unpublished result of Koksma mentioned in Koksma [16]. In an arbitrary number k of dimensions, an upper bound for the integration error of the form

$$(2^{2k-1} + 1)M([D_N^{*-1}]^{-1/k}),$$

where M is the modulus of continuity of the continuous integrand f, was established by Hlawka [26], who also has error estimates for arbitrary Riemann-integrable functions in $\bar{I}^k$. See also Niederreiter [5, 13]. Error estimates for special sequences can be found in Pólya and Szegö [1, II. Abschn., Aufg. 10–12] and Chui [1]. Some extensions of Koksma's inequality in the one-dimensional case were shown by Helmberg [7] and Hlawka and Mück [1].

The generalization of Koksma's inequality to several dimensions was first achieved by Hlawka [12]. The notion of bounded variation employed here was introduced by Hardy [1] and Krause [1] in their work on double Fourier series. A treatment of this concept of variation may also be found in Hobson [1, Sections 253 and 254] and H. Hahn [1, p. 539 ff.]. Our proofs of Lemma 5.2 and Theorem 5.5 follow Zaremba [3]. In a different context—namely, when estimating the error between a multiple sum and its approximation by a simple sum—Korobov [22] arrives by analogous methods at an inequality that has a striking similarity to the Koksma-Hlawka inequality.

Theorem 5.6 was again first proved by Hlawka [11, 12]. Our proof uses the method of Zaremba [3]. A similar inequality was given by Sobol' [3, 7]. Zaremba [3] shows an estimate for the integration error in terms of the L^2 discrepancy. An account of Zaremba's method is also given in Halton [4]. For a special case, see Sobol' [3]. Based on Theorem 5.6, one obtains generalizations of Corollary 5.1 to several dimensions (see Hlawka [11] and Exercise 5.25). For other applications, see Hlawka and Mück [2].

Koksma's inequality may be generalized to more abstract settings. Niederreiter [1] established such an inequality for compact abelian groups with countable base. The special case of the group of p-adic integers was treated earlier by Beer [1]. Using a completely different notion of discrepancy, K. Schmidt [3] has shown analogues of Koksma's inequality for locally compact abelian groups with countable base. Niederreiter [1] also has an estimate for the integration error in terms of the Fourier coefficients of the integrand.

Numerical integration methods based on the sequences of van der Corput-Halton and Hammersley and variants thereof were studied by Sobol' [1, 2, 3, 6, 7]. See Sobol' [4, 7] and K. Schmidt and Zinterhof [1] for numerical integration in the infinite-dimensional unit cube.

Our exposition of Theorem 5.7 and Example 5.4 is based on results of Hlawka [14, 17]. The first investigations along these lines were carried out by Korobov [15, 16]. He restricts his attention to lattice points of the form $\mathbf{g} = (1, a, a^2, \ldots, a^{k-1})$ ("optimal coefficients"). A detailed exposition can be found in Korobov [20]. The order of magnitude of the integration error that he obtains is the same as Hlawka's. There are many Russian papers on optimal coefficients, among them Bahvalov [1, 2], Bahvalov, Korobov, and Čencov [1], Korobov [17, 22], Šahov [2], Saltykov [1], Šarygin [1, 2], Solodov [1, 2, 4], and Stoyantsev [1]. An exposition of the subject is also given in Sobol' [7]. Concerning the explicit construction of good lattice points, see Zaremba [1, 6] and Hua and Wang [2, 3]. Numerical data are compiled in Haber [4] and Maisonneuve [1]. Further theoretical results on good lattice points can be found in Zaremba [4, 9, 10, 11]. Related problems are discussed in Hsu [1], Hua and Wang [1], and Haber and Osgood [1, 2]. A heuristic approach to the subject is presented in Conroy [1].

The basic result for still another number-theoretical integration method is described in Exercise 5.22. The method goes back to Richtmyer [1, 2] and Peck [1], who considered the multidimensional case with an integrand satisfying restrictions as in Example 5.4. Similar ideas were used by Bass and Guillod [1]. For the one-dimensional case, see Bass [2], J.-P. Bertrandias [1], and Couot [1]. For the multidimensional case, see Haselgrove [1], Couot [2], Zinterhof [1] (based on A. Baker [1]), and Niederreiter [11]. The method was improved decisively by Haselgrove [1] and Niederreiter [13]. Numerical data can be found in Davis and Rabinowitz [1] and Roos and Arnold [1]. For further remarks, see Jagerman [3] and Richtmyer, Devaney, and Metropolis [1]. As to the restrictions on the integrand mentioned in Example 5.4, see Haselgrove [1], Hlawka [14], Korobov [20], Niederreiter [11], Šarygin [1], and Zaremba [5] for transforming a nonperiodic integrand into a periodic one, and Zaremba [3] for more tractable sufficient conditions. See also Exercise 5.26.

Hlawka [15, 17], Niederreiter [13], and Zaremba [7] showed that number-theoretical integration methods can be adapted to work for more general classes of integration domains. Sobol' [8] handles certain improper integrals by these methods. The paper of Hlawka and Kuich [1] is a continuation of Hlawka [15]. For a certain general class of integration domains, Solodov [3] shows that the method of optimal coefficients produces satisfactory results.

The above number-theoretical methods have been successfully applied to other areas. A rather immediate application is to integral equations. Russian authors such as Šarygin [1], Šahov [1], and Korobov [19, 20] use optimal coefficients, whereas Hlawka [13, 17] and Hlawka and Kreiter [1] work with low-discrepancy sequences (some of which are constructed by means of good lattice points). Other interesting applications are to interpolation problems: Hlawka [18, 20, 22], Korobov [19], Rjaben'kiĭ [1], Šarygin [2], Smoljak [1], Zinterhof [1]. Korobov [22] uses his method of optimal coefficients to approximately compute multiple sums. Rjaben'kiĭ [2] applies it to the Cauchy problem. In a series of papers, Hlawka [19, 21, 23] studies applications to kinetic gas theory.

Expository accounts of number-theoretical and other integration methods are given in Beresin and Shidkow [1], Davis and Rabinowitz [2], Haber [3], Halton [3], Hammersley and Handscomb [1, Chapter 3], Hlawka [17], Korobov [20], Shreider [1], and Zaremba [2].

Exercises

5.1. Indicate briefly why every function of bounded variation on [0, 1] is Riemann-integrable on [0, 1].

5.2. Carry out the following alternative proof of Koksma's inequality, which avoids the use of Riemann-Stieltjes integrals: Let $0 = t_0 < t_1 < \cdots < t_{k+1} = 1$ be a subdivision of $[0, 1]$ and define $I_j = [t_j, t_{j+1})$ for $0 \leq j \leq k$. Let g be a step function corresponding to that subdivision, that is,

$$g(x) = \sum_{j=0}^{k} a_j c_{I_j}(x) \qquad \text{for } 0 \leq x < 1, \quad g(1) = a_{k+1},$$

where $a_0, \ldots, a_{k+1}$ are arbitrary real numbers. Prove the identity

$$\frac{1}{N} \sum_{n=1}^{N} g(x_n) - \int_0^1 g(x)\, dx$$
$$= \frac{1}{N} \sum_{n=1}^{N} \sum_{j=0}^{k} a_j (c_{[0,t_{j+1})}(x_n) - c_{[0,t_j)}(x_n) - t_{j+1} + t_j)$$

for any N given points $x_1, \ldots, x_N$ in I.

5.3. Using Abel's summation formula in the identity of Exercise 5.2, prove that Koksma's inequality holds for the step function g.

5.4. Let f be a function of bounded variation $V(f)$ on $[0, 1]$, and consider a subdivision of $[0, 1]$ as in Exercise 5.2. Let $f_1(x)$ be the lower Darboux step function corresponding to f and the subdivision, that is, $f_1(x) = \sum_{j=0}^{k} a_j c_{I_j}(x)$ for $0 \leq x < 1$, and $f_1(1) = f(1)$, where $a_j = \inf_{x \in I_j} f(x)$ for $0 \leq j \leq k$. Prove that $V(f_1) \leq V(f)$.

5.5. Prove the same assertion as in the preceding exercise for the upper Darboux step function f_2.

5.6. Based on the results of Exercises 5.3, 5.4, and 5.5, prove Koksma's inequality for f.

5.7. For $N \geq 1$, define $c_N = \sup_\omega (1/ND_N^*(\omega)) \left|\sum_{n=1}^{N} e^{2\pi i x_n}\right|$, where the supremum is extended over all finite sequences ω of N real numbers $x_1, \ldots, x_N$. Prove that $2 \leq c_N \leq 4$ for all $N \geq 1$.

5.8. In the notation of the preceding exercise, show that $c_1 = 2$.

5.9. In the notation of Exercise 5.7, show that

$$c_2 = \max_{0 \leq x \leq 1/2} \frac{4 \sin \pi x}{1 + 2x} = 2.11 \cdots.$$

5.10. Prove that if f has a continuous derivative on $[0, 1]$, then f is of bounded variation on $[0, 1]$ and $V(f) = \int_0^1 |f'(t)|\, dt$. Give a detailed argument not using Riemann-Stieltjes integrals.

5.11. Prove that for any constant $c < 1$ there exists a finite sequence $x_1, \ldots, x_N$ in I, where N may depend on c, such that $|(1/N)\sum_{n=1}^{N} x_n - 1/2| \geq cD_N^*$. *Hint:* Consider sequences of the form

$$\underbrace{0, \ldots, 0}_{m}, \frac{1}{N}, \frac{2}{N}, \ldots, \frac{N-m}{N}$$

with $1 \leq m \leq N$.

5.12. For given points $x_1 \leq x_2 \leq \cdots \leq x_N$ in I, let $R_N(t)$ be defined as in Section 2. Verify the identity

$$\int_0^1 R_N^2(t)\,dt = N\sum_{n=1}^{N}\left(x_n - \frac{2n-1}{2N}\right)^2 + \frac{1}{12}.$$

5.13. Let $x_1, \ldots, x_N$ be as in the preceding exercise, let f be a continuous function on $[0, 1]$, and put $F(t) = \int_0^t f(u)\,du$ for $0 \leq t \leq 1$. Then

$$\int_0^1 R_N^2(t)\,df(t) = 2N^2\int_0^1 F(t)\,dt + 2N\sum_{n=1}^{N} x_n f(x_n) - 2N\sum_{n=1}^{N} F(x_n) - 2\sum_{n=1}^{N} nf(x_n) + \sum_{n=1}^{N} f(x_n).$$

5.14. Prove that the identity in Exercise 5.13 includes the identity in Example 5.2 as a special case.

5.15. Prove in detail that for a continuous function f on $[0, 1]$ its modulus of continuity M satisfies $\lim_{h\to 0+0} M(h) = 0$.

5.16. Verify the following property of the modulus of continuity M of a function f: If $M(h) = o(h)$ as $h \to 0 + 0$, then f is a constant function.

5.17. Exhibit a class of nonconstant functions for which $M(h) = O(h)$ as $h \to 0 + 0$.

5.18. Let f be an arbitrary continuous function on $[0, 1]$ with modulus of continuity M_1, and let g be a continuous function from $[0, 1]$ onto $[0, 1]$ with modulus of continuity M_2. Show that the modulus of continuity M of the composite function $f \circ g$ satisfies $M \leq M_1 \circ M_2$. Give an example which shows that in general $M \neq M_1 \circ M_2$.

5.19. Construct an example that shows that the bound $3ND_N^*M(1/N)$ established in Corollary 5.2 may tend to infinity as $N \to \infty$, even though the sequence (x_n) is u.d. mod 1.

5.20. Let f be a continuous function on $[0, 1]$ with Fourier coefficients $a_h = \int_0^1 f(x)e^{-2\pi ihx}\,dx$, $h \in \mathbb{Z}$, such that the infinite series $\sum_{h=1}^{\infty} h\,|a_h|$ is convergent. Prove that $f(x) = \sum_{h=-\infty}^{\infty} a_h e^{2\pi ihx}$.

5.21. For a function f such as in the preceding exercise and for a finite sequence $x_1, \ldots, x_N$ in I with discrepancy D_N^*, show that the following inequality holds:

$$\left|\frac{1}{N}\sum_{n=1}^{N} f(x_n) - \int_0^1 f(x)\,dx\right| \leq 8D_N^*\sum_{h=1}^{\infty} h\,|a_h|.$$

5.22. Let α be an irrational number of finite type $\eta = s$. Let f be a periodic function with period 1, and suppose that f is represented by the Fourier series $f(t) = \sum_{h=-\infty}^{\infty} c_h e^{2\pi iht}$ for which there exist positive constants M and λ such that $|c_h| \leq M\,|h|^{-s-\lambda}$ for all $h \neq 0$. Prove that

$$\frac{1}{N}\sum_{n=1}^{N} f(n\alpha) - \int_0^1 f(t)\,dt = O\left(\frac{1}{N}\right).$$

5.23. Prove in detail that a function f on $\bar{I}^k$, $k \geq 2$, that depends on less than k variables satisfies $V^{(k)}(f) = 0$.

5.24. Let $\mathbf{g} = (g^{(1)}, \ldots, g^{(k)})$ be a lattice point distinct from the origin. Then we have

$$\int_0^1 \cdots \int_0^1 \left| \begin{matrix}\cos\\ \sin\end{matrix} \, 2\pi(g^{(1)}x^{(1)} + \cdots + g^{(k)}x^{(k)}) \right| dx^{(1)} \cdots dx^{(k)} = \frac{2}{\pi}.$$

5.25. Suppose $\mathbf{g} = (g^{(1)}, \ldots, g^{(k)})$ is a lattice point distinct from the origin and that $\mathbf{x}_1, \ldots, \mathbf{x}_N$ are N points in I^k with discrepancy D_N^*. Using the result of the preceding exercise, prove that

$$\left| \frac{1}{N} \sum_{n=1}^{N} e^{2\pi i \langle \mathbf{g}, \mathbf{x}_n \rangle} \right| \leq \frac{2}{\pi} \left(\prod_{j=1}^{k} (1 + 2\pi \, |g^{(j)}|) - 1 \right) D_N^*.$$

5.26. Let f be an arbitrary Riemann-integrable function on $\bar{I}^k$. Prove that the function g on $\bar{I}^k$ defined by

$$g(x_1, \ldots, x_k) = 2^{-k} \sum_{\substack{\varepsilon_1, \ldots, \varepsilon_k \\ \varepsilon_j = 0,1}} f(\varepsilon_1 + (-1)^{\varepsilon_1} x_1, \ldots, \varepsilon_k + (-1)^{\varepsilon_k} x_k)$$

has period 1 in each variable, that is, $g(x_1, \ldots, x_{j-1}, 0, x_{j+1}, \ldots, x_k) = g(x_1, \ldots, x_{j-1}, 1, x_{j+1}, \ldots, x_k)$ for all $x_1, \ldots, x_{j-1}, x_{j+1}, \ldots, x_k$ and all $j = 1, \ldots, k$. Show also that $\int_{\bar{I}^k} f(\mathbf{x})\, d\mathbf{x} = \int_{\bar{I}^k} g(\mathbf{x})\, d\mathbf{x}$.

5.27. Let $P^{(q)}(\mathbf{g})$ be the sum occurring in Example 5.4, namely,

$$P^{(q)}(\mathbf{g}) = \sum_{\substack{\|\mathbf{h}\| > 0 \\ \langle \mathbf{h}, \mathbf{g} \rangle \equiv 0 \pmod{p}}} r^{-q}(\mathbf{h}).$$

Let F be the function on $\bar{I}^k$ defined by

$$F(\mathbf{x}) = F(x_1, \ldots, x_k) = \prod_{j=1}^{k} \left(1 + \frac{\pi^2}{3} - 2\pi^2 x_j + 2\pi^2 x_j^2 \right).$$

Show first that F can be continued to $\mathbb{R}^k$ so as to have period 1 in each variable. Prove that $P^{(2)}(\mathbf{g})$ is just the error committed when integrating $F(\mathbf{x})$ over $\bar{I}^k$ by means of the sequence generated by $\mathbf{g}$, that is,

$$P^{(2)}(\mathbf{g}) = \left| \frac{1}{p} \sum_{n=1}^{p} F\left(\frac{n}{p} \mathbf{g} \right) - \int_{\bar{I}^k} F(\mathbf{x})\, d\mathbf{x} \right|.$$

5.28. Find an interpretation for $P^{(4)}(\mathbf{g})$ similar to the one given for $P^{(2)}(\mathbf{g})$ in the preceding exercise.

5.29. Show that $\sum_{n=1}^{N} (\{x + (n/N)\} - \frac{1}{2}) = \{Nx\} - \frac{1}{2}$ holds for any real number x and for any positive integer N.

5.30. Let a and b be two positive integers which are relatively prime. Prove that $\int_0^1 (\{ax\} - \frac{1}{2})(\{bx\} - \frac{1}{2})\, dx = 1/12ab$.

5.31. Let $a_1, \ldots, a_N$ be positive integers, and set $S_N(x) = \sum_{n=1}^{N} (\{a_n x\} - \frac{1}{2})$. Evaluate $\int_0^1 S_N(x)\,dx$.

5.32. With $a_1, \ldots, a_N$ and $S_N(x)$ as in the preceding exercise, prove the identity

$$\int_0^1 S_N^2(x)\,dx = \frac{1}{12} \sum_{m,n=1}^{N} \frac{(a_m, a_n)}{[a_m, a_n]}.$$

5.33. For an integer $a \geq 2$, put $S_N(x) = \sum_{n=1}^{N} (\{a^n x\} - \frac{1}{2})$. Prove that

$$\int_0^1 S_N^2(x)\,dx = \frac{a+1}{12(a-1)} N - \frac{a}{6(a-1)^2}\left(1 - \frac{1}{a^N}\right).$$

5.34. Prove that for any constant $c < 4$ there exists a finite sequence $x_1, \ldots, x_N$ in I, where N may depend on c, such that $|(1/N)\sum_{n=1}^{N} e^{2\pi i x_n}| \geq cD_N^*$. *Hint:* Consider sequences of the form

$$\underbrace{\varepsilon, \ldots, \varepsilon}_{k}, \frac{1}{N}, \frac{2}{N}, \ldots, \frac{1}{2} - \frac{k}{N}, \frac{1}{2} + \frac{k}{N}, \ldots,$$
$$\frac{N-2}{N}, \frac{N-1}{N}, \underbrace{1-\varepsilon, \ldots, 1-\varepsilon}_{k},$$

with $N \geq 4$ even, $1 \leq k < N/2$, and $\varepsilon > 0$ sufficiently small.

6. QUANTITATIVE DIFFERENCE THEOREMS

Discrepancy of Difference Sequences

Let ω be a finite sequence consisting of the N real numbers $x_1, \ldots, x_N$. Let τ be the finite sequence consisting of the N^2 differences $x_k - x_j$, $1 \leq k, j \leq N$, in some order. We are interested in the relation between the discrepancies of the two sequences.

THEOREM 6.1. If $D = D_N(\omega)$ denotes the discrepancy of ω, and $F = D_{N^2}(\tau)$ denotes the discrepancy of τ, then

$$D \leq c\sqrt{F}(1 + |\log F|) \tag{6.1}$$

holds with an absolute constant c.

PROOF. For a positive integer r, we put $S_r(\omega) = \sum_{j=1}^{N} \exp(rx_j)$ and $S_r(\tau) = \sum_{k=1}^{N}\sum_{j=1}^{N} \exp(r(x_k - x_j))$, where $\exp(t) = e^{2\pi i t}$ for $t \in \mathbb{R}$. We have

$$|S_r(\omega)|^2 = S_r(\omega)\overline{S_r(\omega)} = \left(\sum_{j=1}^{N} \exp(rx_j)\right)\left(\sum_{j=1}^{N} \exp(-rx_j)\right) = S_r(\tau).$$

Let m be a positive integer that will be specified later on. By (2.42), we obtain

$$D \leq c_1\left(\frac{1}{m} + \frac{1}{N}\sum_{r=1}^{m}\frac{1}{r}\,|S_r(\omega)|\right)$$

with an absolute constant c_1. Now apply the Cauchy-Schwarz inequality to the sum $\sum_{r=1}^{m}(1/r)\,|S_r(\omega)| = \sum_{r=1}^{m}(1/r)^{1/2}((1/r)\,|S_r(\omega)|^2)^{1/2}$. Then

$$D \leq \frac{c_1}{m} + \frac{c_1}{N}\left(\sum_{r=1}^{m}\frac{1}{r}\right)^{1/2}\left(\sum_{r=1}^{m}\frac{1}{r}\,|S_r(\omega)|^2\right)^{1/2}.$$

But $\sum_{r=1}^{m}(1/r) \leq 1 + \log m$ and

$$\sum_{r=1}^{m}\frac{1}{r}\,|S_r(\omega)|^2 = \sum_{r=1}^{m}\frac{1}{r}\,S_r(\tau) = \sum_{k=1}^{N}\sum_{j=1}^{N}\left(\sum_{r=1}^{m}\frac{1}{r}\exp\left(r(x_k - x_j)\right)\right) \leq VN^2F$$

by Example 5.1, where $V = \int_0^1 |f'(x)|\,dx$ with $f(x) = \sum_{r=1}^{m}(1/r)\exp(rx)$. Therefore,

$$D \leq \frac{c_1}{m} + c_1(1 + \log m)^{1/2}(VF)^{1/2}. \tag{6.2}$$

We estimate now $V = 2\pi\int_0^1 |\sum_{r=1}^{m}\exp(rx)|\,dx$ for $m \geq 2$. We write

$$\begin{aligned}\int_0^1\left|\sum_{r=1}^{m}\exp(rx)\right|dx &= \int_0^{1/m}\left|\sum_{r=1}^{m}\exp(rx)\right|dx\\ &\quad + \int_{1/m}^{1-1/m}\left|\sum_{r=1}^{m}\exp(rx)\right|dx\\ &\quad + \int_{1-1/m}^{1}\left|\sum_{r=1}^{m}\exp(rx)\right|dx.\end{aligned}$$

For the first and the last integral, we use the trivial estimate $|\sum_{r=1}^{m}\exp(rx)| \leq m$, and so, these integrals are each at most 1. We note that $|\sum_{r=1}^{m}\exp(rx)| \leq 1/\sin \pi x$ for $0 < x < 1$. But $\sin \pi x \geq 2x$ for $0 \leq x \leq \frac{1}{2}$, and $\sin \pi x \geq 2 - 2x$ for $\frac{1}{2} \leq x \leq 1$, and therefore,

$$\int_{1/m}^{1-1/m}\left|\sum_{r=1}^{m}\exp(rx)\right|dx \leq \frac{1}{2}\int_{1/m}^{1/2}\frac{dx}{x} + \frac{1}{2}\int_{1/2}^{1-1/m}\frac{dx}{1-x} = \log\tfrac{1}{2} + \log m.$$

Altogether, we have shown $V \leq 2\pi(2 + \log\frac{1}{2} + \log m)$, and so,

$$V \leq 4\pi(1 + \log m). \tag{6.3}$$

The inequality (6.3) holds for $m = 1$ as well. Going back to (6.2), we derive the inequality

$$D \leq \frac{c_1}{m} + c_2(1 + \log m)\sqrt{F}.$$

Now we put $m = [F^{-1/2}]$, and we arrive at (6.1). ∎

Using some ideas from the above proof, we can now establish a quantitative version of van der Corput's theorem (see Chapter 1, Theorem 3.1). Let ω be a finite sequence of N real numbers $x_1, \ldots, x_N$. For $1 \leq j \leq N-1$, let ω_j be the sequence of $N-j$ differences $x_{j+1}-x_1, x_{j+2}-x_2, \ldots, x_N - x_{N-j}$. We estimate the discrepancy of ω in terms of the discrepancies of the sequences ω_j.

THEOREM 6.2. Let $D = D_N(\omega)$ be the discrepancy of ω, and let $D^{(j)} = D_{N-j}(\omega_j)$ be the discrepancy of ω_j. Then, for every integer H with $1 \leq H \leq N$, we have

$$D \leq cB(1 + |\log B|), \tag{6.4}$$

where $B = H^{-1/2}(1 + (1/N)\sum_{j=1}^{H-1}(N-j)D^{(j)})^{1/2}$ and c is an absolute constant.

PROOF. We have shown in the proof of the preceding theorem that

$$D \leq \frac{c_1}{m} + \frac{c_1}{N}(1 + \log m)^{1/2}\left(\sum_{r=1}^{m}\frac{1}{r}|S_r(\omega)|^2\right)^{1/2} \tag{6.5}$$

for any positive integer m. To estimate $|S_r(\omega)|^2$, we use the fundamental inequality (see Chapter 1, Lemma 3.1) with $u_n = \exp(rx_n)$. We obtain for every H with $1 \leq H \leq N$:

$$\begin{aligned} H^2|S_r(\omega)|^2 &\leq (H+N-1)HN \\ &\quad + 2(H+N-1)\sum_{h=1}^{H-1}(H-h)\sum_{n=1}^{N-h}\operatorname{Re}\exp(r(x_{n+h}-x_n)) \\ &\leq 2HN^2 + 2(H+N-1)\sum_{h=1}^{H-1}(H-h)\operatorname{Re} S_r(\omega_h), \end{aligned}$$

where $S_r(\omega_h) = \sum_{n=1}^{N-h}\exp(r(x_{n+h}-x_n))$. It follows that

$$|S_r(\omega)|^2 \leq 2N^2H^{-1} + 2(H+N-1)H^{-2}\sum_{h=1}^{H-1}(H-h)\operatorname{Re} S_r(\omega_h).$$

Using $\sum_{r=1}^{m}(1/r) \leq 1 + \log m$, we arrive at

$$\begin{aligned} \sum_{r=1}^{m}\frac{1}{r}|S_r(\omega)|^2 &\leq 2N^2H^{-1}(1+\log m) \\ &\quad + 2(H+N-1)H^{-2}\sum_{h=1}^{H-1}(H-h)\sum_{r=1}^{m}\operatorname{Re}\frac{1}{r}S_r(\omega_h). \end{aligned} \tag{6.6}$$

By Example 5.1, we have

$$\begin{aligned} \sum_{r=1}^{m}\operatorname{Re}\frac{1}{r}S_r(\omega_h) &= \operatorname{Re}\sum_{n=1}^{N-h}\sum_{r=1}^{m}\frac{1}{r}\exp(r(x_{n+h}-x_n)) \\ &\leq \left|\sum_{n=1}^{N-h}f(x_{n+h}-x_n)\right| \leq (N-h)VD^{(h)}, \end{aligned} \tag{6.7}$$

where f and V have the same meaning as in the proof of Theorem 6.1. Combining (6.3), (6.6), and (6.7), we obtain

$$\sum_{r=1}^{m} \frac{1}{r} |S_r(\omega)|^2 \leq 2N^2H^{-1}(1 + \log m)$$
$$+ 16\pi NH^{-1}(1 + \log m) \sum_{h=1}^{H-1} (N - h)D^{(h)}$$
$$\leq 16\pi N^2(1 + \log m)B^2.$$

Returning to (6.5), we arrive at

$$D \leq \frac{c_1}{m} + c_3 B(1 + \log m).$$

If $B \leq 1$, we put $m = [1/B]$, and the result follows. If $B > 1$, then we put $m = [B]$, and the proof is complete. ■

An Integral Identity

In connection with Theorem 6.1, the following integral identity may also be of interest. Let $x_1, \ldots, x_N$ be N real numbers. For two real numbers α and β with $\alpha \leq \beta \leq \alpha + 1$, let $A(\alpha, \beta)$ be the number of x_n, $1 \leq n \leq N$, which lie mod 1 in $[\alpha, \beta)$. Define the error term $R_N(\alpha, \beta)$ by $R_N(\alpha, \beta) = A(\alpha, \beta) - N(\beta - \alpha)$. Likewise, let $R^*_{N^2}(\alpha, \beta)$ be the error term relative to the N^2 real numbers $x_k - x_j$ with $1 \leq k, j \leq N$, that is, $R^*_{N^2}(\alpha, \beta) = A^*(\alpha, \beta) - N^2(\beta - \alpha)$ with $A^*(\alpha, \beta)$ having the obvious meaning.

THEOREM 6.3. With the above notations, the identity

$$\int_0^1 R_N^2(\alpha - t, \alpha + t)\, d\alpha = \int_0^{2t} R^*_{N^2}(-\alpha, \alpha)\, d\alpha \tag{6.8}$$

holds for all t with $0 \leq t \leq \frac{1}{4}$.

PROOF. For α and β with $\alpha \leq \beta \leq \alpha + 1$, let $c(\alpha, \beta, u)$ be the characteristic function of $[\alpha, \beta)$ mod 1; that is, $c(\alpha, \beta, u) = 1$ if $\alpha \leq u < \beta$ (mod 1), and $= 0$ otherwise. Then for any t, $0 \leq t \leq \frac{1}{2}$, and for any u we have

$$\int_0^1 c(\alpha - t, \alpha + t, u)\, d\alpha = 2t, \tag{6.9}$$

and also

$$\int_0^1 c(\alpha - t, \alpha + t, u) c(\alpha - t, \alpha + t, v)\, d\alpha$$
$$= \begin{cases} 2t - \langle u - v \rangle & \text{if } \langle u - v \rangle \leq 2t \\ 0 \text{ otherwise.} \end{cases} \tag{6.10}$$

Furthermore, if $0 \leq \tau \leq \frac{1}{2}$, we have evidently

$$\int_0^\tau c(-\alpha, \alpha, w)\, d\alpha = \begin{cases} \tau - \langle w \rangle & \text{if } \langle w \rangle \leq \tau \\ 0 & \text{otherwise.} \end{cases} \tag{6.11}$$

If $0 \leq t \leq \frac{1}{4}$, then we obtain from (6.10) and (6.11) the identity

$$\int_0^1 c(\alpha - t, \alpha + t, u) c(\alpha - t, \alpha + t, v)\, d\alpha = \int_0^{2t} c(-\alpha, \alpha, u - v)\, d\alpha. \tag{6.12}$$

Now $R_N(\alpha - t, \alpha + t) = \sum_{n=1}^N c(\alpha - t, \alpha + t, x_n) - 2tN$; hence,

$$\begin{aligned}\int_0^1 R_N^2(\alpha - t, \alpha + t)\, d\alpha &= \sum_{k,j=1}^N \int_0^1 c(\alpha - t, \alpha + t, x_k) c(\alpha - t, \alpha + t, x_j)\, d\alpha \\ &\quad - 4tN \sum_{n=1}^N \int_0^1 c(\alpha - t, \alpha + t, x_n)\, d\alpha + 4t^2N^2 \\ &= \sum_{k,j=1}^N \int_0^{2t} c(-\alpha, \alpha, x_k - x_j)\, d\alpha - 4t^2N^2,\end{aligned}$$

according to (6.9) and (6.12). So

$$\begin{aligned}\int_0^1 R_N^2(\alpha - t, \alpha + t)\, d\alpha &= \int_0^{2t} \left(\sum_{k,j=1}^N c(-\alpha, \alpha, x_k - x_j) - 2\alpha N^2 \right) d\alpha \\ &= \int_0^{2t} R^*_{N^2}(-\alpha, \alpha)\, d\alpha. \quad \blacksquare\end{aligned}$$

Notes

Theorems 6.1 and 6.2, which are the strongest of their kind presently known, are from Cassels [8]. In fact, Cassels shows that Theorem 6.1 is best possible in the following sense: Let $\varphi(F)$ be any function of F tending arbitrarily slowly to zero as F tends to zero. Then there exists a finite sequence ω such that $D \leq \varphi(F)\sqrt{F}(1 + |\log F|)$ does not hold true. The author also derives a result on the distribution of quadratic residues modulo a prime from Theorem 6.1.

Earlier results in the direction of Theorem 6.1 are due to (in chronological order) Vinogradov [1], van der Corput and Pisot [1], Koksma [5], and Cassels [5]. For Vinogradov's result, see also Gel'fond and Linnik [1, Chapter 7]. A predecessor of Theorem 6.2 is also contained in the paper of van der Corput and Pisot [1].

Our proofs follow the line of thought of Hlawka [11], who generalized both theorems to several dimensions. These generalizations are obtained by exactly the same method, namely, using the theorem of Erdös-Turán-Koksma (see notes in Section 2) and the analogue of Example 5.1 in several dimensions (see Theorem 5.6). This method also yields a generalization of a result of Coles [1], who compared the discrepancy of a two-dimensional sequence $((x_n, y_n))$ with the discrepancies of the one-dimensional sequences $(hx_n + ky_n)$, where h and k are integers not both zero. For details, see Hlawka [11]. An expository account of these results was also given by Hlawka in [16]. A simplified and improved version of Coles's result can be found in Ungar [1]. Moreover, the above mentioned multidimensional estimates of Hlawka were improved by Helmberg [9].

The integral identity in Theorem 6.3 is from van der Corput and Pisot [1]. Our proof follows Koksma [5].

Exercises

6.1. Let $x_1, \ldots, x_N$ be N real numbers. For two real numbers α and β with $\alpha \le \beta \le \alpha + 1$, define $R_N(\alpha, \beta)$ as in Theorem 6.3. Furthermore, let f be a continuous function on $[0, 1]$. Evaluate the following integrals:

$$\int_0^1 R_N(\alpha - t, \alpha + t)\, d\alpha$$

$$\int_0^1 R_N^2(\alpha - t, \alpha + t)\, d\alpha$$

$$\int_0^1 R_N(\alpha, \alpha + t)\, df(t)$$

$$\int_0^1 R_N^2(\alpha, \alpha + t)\, df(t)$$

$$\int_0^1 R_N(\alpha - t, \alpha)\, df(t)$$

$$\int_0^1 R_N^2(\alpha - t, \alpha)\, df(t)$$

$$\int_0^{1/2} R_N(\alpha - t, \alpha + t)\, df(t)$$

$$\int_0^{1/2} R_N^2(\alpha - t, \alpha + t)\, df(t).$$

6.2. Prove in detail the identity (6.9) in the proof of Theorem 6.3.

6.3. Prove in detail the identity (6.10) in the proof of Theorem 6.3.

6.4. Let $x_1 \le x_2 \le \cdots \le x_N$ be in I, and define $R_N(\alpha, \beta)$ as in Theorem 6.3. Let Δ be the area $\Delta = \{(\alpha, \beta) \in \mathbb{R}^2 : 0 \le \alpha \le 1, 0 \le \beta \le 1, \alpha \le \beta\}$. Prove that

$$\iint_\Delta R_N^2(\alpha, \beta)\, d\alpha\, d\beta = N \sum_{n=1}^{N} \left(x_n - \frac{n}{N}\right)^2 - B_2\left(\sum_{n=1}^{N} (x_n - \tfrac{1}{2})\right),$$

where $B_2(t) = t^2 - t + (1/6)$ is the second Bernoulli polynomial.

6.5. Let $x_1, \ldots, x_N$ be in I, and let $y_1, \ldots, y_{N^2}$ be the sequence of all differences $x_k - x_j \bmod 1$ arranged in some order. For $0 \le t \le 1$, set $R_N(t) = R_N(0, t)$ and $R_{N^2}^*(t) = R_{N^2}^*(0, t)$ (see Theorem 6.3), and extend these functions with period 1 to $\mathbb{R}$. Prove that the identity

$$\int_0^1 (R_N(t+\alpha) - R_N(t))^2\,dt = \int_0^{\langle\alpha\rangle} (R^*_{N^2}(t) - R^*_{N^2}(-t))\,dt$$

holds for all real numbers α, by showing that both sides are equal to

$$-\langle\alpha\rangle^2 N^2 + \sum_{n=1}^{N^2} \max\,(0, \langle\alpha\rangle - \langle y_n\rangle).$$

6.6. Prove that Theorem 6.2 implies van der Corput's difference theorem.

3

UNIFORM DISTRIBUTION IN COMPACT SPACES

In the first two chapters we studied distribution properties of sequences of real numbers with the proviso that two numbers differing by an integer were considered "equivalent." This notion of "equivalence" is indeed an equivalence relation in the rigorous sense; call the set of the resulting equivalence classes the reals mod 1 (or $\mathbb{R}$ mod 1). There is a simple model for $\mathbb{R}$ mod 1, obtained as follows.

Let U be the unit circle in the complex z-plane, $U = \{z \in \mathbb{C}: |z| = 1\}$. The mapping h: $\mathbb{R} \mapsto U$, defined by $h(x) = e^{2\pi i x}$ for $x \in \mathbb{R}$, is constant on the equivalence classes comprising $\mathbb{R}$ mod 1 and can thus be considered as a mapping from $\mathbb{R}$ mod 1 onto U. If we give $\mathbb{R}$ mod 1 the natural topology (open sets in $\mathbb{R}$ mod 1 are open sets in $\mathbb{R}$ after identification mod 1) and furnish U with the relative topology in the plane, then h is a homeomorphism from $\mathbb{R}$ mod 1 onto U. Thus, we could as well consider uniform distribution of sequences in U. For our purposes, it is pertinent that U is a compact Hausdorff space with countable base.

It will turn out that a satisfactory theory of uniform distribution can be developed in the more abstract setting of an arbitrary compact Hausdorff space with countable base. Some important facts will even hold true without requiring the second axiom of countability.

1. DEFINITION AND IMPORTANT PROPERTIES

Definition

Let X be a compact Hausdorff space. The elements of the σ-algebra generated by the open sets in X are called the *Borel sets* in X. Let μ be a *nonnegative regular normed Borel measure* in X, that is, a nonnegative measure μ defined on the class of Borel sets with $\mu(X) = 1$ and $\mu(E) = \sup\{\mu(C): C \subseteq E, C \text{ closed}\} = \inf\{\mu(D): E \subseteq D, D \text{ open}\}$ for all Borel sets E in X.

Suppose we are given a sequence (x_n), $n = 1, 2, \ldots$, of elements $x_n \in X$. Among the various characterizations that we had for u.d. mod 1, the one most easily adaptable to a more general situation seems to be Theorem 1.1 of Chapter 1. It is convenient to have the following notions available: By $\mathscr{B}(X)$ we mean the set of all bounded real-valued Borel-measurable functions on X. Under the norm $\|f\| = \sup_{x\in X} |f(x)|$ for $f \in \mathscr{B}(X)$, the set $\mathscr{B}(X)$ forms a Banach space, and even a Banach algebra if algebraic operations for functions are defined in the usual way. The subset $\mathscr{R}(X)$ of $\mathscr{B}(X)$ consisting of all real-valued continuous functions on X is then a Banach subalgebra of $\mathscr{B}(X)$.

DEFINITION 1.1. The sequence (x_n), $n = 1, 2, \ldots$, of elements in X is called *μ-u.d. in X* if

$$\lim_{N\to\infty} \frac{1}{N} \sum_{n=1}^{N} f(x_n) = \int_X f\, d\mu \qquad \text{for all } f \in \mathscr{R}(X). \tag{1.1}$$

If $X = U$, μ = normed Lebesgue measure on U, then the notions of u.d. in U (or, equivalently, u.d. mod 1) and μ-u.d. in U coincide. The first and most natural question to ask is of course whether u.d. sequences exist at all for an arbitrary X. One can easily show that for each X there is a μ with corresponding μ-u.d. sequences (see Exercise 1.1). As to the tougher problem of the existence of a μ-u.d. sequence for any given X and μ, we refer to the notes.

In this chapter, we will sometimes take recourse to stronger topological conditions on X. We shall see in Section 2 that, under the additional assumption of a countable base for the topology in X, there is really an abundance of u.d. sequences. Fortunately enough, it will also turn out that requiring a countability condition of some type does not lead to an actual loss of generality (see Exercise 2.1).

It is sometimes convenient to consider also complex-valued functions f on X. Such a function can be written in the form $f = f_1 + if_2$ where f_1 and f_2 are uniquely determined real-valued functions called the *real part* and *imaginary part* of f, respectively. If both f_1 and f_2 are in $\mathscr{B}(X)$, then f is again called a bounded (complex-valued) Borel-measurable function. In this case, we define $\int_X f\,d\mu = \int_X f_1\,d\mu + i\int_X f_2\,d\mu$. We shall denote by $\mathscr{C}(X)$ the set of all continuous complex-valued functions on X. Clearly, a function is in $\mathscr{C}(X)$ if and only if its real and imaginary parts both belong to $\mathscr{R}(X)$. The set $\mathscr{C}(X)$ forms a (complex) Banach algebra under the supremum norm and with the usual algebraic operations. It is an easy exercise to show that a sequence (x_n) is μ-u.d. in X if and only if (1.1) holds for all $f \in \mathscr{C}(X)$.

Convergence-Determining Classes

In the classical theory we noticed that we need not consider all f from $\mathscr{R}(X)$ in (1.1) to guarantee uniform distribution. This suggests the following definition.

DEFINITION 1.2. A class $\mathscr{V}$ of functions from $\mathscr{B}(X)$ is called *convergence-determining* (with respect to μ) if for any sequence (x_n) in X, the validity of the equation

$$\lim_{N\to\infty} \frac{1}{N} \sum_{n=1}^{N} f(x_n) = \int_X f\,d\mu \qquad \text{for all } f \in \mathscr{V} \tag{1.2}$$

already implies the μ-u.d. of (x_n).

EXAMPLE 1.1. In $X = U$, the class of all characteristic functions of half-open (open, closed) intervals is convergence-determining with respect to the normed Lebesgue measure. ■

For a class $\mathscr{V}$ of functions from $\mathscr{B}(X)$, let sp $(\mathscr{V})$ denote the linear subspace of $\mathscr{B}(X)$ generated by $\mathscr{V}$. In other words, sp $(\mathscr{V})$ consists of all finite linear combinations of elements from $\mathscr{V}$ with real coefficients. The construction of many important convergence-determining classes is based on the following theorem.

THEOREM 1.1. If $\mathscr{V}$ is a class of functions from $\mathscr{B}(X)$ such that sp $(\mathscr{V})$ is dense in $\mathscr{R}(X)$, that is, $\overline{\text{sp}\ (\mathscr{V})} \supseteq \mathscr{R}(X)$, then $\mathscr{V}$ is convergence-determining with respect to any μ in X.

PROOF. Let us first show that (1.1) holds for all $g \in$ sp $(\mathscr{V})$ provided that (1.1) holds for all $f \in \mathscr{V}$. For, in this case, $g = \alpha_1 f_1 + \cdots + \alpha_k f_k$ for some $f_i \in \mathscr{V}$ and $\alpha_i \in \mathbb{R}$, $1 \leq i \leq k$. Then,

$$\lim_{N\to\infty} \frac{1}{N} \sum_{n=1}^{N} g(x_n) = \int_X g\,d\mu$$

follows from the linearity of both the right-hand side and the left-hand side in (1.1). Turning to an arbitrary $f \in \mathscr{R}(X)$, we choose $\varepsilon > 0$, and by hypothesis, we can find $h \in \text{sp}(\mathscr{V})$ with $\|f - h\| < \varepsilon$. Then, we have

$$\begin{aligned}\left|\frac{1}{N}\sum_{n=1}^{N} f(x_n) - \int_X f\,d\mu\right| &\leq \left|\frac{1}{N}\sum_{n=1}^{N}(f-h)(x_n) - \int_X (f-h)\,d\mu\right| \\ &\quad + \left|\frac{1}{N}\sum_{n=1}^{N} h(x_n) - \int_X h\,d\mu\right| \\ &\leq \frac{1}{N}\sum_{n=1}^{N}|(f-h)(x_n)| + \int_X |f-h|\,d\mu \\ &\quad + \left|\frac{1}{N}\sum_{n=1}^{N} h(x_n) - \int_X h\,d\mu\right| \\ &\leq 2\,\|f-h\| + \left|\frac{1}{N}\sum_{n=1}^{N} h(x_n) - \int_X h\,d\mu\right| < 3\varepsilon\end{aligned}$$

for sufficiently large N. ■

Using a well-known and fundamental theorem that we quote below for easy reference, we obtain then a useful corollary.

LEMMA 1.1: Stone-Weierstrass Theorem. Let X be a compact Hausdorff space, and let $\mathscr{A}$ be a subalgebra of $\mathscr{R}(X)$ that contains the constant functions and that separates points; that is, for any two distinct points $x_1, x_2 \in X$ there exists $f \in \mathscr{A}$ with $f(x_1) \neq f(x_2)$. Then $\mathscr{A}$ is dense in $\mathscr{R}(X)$. If $\mathscr{B}$ is a subalgebra of $\mathscr{C}(X)$ that, in addition to the above properties, is closed under complex conjugation (i.e., $f \in \mathscr{B}$ implies $\bar{f} \in \mathscr{B}$), then $\mathscr{B}$ is dense in $\mathscr{C}(X)$.

COROLLARY 1.1. If $\text{sp}(\mathscr{V})$ is a subalgebra of $\mathscr{R}(X)$ that separates points and contains the constant functions, then $\mathscr{V}$ is a convergence-determining class with respect to any μ in X.

PROOF. This follows from Theorem 1.1 and Lemma 1.1. ■

The notion of convergence-determining class may of course also be defined for classes of bounded complex-valued Borel-measurable functions. A result analogous to Theorem 1.1 can be shown for this case as well, since all the arguments in the proof go through without any change. In fact, $\text{sp}(\mathscr{V})$ may now even be taken as the subspace of the complex vector space $\mathscr{C}(X)$ generated by $\mathscr{V}$. Using the complex version of the Stone-Weierstrass theorem, we arrive then at the following result.

COROLLARY 1.2. Let $\mathscr{V}$ be a set of functions from $\mathscr{C}(X)$ such that the subspace of $\mathscr{C}(X)$ generated by $\mathscr{V}$ is a subalgebra of $\mathscr{C}(X)$ that separates points, contains the constant functions, and is closed under complex conjugation. Then $\mathscr{V}$ is a convergence-determining class with respect to any μ in X.

Continuity Sets

For a subset M of X, let int M stand for the interior of M, and let $\partial M = \bar{M}\backslash \text{int } M$ be the boundary of M. Furthermore, we shall use M' for the complement of M in X.

DEFINITION 1.3. A Borel set $M \subseteq X$ is called a *μ-continuity set* if $\mu(\partial M) = 0$.

Note that ∂M is closed and is therefore again a Borel set. Based on this concept of μ-continuity set, we can then exhibit an important convergence-determining class, namely, the class of all characteristic functions of μ-continuity sets. Before we endeavor to do this, we first give another example of a nontrivial convergence-determining class consisting entirely of continuous functions. This example will also show some of the common techniques.

EXAMPLE 1.2. Since X is normal, there will exist, for any two disjoint closed subsets A and B of X, a so-called *Urysohn function*, that is, a function $f \in \mathscr{R}(X)$ with $0 \leq f(x) \leq 1$ for all $x \in X$, $f(x) = 0$ for all $x \in A$, and $f(x) = 1$ for all $x \in B$. Now, for fixed $\varepsilon > 0$, consider all ordered pairs (C, D) of closed μ-continuity sets C and open sets D with $D \supseteq C$ and $\mu(D\backslash C) < \varepsilon$. For each ordered pair (C, D), choose one Urysohn function f with $f(x) = 1$ for $x \in C$ and $f(x) = 0$ for $x \in D'$. Let $\mathscr{U}_\varepsilon$ be the collection of the functions f obtained in this way. Furthermore, let $\mathscr{P}_\varepsilon$ denote the set of all finite products of functions from $\mathscr{U}_\varepsilon$. We claim that $\mathscr{P}_\varepsilon$ is convergence-determining with respect to any μ in X.

We can quickly convince ourselves that sp $(\mathscr{P}_\varepsilon)$ is a subalgebra of $\mathscr{R}(X)$. Therefore, the use of Corollary 1.1 is suggested. Now sp $(\mathscr{P}_\varepsilon)$ contains the constant functions, since $\mathscr{U}_\varepsilon$ contains the function $f(x) \equiv 1$ (take $C = D = X$). The proof is completed by showing that $\mathscr{U}_\varepsilon$ separates points. Take $x_0, y_0 \in X$, $x_0 \neq y_0$, and choose an open neighborhood D of x_0 that does not contain y_0. Since μ is regular, there exists a closed $B \subseteq D$ with $\mu(D\backslash B) < \varepsilon$. We may assume $x_0 \in B$, for otherwise we look at $B \cup \{x_0\}$. Of course, B need not be a μ-continuity set. But, as we shall see, there is such an abundance of closed μ-continuity sets that we can easily find one between B and D. Consider a Urysohn function g corresponding to the disjoint closed sets D' and B, that is, $0 \leq g(x) \leq 1$ for all $x \in X$, $g(x) = 0$ for $x \in D'$, $g(x) = 1$ for $x \in B$. For $0 \leq \alpha \leq 1$, put $G_\alpha = \{x \in X\colon g(x) = \alpha\}$. Since $X = \bigcup_{0 \leq \alpha \leq 1} G_\alpha$ has finite μ-measure, there are at most countably many G_α with $\mu(G_\alpha) > 0$ (see Exercise 1.21). Thus, there exists an α, $0 < \alpha < 1$, with $\mu(G_\alpha) = 0$. Now let C be the closed set $C = \{x \in X\colon g(x) \geq \alpha\}$. Since $\{x \in X\colon g(x) > \alpha\}$ is open, we have $\partial C \subseteq G_\alpha$, and C is a μ-continuity set. From $B \subseteq C \subseteq D$ we can infer

$\mu(D\backslash C) < \varepsilon$. To the pair (C, D), there corresponds a Urysohn function $f \in \mathscr{U}_\varepsilon$ that satisfies $f(x_0) = 1$ and $f(y_0) = 0$. By using a direct approximation technique, one can even show that $\mathscr{U}_\varepsilon$ is already convergence-determining (see Exercise 1.16). ■

EXAMPLE 1.3. If X is a compact metric space, then a crucial step in the above argument can be carried out using the metric d of X. In particular, we show the following: Let $B(x, \delta_1) = \{y \in X\colon d(x, y) < \delta_1\}$ and $B(x, \delta_2) = \{y \in X\colon d(x, y) < \delta_2\}$ be open balls with center x and $0 < \delta_1 < \delta_2$; then there are "many" open balls between $B(x, \delta_1)$ and $B(x, \delta_2)$ that are μ-continuity sets. To see this, put $S(x, \delta) = \{y \in X\colon d(x, y) = \delta\}$ and $B(x, \delta) = \{y \in X\colon d(x, y) < \delta\}$ for $\delta > 0$. Then $\partial B(x, \delta) \subseteq S(x, \delta)$. But $B(x, \delta_2)\backslash B(x, \delta_1) = \bigcup_{\delta_1 \leq \delta < \delta_2} S(x, \delta)$ has finite μ-measure; therefore, at most countably many of the $S(x, \delta)$ can have positive μ-measure. Thus, apart from at most countably many exceptions, the open balls $B(x, \delta)$ with $\delta_1 \leq \delta \leq \delta_2$ are μ-continuity sets. ■

For a given sequence (x_n) in X, a subset M of X, and a natural number N, we define the counting function $A(M; N)$ by $A(M; N) = \sum_{n=1}^{N} c_M(x_n) =$ number of x_n, $1 \leq n \leq N$, with $x_n \in M$, where c_M denotes the characteristic function of M. We shall now prove the result we announced earlier, and even more.

THEOREM 1.2. The sequence (x_n) is μ-u.d. in X if and only if

$$\lim_{N\to\infty} \frac{A(M; N)}{N} = \mu(M) \tag{1.3}$$

holds for all μ-continuity sets $M \subseteq X$. In particular, the class $\mathscr{I} = \{c_M\colon M$ is a μ-continuity set in $X\}$ is convergence-determining with respect to μ.

PROOF. Suppose first that (x_n) is μ-u.d. and let M be a μ-continuity set. It suffices to construct for every $\varepsilon > 0$ two functions f_ε and g_ε from $\mathscr{R}(X)$ with $f_\varepsilon(x) \leq c_M(x) \leq g_\varepsilon(x)$ for all $x \in X$, and $\int_X (g_\varepsilon - f_\varepsilon)\, d\mu < \varepsilon$. Then the proof of Theorem 1.1 in Chapter 1 can be reproduced *verbatim* to yield the desired result. By the regularity of μ, there exists a closed $C \subseteq \operatorname{int} M$ with $\mu((\operatorname{int} M)\backslash C) < \varepsilon/2$, and an open $D \supseteq \bar{M}$ with $\mu(D\backslash\bar{M}) < \varepsilon/2$. Let $f_\varepsilon \in \mathscr{R}(X)$ be a Urysohn function with $f_\varepsilon(x) = 1$ for $x \in C$, $f_\varepsilon(x) = 0$ for $x \in (\operatorname{int} M)'$, and let $g_\varepsilon \in \mathscr{R}(X)$ be a Urysohn function with $g_\varepsilon(x) = 1$ for $x \in \bar{M}$, $g_\varepsilon(x) = 0$ for $x \in D'$. Then $f_\varepsilon(x) \leq c_M(x) \leq g_\varepsilon(x)$ for all $x \in X$. Furthermore, f_ε and g_ε coincide on C and on D', and $0 \leq g_\varepsilon(x) - f_\varepsilon(x) \leq 1$ for all $x \in X$; therefore,

$$\begin{aligned}\int_X (g_\varepsilon - f_\varepsilon)\, d\mu &= \int_{D\backslash C} (g_\varepsilon - f_\varepsilon)\, d\mu \leq \mu(D\backslash C) \\ &= \mu(D\backslash M) + \mu(\partial M) + \mu((\operatorname{int} M)\backslash C) < \varepsilon. \end{aligned} \tag{1.4}$$

Conversely, suppose that the condition of the theorem is satisfied. By Theorem 1.1, it suffices to show that $\overline{\mathrm{sp}\,(\mathscr{I})} \supseteq \mathscr{R}(X)$. Choose $f \in \mathscr{R}(X)$, and since $\overline{\mathrm{sp}\,(\mathscr{I})}$ is a linear subspace of $\mathscr{B}(X)$ containing the constant functions, we may assume without loss of generality that $0 \leq f(x) < 1$ for all $x \in X$, by possibly transforming f into $af + b$, $a, b \in \mathbb{R}$, $a \neq 0$. As in Example 1.2, we see that apart from at most countably many exceptions, the sets $E_\alpha = \{x \in X: f(x) \geq \alpha\}$ are μ-continuity sets. Consequently, for a given $\varepsilon > 0$ there will exist a sequence of numbers $0 = \alpha_0 < \alpha_1 < \cdots < \alpha_n = 1$ such that for all $0 \leq i \leq n - 1$ the following holds: $\alpha_{i+1} - \alpha_i \leq \varepsilon$, and $F_i = \{x \in X: f(x) \geq \alpha_i\}$ is a μ-continuity set. We assert that

$$\left\| \sum_{i=0}^{n-1} (\alpha_{i+1} - \alpha_i) c_{F_i} - f \right\| \leq \varepsilon. \tag{1.5}$$

For each $x \in X$, there exists an integer k, $0 \leq k \leq n - 1$, with $\alpha_k \leq f(x) < \alpha_{k+1}$. Then

$$\begin{aligned} \left| \sum_{i=0}^{n-1} (\alpha_{i+1} - \alpha_i) c_{F_i}(x) - f(x) \right| &= \left| \sum_{i=0}^{k} (\alpha_{i+1} - \alpha_i) - f(x) \right| \\ &= |\alpha_{k+1} - f(x)| \leq \varepsilon, \end{aligned} \tag{1.6}$$

and (1.5) follows. ∎

We remark that since the F_i in the foregoing argument are closed μ-continuity sets, the validity of (1.3) for all closed μ-continuity sets M will already guarantee μ-u.d. in X.

There are certain pathological cases where (1.3) might even hold for all Borel sets in X (consider the example given in Exercise 1.1). However, under very mild conditions on the measure μ, one can, for each given μ-u.d. sequence, easily construct closed (open) sets for which (1.3) fails drastically (see Exercise 1.7).

Support

DEFINITION 1.4. The *support* K of the measure μ in X is defined to be the set $K = \{x \in X: \mu(D) > 0$ for all open neighborhoods D of $x\}$.

LEMMA 1.2. The support K of μ has the following important properties:

i. K is closed

ii. $\mu(K) = 1$

iii. If $\mu(C) = 1$ for a closed set $C \subseteq X$, then $C \supseteq K$.

PROOF. i. We show that K' is open. Take $x \in K'$; then there exists an open neighborhood D of x with $\mu(D) = 0$. But none of the points of D can belong to K, and thus, $D \subseteq K'$.

ii. Choose $\varepsilon > 0$. Since μ is regular, there exists a closed $B \subseteq K'$ with $\mu(K' \backslash B) < \varepsilon$. For every $x \in B$, there exists an open neighborhood D_x of x with $\mu(D_x) = 0$. B is compact; therefore, finitely many D_x suffice to cover B. It follows that $\mu(B) = 0$; hence, $\mu(K') < \varepsilon$. Since ε was arbitrary, we obtain $\mu(K') = 0$, or $\mu(K) = 1$.

iii. Let C be a closed set with $\mu(C) = 1$. If $C = X$, then there is nothing to show. If $C \neq X$, then for every $x \in C'$ there exists an open neighborhood D of x with $\mu(D) = 0$, namely, $D = C'$. Thus, $C' \subseteq K'$, or $C \supseteq K$. ■

The following theorem will show the usefulness of the notion of support. Let μ be a nonnegative regular normed Borel measure in X with support K. Let μ^* be the restriction of μ to K; that is, the measure μ^* in K defined by $\mu^*(A) = \mu(A)$ for all μ-measurable sets $A \subseteq K$. Then μ^* is a nonnegative regular normed Borel measure in K in the relative topology, since the Borel sets in the space K are exactly those of the form $B \cap K$ with B a Borel set in X (see Halmos [1, p. 25]). We shall also need the following classical result.

LEMMA 1.3: Tietze's Extension Theorem. If C is a closed subset of a normal space X, then every real-valued continuous function on C can be extended to a real-valued continuous function on the entire space X.

THEOREM 1.3. The sequence (x_n) in K is μ-u.d. in X if and only if (x_n) is μ^*-u.d. in K.

PROOF. By Tietze's extension theorem, the space $\mathscr{R}(K)$ just consists of the restrictions of all functions $g \in \mathscr{R}(X)$ to K. Let $g \mid K$ denote the restriction of g to K. Then (x_n) is μ^*-u.d. in K if and only if $\lim_{N\to\infty} (1/N) \sum_{n=1}^{N} (g \mid K)(x_n) = \int_K (g \mid K)\, d\mu^*$ for all $g \in \mathscr{R}(X)$. But $(g \mid K)(x_n) = g(x_n)$ for all $n \geq 1$, and $\int_K (g \mid K)\, d\mu^* = \int_K g\, d\mu = \int_K g\, d\mu + \int_{K'} g\, d\mu = \int_X g\, d\mu$. Therefore, (x_n) is μ^*-u.d. in K if and only if $\lim_{N\to\infty} (1/N) \sum_{n=1}^{N} g(x_n) = \int_X g\, d\mu$ holds for all $g \in \mathscr{R}(X)$, and the proof is complete. ■

Notes

For proofs of the Stone-Weierstrass theorem and Tietze's extension theorem, see Gaal [1] and Hewitt and Stromberg [1].

Some even more general concepts of u.d. than the one presented here have been considered in the literature. In particular, we mention a notion of u.d. introduced by Helmberg [3], which is also briefly discussed in Cigler and Helmberg [1] and Helmberg [5]. A variant of Helmberg's concept was considered by Kemperman [3]. Auslander and Brezin [1] employ a special case of Kemperman's definition, with X being a certain compact quotient of a connected, simply connected, solvable Lie group. Their Weyl criterion (loc. cit., Theorem 3.2 (*a*), (*b*)) holds in fact for any compact Hausdorff space X.

The μ-u.d. of a sequence in X can be viewed as a special case of *weak convergence of measures*. A sequence (μ_n) of nonnegative regular normed Borel measures on X is said to converge weakly to μ, written $\lim_{n\to\infty} \mu_n = \mu$, if $\lim_{n\to\infty} \mu_n(M) = \mu(M)$ for all μ-continuity sets M in X. The sequence (x_n) in X is μ-u.d. if and only if $\lim_{n\to\infty} (1/n)(\mu_{x_1} + \cdots + \mu_{x_n}) = \mu$, where μ_x is the point measure at x (compare with Exercise 1.1). For an exposition of the theory of weak convergence, see Billingsley [2], Parthasarathy [1, Section 2.6], and Topsøe [2]. The following general existence theorem was shown by Niederreiter [9]. The measure μ on X admits a μ-u.d. sequence if and only if μ lies in the weak sequential closure of the convex hull of point measures on X. A class $\mathscr{M}$ of Borel sets in X is called a *μ-uniformity class* if $\lim_{n\to\infty} \mu_n(M) = \mu(M)$ uniformly in $M \in \mathscr{M}$ whenever $\lim_{n\to\infty} \mu_n = \mu$. This notion is of interest in relation to the theory of discrepancy. For significant work on uniformity classes, see Billingsley and Topsøe [1], Ranga Rao [1], and Topsøe [1, 2]. Based on ideas of Cigler [6], K. Schmidt [1] introduced and studied a notion of u.d. for sequences of measures on a compact Hausdorff space.

The study of u.d. sequences in compact spaces was initiated by Hlawka [3, 6]. In this approach, Hlawka also replaced arithmetic means by a larger class of summation methods (see Section 4). The second axiom of countability is presupposed in both papers. Remarks on Theorem 1.2 were made by Paganoni [1].

U.d. in locally compact spaces can be defined in a natural way. If X is a locally compact Hausdorff space with countable base and μ is a nonnegative regular normed Borel measure in X, then the sequence (x_n) in X is called μ-u.d. in X provided that

$$\lim_{N\to\infty} \frac{1}{N} \sum_{n=1}^{N} f(x_n) = \int_X f\, d\mu$$

holds for all real-valued continuous functions f on X with compact support (Stapleton [1]). In case X is compact, this coincides with Definition 1.1. The same author shows that (x_n) is μ-u.d. if and only if $\lim_{N\to\infty} A(M; N)/N = \mu(M)$ holds for all relatively compact μ-continuity sets M and constructs μ-u.d. sequences. The subject was taken up again by Helmberg [7], who dropped the second axiom of countability. The relation between the noncompact case and the compact case is quite close. For noncompact X, let $\tilde{X} = X \cup \{\infty\}$ be the one-point compactification of X, and extend μ to a measure $\tilde{\mu}$ on $\tilde{X}$ by $\tilde{\mu}(E) = \tilde{\mu}(E \cup \{\infty\}) = \mu(E)$ for all E in the σ-algebra generated by the compact sets in X. Then the sequence (x_n) in X is μ-u.d. if and only if (x_n) is $\tilde{\mu}$-u.d. in $\tilde{X}$ (Helmberg [7]). See also Lesca [4]. Interesting new aspects arise if one considers unbounded continuous functions on a noncompact X. Post [1] has shown the following for such X with countable base: For any μ-u.d. sequence (x_n) in X without repetitions and any three numbers α, β, γ, with $0 < \alpha \leq \beta \leq \gamma \leq \infty$, there exists a nonnegative continuous function f on X such that $\int_X f\, d\mu = \alpha$, $\underline{\lim}_{N\to\infty} (1/N) \sum_{n=1}^{N} f(x_n) = \beta$, and $\overline{\lim}_{N\to\infty} (1/N) \sum_{n=1}^{N} f(x_n) = \gamma$. Dupain and Lesca [1] study u.d. subsequences of sequences in locally compact spaces.

An interesting notion of u.d. in locally compact spaces was introduced by Gerl [7, 8]: Let X be a locally compact Hausdorff space with countable base, and let μ be a positive Radon measure in X (it need not be finite). A sequence (x_n) in X is called *μ-relatively equidistributed* (*μ-relativ gleichverteilt*) if $\lim_{N\to\infty} A(C; N)/A(D; N) = \mu(C)/\mu(D)$ holds for all relatively compact μ-continuity sets C and D in X with $\mu(D) > 0$. Under the stated hypotheses on X and μ, the author can show the existence of μ-relatively equidistributed sequences. If μ is normed, then a μ-u.d. sequence in X (in the sense of Stapleton) is also μ-relatively equidistributed. On the other hand, if X is noncompact and μ is normed with compact support, then there are μ-relatively equidistributed sequences in X that are not μ-u.d. For u.d. in locally compact groups, see Section 5 of Chapter 4.

Exercises

In all the subsequent problems, X denotes an arbitrary compact Hausdorff space and μ a nonnegative regular normed Borel measure in X unless stated otherwise.

1.1. Let $x \in X$ be fixed. Prove that the measure μ_x defined for all subsets E of X by $\mu_x(E) = 1$ if $x \in E$, $\mu_x(E) = 0$ if $x \notin E$, is a nonnegative regular normed Borel measure (called the *point measure at* x). The sequence (x_n), $x_n = x$ for all $n \geq 1$, is μ_x-u.d. in X.

1.2. In $X = [0, 1]$ with the usual topology and Lebesgue measure, the functions $f_0(x) \equiv 1, f_k(x) = x^k$ for $k = 1, 2, \ldots$, form a convergence-determining class. Does this hold for other measures in X as well?

1.3. Show that the family of μ-continuity sets in X forms a field, that is, that it is closed under finite unions, finite intersections, and complementation.

1.4. Give an example of a space X and a measure μ for which the μ-continuity sets do not form a σ-algebra.

1.5. Prove that every nonvoid open set in X contains a nonvoid open μ-continuity set.

1.6. The following criterion holds: (x_n) is μ-u.d. in X if and only if

$$\varliminf_{N\to\infty} \frac{A(D; N)}{N} \geq \mu(D)$$

for all open $D \subseteq X$. Consequently, $\varlimsup_{N\to\infty} A(C; N)/N \leq \mu(C)$ for all closed $C \subseteq X$ is also a necessary and sufficient condition for μ-u.d. in X.

1.7. If μ is not concentrated on a countable set (i.e., if $\mu(E) < 1$ for all countable $E \subseteq X$) and if (x_n) is μ-u.d. in X, then there exists a closed $C \subseteq X$ with $\lim_{N\to\infty} A(C; N)/N < \mu(C)$.

1.8. Prove that a μ-u.d. sequence is dense in the support of μ.

1.9. Suppose X contains μ-u.d. sequences. The statement "All μ-u.d. sequences are everywhere dense in X" holds true if and only if the support of μ is X.

1.10. Let X and Y be compact Hausdorff spaces, let $T\colon X \mapsto Y$ be continuous, and let (x_n) be μ-u.d. in X. Then (Tx_n) is $T^{-1}\mu$-u.d. in Y.

1.11. Let μ satisfy the same condition as in Exercise 1.7. Show that for every μ-u.d. sequence (x_n) in X, there exists a compact Hausdorff space Y and a measurable function $T\colon X \mapsto Y$ such that (Tx_n) is not $T^{-1}\mu$-u.d. in Y.

1.12. If (x_n) is μ-u.d. in X, then $\lim_{N\to\infty} (1/N) \sum_{n=1}^{N} f(x_n) = \int_X f\, d\mu$ holds for functions $f \in \mathscr{B}(X)$ whose discontinuities are contained in a μ-null set. *Hint:* For each $\varepsilon > 0$, construct step functions f_1 and f_2 as in (1.5) with $f_1 \leq f \leq f_2$ and $\int_X (f_2 - f_1)\, d\mu < \varepsilon$.

1.13. Let M be a closed μ-continuity set in X with $\mu(M) = \alpha > 0$. If $\omega = (x_n)$ is μ-u.d. in X, then the subsequence of ω consisting of all elements lying in M is $(1/\alpha)\mu^*$-u.d. in M with the relative topology, where μ^* is the restriction of μ to M. *Hint:* Use Tietze's extension theorem and Exercise 1.12.

1.14. Prove the inclusion-exclusion formula: If $A_1, \ldots, A_n$ are Borel sets in X, then

$$\mu\left(\bigcup_{i=1}^{n} A_i\right) = \sum_{k=1}^{n}(-1)^{k-1} \sum_{1 \le j_1 < j_2 < \cdots < j_k \le n} \mu(A_{j_1} \cap A_{j_2} \cap \cdots \cap A_{j_k}).$$

1.15. Use Exercises 1.6 and 1.14 to show that if $\mathcal{M}$ is a class of Borel sets in X closed under finite intersections and such that each open set in X can be approximated from below in μ-measure by finite unions of elements of $\mathcal{M}$, then the collection of characteristic functions of elements of $\mathcal{M}$ forms a convergence-determining class with respect to μ. (Note: This generalizes Example 1.1.)

1.16. Using the approximation technique employed in the second part of the proof of Theorem 1.2, show that the class $\mathcal{U}_\varepsilon$ in Example 1.2 is a convergence-determining class with respect to any μ in X.

1.17. Let (x_n) be μ-u.d. in X, and let $k(1), k(2), \ldots, k(N), \ldots$ be an increasing sequence of positive integers with $\lim_{N\to\infty} N/k(N) = 0$. Construct a new sequence (y_n) by inserting between any two terms $x_{k(N)}$ and $x_{k(N)+1}$ an arbitrary element of X. Show that (y_n) is again μ-u.d. in X.

1.18. Let $(x_n^{(i)})$, $i = 1, 2, \ldots, s$, be finitely many μ-u.d. sequences in X. Furthermore, let $P_1, P_2, \ldots, P_s$ be a partition of the set of positive integers into s infinite subsets P_i whose elements are arranged in natural order. For given $n \ge 1$, the integer n lies in a unique P_i. Define $y_n = x_k^{(i)}$ if n is the kth element of P_i. Show that (y_n) is μ-u.d. in X.

1.19. Let the support of the measure μ be a μ-continuity set. Then there exist μ-u.d. sequences in X if and only if there exist μ^*-u.d. sequences in the support of μ, where μ^* has the same meaning as in Theorem 1.3.

1.20. Let K be the support of μ. For every μ-continuity set M in X, the set $M \cap K$ is a μ^*-continuity set in K, where μ^* is the restriction of μ to K.

1.21. Let $(Y, \mathcal{F}, \nu)$ be a σ-finite measure space (i.e., Y is the countable union of measurable sets of finite ν-measure). Then in any family of pairwise disjoint measurable sets at most countably many sets can have positive ν-measure.

2. SPACES WITH COUNTABLE BASE

Weyl Criterion

Let now X be a compact Hausdorff space satisfying the second axiom of countability. This additional condition leads to several simplifications relative to the considerations in the preceding section. First of all, we need not require explicitly that μ be regular. Every nonnegative normed Borel measure in X is automatically regular (Halmos [1, Sections 50–52]). Furthermore, by invoking the Urysohn metrization theorem, we obtain that X is metrizable (and, conversely, every compact metric space has a countable base). There is still another topological property of X that is in fact the most pertinent one for the theory of uniform distribution. Let us first note that we have the following corollary to the classical Weyl criterion (see Chapter 1, Theorem 2.1): For the unit circle U with normed Lebesgue measure, there exists a countable convergence-determining class. This statement holds true for all spaces X under consideration, by virtue of the following general result: For a compact Hausdorff space X, $\mathscr{R}(X)$ is separable if and only if X is metrizable (Kelley [1, p. 245]). We shall prefer to construct explicitly a countable convergence-determining class for any X and μ, and we thereby give a more self-contained proof of the general Weyl criterion.

THEOREM 2.1: Weyl Criterion. In a compact Hausdorff space X with countable base, there exists a countable convergence-determining class of real-valued continuous functions with respect to any nonnegative normed Borel measure in X.

PROOF. Since X is separable, there will exist an everywhere dense sequence (x_k) in X. Let d be a metric in X. For each x_k and every integer $n \geq 1$, choose a Urysohn function $f_{k,n}$ with $f_{k,n}(x) = 1$ for $d(x_k, x) \leq 2^{-n-1}$ and $f_{k,n}(x) = 0$ for $d(x_k, x) \geq 2^{-n}$. Let $\mathscr{U}$ be the class of all functions $f_{k,n}$, and let $\mathscr{P}$ be the class consisting of the function $f \equiv 1$ and all finite products of functions from $\mathscr{U}$. We shall show that the countable class $\mathscr{P}$ is convergence-determining. We proceed by Corollary 1.1. It is easily seen that sp ($\mathscr{P}$) is a subalgebra of $\mathscr{R}(X)$. By the construction of $\mathscr{P}$, sp ($\mathscr{P}$) contains the constant functions. We conclude the proof by showing that the functions from $\mathscr{U}$ already separate points. Let $a, b \in X$ with $a \neq b$. Choose an integer $n \geq 0$ with $2^{-n} \leq d(a, b)$. For an x_k with $d(x_k, a) \leq 2^{-n-2}$, we look at the function $f_{k,n+1} \in \mathscr{U}$. We certainly have $f_{k,n+1}(a) = 1$. On the other hand, $d(x_k, b) \geq d(a, b) - d(x_k, a) \geq 2^{-n} - 2^{-n-2} > 2^{-n-1}$, and therefore, $f_{k,n+1}(b) = 0$. ∎

Metric Theorems

The above theorem has many important consequences. For instance, we can show that in a certain technical sense, almost all sequences in X are μ-u.d. We note that a sequence (x_n) in X can be viewed as a point in the Cartesian product X^∞ of denumerably many copies of X; that is, $X^\infty = \prod_{i=1}^{\infty} X_i$, with $X_i = X$ for all i. Of course, the sequence (x_n) is identified with the point $\xi = (x_1, x_2, \ldots, x_n, \ldots) \in X^\infty$. Furnished with the product topology, X^∞ is again a compact Hausdorff space with countable base. The given measure μ on X induces the product measure μ_∞ in X^∞, which we may assume to be complete. We first prove an auxiliary result that is a special case of the so-called law of large numbers in probability theory (Feller [2, p. 233], Loève [1, p. 239], Rényi [2, p. 332]).

LEMMA 2.1. For given $f \in \mathscr{B}(X)$, we have

$$\lim_{N\to\infty} \frac{1}{N} \sum_{n=1}^{N} f(x_n) = \int_X f\,d\mu \tag{2.1}$$

for μ_∞-almost all points $(x_1, x_2, \ldots, x_n, \ldots) \in X^\infty$.

PROOF. It suffices to prove the assertion for functions $f \in \mathscr{B}(X)$ with $\int_X f\,d\mu = 0$. For $N \geq 1$, let F_N be the function on X^∞ defined by

$$F_N(x_1, x_2, \ldots, x_n, \ldots) = \frac{1}{N} \sum_{n=1}^{N} f(x_n)$$

for $(x_1, x_2, \ldots, x_n, \ldots) \in X^\infty$. Then,

$$\int_{X^\infty} F_N^2\,d\mu_\infty = \frac{1}{N^2} \sum_{n=1}^{N} \int_{X^\infty} f^2(x_n)\,d\mu_\infty + \frac{2}{N^2} \sum_{1\leq i<j\leq N} \int_{X^\infty} f(x_i)f(x_j)\,d\mu_\infty = \frac{1}{N} \int_X f^2\,d\mu. \tag{2.2}$$

Therefore,

$$\sum_{m=1}^{\infty} \int_{X^\infty} (F_{m^2})^2\,d\mu_\infty = \left(\int_X f^2\,d\mu\right) \sum_{m=1}^{\infty} \frac{1}{m^2} < \infty. \tag{2.3}$$

It follows from Levi's theorem that $\lim_{m\to\infty} F_{m^2} = 0$ μ_∞-a.e. For arbitrary $N \geq 1$, there exists $m \geq 1$ with $m^2 \leq N < (m+1)^2$. Then

$$|F_N| = \left| \frac{m^2}{N} F_{m^2} + \frac{1}{N}(f(x_{m^2+1}) + \cdots + f(x_N)) \right| \leq |F_{m^2}| + \frac{2}{m} \|f\|. \tag{2.4}$$

Since the right-hand side of (2.4) tends to zero μ_∞-a.e. as $N \to \infty$ (or, equivalently, as $m \to \infty$), we are done. ∎

THEOREM 2.2. Let S be the set of all μ-u.d. sequences in X, viewed as a subset of X^∞. Then $\mu_\infty(S) = 1$.

PROOF. Let $f_1, f_2, \ldots, f_k, \ldots$ be a countable convergence-determining class of functions from $\mathscr{B}(X)$. For $k = 1, 2, \ldots$, let B_k be the exceptional μ_∞-null set corresponding to f_k according to Lemma 2.1. Then, for all $(x_1, x_2, \ldots, x_n, \ldots) \in X^\infty$ lying in the complement of the μ_∞-null set $B = \bigcup_{k=1}^\infty B_k$, we have

$$\lim_{N\to\infty} \frac{1}{N} \sum_{n=1}^{N} f_k(x_n) = \int_X f_k \, d\mu \qquad \text{for all } k = 1, 2, \ldots. \tag{2.5}$$

In other words, all sequences (x_n) corresponding to points in X^∞ outside of B are μ-u.d. ∎

We indicate now how Lemma 2.1 may also be deduced from a very important general principle, namely, the individual ergodic theorem. We need some definitions that will be useful in other contexts as well.

DEFINITION 2.1. Let $(Y, \mathscr{F}, \nu)$ be a measure space with ν being a nonnegative normed measure. A measurable transformation $T\colon Y \mapsto Y$ is called *measure-preserving* (with respect to ν) if $\nu(T^{-1}F) = \nu(F)$ holds for all $F \in \mathscr{F}$. A measure-preserving transformation T of Y is called *ergodic* (with respect to ν) if for all sets $F \in \mathscr{F}$ with $T^{-1}F = F$, we have $\nu(F) = 0$ or 1.

The mapping $T\colon X^\infty \mapsto X^\infty$, defined by $T(x_1, x_2, x_3, \ldots) = (x_2, x_3, \ldots)$ for $(x_1, x_2, x_3, \ldots) \in X^\infty$, is called the *one-sided shift in* X^∞. T is ergodic with respect to μ_∞ (see Exercise 2.20).

LEMMA 2.2: Individual Ergodic Theorem. Let $(Y, \mathscr{F}, \nu)$ be a measure space with ν being a nonnegative normed measure, and let T be an ergodic transformation of Y with respect to ν. Then, for any ν-integrable function f on Y, we have

$$\lim_{N\to\infty} \frac{1}{N} \sum_{n=0}^{N-1} f(T^n y) = \int_Y f \, d\nu \qquad \text{for } \nu\text{-almost all } y \in Y. \tag{2.6}$$

For the alternative proof of Lemma 2.1, we consider the projection map $p_1\colon X^\infty \mapsto X$, defined by

$$p_1(x_1, x_2, \ldots, x_n, \ldots) = x_1 \qquad \text{for } (x_1, x_2, \ldots, x_n, \ldots) \in X^\infty. \tag{2.7}$$

With $f \in \mathscr{B}(X)$ being given, the composite function $f \circ p_1$ will be bounded and measurable and, hence, integrable on X^∞. Applying the individual ergodic theorem to the function $f \circ p_1$ and the one-sided shift T in X^∞, we get that for μ_∞-almost all $\xi \in X^\infty$,

$$\lim_{N\to\infty} \frac{1}{N} \sum_{n=0}^{N-1} (f \circ p_1)(T^n \xi) = \int_{X^\infty} (f \circ p_1)\, d\mu_\infty. \tag{2.8}$$

Let us observe that, with $\xi = (x_1, x_2, \ldots, x_n, \ldots)$, we have $(f \circ p_1)(T^n\xi) = f(x_{n+1})$ for all $n \geq 0$. Therefore, (2.8) is equivalent to

$$\lim_{N\to\infty} \frac{1}{N} \sum_{n=1}^{N} f(x_n) = \int_{X^\infty} (f \circ p_1)\, d\mu_\infty = \int_X f\, d\mu \qquad \mu_\infty\text{-a.e.} \tag{2.9}$$

But (2.9) is identical with the assertion of Lemma 2.1.

From the viewpoint of measure theory, the μ-u.d. sequences constitute a large subset of X^∞. Therefore, it may come as a surprise that, topologically speaking, the μ-u.d. sequences form a rather small set. There is one trivial exception, namely, if X contains only one element. Then, of course, the set of u.d. sequences (there is only one possible μ) is identical with X^∞.

THEOREM 2.3. If X contains more than one element, then the set S from Theorem 2.2 is a set of the first category in X^∞.

PROOF. Let $\mathscr{V} = \{f_1, f_2, \ldots, f_k, \ldots\}$ be a countable convergence-determining class of real-valued continuous functions on X with $\overline{\text{sp}(\mathscr{V})} = \mathscr{R}(X)$ (such a class was constructed in the proof of Theorem 2.1). For given positive integers k, m, and t, let S_{kmt} be the set of all points $(x_1, x_2, \ldots, x_n, \ldots) \in X^\infty$ for which $|(1/N)\sum_{n=1}^N f_k(x_n) - \int_X f_k\, d\mu| \leq 1/m$ holds for all $N \geq t$. Since, for each fixed N, the function $F_{k,N}$ on X^∞ defined by

$$F_{k,N}(x_1, x_2, \ldots, x_n, \ldots) = \frac{1}{N} \sum_{n=1}^{N} f_k(x_n)$$

is continuous, S_{kmt} is closed. Furthermore, we have $S = \bigcap_{k=1}^\infty \bigcap_{m=1}^\infty \bigcup_{t=1}^\infty S_{kmt}$. The hypothesis that X contains at least two points implies that there exist nonconstant continuous functions on X (e.g., take a Urysohn function corresponding to two distinct points). Consequently, there exists a nonconstant function f_{k_0} in $\mathscr{V}$ (otherwise $\overline{\text{sp}(\mathscr{V})}$ would consist entirely of constant functions). In particular, there is a $y_0 \in X$ with $f_{k_0}(y_0) \neq \int_X f_{k_0}\, d\mu$. Put $c = |f_{k_0}(y_0) - \int_X f_{k_0}\, d\mu|$.

We claim that for $m > 2/c$ and all t, S_{k_0mt} is nowhere dense in X^∞. Assume, on the contrary, that we can find a nonvoid open set D in X^∞ with $D \subseteq S_{k_0mt}$. The open set D contains a nonvoid cylinder $E = \prod_{i=1}^\infty E_i$ with $E_i = X$ for i greater than some r. Consider a point $\zeta = (z_1, z_2, \ldots, z_n, \ldots) \in E$ with

$z_n = y_0$ for $n > r$. Then,

$$\begin{aligned}\left|\frac{1}{N}\sum_{n=1}^{N} f_{k_0}(z_n) - \int_X f_{k_0}\, d\mu\right| &= \left|\frac{1}{N}\sum_{n=1}^{N} f_{k_0}(y_0) - \int_X f_{k_0}\, d\mu + \frac{1}{N}\sum_{n=1}^{r}(f_{k_0}(z_n) - f_{k_0}(y_0))\right| \\ &\geq \left|f_{k_0}(y_0) - \int_X f_{k_0}\, d\mu\right| - \frac{1}{N}\sum_{n=1}^{r}|f_{k_0}(z_n) - f_{k_0}(y_0)| \\ &\geq c - \frac{r}{N} 2\, \|f_{k_0}\| \geq \frac{c}{2} > \frac{1}{m} \qquad \text{for sufficiently large } N. \quad (2.10)\end{aligned}$$

Thus, $\zeta \notin S_{k_0mt}$, a contradiction. We infer from the preceding that for sufficiently large m, the set $\bigcup_{t=1}^{\infty} S_{k_0mt}$ is of the first category in X^∞. Hence, S, being contained in this set, is also of the first category in X^∞. ∎

The above proof might even suggest that S itself is nowhere dense in X^∞. But, quite on the contrary, we have the following result.

THEOREM 2.4. The set S is everywhere dense in X^∞.

PROOF. We show that an arbitrary point $\xi = (x_1, x_2, \ldots, x_n, \ldots) \in X^\infty$ can be approximated (in the topology of X^∞) by points from S. Since $\mu_\infty(S) = 1$, there exists a μ-u.d. sequence (y_n) in X. Consider the points $\xi^{(k)} = (z_1^{(k)}, z_2^{(k)}, \ldots, z_n^{(k)}, \ldots) \in X^\infty$ defined by $z_i^{(k)} = x_i$ for $1 \leq i \leq k$ and $z_i^{(k)} = y_{i-k}$ for $i > k$. The points $\xi^{(k)}$ certainly correspond to μ-u.d. sequences in X; therefore, $\xi^{(k)} \in S$ for all $k \geq 1$. Moreover, $\lim_{k\to\infty} \xi^{(k)} = \xi$, and the proof is complete. ∎

Rearrangement of Sequences

We have seen in Section 4 of Chapter 2 that an everywhere dense sequence in the unit interval $[0, 1)$ can be rearranged so as to yield a u.d. sequence. A similar result holds in our present, more general setting. We can give a necessary and sufficient condition for a sequence to possess a μ-u.d. rearrangement.

THEOREM 2.5. The sequence (x_n) in X has a μ-u.d. rearrangement if and only if all open neighborhoods of points in the support of μ contain infinitely many terms of the sequence (x_n).

PROOF. Since $\mu(D) > 0$ for every open neighborhood D of a point in the support of μ, the condition is easily seen to be necessary (compare with Exercises 1.5 and 1.6).

To prove the converse, we note that Theorems 2.2 and 1.3 imply the existence of a μ-u.d. sequence (y_k) that is entirely contained in the support of μ. We choose a countable convergence-determining class of functions $f_1, f_2, \ldots, f_r, \ldots$ from $\mathscr{R}(X)$. For each $k \geq 1$, the set $D_k = \{x \in X\colon |f_r(y_k) - f_r(x)| < 1/k \text{ for } 1 \leq r \leq k\}$ is an open neighborhood of y_k. By the condition on the sequence (x_n), we can choose for each $k \geq 1$ an $x_{n_k} \in D_k$ in such a way that $n_k \neq n_m$ whenever $k \neq m$. Now consider a fixed f_r. By the construction of the x_{n_k}, we have $|f_r(y_k) - f_r(x_{n_k})| < 1/k$ for all $k \geq r$. Therefore, $(f_r(y_k) - f_r(x_{n_k})) \to 0$ as $k \to \infty$. By Cauchy's theorem, we get then

$$\lim_{N\to\infty} \frac{1}{N} \sum_{k=1}^{N} (f_r(y_k) - f_r(x_{n_k})) = 0. \tag{2.11}$$

We rearrange the given sequence (x_n) in the following way. We first enumerate those subscripts n, which are either no n_k or else are n_k with k being a perfect square, in an arbitrary fashion: $m_1, m_2, \ldots$. Then we define a sequence (u_k) by $u_k = x_{n_k}$ if k is not a perfect square and by $u_k = x_{m_p}$ if $k = p^2$. Clearly, (u_k) is a rearrangement of (x_n). Furthermore, for fixed r:

$$\left| \frac{1}{N} \sum_{k=1}^{N} (f_r(u_k) - f_r(x_{n_k})) \right| = \left| \frac{1}{N} \sum_{\substack{k=1 \\ k=p^2}}^{N} (f_r(u_k) - f_r(x_{n_k})) \right| \leq \frac{\sqrt{N}}{N} 2 \, \|f_r\|,$$

and so,

$$\lim_{N\to\infty} \frac{1}{N} \sum_{k=1}^{N} (f_r(u_k) - f_r(x_{n_k})) = 0. \tag{2.12}$$

Using (2.11) and (2.12), we have for all r,

$$\begin{aligned} \lim_{N\to\infty} \frac{1}{N} \sum_{k=1}^{N} f_r(u_k) &= \lim_{N\to\infty} \frac{1}{N} \sum_{k=1}^{N} (f_r(u_k) - f_r(x_{n_k})) \\ &\quad + \lim_{N\to\infty} \frac{1}{N} \sum_{k=1}^{N} (f_r(x_{n_k}) - f_r(y_k)) \\ &\quad + \lim_{N\to\infty} \frac{1}{N} \sum_{k=1}^{N} f_r(y_k) = \int_X f_r \, d\mu; \end{aligned}$$

hence, (u_k) is μ-u.d. in X. ∎

This theorem has some easy consequences concerning the rearrangement of everywhere dense sequences. For instance, if the support of μ does not contain any isolated points, then any everywhere dense sequence in X can be rearranged to a μ-u.d. sequence (see Exercise 2.11). However, if X has at least two points and the support of μ contains isolated points, then the foregoing statement need not be true any more (see Exercise 2.12). In particular, the statement does not hold for finite X having at least two points. Again, the space X consisting of only one point plays an exceptional role: for this space the above statement is true in a very trivial way.

We may further exploit the method in the proof of Theorem 2.5 in order to arrive at a quantitative result that bears some resemblance to Theorem 4.2 of Chapter 2. The restriction imposed on the sequence (x_n) below will be clear from Theorem 2.5.

THEOREM 2.6. Let (y_k) be a given sequence in X, let $f_1, f_2, \ldots, f_r, \ldots$ be a sequence of functions from $\mathscr{R}(X)$ with $\lim_{r\to\infty} \|f_r\| = 0$, and let $a_1, a_2, \ldots, a_N, \ldots$ be an increasing sequence of positive real numbers with $\lim_{N\to\infty} a_N = \infty$. Suppose the sequence (x_n) in X satisfies the following condition: Every open neighborhood of each y_k contains infinitely many terms of the sequence (x_n). Then (x_n) can be rearranged into a sequence (u_k) with

$$\sup_{r=1,2,\ldots} \left| \frac{1}{N} \sum_{k=1}^{N} (f_r(y_k) - f_r(u_k)) \right| \leq \frac{a_N}{N} \tag{2.13}$$

for all sufficiently large N.

PROOF. We can easily find a sequence of positive real numbers $\varepsilon_1, \varepsilon_2, \ldots$ such that $\sum_{k=1}^{N} \varepsilon_k \leq \frac{1}{3} a_N$ for all $N \geq 1$; for example, choose $\varepsilon_1 \leq \frac{1}{3} a_1$ and $\varepsilon_i \leq \frac{1}{3}(a_i - a_{i-1})$ for $i \geq 2$. By the assumption on $\|f_r\|$, there exists, for each $k \geq 1$, a least nonnegative integer $h(k)$ such that $\|f_r\| \leq 1/k$ for all $r > h(k)$. Let us note for later use that the $h(k)$ form a nondecreasing sequence. The set $E_k = \{x \in X\colon |f_r(y_k) - f_r(x)| < \varepsilon_k \text{ for } 1 \leq r \leq h(k)\}$ is an open neighborhood of y_k. In the same way as in the proof of Theorem 2.5 we can construct a subsequence (x_{n_k}) of (x_n) such that $x_{n_k} \in E_k$ for all $k \geq 1$ and $n_k \neq n_m$ for $k \neq m$.

For the following discussion, it will be useful to define $h(0) = 0$. Consider a fixed f_r. If $h(k) < r$ for all $k \geq 1$, then $\|f_r\| \leq 1/k$ for all k, or $f_r \equiv 0$. In this case, the subsequent estimates will be trivial. Otherwise, let k_0 be the largest $k \geq 0$ such that $h(k) < r$. Then, for all $k > k_0$, we have $r \leq h(k)$; hence, $|f_r(y_k) - f_r(x_{n_k})| < \varepsilon_k$. Therefore,

$$\begin{aligned} \left| \frac{1}{N} \sum_{k=1}^{N} (f_r(y_k) - f_r(x_{n_k})) \right| &\leq \left| \frac{1}{N} \sum_{k=1}^{k_0} (f_r(y_k) - f_r(x_{n_k})) \right| \\ &\quad + \left| \frac{1}{N} \sum_{k=k_0+1}^{N} (f_r(y_k) - f_r(x_{n_k})) \right| \\ &\leq \frac{1}{N} 2k_0 \|f_r\| + \frac{1}{N} \sum_{k=1}^{N} \varepsilon_k \\ &\leq \frac{1}{N} 2k_0 \|f_r\| + \frac{a_N}{3N}. \end{aligned} \tag{2.14}$$

If $k_0 > 0$, then $r > h(k_0)$ implies that $\| f_r \| \le 1/k_0$. Then

$$\left| \frac{1}{N} \sum_{k=1}^{N} (f_r(y_k) - f_r(x_{n_k})) \right| \le \frac{2}{N} + \frac{a_N}{3N}. \tag{2.15}$$

This inequality is trivially true for $k_0 = 0$ as well.

The sequence of $\| f_r \|$ is bounded; therefore, $\| f_r \| \le M$ for some positive constant M. In the same vein as in the proof of Theorem 2.5, we now enumerate those subscripts n that are either no n_k or else are n_k with $k \ge 2$ and $[a_{k-1}/6M] < [a_k/6M]$, in an arbitrary fashion: $m_1, m_2, \ldots$. The sequence (u_k) is then defined in the following way: $u_k = x_{n_k}$ if $k = 1$ or $[a_{k-1}/6M] = [a_k/6M]$; $u_k = x_{m_p}$ if $[a_{k-1}/6M] < [a_k/6M]$ and $k \ge 2$ is the pth subscript for which this happens. Again, it is easily seen that (u_k) is a rearrangement of (x_n). For fixed r, we have

$$\begin{aligned}
&\left| \frac{1}{N} \sum_{k=1}^{N} (f_r(u_k) - f_r(x_{n_k})) \right| \\
&\quad = \left| \frac{1}{N} \sum_{\substack{k=2 \\ [a_{k-1}/6M] < [a_k/6M]}}^{N} (f_r(u_k) - f_r(x_{n_k})) \right| \\
&\quad \le \frac{1}{N} 2 \, \| f_r \| \\
&\quad \cdot \left(\text{number of } k,\ 2 \le k \le N, \text{ such that } \left[\frac{a_{k-1}}{6M} \right] < \left[\frac{a_k}{6M} \right] \right) \\
&\quad \le \frac{1}{N} 2 \, \| f_r \| \left(\left[\frac{a_N}{6M} \right] - \left[\frac{a_1}{6M} \right] \right) \le \frac{1}{N} 2 \, \| f_r \| \frac{a_N}{6M} \le \frac{a_N}{3N}.
\end{aligned} \tag{2.16}$$

Together with (2.15) we get

$$\left| \frac{1}{N} \sum_{k=1}^{N} (f_r(y_k) - f_r(u_k)) \right| \le \frac{2a_N}{3N} + \frac{2}{N} \qquad \text{for all } r = 1, 2, \ldots. \tag{2.17}$$

Since this upper bound does not depend on r, we obtain

$$\sup_{r=1,2,\ldots} \left| \frac{1}{N} \sum_{k=1}^{N} (f_r(y_k) - f_r(u_k)) \right| \le \frac{2a_N}{3N} + \frac{2}{N}. \tag{2.18}$$

But, for sufficiently large N, we have $2 \le \frac{1}{3} a_N$, and we are done. ∎

An interesting special case occurs when the sequence (y_k) consists of nonisolated points. For (x_n) one may then take, for instance, any everywhere dense sequence in X. Theorem 2.6 is, of course, only of interest if $a_N/N \to 0$ as $N \to \infty$. In this case, sequences (y_k) containing isolated points have to be treated with care (see Exercise 2.13).

A countable convergence-determining class of continuous functions $f_1, f_2, \ldots, f_r, \ldots$ satisfying the hypothesis of Theorem 2.6 can be easily constructed. Just take an arbitrary countable convergence-determining class of continuous functions $g_1, g_2, \ldots, g_r, \ldots$ and replace those g_r with $\|g_r\| > 0$ by $f_r = (1/r\, \|g_r\|)g_r$.

Quantitative Theory

We can make the resemblance between Theorem 2.6 and Theorem 4.2 of Chapter 2 even more conspicuous by introducing the following definition.

DEFINITION 2.2. Let $f_1, f_2, \ldots, f_r, \ldots$ be a countable convergence-determining class of functions from $\mathscr{R}(X)$ with $\lim_{r\to\infty} \|f_r\| = 0$. For a sequence (x_n) in X and a natural number N, the *maximal deviation* M_N is defined to be

$$M_N = \sup_{r=1,2,\ldots} \left| \frac{1}{N} \sum_{n=1}^{N} f_r(x_n) - \int_X f_r\, d\mu \right|. \tag{2.19}$$

We have the following criterion analogous to Theorem 1.1 of Chapter 2.

LEMMA 2.3. The sequence (x_n) is μ-u.d. in X if and only if $\lim_{N\to\infty} M_N = 0$.

PROOF. It is obvious that (M_N) being a null sequence implies that (x_n) is μ-u.d. Conversely, suppose (x_n) is μ-u.d., and choose $\varepsilon > 0$. There is an R such that $\|f_r\| \le \varepsilon/2$ for all $r > R$. For those r, we have for all $N \ge 1$:

$$\left| \frac{1}{N} \sum_{n=1}^{N} f_r(x_n) - \int_X f_r\, d\mu \right| \le 2\, \|f_r\| \le \varepsilon.$$

For the finitely many functions $f_1, \ldots, f_R$, there exists N_0 such that $|(1/N)\sum_{n=1}^{N} f_r(x_n) - \int_X f_r\, d\mu| \le \varepsilon$ for all $N > N_0$ and all $r = 1, 2, \ldots, R$. It follows that $M_N \le \varepsilon$ for all $N > N_0$, and this is what we had to show. ■

From Theorem 2.6 and Definition 2.2 we can easily infer the following consequence.

COROLLARY 2.1. Let (y_k) be a sequence in X with maximal deviation M_N and let $a_1, a_2, \ldots, a_N, \ldots$ be an increasing sequence of positive real numbers with $\lim_{N\to\infty} a_N = \infty$. Then any sequence (x_n) in X satisfying the condition in Theorem 2.6 can be rearranged into a sequence (u_k) whose maximal deviation M_N^*, based on the same convergence-determining class as M_N, satisfies $|M_N - M_N^*| \le a_N/N$ for sufficiently large N.

EXAMPLE 2.1. Consider $X = [0, 1]$ with the relative topology of the reals and Lebesgue measure λ on X. It follows from Exercise 1.2 that the functions $f_0(x) \equiv 1$ and $f_r(x) = x^r/r$, $r = 1, 2, \ldots$, form a convergence-determining class with $\lim_{r\to\infty} \|f_r\| = 0$. Let M_N be the maximal deviation defined in terms of this system, and let D_N^* denote the discrepancy as defined in Definition 1.2 of Chapter 2. For $r \geq 1$ and $\omega = (x_n)$ in $[0, 1)$, we have $|(1/N)\sum_{n=1}^N f_r(x_n) - \int_X f_r d\lambda| \leq V(f_r)D_N^*(\omega) = (1/r)D_N^*(\omega)$ by Koksma's inequality (see Chapter 2, Theorem 5.1). Therefore, $\sup_{r=1,2,\ldots} |(1/N) \sum_{n=1}^N f_r(x_n) - \int_X f_r\, d\lambda| \leq D_N^*(\omega)$, and this remains valid if we include $r = 0$. Thus, $M_N \leq D_N^*(\omega)$ for all N and all sequences $\omega = (x_n)$. ∎

Notes

For a general survey of ergodic theory and proofs of the individual ergodic theorem, we refer to Halmos [2] and Billingsley [1]. For relations between laws of large numbers and ergodic theorems, see Loève [1, Chapter 9]. An in-depth study of laws of large numbers is carried out in Révész [1]. Further applications of the individual ergodic theorem related to problems discussed in this section were exhibited by Cigler [2, 9].

Theorems 2.1, 2.2, and 2.3 were shown in the fundamental paper of Hlawka [3], and even in the more general setting of summation methods (compare with Section 4). We have preferred to use the term *convergence-determining class* also in the context of this section, since it is more suggestive than Hlawka's *Hauptsystem* (Hlawka [3]).

In recognition of the pioneer work of Borel [1], the statement in Theorem 2.2 is sometimes referred to as the Borel property of μ-u.d. (see also Section 4). In conjunction with Theorem 2.2, the following zero-one law from probability theory is of interest: Let $(Y, \mathscr{F}, \nu)$ be an arbitrary probability space, and let A be a homogeneous set in Y^∞, that is, a ν_∞-measurable subset of Y^∞ that is independent of all cylinder sets; then $\nu_\infty(A) = 0$ or 1 (Kolmogorov [1, p. 60], Feller [2, p. 122]). Moreover, if A is a ν_∞-measurable subset of Y^∞ for which $(y_1, y_2, \ldots) \in A$ remains true whenever a finite number of y_i is replaced by arbitrary elements from Y, then A is homogeneous (Kolmogorov [1, p. 60], Visser [1]). It follows then, in particular, that if X is an arbitrary compact Hausdorff space and if the set S of μ-u.d. sequences in X is μ_∞-measurable, then $\mu_\infty(S)$ can only be 0 or 1. More general laws of the above type are the zero-one laws of Hewitt and Savage [1] and of Horn and Schach [1]. For arbitrary compact Hausdorff spaces, one does not know how to characterize the measures μ for which S is μ_∞-measurable and $\mu_\infty(S) = 1$.

By the law of the iterated logarithm (Loève [1, p. 260], Rényi [2, p. 337]), we have the following quantitative version of Lemma 2.1: If $f \in \mathscr{B}(X)$, then $(1/N) \sum_{n=1}^N f(x_n) = \int_X f\, d\mu + O(\sqrt{(\log\log N)/N})$ for μ_∞-almost all sequences (x_n) in X, the exceptional set depending on f. Thus, for a countable convergence-determining class $f_1, f_2, \ldots, f_r, \ldots$, we obtain $(1/N) \sum_{n=1}^N f_r(x_n) = \int_X f_r\, d\mu + O(\sqrt{(\log\log N)/N})$ for all $r = 1, 2, \ldots$ (μ_∞-a.e.), and Theorem 2.2 follows. A survey of probability methods in the theory of u.d. and some general results can be found in Kemperman [2].

Theorem 2.5 was enunciated in this form by Descovich [1]. The basic ideas can be traced back to Hlawka [3]. Both authors prove their results for certain classes of summation methods. Stapleton [1] has a very similar rearrangement result for locally compact Hausdorff spaces with countable base. Gerl [7] shows a rearrangement theorem for μ-relatively equidistributed sequences (compare with the notes in Section 1).

A technique of "lifting" sequences from the unit interval was used by Hedrlín [1, 2] to construct u.d. sequences in compact metric spaces. The method was in fact already employed earlier by Stapleton [1] in order to show the existence of u.d. sequences in locally compact Hausdorff spaces with countable base. For another application of the method, see Baayen and Hedrlín [1]. The possibility of defining a discrepancy in a compact space was investigated by Grassini [1]. An error in this paper was pointed out by Post [2].

A notion of discrepancy in separable metric spaces based on the Prohorov distance of probability measures was introduced by Bauer [1]. See also Mück and Philipp [1]. *Suites eutaxiques* in compact metric spaces were studied by de Mathan [3]. For quantitative theory in compact groups, see the notes in Section 1 of Chapter 4. For u.d. in Cartesian products of compact Hausdorff spaces, see Christol [1].

Exercises

2.1. Let X be an arbitrary compact Hausdorff space. A Borel set B in X is called a *carrier* of the measure μ if $\mu(B) = 1$. Prove that if there exist μ-u.d. sequences in X, then μ has a compact separable carrier. (Note: The support of μ is of course also a carrier of μ.)

2.2. For the same situation as in Exercise 2.1, prove that if there exists a carrier of μ that is compact with countable base in the relative topology, then there exist μ-u.d. sequences in X. Show that even μ_∞-almost all sequences in X are μ-u.d. (Note: Exercises 2.1 and 2.2 together still do not yield a necessary and sufficient condition for the measure μ to allow a μ-u.d. sequence, because there are compact separable Hausdorff spaces that do not possess a countable base; see Kelley [1, p. 164]. For a complete characterization of those measures μ, see the notes to Section 1. Compare also with Exercise 2.15.)

2.3. Let (X, d) be a compact metric space with an everywhere dense sequence (x_n). Show that the open balls $B_{nk} = \{x \in X: d(x_n, x) < 1/k\}$, $n = 1, 2, \ldots, k = 1, 2, \ldots$, form a countable base.

2.4. In a compact metric space with nonnegative normed Borel measure μ, there exists a countable base of open balls that are μ-continuity sets. *Hint:* Use Exercise 2.3.

2.5. Consider the discrete space $X = \{0, 1\}$ with measure μ defined by $\mu(\{0\}) = \alpha$, $\mu(\{1\}) = 1 - \alpha$, $0 \leq \alpha \leq 1$. Construct a μ-u.d. sequence in X.

2.6. Let $X = \{0, 1, 2, \ldots, k\}$ be a discrete space with measure μ defined by $\mu(\{i\}) = \lambda_i \geq 0$ for $0 \leq i \leq k$, $\lambda_0 + \lambda_1 + \cdots + \lambda_k = 1$. Construct a μ-u.d. sequence in X. *Hint:* Use recursion on k.

2.7. Use the individual ergodic theorem to prove that if T: $X \mapsto X$ is an ergodic transformation (with respect to μ) on a compact Hausdorff space X with countable base, then the sequence $(T^n x)$ is μ-u.d. in X for μ-almost all $x \in X$.

2.8. Prove without using Baire's category theorem that if the compact Hausdorff space X has at least two points, then the set S from Theorem 2.2 has a void interior in X^∞.

2.9. Give a detailed proof of the following fact: If the Hausdorff space X has no isolated points and (x_n) is everywhere dense in X, then the sequence remains everywhere dense after deletion of finitely many terms. Why is this false if X contains isolated points?

2.10.–2.13. In these four exercises, X denotes a compact Hausdorff space with countable base, and μ is a nonnegative normed Borel measure in X.

2.10. Prove that if $\mu(\{x\}) = 0$ for all isolated points $x \in X$, then there exists a μ-u.d. sequence in X consisting entirely of nonisolated points. *Hint:* Since X has a countable base, at most countably many points in X can be isolated.

2.11. Any everywhere dense sequence in X can be rearranged to a μ-u.d. sequence if we require that $\mu(\{x\}) = 0$ for all isolated points $x \in X$.

2.12. Prove that if X has at least two points and if there is an isolated point $x \in X$ with $\mu(\{x\}) > 0$, then there are everywhere dense sequences in X that cannot be rearranged to a μ-u.d. sequence.

2.13. Prove that Theorem 2.6 is best possible in the following sense: If X has at least two points some of which are isolated and if $a_N/N \to 0$ as $N \to \infty$, then one can construct a sequence (y_k) containing isolated points (even a μ-u.d. one), a countable convergence-determining class $f_1, f_2, \ldots, f_r, \ldots$ of functions from $\mathscr{R}(X)$ with $\|f_r\| \to 0$ as $r \to \infty$, and an everywhere dense sequence (x_n) in X such that the assertion of Theorem 2.6 does not hold. *Hint:* If $y \in X$ is isolated, then the function $f(x) = 1$ for $x = y$, $f(x) = 0$ for $x \neq y$, is continuous; use this function in the system $f_1, f_2, \ldots$; take a sequence (y_k) that contains the point y with a relative frequency larger than a_N/N for sufficiently large N.

2.14. For the space $X = \{0\} \cup \{1/n\colon n = 1, 2, \ldots\}$ in the relative topology of the reals and the point measure μ at $x = 0$, show that the functions f_r, $r = 1, 2, \ldots$, defined by $f_r(x) = 1/r$ for $x \leq 1/r$ and $f_r(x) = 0$ otherwise, form a convergence-determining class with $\lim_{r\to\infty} \|f_r\| = 0$. Let M_N be the maximal deviation defined in terms of the f_r. Show that for the sequence $1, \frac{1}{2}, \frac{1}{3}, \ldots$ we have $NM_N < 1$ for all $N \geq 1$.

2.15. Let X be a compact separable Hausdorff space but not necessarily with countable base. Let $x_0, x_1, x_2, \ldots$ be everywhere dense in X. Define the measure μ by

$$\mu(M) = \sum_{\substack{i=0 \\ x_i \in M}}^{\infty} 2^{-i-1} \qquad \text{for all } M \subseteq X.$$

Show that μ is regular and that its only compact carrier is X (in the sense of Exercise 2.1). Construct a sequence (y_k), $k \geq 1$, in the following way: The integer k has a unique dyadic representation $k = 2^{j_1} + 2^{j_2} + \cdots + 2^{j_s}$, $j_1 > j_2 > \cdots > j_s \geq 0$; define $y_k = x_{j_s}$. Prove that (y_k) is μ-u.d. in X. *Hint:* Show first that each x_i occurs with the proper frequency.

This example shows that even if μ has no compact carrier with countable base there may exist μ-u.d. sequences.

2.16. Let X be a compact metric space with metric d, and let μ be a nonnegative normed Borel measure in X with support K. Suppose (x_n) is a μ-u.d. sequence in X. For each n, let y_n be a point in K that is closest to x_n; that is, $d(x_n, y_n) = d(x_n, K) = \min_{z \in K} d(x_n, z)$. Prove that (y_n) is again μ-u.d. in X.

2.17. Let $b \geq 2$ be an integer. Prove that the transformation T of $[0, 1)$ defined by $Tx = \{bx\}$ (= fractional part of bx) is ergodic with respect to the Lebesgue measure λ in $[0, 1)$. *Hint:* Show first that an invariant Borel set A satisfies $\lambda(A)\lambda(J) = \lambda(A \cap J)$ for every interval J of the form $J = [m/b^k, (m + 1)/b^k)$, where k and m are integers with $k \geq 1$ and $0 \leq m < b^k$.

Deduce from the individual ergodic theorem that almost all real numbers are normal to the base b.

2.18. Why does Theorem 2.3 imply the existence of non-μ-u.d. sequences in X?

2.19. Prove that for any compact Hausdorff space X containing at least two points and for any nonnegative regular normed Borel measure μ in X, there exist sequences in X that are not μ-u.d. in X.

2.20. Prove that the one-sided shift T in X^∞ is ergodic with respect to μ_∞. *Hint:* Show first that $\mu_\infty(T^{-1}A) = \mu_\infty(A)$ and $\lim_{n\to\infty} \mu_\infty(A \cap T^{-n}B) = \mu_\infty(A)\mu_\infty(B)$ for cylinder sets A, B in X^∞.

3. EQUI-UNIFORM DISTRIBUTION

Basic Results

Let X again be a compact Hausdorff space but not necessarily with countable base. As usual, μ shall denote a nonnegative regular normed Borel measure in X. Instead of looking at one given sequence in X, we now consider a whole family of such sequences. We are interested in the extent to which we can expect a uniform approximation of the μ-integral by arithmetic means in the given family of sequences. To this end, we introduce the notion of a family of equi-μ-u.d. sequences.

DEFINITION 3.1. Let $\mathscr{S} = \{(x_{n,\sigma}) : \sigma \in J\}$ be a family of sequences in X, where J denotes an arbitrary index set. $\mathscr{S}$ is called a *family of equi-μ-u.d. sequences in X* if for every $f \in \mathscr{R}(X)$, we have $\lim_{N\to\infty} (1/N) \sum_{n=1}^{N} f(x_{n,\sigma}) = \int_X f \, d\mu$ uniformly in σ; that is, if for every $f \in \mathscr{R}(X)$ and for every $\varepsilon > 0$ there exists an integer $N(f, \varepsilon)$, independent of σ, such that

$$\left| \frac{1}{N} \sum_{n=1}^{N} f(x_{n,\sigma}) - \int_X f \, d\mu \right| \leq \varepsilon \tag{3.1}$$

for all $N \geq N(f, \varepsilon)$ and for all $\sigma \in J$.

EXAMPLE 3.1. Whenever $\mathscr{S}$ consists of finitely many μ-u.d. sequences in X, then $\mathscr{S}$ is a family of equi-μ-u.d. sequences in X. ■

EXAMPLE 3.2. Let X be a compact Hausdorff uniform space with uniformity $\mathscr{U}$. Actually, the assumption that X is a uniform space is no restriction at all. If $\Delta = \{(x, x): x \in X\}$ denotes the diagonal of $X \times X$, where X is an arbitrary compact Hausdorff space, then the family of all neighborhoods of Δ (in the product topology) is a uniformity for X and the corresponding uniform topology is identical with the original topology in X (Kelley [1, p. 198]). For the simple facts about uniform spaces that we need here, the reader is referred to Kelley [1, Chapter 6], Gaal [1], and Isbell [1]. We use the following well-known notions: If U and V are two subsets of $X \times X$, then $U \circ V = \{(x, z) \in X \times X: (x, y) \in U \text{ and } (y, z) \in V \text{ for some } y \in X\}$; moreover, U is called symmetric if $(x, y) \in U$ implies $(y, x) \in U$. A family $\{P_\sigma: \sigma \in J\}$ of transformations $P_\sigma: X \mapsto X$ is said to be equicontinuous at the point $x \in X$ if for each $U \in \mathscr{U}$, there exists a neighborhood R of x such that $(P_\sigma x, P_\sigma y) \in U$ for all $y \in R$ and for all $\sigma \in J$.

Now suppose that $\{P_\sigma: \sigma \in J\}$ is a family of measure-preserving transformations on X (with respect to the given nonnegative regular normed Borel measure μ) that is equicontinuous at every point $x \in X$, and let (x_n) be a given μ-u.d. sequence in X. Then we claim that $\{(P_\sigma x_n): \sigma \in J\}$ is a family of equi-μ-u.d. sequences in X. We have to start from a function $f \in \mathscr{R}(X)$ and an $\varepsilon > 0$. Since a continuous function on a compact uniform space is uniformly continuous, there exists $W \in \mathscr{U}$ such that $|f(u) - f(v)| < \varepsilon$ whenever $(u, v) \in W$. We choose a symmetric $V \in \mathscr{U}$ so that $V \circ V \subseteq W$. Then, by the equicontinuity of the family $\{P_\sigma: \sigma \in J\}$, for every $y \in X$ there is a neighborhood $R(y)$ of y such that $z \in R(y)$ implies $(P_\sigma y, P_\sigma z) \in V$ for all $\sigma \in J$. By the usual argument—take a Urysohn function corresponding to $\{y\}$ and the complement of the interior of $R(y)$—we can show that every $R(y)$ contains an open μ-continuity set $S(y)$ with $y \in S(y)$. Since X is compact, finitely many $S(y)$, say $S_1 = S(y_1), \ldots, S_m = S(y_m)$, will already cover X. Define $B_1 = S_1$ and $B_i = S_i \cap S_1' \cap \cdots \cap S_{i-1}'$ for $2 \leq i \leq m$. Without loss of generality, all B_i are nonvoid; otherwise, we just delete those B_i that are void. The B_i, $1 \leq i \leq m$, are pairwise disjoint μ-continuity sets that cover X. Moreover, since $B_i \subseteq S_i \subseteq R(y_i)$, we have for all $x, y \in B_i$: $(P_\sigma y_i, P_\sigma x) \in V$ and $(P_\sigma y_i, P_\sigma y) \in V$ for all $\sigma \in J$, which, in turn, implies $(P_\sigma x, P_\sigma y) \in W$ for all $\sigma \in J$. Therefore,

$$|f(P_\sigma x) - f(P_\sigma y)| < \varepsilon \tag{3.2}$$

for all x and y in the same B_i and all $\sigma \in J$. From each B_i, $1 \leq i \leq m$, we choose a point z_i that will be fixed throughout the remainder of the proof. Let $L > 0$ be such that $|f(x)| \leq L$ for all $x \in X$. The sequence (x_n) is μ-u.d.

and there are only finitely many B_i; therefore, we can find an integer $N(f, \varepsilon)$ such that

$$\left|\frac{A(B_i; N)}{N} - \mu(B_i)\right| \leq \frac{\varepsilon}{Lm} \qquad \text{for all } N \geq N(f, \varepsilon) \text{ and all } B_i. \tag{3.3}$$

We complete the proof by showing that

$$\left|\frac{1}{N}\sum_{n=1}^{N} f(P_\sigma x_n) - \int_X f\,d\mu\right| \leq 3\varepsilon \tag{3.4}$$

for all $N \geq N(f, \varepsilon)$ and all $\sigma \in J$. The basic idea is that each x_n lies in a unique B_i; we therefore approximate $f(P_\sigma x_n)$ by $f(P_\sigma z_i)$ and estimate the error. Thus, we write $\sum_{n=1}^{N} f(P_\sigma x_n) = \sum_{i=1}^{m} \sum_{\substack{n=1\\ x_n \in B_i}}^{N} f(P_\sigma x_n)$, which will then be approximated by $\sum_{i=1}^{m} \sum_{\substack{n=1\\ x_n \in B_i}}^{N} f(P_\sigma z_i)$, or $\sum_{i=1}^{m} A(B_i; N) f(P_\sigma z_i)$. For each x_n, $1 \leq n \leq N$, we commit an error of at most ε by (3.2); therefore, for the total error we have

$$\left|\sum_{n=1}^{N} f(P_\sigma x_n) - \sum_{i=1}^{m} A(B_i; N) f(P_\sigma z_i)\right| \leq N\varepsilon. \tag{3.5}$$

We also note that from the assumption that all P_σ are measure-preserving we get $\int_X f(P_\sigma x)\,d\mu(x) = \int_X f\,d\mu$ for all $\sigma \in J$. By using (3.2), (3.3), and (3.5), we have now for all $N \geq N(f, \varepsilon)$ and all $\sigma \in J$:

$$\begin{aligned}
\left|\frac{1}{N}\sum_{n=1}^{N} f(P_\sigma x_n) - \int_X f\,d\mu\right| &= \left|\frac{1}{N}\sum_{n=1}^{N} f(P_\sigma x_n) - \int_X f(P_\sigma x)\,d\mu(x)\right| \\
&\leq \left|\frac{1}{N}\sum_{n=1}^{N} f(P_\sigma x_n) - \frac{1}{N}\sum_{i=1}^{m} A(B_i; N) f(P_\sigma z_i)\right| \\
&\quad + \left|\frac{1}{N}\sum_{i=1}^{m} A(B_i; N) f(P_\sigma z_i) - \sum_{i=1}^{m} \mu(B_i) f(P_\sigma z_i)\right| \\
&\quad + \left|\sum_{i=1}^{m} \mu(B_i) f(P_\sigma z_i) - \int_X f(P_\sigma x)\,d\mu(x)\right| \\
&\leq \varepsilon + \sum_{i=1}^{m} \left|\frac{A(B_i; N)}{N} - \mu(B_i)\right| |f(P_\sigma z_i)| \\
&\quad + \left|\sum_{i=1}^{m} \int_{B_i} (f(P_\sigma z_i) - f(P_\sigma x))\,d\mu(x)\right| \\
&\leq \varepsilon + \varepsilon + \sum_{i=1}^{m} \int_{B_i} |f(P_\sigma z_i) - f(P_\sigma x)|\,d\mu(x) \\
&\leq 2\varepsilon + \sum_{i=1}^{m} \varepsilon\mu(B_i) = 3\varepsilon. \qquad \blacksquare
\end{aligned}$$

It does not require too much optimism to expect that criteria similar to those in Section 1 might hold as well in the present situation if we only make some provisions concerning the uniformity in σ. In fact, the proofs of the subsequent theorems are almost identical with the proofs of the corresponding theorems in Section 1, so that we take the liberty of carrying out the details in one case only.

THEOREM 3.1. Let $\mathscr{V}$ be a class of functions from $\mathscr{B}(X)$ such that $\operatorname{sp}(\mathscr{V})$ is dense in $\mathscr{R}(X)$, and let $\mathscr{S} = \{(x_{n,\sigma}): \sigma \in J\}$ be a family of sequences in X. If for every $f \in \mathscr{V}$, we have

$$\lim_{N\to\infty} \frac{1}{N} \sum_{n=1}^{N} f(x_{n,\sigma}) = \int_X f\, d\mu \qquad \text{uniformly in } \sigma \in J, \tag{3.6}$$

then $\mathscr{S}$ is a family of equi-μ-u.d. sequences in X. If $\mathscr{V} \subseteq \mathscr{R}(X)$, then the converse evidently holds as well.

PROOF. We proceed along the same lines as in the proof of Theorem 1.1. Take $g \in \operatorname{sp}(\mathscr{V})$; that is, $g = \alpha_1 f_1 + \cdots + \alpha_k f_k$ for some $f_i \in \mathscr{V}$ and $\alpha_i \in \mathbb{R}$, $1 \leq i \leq k$. Let M be a positive constant such that $|\alpha_i| \leq M$ for all $1 \leq i \leq k$, and choose $\varepsilon > 0$. Then, for the positive number ε/Mk and each f_i, there exists an integer N_i, independent of σ, such that

$$\left| \frac{1}{N} \sum_{n=1}^{N} f_i(x_{n,\sigma}) - \int_X f_i\, d\mu \right| \leq \frac{\varepsilon}{Mk} \tag{3.7}$$

for all $N \geq N_i$ and all $\sigma \in J$. Hence, for $N \geq \max_{i=1,\ldots,k} N_i$ and all $\sigma \in J$, we obtain

$$\begin{aligned} \left| \frac{1}{N} \sum_{n=1}^{N} g(x_{n,\sigma}) - \int_X g\, d\mu \right| &= \left| \sum_{i=1}^{k} \alpha_i \left(\frac{1}{N} \sum_{n=1}^{N} f_i(x_{n,\sigma}) - \int_X f_i\, d\mu \right) \right| \\ &\leq \sum_{i=1}^{k} |\alpha_i| \frac{\varepsilon}{Mk} \leq \varepsilon. \end{aligned} \tag{3.8}$$

Now, for a given $f \in \mathscr{R}(X)$ and $\varepsilon > 0$, there exists $h \in \operatorname{sp}(\mathscr{V})$ with $\|f - h\| < \varepsilon$. By what we have already seen, there exists an integer $N(h, \varepsilon)$ such that $|(1/N) \sum_{n=1}^{N} h(x_{n,\sigma}) - \int_X h\, d\mu| \leq \varepsilon$ for all $N \geq N(h, \varepsilon)$ and all $\sigma \in J$. Then, in exactly the same way as in the proof of Theorem 1.1, we get for all $N \geq N(h, \varepsilon)$ and all $\sigma \in J$:

$$\begin{aligned} &\left| \frac{1}{N} \sum_{n=1}^{N} f(x_{n,\sigma}) - \int_X f\, d\mu \right| \\ &\qquad \leq 2\|f - h\| + \left| \frac{1}{N} \sum_{n=1}^{N} h(x_{n,\sigma}) - \int_X h\, d\mu \right| < 3\varepsilon. \quad \blacksquare \end{aligned} \tag{3.9}$$

For X with countable base, we constructed in Theorem 2.1 a countable class $\mathscr{V} \subseteq \mathscr{R}(X)$ with $\overline{\mathrm{sp}\,(\mathscr{V})} = \mathscr{R}(X)$. This, together with Theorem 3.1, yields the following criterion.

THEOREM 3.2: Weyl Criterion. Let X satisfy the second axiom of countability, and let $\mathscr{V} = \{f_1, f_2, \ldots\}$ be a countable class of real-valued continuous functions on X with $\overline{\mathrm{sp}\,(\mathscr{V})} = \mathscr{R}(X)$. Then $\mathscr{S} = \{(x_{n,\sigma})\colon \sigma \in J\}$ is a family of equi-μ-u.d. sequences in X if and only if for all $f_i \in \mathscr{V}$, we have

$$\lim_{N\to\infty} \frac{1}{N} \sum_{n=1}^{N} f_i(x_{n,\sigma}) = \int_X f_i \, d\mu \qquad \text{uniformly in } \sigma \in J. \tag{3.10}$$

Returning to arbitrary X, the following analogue of Theorem 1.2 holds.

THEOREM 3.3. $\mathscr{S} = \{(x_{n,\sigma})\colon \sigma \in J\}$ is a family of equi-μ-u.d. sequences in X if and only if for all (closed) μ-continuity sets $M \subseteq X$, we have

$$\lim_{N\to\infty} \frac{A_\sigma(M; N)}{N} = \mu(M) \qquad \text{uniformly in } \sigma \in J, \tag{3.11}$$

where $A_\sigma(M; N)$ denotes the counting function corresponding to the sequence $(x_{n,\sigma})$. In detail, for every (closed) μ-continuity set $M \subseteq X$ and every $\varepsilon > 0$ there should exist an integer $N(M, \varepsilon)$, independent of σ, such that

$$\left| \frac{A_\sigma(M; N)}{N} - \mu(M) \right| \le \varepsilon \qquad \text{for all } N \ge N(M, \varepsilon) \quad \text{and all } \sigma \in J. \tag{3.12}$$

The Size of Families of Equi-u.d. Sequences

Let $\mathscr{S}$ be a family of equi-μ-u.d. sequences in X, viewed as a subset of X^∞. The subsequent discussions will be in the same vein as those concerning the set S in Section 2. We know at any rate that $\mathscr{S}$ has to be contained in S.

THEOREM 3.4. Suppose that $\mathscr{S} = \{(x_{n,\sigma})\colon \sigma \in J\}$ is a family of equi-μ-u.d. sequences in X. Then so is $\overline{\mathscr{S}}$, the closure taken in X^∞.

PROOF. By hypothesis, for given $f \in \mathscr{R}(X)$ and $\varepsilon > 0$, there exists an integer $N(f, \varepsilon)$ such that $|(1/N)\sum_{n=1}^{N} f(x_{n,\sigma}) - \int_X f\, d\mu| \le \varepsilon$ for all $N \ge N(f, \varepsilon)$ and all $\sigma \in J$. Using an argument similar to that in the proof of Theorem 2.3, we consider the set Q of all points $(y_1, y_2, \ldots, y_n, \ldots) \in X^\infty$ for which $|(1/N)\sum_{n=1}^{N} f(y_n) - \int_X f\, d\mu| \le \varepsilon$ holds for all $N \ge N(f, \varepsilon)$. As in the proof of Theorem 2.3, it follows that Q is a closed subset of X^∞. But $\mathscr{S} \subseteq Q$, and therefore, $\overline{\mathscr{S}} \subseteq Q$, which completes the proof. ■

THEOREM 3.5. Suppose that X contains at least two points and that $\mathscr{S}$ is a family of equi-μ-u.d. sequences in X. Then $\mathscr{S}$ is nowhere dense in X^∞.

PROOF. Assume, on the contrary, that $\bar{\mathscr{S}}$ contains a nonvoid open set D in X^∞. Then D, in turn, contains a nonvoid open set $E = \prod_{i=1}^\infty E_i$ with $E_i = X$ for i greater than some k. For two distinct points a and b in X, let f be a Urysohn function (see Section 1) with $f(a) = 1, f(b) = 0$. In $\bar{\mathscr{S}}$, there is a sequence (y_n) with $y_n = a$ for $n > k$, and a sequence (z_n) with $z_n = b$ for $n > k$. Then, for all $N \geq 3k$, we have $(1/N)\sum_{n=1}^N f(y_n) \geq (N-k)/N \geq \frac{2}{3}$ and $(1/N)\sum_{n=1}^N f(z_n) \leq k/N \leq \frac{1}{3}$. But this is not possible, since by Theorem 3.4, $\bar{\mathscr{S}}$ is a family of equi-μ-u.d. sequences in X. ■

At this stage, it should already be evident that the restriction on X in the foregoing theorem is necessary (compare with Section 2). If X has a countable base, the following alternative proof of Theorem 3.5 can be given: $\bar{\mathscr{S}}$, as a subset of S, is of the first category in X^∞ (see Theorem 2.3); since X^∞ is metrizable, Baire's category theorem applies, and so, $\bar{\mathscr{S}}$ has a void interior; in other words, $\mathscr{S}$ is nowhere dense in X^∞. In this argument, we used the following special form of Baire's category theorem.

LEMMA 3.1: Baire's Category Theorem. Let M be a closed subset of the compact metric space Y. If M is of the first category in Y, then M has a void interior.

The support of μ (see Definition 1.4) will be used in the subsequent theorems, the first of which is a counterpart to Theorem 2.2. But now we do not need a countable base for X. Let μ_∞ be the product measure in X^∞ induced by μ (if X has a countable base, then μ_∞ is again a Borel measure). Furthermore, let $\bar{\mu}_\infty$ be the outer measure in X^∞ defined by $\bar{\mu}_\infty(B) = \inf\{\mu_\infty(A)\colon A \text{ is } \mu_\infty\text{-measurable and } B \subseteq A\}$ for every $B \subseteq X^\infty$.

THEOREM 3.6. Suppose that μ is not a point measure, and let $\mathscr{S}$ be a family of equi-μ-u.d. sequences in X. Then $\bar{\mu}_\infty(\mathscr{S}) < 1$.

PROOF. It suffices to show that $\bar{\mu}_\infty(\bar{\mathscr{S}}) < 1$. Suppose, on the contrary, that $\bar{\mu}_\infty(\bar{\mathscr{S}}) = 1$. Then, for any μ_∞-measurable subset A of $X^\infty \backslash \bar{\mathscr{S}}$, we have $\mu_\infty(A) = 0$. By Theorems 3.4 and 3.5, $X^\infty \backslash \bar{\mathscr{S}}$ is a nonvoid open set in X^∞. Choose an arbitrary $\xi = (x_1, x_2, \ldots) \in X^\infty \backslash \bar{\mathscr{S}}$, and a neighborhood E of ξ that is a cylinder set $E = \prod_{i=1}^\infty E_i$ contained in $X^\infty \backslash \bar{\mathscr{S}}$, where E_i is open in X for all i and $E_i = X$ for sufficiently large i. Now, $\mu_\infty(E) = 0$; hence, $\mu(E_j) = 0$ for some j. But then $x_j \in K'$, where K is the support of μ. Thus, we have shown $X^\infty \backslash \bar{\mathscr{S}} \subseteq X^\infty \backslash K^\infty$, or $K^\infty \subseteq \bar{\mathscr{S}}$. It follows that K^∞ is a family of equi-μ-u.d. sequences in X.

Since μ is not a point measure, the support K of μ contains at least two points. But then there exist points in K^∞ that do not correspond to μ-u.d.

sequences in X (see Theorem 1.3 and Exercise 2.19), and we arrive at a contradiction. ■

If μ is a point measure (i.e., $\mu(\{x_0\}) = 1$ for some $x_0 \in X$), then the family $\mathscr{S}$ consisting only of the constant sequence $(x_0, x_0, \ldots)$ is a family of equi-μ-u.d. sequences in X satisfying $\mu_\infty(\mathscr{S}) = 1$. In case the support of μ has a countable base (in the relative topology), we can prove a positive result on the μ_∞-measure of $\overline{\mathscr{S}}$ supplementing Theorem 3.6, namely, that $\mu_\infty(\overline{\mathscr{S}})$ may come arbitrarily close to 1. We shall need the following important fact from measure theory.

LEMMA 3.2: Egoroff's Theorem. Let $(Y, \mathscr{F}, \nu)$ be a measure space with ν being a nonnegative normed measure. Suppose that f_n, $n = 1, 2, \ldots$, and f are $\mathscr{F}$-measurable functions on Y that are finite ν-a.e. and for which $\lim_{n\to\infty} f_n(y) = f(y)$ ν-a.e. Then for every $\varepsilon > 0$ there exists a set $M \in \mathscr{F}$ with $\nu(M) \geq 1 - \varepsilon$ such that $\lim_{n\to\infty} f_n(y) = f(y)$ uniformly on M.

THEOREM 3.7. Suppose the support of μ has a countable base, and let a real number δ with $0 < \delta < 1$ be given. Then there exists a closed family $\mathscr{S}$ of equi-μ-u.d. sequences in X such that $\mu_\infty(\mathscr{S}) \geq 1 - \delta$.

PROOF. Let K be the support of μ and let μ^* be the restriction of μ to K. For simplicity, we write μ_∞^* instead of $(\mu^*)_\infty$. Let $\mathscr{V} = \{f_1, f_2, \ldots\}$ be a countable class of functions from $\mathscr{R}(K)$ with $\overline{\text{sp}(\mathscr{V})} = \mathscr{R}(K)$. As we did in the proof of Theorem 2.3, we look, for fixed $k \geq 1$ and $N \geq 1$, at the function $F_{k,N} \in \mathscr{R}(K^\infty)$, defined by $F_{k,N}(x_1, x_2, \ldots) = (1/N) \sum_{n=1}^N f_k(x_n)$ for $(x_1, x_2, \ldots) \in K^\infty$. From Lemma 2.1, we get

$$\lim_{N\to\infty} F_{k,N}(x_1, x_2, \ldots) = \int_K f_k \, d\mu^* \qquad \mu_\infty^*\text{-a.e.}$$

Therefore, by Egoroff's theorem, there exists a μ_∞^*-measurable subset $\mathscr{S}_k$ of K^∞ of measure $\mu_\infty^*(\mathscr{S}_k) \geq 1 - (\delta/2^k)$ such that $\lim_{N\to\infty} F_{k,N}(x_1, x_2, \ldots) = \int_K f_k \, d\mu^*$ uniformly on $\mathscr{S}_k$. Consequently, for $\mathscr{S}_0 = \bigcap_{k=1}^\infty \mathscr{S}_k$ the following holds: For all $f_k \in \mathscr{V}$ we have

$$\lim_{N\to\infty} \frac{1}{N} \sum_{n=1}^N f_k(x_n) = \int_K f_k \, d\mu^* \qquad \text{uniformly on } \mathscr{S}_0.$$

By Theorem 3.2, $\mathscr{S}_0$ is a family of equi-μ^*-u.d. sequences in K. Applying Theorem 3.4, we get a closed family $\mathscr{S} = \overline{\mathscr{S}}_0$ (closure in K^∞) of equi-μ^*-u.d. sequences in K satisfying $\mu_\infty^*(\mathscr{S}) \geq \mu_\infty^*(\mathscr{S}_0) \geq 1 - \sum_{k=1}^\infty (1 - \mu_\infty^*(\mathscr{S}_k)) \geq 1 - \delta$; hence, $\mu_\infty(\mathscr{S}) \geq 1 - \delta$. Note that K^∞ is closed in X^∞, and so, $\mathscr{S}$ is closed in X^∞. It is easy to see that $\mathscr{S}$ is also a family of equi-μ-u.d. sequences in X, which completes the proof. ■

Well-Distributed Sequences

DEFINITION 3.2. The sequence (x_n) in X is said to be *μ-well distributed in X* if for every $f \in \mathscr{R}(X)$, we have

$$\lim_{N\to\infty} \frac{1}{N} \sum_{n=1+h}^{N+h} f(x_n) = \int_X f\,d\mu \qquad \text{uniformly in } h = 0, 1, 2, \ldots. \tag{3.13}$$

Thus, $\xi = (x_n)$ is μ-well distributed in X if ξ, $T\xi$, $T^2\xi, \ldots$ form a family of equi-μ-u.d. sequences in X, where T is the one-sided shift in X^∞. As immediate consequences of Theorems 3.1, 3.2, and 3.3, respectively, we get the following criteria.

COROLLARY 3.1. Let $\mathscr{V}$ be a class of functions from $\mathscr{R}(X)$ with $\overline{\operatorname{sp}(\mathscr{V})} = \mathscr{R}(X)$. Then (x_n) is μ-well distributed in X if and only if for all $f \in \mathscr{V}$, we have $\lim_{N\to\infty} (1/N) \sum_{n=1+h}^{N+h} f(x_n) = \int_X f\,d\mu$ uniformly in $h = 0, 1, 2, \ldots$. If X satisfies the second axiom of countability, then there exists a countable class $\mathscr{V}$ satisfying the above conditions.

COROLLARY 3.2. The sequence (x_n) is μ-well distributed in X if and only if for all (closed) μ-continuity sets $M \subseteq X$, we have

$$\lim_{N\to\infty} \frac{1}{N} \sum_{n=1+h}^{N+h} c_M(x_n) = \mu(M) \qquad \text{uniformly in } h = 0, 1, 2, \ldots, \tag{3.14}$$

where c_M denotes the characteristic function of M.

A simple, but interesting, consequence of the preceding result that sheds some more light on the property of μ-well-distributivity is the following.

LEMMA 3.3. If (x_n) is μ-well distributed in X, then for every μ-continuity set M with $\mu(M) > 0$, there exists a natural number $N_0 = N_0(M)$ such that at least one of any N_0 consecutive elements from (x_n) lies in M.

PROOF. Choose $\varepsilon = \frac{1}{2}\mu(M)$. Then, by Corollary 3.2, there exists N_0 such that

$$\left| \frac{1}{N} \sum_{n=1+h}^{N+h} c_M(x_n) - \mu(M) \right| \leq \tfrac{1}{2}\mu(M)$$

for all $N \geq N_0$ and all $h = 0, 1, 2, \ldots$. In particular,

$$\frac{1}{N_0} \sum_{n=1+h}^{N_0+h} c_M(x_n) \geq \tfrac{1}{2}\mu(M) > 0 \qquad \text{for all } h = 0, 1, 2, \ldots,$$

which already proves our assertion. ∎

Not too surprisingly, well-distributed sequences are rather scarce. To prove a precise version of this statement, the following measure-theoretic lemma will be useful.

LEMMA 3.4. If μ is not a point measure in X, then there exists a μ-continuity set $M \subseteq X$ with $0 < \mu(M) < 1$.

PROOF. By the assumption on μ, there exist at least two distinct points a and b in the support K of μ. Let f be a Urysohn function on X with $f(a) = 1$ and $f(b) = 0$. By the usual argument (see Example 1.2), there is a μ-continuity set M of the form $M = \{x \in X: f(x) > \varepsilon\}$ for some ε with $0 < \varepsilon < 1$. Since M is an open neighborhood of $a \in K$, we have $\mu(M) > 0$. From $\bar{M} \subseteq \{x \in X: f(x) \geq \varepsilon\}$ we infer that $\bar{M}'$ is an open neighborhood of $b \in K$; thus, $\mu(\bar{M}') > 0$. This implies $\mu(\bar{M}) < 1$, and so, $\mu(M) < 1$. ∎

THEOREM 3.8. Let W be the set of all μ-well distributed sequences in X, viewed as a subset of X^∞. If μ is not a point measure, then $\mu_\infty(W) = 0$, where μ_∞ is the complete product measure in X^∞.

PROOF. Let M be a fixed μ-continuity set with $0 < \mu(M) < 1$, which exists by Lemma 3.4. For $N \geq 1$, let W_N be the set of all μ-well distributed sequences in X for which at least one of any N consecutive elements lies in M. By Lemma 3.3, we have $W = \bigcup_{N=1}^\infty W_N$. Thus, it suffices to show that $\mu_\infty(W_N) = 0$ for all $N \geq 1$. For given $N \geq 1$, let X^N be the Cartesian product of N copies of X, and let μ_N be the product measure on X^N. Define F_N to be the set consisting of all points of X^N for which at least one coordinate belongs to M; that is, $F_N = X^N \backslash \prod_{i=1}^N M_i'$ with $M_i = M$ for all i. Note that $\mu_N(F_N) = 1 - (1 - \alpha)^N$, where $\alpha = \mu(M)$. For $k \geq 0$, put

$$F_N^{(k)} = \{(x_1, x_2, \ldots) \in X^\infty: (x_{jN+1}, \ldots, x_{jN+N}) \in F_N \quad \text{for } 0 \leq j \leq k\}.$$

It follows from the definition of W_N that $W_N \subseteq \bigcap_{k=0}^\infty F_N^{(k)}$. Now $\mu_\infty(F_N^{(k)}) = (1 - (1 - \alpha)^N)^{k+1}$, and so, $0 < 1 - (1 - \alpha)^N < 1$ implies $\mu_\infty(\bigcap_{k=0}^\infty F_N^{(k)}) = 0$. Thus, *a fortiori*, $\mu_\infty(W_N) = 0$. ∎

A question that poses itself naturally is whether there exist μ-well distributed sequences at all. In case X has a countable base, a more or less explicit construction of μ-well distributed sequences is available (see notes). But μ-well distributed sequences may also exist for measures whose support does not have a countable base (see Exercise 3.3). For arbitrary compact Hausdorff spaces, a characterization of the measures μ for which μ-well distributed sequences exist is not known. In particular, one does not know whether there are measures μ for which μ-u.d. sequences, but no μ-well distributed sequences, exist.

The next result is very useful in that it enables us to find new μ-well distributed sequences from a given one.

THEOREM 3.9. Let $\xi = (x_n)$ be μ-well distributed in X. Then so is every sequence in the closure (in X^∞) of the set $\mathscr{S} = \{\xi, T\xi, T^2\xi, \ldots\}$.

PROOF. By the definition of μ-well-distributivity, $\mathscr{S}$ is a family of equi-μ-u.d. sequences in X. Then so is $\overline{\mathscr{S}}$, by Theorem 3.4. Now take a sequence $\eta \in \overline{\mathscr{S}}$. We claim that $T\eta \in \overline{\mathscr{S}}$. For if $T\eta \notin \overline{\mathscr{S}}$, then the continuity of T implies the existence of an open neighborhood D of η such that $T\zeta \notin \overline{\mathscr{S}}$ whenever $\zeta \in D$. But $T^m\xi \in D$ for some $m \geq 0$, and so, it would follow that $T^{m+1}\xi \notin \overline{\mathscr{S}}$, a contradiction. By induction, we get then $T^r\eta \in \overline{\mathscr{S}}$ for all $r \geq 0$. Thus, $\{\eta, T\eta, T^2\eta, \ldots\} \subseteq \overline{\mathscr{S}}$ is a family of equi-μ-u.d. sequences in X; in other words, η is μ-well distributed in X. ∎

THEOREM 3.10. Suppose X has a countable base and contains no isolated points. Then any everywhere dense sequence (x_n) in X can be rearranged so as to yield a μ-well distributed sequence in X.

PROOF. We proceed as in the proof of Theorem 2.5. We start out from a μ-well distributed sequence (y_k) in X, and we show that the pertinent sums $\sum_{k=1}^{N}$ occurring there, if replaced by $\sum_{k=1+h}^{N+h}$, can be estimated uniformly in h. To simplify the computations, we consider slightly altered sets D_k, namely, $D_k = \{x \in X\colon |f_r(y_k) - f_r(x)| < 2^{-k} \text{ for } 1 \leq r \leq k\}$. The sequence (x_{n_k}) is then constructed in exactly the same way, and we have for fixed f_r:

$$\left|\frac{1}{N}\sum_{k=1+h}^{N+h}(f_r(y_k) - f_r(x_{n_k}))\right| \leq \frac{1}{N}\sum_{k=1+h}^{N+h}|f_r(y_k) - f_r(x_{n_k})|$$
$$\leq \frac{r-1}{N}2\,\|f_r\| + \frac{1}{N}\sum_{k=1+h}^{N+h}2^{-k}$$
$$\leq \frac{r-1}{N}2\,\|f_r\| + \frac{1}{N},$$

which can be made arbitrarily small independent of h. The rearrangement (u_k) of (x_n) is again defined in the same way as in Theorem 2.5. For a fixed f_r, we get

$$\left|\frac{1}{N}\sum_{k=1+h}^{N+h}(f_r(u_k) - f_r(x_{n_k}))\right| = \left|\frac{1}{N}\sum_{\substack{k=1+h\\k=p^2}}^{N+h}(f_r(u_k) - f_r(x_{n_k}))\right|$$
$$\leq \frac{\sqrt{N+h} - \sqrt{h} + 1}{N}2\,\|f_r\|$$
$$= \frac{2\,\|f_r\|}{\sqrt{N+h} + \sqrt{h}} + \frac{2\,\|f_r\|}{N}$$
$$\leq \frac{2\,\|f_r\|}{\sqrt{N}} + \frac{2\,\|f_r\|}{N},$$

which can again be made arbitrarily small independent of h. Thus,

$$\left|\frac{1}{N}\sum_{k=1+h}^{N+h} f_r(u_k) - \int_X f_r\,d\mu\right| \leq \left|\frac{1}{N}\sum_{k=1+h}^{N+h} (f_r(u_k) - f_r(x_{n_k}))\right| + \left|\frac{1}{N}\sum_{k=1+h}^{N+h} (f_r(x_{n_k}) - f_r(y_k))\right| + \left|\frac{1}{N}\sum_{k=1+h}^{N+h} f_r(y_k) - \int_X f_r\,d\mu\right| \leq \varepsilon$$

for $N \geq N(f_r, \varepsilon)$, independent of h. Corollary 3.1 implies the μ-well-distributivity of (u_k). ■

We give a sample result to show how the concept of μ-well-distributivity can be used in analysis.

THEOREM 3.11. Let (x_n) be a μ-well distributed sequence in X. Let f be a function on X that is bounded away from 0 on an open set of positive measure; that is, there exists an open set E with $\mu(E) > 0$ and a positive constant c such that $|f(x)| \geq c$ for all $x \in E$. Furthermore, we are given a sequence (α_n) of positive real numbers with $\alpha_{n+1} \leq \sigma\alpha_n$ for all $n = 1, 2, \ldots$ and some fixed $\sigma > 0$. Then the absolute convergence of $\sum_{n=1}^{\infty} \alpha_n f(x_n)$ implies the convergence of $\sum_{n=1}^{\infty} \alpha_n$.

PROOF. The open set E contains a μ-continuity set M with $\mu(M) > 0$. To see this, we observe that, by regularity, there is a closed $C \subseteq E$ with $\mu(C) > 0$. As we have shown in Section 1, a Urysohn function corresponding to the disjoint closed sets C and E' then enables us to construct a μ-continuity set M with $C \subseteq M \subseteq E$. By Lemma 3.3, the μ-well-distributivity of (x_n) implies the existence of a positive integer N such that at least one of any N consecutive elements of (x_n) lies in M. We shall use that $|f(x)| \geq c$ for all $x \in M$. Without loss of generality, we may assume $\sigma \geq 1$, since for $\sigma < 1$ the theorem is obvious. For every positive integer s and every k with $1 \leq k \leq N$, we have $\alpha_{s+k} \geq \alpha_{s+N}\sigma^{k-N} \geq \alpha_{s+N}\sigma^{-N}$. We write

$$\sum_{n=1}^{\infty} \alpha_n\,|f(x_n)| = \sum_{q=0}^{\infty}\sum_{k=1}^{N} \alpha_{qN+k}\,|f(x_{qN+k})|$$

and note that in the inner sum for at least one term $|f(x_{qN+k})| \geq c$. Therefore,

$$\sum_{n=1}^{\infty} \alpha_n\,|f(x_n)| \geq c\sigma^{-N}\sum_{q=0}^{\infty} \alpha_{(q+1)N},$$

and $\sum_{q=0}^{\infty} \alpha_{(q+1)N}$ converges. Finally, for $L \geq 2$, we get

$$\sum_{n=1}^{LN} \alpha_n = \sum_{n=1}^{N} \alpha_n + \sum_{q=0}^{L-2}\sum_{k=1}^{N} \alpha_{(q+1)N+k} \leq \sum_{n=1}^{N} \alpha_n + N\sigma^N \sum_{q=0}^{L-2} \alpha_{(q+1)N} \leq \sum_{n=1}^{N} \alpha_n + N\sigma^N \sum_{q=0}^{\infty} \alpha_{(q+1)N},$$

and so $\sum_{n=1}^{\infty} \alpha_n$ converges. ■

Complete Uniform Distribution

A notion complementary in a sense to μ-well-distributivity is the so-called *complete μ-uniform distribution.* Throughout the remainder of this section, we suppose that X has a countable base.

DEFINITION 3.3. The sequence $\xi = (x_n)$ in X is called *completely μ-u.d. in X* if the sequence $(T^n\xi)$, $n = 0, 1, 2, \ldots$, is μ_∞-u.d. in X^∞.

It can be easily seen that a completely μ-u.d. sequence is in particular μ-u.d. (see Exercise 3.6). On the other hand, a completely μ-u.d. sequence cannot be μ-well distributed, apart from an obvious exception.

THEOREM 3.12. If μ is not a point measure, then a completely μ-u.d. sequence in X cannot be μ-well distributed in X.

PROOF. By Lemma 3.4, we can find a μ-continuity set $M \subseteq X$ with $0 < \mu(M) < 1$. Now suppose that $\xi = (x_n)$ is a sequence that is both completely μ-u.d. and μ-well distributed in X. Then, by Lemma 3.3, there exists a natural number N such that at least one of any N consecutive elements from (x_n) lies in M. On the other hand, consider the following set in X^∞: $F_\infty^{(N)} = \prod_{i=1}^{\infty} F_i$ with $F_i = M'$ for $1 \leq i \leq N$ and $F_i = X$ for $i > N$. Since M' is a μ-continuity set with $0 < \mu(M') < 1$, $F_\infty^{(N)}$ is a μ_∞-continuity set satisfying $0 < \mu_\infty(F_\infty^{(N)}) < 1$. But the sequence $(T^n\xi)$ is μ_∞-u.d. in X^∞, so, in particular, some elements of this sequence will lie in $F_\infty^{(N)}$. Thus, $T^h\xi \in F_\infty^{(N)}$ for some $h \geq 0$. In other words, $(x_{h+1}, x_{h+2}, \ldots) \in F_\infty^{(N)}$, or $x_{h+i} \in M'$ for $1 \leq i \leq N$. This is a contradiction to the construction of N. ∎

THEOREM 3.13. μ_∞-almost all sequences are completely μ-u.d. in X.

PROOF. The arguments are very similar to those employed in the proof of Theorem 2.2. Given a function $g \in \mathscr{B}(X^\infty)$, the individual ergodic theorem yields $\lim_{N\to\infty} (1/N) \sum_{n=0}^{N-1} g(T^n\xi) = \int_{X^\infty} g \, d\mu_\infty$ for μ_∞-almost all $\xi \in X^\infty$. Since X^∞ has also a countable base, there exists a countable convergence-determining class $g_1, g_2, \ldots$ of functions from $\mathscr{B}(X^\infty)$. Hence, for μ_∞-almost all $\xi \in X^\infty$,

$$\lim_{N\to\infty} \frac{1}{N} \sum_{n=0}^{N-1} g_i(T^n\xi) = \int_{X^\infty} g_i \, d\mu_\infty \qquad \text{for all } i = 1, 2, \ldots.$$

This means that $(T^n\xi)$ is μ_∞-u.d. in X^∞ for μ_∞-almost all $\xi \in X^\infty$, and the proof is complete. ∎

For X with countable base, Theorems 3.12 and 3.13 together provide an alternative proof for Theorem 3.8.

Notes

Proofs of Baire's category theorem can be found in Kelley [1] and Hewitt and Stromberg [1]. For proofs of Egoroff's theorem we refer to Halmos [1] and Hewitt and Stromberg [1].

Most of the results in this section are from Baayen and Helmberg [1] and Helmberg and Paalman-de Miranda [1]. Theorem 3.8, in this general form, is from Niederreiter [9]. It was shown earlier for X with countable base by Helmberg and Paalman-de Miranda [1]. For compact groups, several results were already shown earlier by Hlawka [1]. The explicit construction of μ-well distributed sequences that we mentioned was given by Baayen and Hedrlín [1]. An interesting measure-theoretic characterization of well-distributed sequences is in Cigler [12, 15].

A very thorough analysis of the relation between various concepts like uniformly distributed, well-distributed, almost well-distributed, weakly well-distributed, and completely uniformly distributed sequences with respect to a measure μ was carried out by Baayen and Helmberg [1]. One question left open in this paper was settled by Zame [3]. For a discussion of complete u.d. in compact spaces, see also Cigler [9] and Kemperman [2]. We remark that a sequence $\xi = (x_n)$ is completely μ-u.d. in X if and only if ξ is a generic point (in the sense of Furstenberg [1, 2]) with respect to the triple $(X^\infty, T, \mu_\infty)$, where T is the one-sided shift in X^∞.

In the mod 1 case, the definition of complete u.d. is due to Korobov [1]. A sequence (x_n) of real numbers is completely u.d. mod 1 if and only if for all $s \geq 1$ and all lattice points $(h_1, \ldots, h_s) \neq (0, \ldots, 0)$, the sequence $(h_1x_n + h_2x_{n+1} + \cdots + h_sx_{n+s-1})$, $n = 1, 2, \ldots$, is u.d. mod 1. For results on complete u.d. mod 1, see Cigler [2, 9], Franklin [2], Haber [2], Hlawka [4, 8], Knuth [1], Korobov [1, 2, 5, 7, 8, 10, 14, 18], Postnikov [4], Postnikova [1], and Starčenko [1, 2]. Completely u.d. mod 1 sequences are important random number generators (see Knuth [2, Chapter 3]).

Theorem 3.11 is essentially due to Hlawka [1] and has its origin in a theorem of Fatou (Zygmund [1, p. 232]). Hlawka [2] showed that the theorem remains true if "μ-well distributed" is replaced by "weakly μ-well distributed" (*schwach gleichmässig gleichverteilt*), a concept defined in Exercise 3.13. This is even more interesting, since almost all sequences (in the usual sense) are weakly μ-well distributed (Hlawka [2], Baayen and Helmberg [1]). Hlawka proved these results for compact groups with countable base, but they hold true as well for arbitrary compact Hausdorff spaces (only for the metric result one needs the second axiom of countability).

Exercises

3.1. Let (x_n) be u.d. mod 1. Prove that the family of sequences $\{(x_n + \sigma): 0 \leq \sigma < 1\}$ is equi-u.d. mod 1. *Hint:* Use Example 3.2.

3.2. Generalize Exercise 3.1 to several dimensions.

3.3. Prove that the sequence from Exercise 2.15 is μ-well distributed. *Hint:* Show first that the approximation of μ-measure by relative frequencies is uniform in h for a singleton $\{x_i\}$.

3.4. Let X be a discrete space with k elements, and let μ be the measure defined by $\mu(B) = (1/k)\,\mathrm{card}\,B$ for all $B \subseteq X$. Construct explicitly a μ-well distributed sequence in X.

3.5. Let (x_n) be a sequence in a compact Hausdorff space X such that for every μ-continuity set M, there exists a positive constant $C(M)$ such that $|A(M; N)/N - \mu(M)| \leq C(M)/N$ holds for all $N \geq 1$. Prove that (x_n) is μ-well distributed in X.

3.6. Here and in the three following exercises, suppose X has a countable base. Prove that if (x_n) is completely μ-u.d. in X, then (x_n) is μ-u.d. in X.

3.7. Let $\xi = (x_n)$ be a sequence in X. Prove that if $(T^n\xi)$, $n = 0, 1, 2, \ldots$, is μ_∞-well distributed in X^∞, then (x_n) is μ-well distributed in X.

3.8. Deduce from Exercise 3.7 that if μ is not a point measure, then no sequence of the form $(T^n\xi)$ is μ_∞-well distributed in X^∞.

3.9. Let (ξ_n) be a sequence in X^∞, $\xi_n = (x_{n1}, x_{n2}, \ldots, x_{nk}, \ldots)$, $n = 1, 2, \ldots$. Prove that (ξ_n) is μ_∞-u.d. in X^∞ if and only if for each $k \geq 1$, the sequence $(x_{n1}, \ldots, x_{nk})$, $n = 1, 2, \ldots$, is u.d. in $X^k (= \prod_{i=1}^k X_i$ with $X_i = X$ for all $i)$ with respect to the projection μ_k of μ_∞ on X^k; that is, $\mu_k(B) = \mu_\infty(B \times X \times X \times \cdots)$ for Borel sets B in X^k. *Hint:* Show first that the continuous functions on X^∞ that depend only on finitely many coordinates are dense in $\mathscr{R}(X^\infty)$.

3.10. Let $b \geq 2$ be an integer, and let $x = \sum_{n=1}^\infty x_n/b^n$ be the b-adic representation of $x \in [0, 1)$ (see Chapter 1, (8.1)). Consider the discrete space $X = \{0, 1, \ldots, b - 1\}$ with measure μ defined as in Exercise 3.4. Prove that x is normal to the base b if and only if the sequence (x_n) is completely μ-u.d. in X. *Hint:* Use Exercise 3.9.

3.11. Let X have a countable base, and let μ^* be the restriction of μ to its support K. Prove that a sequence in K is completely μ-u.d. in X if and only if it is completely μ^*-u.d. in K (in the relative topology).

3.12. Similar to Exercise 3.11, but with "completely u.d." replaced by "well-distributed."

3.13. A sequence (x_n) in X is called *weakly μ-well distributed in X* if

$$\lim_{N\to\infty} \overline{\lim_{H\to\infty}} \frac{1}{H} \sum_{h=0}^{H-1} \left| \frac{1}{N} \sum_{n=hN+1}^{hN+N} f(x_n) - \int_X f\, d\mu \right| = 0 \tag{3.15}$$

holds for every $f \in \mathscr{R}(X)$. Prove that every weakly μ-well distributed sequence in X is μ-u.d. in X.

3.14. Prove that every μ-well distributed sequence in X is weakly μ-well distributed in X.

3.15. Let $\mathscr{V}$ be a class of continuous functions on X with $\overline{\text{sp}(\mathscr{V})} = \mathscr{R}(X)$. Prove that (x_n) is weakly μ-well distributed in X if and only if (3.15) holds for all $f \in \mathscr{V}$.

3.16. Let X have a countable base. Prove that the support of the Borel measure μ_∞ in X^∞ is K^∞, where K is the support of μ.

4. SUMMATION METHODS

Matrix Methods

Let X be a compact Hausdorff space, and let μ be a nonnegative regular normed Borel measure on X. Furthermore, we take an infinite real matrix $\mathbf{A} = (a_{nk})$, $n = 1, 2, \ldots, k = 1, 2, \ldots$, satisfying the conditions

i. $$\|\mathbf{A}\| \overset{\text{def}}{=} \sup_n \sum_{k=1}^{\infty} |a_{nk}| < \infty$$

ii. $$\lim_{n\to\infty} \sum_{k=1}^{\infty} a_{nk} = 1.$$

In the sequel, if we speak of a *matrix method*, it will be tacitly assumed that these two conditions are satisfied. For important facts on matrix methods and some examples, see Section 7 of Chapter 1.

DEFINITION 4.1. The sequence (x_n) in X is said to be $(\mathbf{A}, \mu)$-*u.d. in* X if

$$\lim_{n\to\infty} \sum_{k=1}^{\infty} a_{nk} f(x_k) = \int_X f\, d\mu \qquad \text{for all } f \in \mathscr{R}(X). \tag{4.1}$$

If we choose the matrix method of arithmetic means (i.e., $a_{nk} = 1/n$ for $1 \leq k \leq n$ and $a_{nk} = 0$ for $k > n$), then we are back safely at Definition 1.1. As in Section 1, it suffices to require (4.1) only for a rather small class of continuous functions in order to guarantee $(\mathbf{A}, \mu)$-u.d.

THEOREM 4.1. Let $\mathscr{V}$ be a class of continuous functions on X with $\overline{\text{sp}(\mathscr{V})} = \mathscr{R}(X)$. Then (x_n) is $(\mathbf{A}, \mu)$-u.d. in X if and only if

$$\lim_{n\to\infty} \sum_{k=1}^{\infty} a_{nk} f(x_k) = \int_X f\, d\mu \qquad \text{for all } f \in \mathscr{V}. \tag{4.2}$$

PROOF. Necessity is clear. By linearity, (4.2) holds for all $g \in \text{sp}(\mathscr{V})$. With $h \in \mathscr{R}(X)$ and $\varepsilon > 0$ being given, we can find $g \in \text{sp}(\mathscr{V})$ with $\|h - g\| < \varepsilon$. Then, for sufficiently large n,

$$\begin{aligned}\left|\sum_{k=1}^{\infty} a_{nk} h(x_k) - \int_X h\, d\mu\right| &\leq \left|\sum_{k=1}^{\infty} a_{nk}(h - g)(x_k) - \int_X (h - g)\, d\mu\right| \\ &\quad + \left|\sum_{k=1}^{\infty} a_{nk} g(x_k) - \int_X g\, d\mu\right| \\ &\leq \|\mathbf{A}\|\, \|h - g\| + \|h - g\| \\ &\quad + \left|\sum_{k=1}^{\infty} a_{nk} g(x_k) - \int_X g\, d\mu\right| \\ &< (\|\mathbf{A}\| + 2)\varepsilon. \quad \blacksquare\end{aligned} \tag{4.3}$$

This proof is of course just an adaptation of the proof of Theorem 1.1 to the present situation. Under the additional condition that $a_{nk} \geq 0$ for all $n, k = 1, 2, \ldots$, we can also adapt the proof of Theorem 1.2 to yield an analogous result for $(\mathbf{A}, \mu)$-u.d. (see Exercise 4.1).

The Borel Property

We enter one of the most interesting aspects of the theory of $(\mathbf{A}, \mu)$-u.d. when we ask whether the metric result enunciated in Theorem 2.2 also holds for the matrix methods under consideration. Unfortunately, no conditions on the matrix $\mathbf{A}$ are known that are both necessary and sufficient for the analogue of Theorem 2.2 to be valid. However, we can prove a useful sufficient condition that is satisfactory in many important special cases. Example 4.4 will show that the condition is not necessary.

THEOREM 4.2. Let X have a countable base, and let $\mathbf{A} = (a_{nk})$ be a matrix method. Put $a_n = \sum_{k=1}^{\infty} a_{nk}^2$ for $n \geq 1$. If

$$\sum_{n=1}^{\infty} e^{-\delta/a_n} < \infty \qquad \text{for all } \delta > 0, \tag{4.4}$$

then μ_∞-almost all sequences are $(\mathbf{A}, \mu)$-u.d. in X.

In case $a_n = 0$, the term $e^{-\delta/a_n}$ is interpreted to be zero; because of condition (ii), this can happen for only finitely many n. For the proof of this theorem, we need a few auxiliary results. Throughout this discussion, Exp (v) will stand for the exponential function e^v with real v.

LEMMA 4.1. For $s \in \mathscr{B}(X)$, put $\alpha = \int_X s\, d\mu$. Then for all real numbers u we have

$$\int_X \operatorname{Exp}(us)\, d\mu \leq \operatorname{Exp}(\alpha u + \tfrac{1}{2}\, \|s\|^2\, u^2). \tag{4.5}$$

PROOF. As the composition of s with a continuous function on $\mathbb{R}$, Exp (us) is Borel-measurable on X. For each u, we have

$$\operatorname{Exp}(us) \leq \operatorname{Exp}(|us|) \leq \operatorname{Exp}(|u|\, \|s\|); \tag{4.6}$$

therefore, Exp (us) is integrable. Let ψ be the function on $\mathbb{R}$ defined by $\psi(u) = \int_X \operatorname{Exp}(us)\, d\mu$. We shall compute the first and second derivative of ψ. By (4.6), we may apply the dominated convergence theorem to

$$\psi(u) = \int_X \left(\sum_{j=0}^{\infty} \frac{u^j}{j!} s^j \right) d\mu,$$

and obtain

$$\psi(u) = \sum_{j=0}^{\infty} \frac{u^j}{j!} \int_X s^j\, d\mu.$$

Differentiation yields

$$\psi'(u) = \sum_{j=1}^{\infty} \frac{ju^{j-1}}{j!} \int_X s^j \, d\mu = \sum_{j=0}^{\infty} \frac{u^j}{j!} \int_X s^{j+1} \, d\mu = \int_X s \operatorname{Exp}(us) \, d\mu, \tag{4.7}$$

and similarly,

$$\psi''(u) = \int_X s^2 \operatorname{Exp}(us) \, d\mu. \tag{4.8}$$

With $\varphi(u) = \log \psi(u)$, we get

$$\varphi(0) = 0, \qquad \varphi'(0) = \frac{\psi'(0)}{\psi(0)} = \alpha, \tag{4.9}$$

and

$$\varphi''(u) = \frac{\psi''(u)\psi(u) - (\psi'(u))^2}{\psi^2(u)} \leq \frac{\psi''(u)}{\psi(u)} \leq \frac{\|s\|^2 \int_X \operatorname{Exp}(us) \, d\mu}{\int_X \operatorname{Exp}(us) \, d\mu} = \|s\|^2 \quad \text{for all } u \in \mathbb{R}. \tag{4.10}$$

Then, by (4.9), (4.10), and Taylor's theorem,

$$\varphi(u) \leq \alpha u + \tfrac{1}{2} \|s\|^2 u^2 \qquad \text{for all } u \in \mathbb{R}, \tag{4.11}$$

and the desired result follows. ■

LEMMA 4.2. Let $f \in \mathscr{B}(X)$ with $\int_X f \, d\mu = 0$ and $\|f\| = 1$. For $n \geq 1$, let S_n be the bounded measurable function on X^∞ defined by $S_n(x_1, x_2, \ldots) = \sum_{k=1}^{\infty} a_{nk} f(x_k)$. Then for all $\eta > 0$,

$$\mu_\infty(\{\xi \in X^\infty : |S_n(\xi)| > \eta\}) \leq 2 \operatorname{Exp}\left(-\frac{\eta^2}{2a_n}\right). \tag{4.12}$$

PROOF. We first estimate $\int_{X^\infty} \operatorname{Exp}(uS_n) \, d\mu_\infty$ for $u \in \mathbb{R}$. Since

$$\left| \sum_{k=1}^{N} a_{nk} f(x_k) \right| \leq \|\mathbf{A}\| \qquad \text{for all } N \geq 1,$$

we can use the dominated convergence theorem:

$$\begin{aligned}
\int_{X^\infty} \operatorname{Exp}(uS_n) \, d\mu_\infty &= \int_{X^\infty} \left(\lim_{N \to \infty} \operatorname{Exp}\left(\sum_{k=1}^{N} u a_{nk} f(x_k) \right) \right) d\mu_\infty \\
&= \lim_{N \to \infty} \int_{X^\infty} \operatorname{Exp}\left(\sum_{k=1}^{N} u a_{nk} f(x_k) \right) d\mu_\infty \\
&= \lim_{N \to \infty} \int_{X^\infty} \prod_{k=1}^{N} \operatorname{Exp}(u a_{nk} f(x_k)) \, d\mu_\infty \\
&= \lim_{N \to \infty} \prod_{k=1}^{N} \int_X \operatorname{Exp}(u a_{nk} f) \, d\mu.
\end{aligned} \tag{4.13}$$

In the last step, we used Fubini's theorem. Now, by Lemma 4.1,

$$\int_X \operatorname{Exp}(ua_{nk}f)\,d\mu \le \operatorname{Exp}(\tfrac{1}{2}u^2 a_{nk}^{\ 2}), \tag{4.14}$$

and so,

$$\int_{X^\infty} \operatorname{Exp}(uS_n)\,d\mu_\infty \le \lim_{N\to\infty} \prod_{k=1}^{N} \operatorname{Exp}(\tfrac{1}{2}u^2 a_{nk}^{\ 2})$$
$$= \lim_{N\to\infty} \operatorname{Exp}\left(\tfrac{1}{2}u^2 \sum_{k=1}^{N} a_{nk}^{\ 2}\right) = \operatorname{Exp}(\tfrac{1}{2}u^2 a_n). \tag{4.15}$$

For $u > 0$, we have by (4.15),

$$\operatorname{Exp}(\tfrac{1}{2}u^2 a_n) \ge \int_{X^\infty} \operatorname{Exp}(uS_n)\,d\mu_\infty \ge \operatorname{Exp}(u\eta)\mu_\infty(\{\xi \in X^\infty\colon\ S_n(\xi) > \eta\}). \tag{4.16}$$

We note that (4.12) is trivial for $a_n = 0$. So suppose $a_n > 0$. Choose $u = \eta/a_n$, and it follows from (4.16) that

$$\mu_\infty(\{\xi \in X^\infty\colon\ S_n(\xi) > \eta\}) \le \operatorname{Exp}\left(-\frac{\eta^2}{2a_n}\right). \tag{4.17}$$

For $u < 0$, we have by (4.15),

$$\operatorname{Exp}(\tfrac{1}{2}u^2 a_n) \ge \int_{X^\infty} \operatorname{Exp}(uS_n)\,d\mu_\infty \ge \operatorname{Exp}(-u\eta)\mu_\infty(\{\xi \in X^\infty\colon\ S_n(\xi) < -\eta\}). \tag{4.18}$$

Choose $u = -\eta/a_n$, and it follows from (4.18) that

$$\mu_\infty(\{\xi \in X^\infty\colon\ S_n(\xi) < -\eta\}) \le \operatorname{Exp}\left(-\frac{\eta^2}{2a_n}\right). \tag{4.19}$$

(4.17) and (4.19) together imply (4.12). ■

PROOF OF THEOREM 4.2. Let $f \in \mathscr{B}(X)$ and S_n be as in Lemma 4.2. For fixed $\eta > 0$, put

$$L(\eta) = \{\xi \in X^\infty\colon \text{there exists } n_0(\xi) \text{ such that } |S_n(\xi)| \le \eta \text{ for all } n \ge n_0(\xi)\}. \tag{4.20}$$

With $B_n = \{\xi \in X^\infty\colon\ |S_n(\xi)| \le \eta\}$, we have $L(\eta) = \bigcup_{N=1}^\infty \bigcap_{n=N}^\infty B_n$. Since $\sum_{n=1}^\infty \operatorname{Exp}(-\eta^2/2a_n)$ converges by assumption, there exists, for given $\varepsilon > 0$, a positive integer N_0 with $\sum_{n=N_0}^\infty \operatorname{Exp}(-\eta^2/2a_n) \le \varepsilon/2$. Then, by Lemma 4.2,

$$\mu_\infty(L(\eta)) \ge \mu_\infty\left(\bigcap_{n=N_0}^\infty B_n\right) = 1 - \mu_\infty\left(\bigcup_{n=N_0}^\infty B_n'\right) \ge 1 - \sum_{n=N_0}^\infty \mu_\infty(B_n')$$
$$\ge 1 - 2\sum_{n=N_0}^\infty \operatorname{Exp}\left(-\frac{\eta^2}{2a_n}\right) \ge 1 - \varepsilon; \tag{4.21}$$

hence, $\mu_\infty(L(\eta)) = 1$. Then, $\mu_\infty(\bigcap_{r=1}^{\infty} L(1/r)) = 1$, and therefore, for μ_∞-almost all sequences (x_n), we have

$$\lim_{n\to\infty} S_n = \lim_{n\to\infty} \sum_{k=1}^{\infty} a_{nk} f(x_k) = \int_X f\, d\mu. \tag{4.22}$$

Now there is a countable system $\mathscr{V} = \{f_1, f_2, \ldots\}$ in $\mathscr{R}(X)$ with $\overline{\text{sp}\,(\mathscr{V})} = \mathscr{R}(X)$ and $f_1 \equiv 1$. By replacing each nonconstant f_i by $f_i - \int_X f_i\, d\mu$ and multiplying by a suitable constant, we get a countable system $\mathscr{V}^* = \{g_1, g_2, \ldots\}$ in $\mathscr{R}(X)$ with $\overline{\text{sp}\,(\mathscr{V}^*)} = \mathscr{R}(X)$ and where each nonconstant g_i satisfies $\int_X g_i\, d\mu = 0$ and $\|g_i\| = 1$. By what we have already shown, (4.22) holds for the nonconstant g_i in place of f and is trivial for the constant g_i. Then, by Theorem 4.1 and the same argument as in the proof of Theorem 2.2, we may conclude that μ_∞-almost all sequences are $(\mathbf{A}, \mu)$-u.d. in X. ■

In the first part of the above proof, we have of course just reproduced the Borel-Cantelli lemma from probability theory.

A matrix $\mathbf{A}$ for which μ_∞-almost all sequences are $(\mathbf{A}, \mu)$-u.d. is said to have the *Borel property* with respect to the measure μ (see also the notes in Section 1). The following condition, although more restrictive than the so-called *Hill condition* (4.4), often suffices to verify the Borel property.

COROLLARY 4.1. If $\lim_{n\to\infty} a_n \log n = 0$ with the notation of Theorem 4.2, then $\mathbf{A}$ has the Borel property with respect to all nonnegative normed Borel measures μ on the compact Hausdorff space X with countable base.

PROOF. We show that $\lim_{n\to\infty} a_n \log n = 0$ implies the Hill condition. Choose $\delta > 0$, and then an ε with $0 < \varepsilon < \delta$. By hypothesis, there exists an integer N such that $a_n \log n < \varepsilon$ for all $n \geq N$. Then $\text{Exp}\,(-\delta/a_n) \leq \text{Exp}\,(-(\delta/\varepsilon)\log n) = n^{-\delta/\varepsilon}$ for all $n \geq N$, with $\delta/\varepsilon > 1$. Therefore the series $\sum_{n=1}^{\infty} \text{Exp}\,(-\delta/a_n)$ converges. ■

EXAMPLE 4.1. Theorem 2.2 is now just a special case of Corollary 4.1. For the summation method of arithmetic means ($a_{nk} = 1/n$ for $1 \leq k \leq n$ and $a_{nk} = 0$ for $k > n$) we get $a_n = 1/n$ for all $n \geq 1$; therefore,

$$\lim_{n\to\infty} a_n \log n = 0. \quad ■$$

We could as well ask for necessary conditions on the matrix $\mathbf{A}$ that are implied by the Borel property. We show one such condition that is closely related to Corollary 4.1. For this theorem, X need not have a countable base.

THEOREM 4.3. If $\mathbf{A}$ has the Borel property with respect to a measure μ that is not a point measure, then $\lim_{n\to\infty} a_n = 0$.

PROOF. Let us first note that the assumption on μ implies that the support of μ contains at least two points. We now want to construct a function

$f \in \mathscr{R}(X)$ with $\int_X f\,d\mu = 0$ and $\int_X f^2\,d\mu > 0$. Take $g \in \mathscr{R}(X)$ such that g is not constant on the support of μ (e.g., take a Urysohn function with respect to two distinct points from the support of μ). Then $f = g - \int_X g\,d\mu$ will do. We certainly have $\int_X f\,d\mu = 0$. Furthermore, by the construction of g, there exists a point b from the support of μ such that $g(b) \neq \int_X g\,d\mu$. Thus, the continuous function f^2 is positive on some open neighborhood of b, which has positive μ-measure, since b is in the support of μ. Therefore, $\int_X f^2\,d\mu > 0$.

With this $f \in \mathscr{R}(X)$, put as in Lemma 4.2: $S_n(\xi) = S_n(x_1, x_2, \ldots) = \sum_{k=1}^{\infty} a_{nk} f(x_k)$. Since $\mathbf{A}$ has the Borel property, we have $\lim_{n\to\infty} S_n(\xi) = \int_X f\,d\mu = 0$ for μ_∞-almost all $\xi \in X^\infty$. Consequently, $\lim_{n\to\infty} S_n^2(\xi) = 0$ for μ_∞-almost all $\xi \in X^\infty$, and so, $\int_{X^\infty} (\lim_{n\to\infty} S_n^2)\,d\mu_\infty = 0$. On the other hand, the inequality $S_n^2 \leq \|\mathbf{A}\|^2\,\|f\|^2$ and the dominated convergence theorem yield

$$\begin{aligned}
\int_{X^\infty} \left(\lim_{n\to\infty} S_n^2\right) d\mu_\infty &= \lim_{n\to\infty} \int_{X^\infty} S_n^2\,d\mu_\infty \\
&= \lim_{n\to\infty} \int_{X^\infty} \left(\sum_{k=1}^{\infty} a_{nk} f(x_k)\right)^2 d\mu_\infty \\
&= \lim_{n\to\infty} \int_{X^\infty} \left(\sum_{i,j=1}^{\infty} a_{ni} a_{nj} f(x_i) f(x_j)\right) d\mu_\infty. \qquad (4.23)
\end{aligned}$$

The absolute convergence of $\sum_{k=1}^{\infty} a_{nk} f(x_k)$ implies that any arrangement of the above double series into a simple series converges to $(\sum_{k=1}^{\infty} a_{nk} f(x_k))^2$. Another application of the dominated convergence theorem yields

$$\begin{aligned}
0 &= \int_{X^\infty} \left(\lim_{n\to\infty} S_n^2\right) d\mu_\infty \\
&= \lim_{n\to\infty} \sum_{i,j=1}^{\infty} a_{ni} a_{nj} \int_{X^\infty} f(x_i) f(x_j)\,d\mu_\infty \\
&= \lim_{n\to\infty} \sum_{k=1}^{\infty} a_{nk}^2 \int_X f^2\,d\mu. \qquad (4.24)
\end{aligned}$$

Thus, $(\int_X f^2\,d\mu) \lim_{n\to\infty} a_n = 0$, and since $\int_X f^2\,d\mu \neq 0$, the proof is complete. ∎

COROLLARY 4.2. *If $\mathbf{A}$ has the Borel property with respect to a measure μ that is not a point measure, then $\mathbf{A}$ is regular; that is, $\lim_{n\to\infty} a_{nk} = 0$ for every fixed k.*

If the matrix method $\mathbf{A}$ includes the matrix method $\mathbf{B}$ (in the sense of Definition 7.4 of Chapter 1) and if (x_n) is $(\mathbf{B}, \mu)$-u.d. in X, then (x_n) is obviously also $(\mathbf{A}, \mu)$-u.d. in X. Hence, if $\mathbf{B}$ has the Borel property with respect to the measure μ on X, and if **A** includes **B**, then **A** has the Borel property with respect to μ.

EXAMPLE 4.2. Consider a Cesàro means $(\mathbf{C}, r)$ with $r \geq 1$ as defined in Example 7.1 of Chapter 1. Since $(\mathbf{C}, r)$ includes $(\mathbf{C}, 1)$ (Zeller and Beekmann [1, p. 104], Peyerimhoff [1, p. 15]), it enjoys the Borel property with respect to any nonnegative normed Borel measure μ on a compact Hausdorff space X with countable base. ■

EXAMPLE 4.3. A so-called *discrete Abel method* is defined as follows: We are given a sequence (c_n) of real numbers with $0 < c_n < 1$ and $\lim_{n\to\infty} c_n = 1$; the matrix $\mathbf{A} = (a_{nk})$ is defined by $a_{nk} = (1 - c_n)c_n^{k-1}$, $n, k = 1, 2, \ldots$. The matrix **A** satisfies conditions (i) and (ii), since $\sum_{k=1}^{\infty} a_{nk} = \sum_{k=1}^{\infty} |a_{nk}| = (1 - c_n) \sum_{k=1}^{\infty} c_n^{k-1} = 1$ for all $n \geq 1$. Moreover, **A** is regular. Now let (s_k) be a sequence of real numbers such that the radius of convergence of the power series $\alpha(x) = (1 - x) \sum_{k=1}^{\infty} s_k x^{k-1}$ is at least 1. Note that $\sum_{k=1}^{\infty} a_{nk} s_k = (1 - c_n) \sum_{k=1}^{\infty} s_k c_n^{k-1} = \alpha(c_n)$; thus, (s_k) will certainly be summable to the value m by the discrete Abel method if $\lim_{x\to 1-0} \alpha(x) = m$. According to the theorem of Frobenius stated in the proof of Theorem 2.4 of Chapter 1, every discrete Abel method includes $(\mathbf{C}, 1)$ and has, therefore, the Borel property (in the same universal sense as in Example 4.2). ■

EXAMPLE 4.4. We exhibit a matrix method that has the Borel property but does not satisfy the Hill condition. The matrix method that we construct will be a discrete Abel method (see preceding example). Let γ be a function with $0 < \gamma(n) < 1$ for positive integers n, that tends to 0 sufficiently slowly as $n \to \infty$; γ will be specified later on. The sequence (c_n), defined by $c_n = (1 - \gamma(n))/(1 + \gamma(n))$ for $n \geq 1$, will then yield a discrete Abel method. For the corresponding matrix **A**, we have then

$$a_n = \sum_{k=1}^{\infty} a_{nk}^{\ 2} = (1 - c_n)^2 \sum_{k=1}^{\infty} (c_n^{\ 2})^{k-1} = \frac{(1 - c_n)^2}{1 - c_n^{\ 2}} = \frac{1 - c_n}{1 + c_n} = \gamma(n).$$

We can now choose $\gamma(n)$ in such a way that $\sum_{n=1}^{\infty} \operatorname{Exp}(-\delta/a_n)$ diverges for every $\delta > 0$. Namely, take $\gamma(n) = 1/(\log\log(n + p))$ with p so large that $0 < \gamma(n) < 1$ is guaranteed (for instance, $p = e^e$). Then $\sum_{n=1}^{\infty} \operatorname{Exp}(-\delta/a_n) = \sum_{n=1}^{\infty} (\log(n + p))^{-\delta}$, which diverges for every $\delta > 0$. ■

EXAMPLE 4.5. Let $(\mathbf{R}, p_n)$ be a simple Riesz (or weighted arithmetical) means (see Example 7.3 of Chapter 1). By Corollary 4.2, a necessary condition for $(\mathbf{R}, p_n)$ to satisfy the Borel property with respect to a measure that is not a point measure is $\lim_{n\to\infty} P_n = \lim_{n\to\infty} (p_1 + \cdots + p_n) = \infty$. If the sequence (p_n) is nonincreasing and $\lim_{n\to\infty} P_n = \infty$, then Lemma 7.1 of Chapter 1 implies that $(\mathbf{R}, p_n)$ includes $(\mathbf{C}, 1)$ and thus has the Borel property in the universal sense of Example 4.2. Moreover, if (p_n) is nondecreasing but not too rapidly (more precisely, if there is an $H > 0$ such that $np_n \leq HP_n$ for all n), then by the same lemma $(\mathbf{R}, p_n)$ includes $(\mathbf{C}, 1)$ and has again the Borel property. ■

A necessary and sufficient condition for a matrix method **A** to include $(\mathbf{C}, 1)$ can be given (Zeller and Beekmann [1, p. 100]). This condition is the following: Every convergent sequence is summable by **A** and

$$\sup_n \sum_{k=1}^{\infty} (k+1)\, |a_{nk} - a_{n,k+1}| < \infty. \tag{4.25}$$

In particular, if **A** is regular, then the validity of the condition (4.25) guarantees that **A** includes $(\mathbf{C}, 1)$.

A Constructive Result

For the following, let X be a compact Hausdorff uniform space with uniformity $\mathscr{U}$ (see the remarks in Example 3.2). We shall present a theorem that will enable us to construct new $(\mathbf{A}, \mu)$-u.d. sequences from a given $(\mathbf{A}, \mu)$-u.d. sequence. To this end, we suppose that we are given, for every $n \geq 1$, an arbitrary transformation T_n on X such that the T_n converge uniformly to a continuous and measure-preserving transformation T on X; that is, for every $U \in \mathscr{U}$ there should exist a positive integer $N(U)$, independent of $x \in X$, such that $(T_n x, Tx) \in U$ for all $n > N(U)$ and every $x \in X$. Having these provisions in mind, we show the following.

THEOREM 4.4. Under the above assumptions, let $\mathbf{A} = (a_{nk})$ be a regular matrix method and let (x_n) be a given $(\mathbf{A}, \mu)$-u.d. sequence in X. Then the sequence $(T_n x_n)$ is again $(\mathbf{A}, \mu)$-u.d. in X.

PROOF. Take $f \in \mathscr{R}(X)$. Then f is uniformly continuous, so for a given $\varepsilon > 0$ there exists $U \in \mathscr{U}$ such that $|f(y) - f(z)| < \varepsilon$ whenever $(y, z) \in U$. It follows that for $k > N(U) = K_0$ we have $|f(T_k x) - f(Tx)| < \varepsilon$ for all $x \in X$. Since **A** is regular, we can find a positive integer N_0 such that

$$\sum_{k=1}^{K_0} |a_{nk}| < \varepsilon$$

for all $n > N_0$. We note also that T being measure-preserving implies $\int_X f(Tx)\, d\mu(x) = \int_X f\, d\mu$. Furthermore, the function $g(x) = f(Tx)$ is in $\mathscr{R}(X)$ and (x_n) is $(\mathbf{A}, \mu)$-u.d.; therefore, there exists a positive integer N_1, so that

$$\left| \sum_{k=1}^{\infty} a_{nk} f(Tx_k) - \int_X f(Tx)\, d\mu(x) \right| < \varepsilon \qquad \text{for } n > N_1.$$

Then, for sufficiently large n—that is, for $n > \max(N_0, N_1)$—we have

$$\begin{aligned}\left|\sum_{k=1}^{\infty} a_{nk} f(T_k x_k) - \int_X f\, d\mu\right| &\leq \left|\sum_{k=1}^{\infty} a_{nk} f(T x_k) - \int_X f(Tx)\, d\mu(x)\right| \\ &\quad + \left|\sum_{k=1}^{\infty} a_{nk}(f(T_k x_k) - f(T x_k))\right| \\ &< \varepsilon + \left|\sum_{k=1}^{K_0} a_{nk}(f(T_k x_k) - f(T x_k))\right| \\ &\quad + \left|\sum_{k=K_0+1}^{\infty} a_{nk}(f(T_k x_k) - f(T x_k))\right| \\ &\leq \varepsilon + 2\,\|f\| \sum_{k=1}^{K_0} |a_{nk}| + \varepsilon \sum_{k=K_0+1}^{\infty} |a_{nk}| \\ &\leq (1 + 2\,\|f\| + \|\mathbf{A}\|)\varepsilon.\end{aligned}$$

Hence, $\lim_{n\to\infty} \sum_{k=1}^{\infty} a_{nk} f(T_k x_k) = \int_X f\, d\mu$, and $(T_n x_n)$ is $(\mathbf{A}, \mu)$-u.d. ■

Almost Convergence

We are now going to discuss a summation method that is pertinent to the theory of well-distributed sequences but is not a matrix method. Essentially, this will be the (C, 1) method with an additional uniformity condition.

DEFINITION 4.2. A sequence (s_n) of real numbers is called *almost convergent to the value* s if

$$\lim_{N\to\infty} \frac{1}{N} \sum_{n=1+h}^{N+h} s_n = s \qquad \text{uniformly in } h = 0, 1, 2, \ldots. \tag{4.26}$$

The resulting summation method of almost convergence will be denoted by F (from the German *Fastkonvergenz*). The relation to μ-well distributed sequences (see Definition 3.2) should be clear.

COROLLARY 4.3. A sequence (x_n) in X is μ-well distributed in X if and only if for every $f \in \mathscr{R}(X)$, the sequence $(f(x_n))$ is almost convergent to the value $\int_X f\, d\mu$.

We want to point out a remarkable connection between the method F and the theory of Banach limits in functional analysis. First of all, it is easily seen that every almost-convergent sequence is bounded (see Exercise 4.7). The set Σ of all bounded sequences $\sigma = (s_n)$ of real numbers is a

Banach space if the linear operations with sequences are defined termwise—that is, $(s_n) + (t_n) = (s_n + t_n)$ and $a(s_n) = (as_n)$ for real a—and if the norm $\|\sigma\|$ of a sequence $\sigma = (s_n)$ is defined by $\|\sigma\| = \sup_n |s_n|$. By a *Banach limit* L in Σ we mean a linear functional $L\colon \Sigma \mapsto \mathbb{R}$ that is

1. normed: $L((1, 1, 1, \ldots)) = 1$
2. positive: If $s_n \geq 0$ for all n, then $L(\sigma) \geq 0$
3. shift-invariant: $L(\sigma) = L(T\sigma)$; that is, $L((s_1, s_2, \ldots)) = L((s_2, s_3, \ldots))$.

The existence of such functionals was shown by Banach and Mazur. We state the following interesting characterization of the method F without proof (for a proof, see Lorentz [1], G. M. Petersen [2, Chapter 3]).

THEOREM 4.5. The sequence $\sigma = (s_n)$ is almost convergent to the value s if and only if $L(\sigma) = s$ holds for every Banach limit L in Σ.

It is natural to ask whether there exists a matrix method **A** that is equivalent to F. Let us first note that if (s_n) converges to s in the ordinary sense, then (s_n) is almost convergent to the same value s (Lorentz [1], G. M. Petersen [2, Chapter 3]). Therefore, a matrix method **A** with the desired property should at least transform every sequence converging to s into another such sequence and should thus be regular. Certainly, it is also necessary that **A** includes F. A matrix method **A** that includes F is called *strongly regular*, and the following criterion holds: A regular matrix method $\mathbf{A} = (a_{nk})$ is strongly regular if and only if $\lim_{n\to\infty} \sum_{k=1}^{\infty} |a_{nk} - a_{n,k+1}| = 0$ (Lorentz [1], G. M. Petersen [2, Chapter 3], Zeller and Beekmann [1, Section 6]). On the other hand, if **A** is strongly regular, then one can construct bounded sequences that are summable by **A** but are not almost-convergent (G. M. Petersen [2, Chapter 3]). Therefore, no matrix method is equivalent to F. But, at least, there are many important examples of matrix methods that include F, for example, the Cesàro means $(\mathbf{C}, r)$ with $r > 0$, some classes of Riesz means (see Exercises 4.10 and 4.11), and all discrete Abel methods. We note a trivial consequence of the definition of strong regularity, namely, that a μ-well distributed sequence is $(\mathbf{A}, \mu)$-u.d. for strongly regular **A**.

We have already mentioned that the method F can be considered as a Cesàro means $(\mathbf{C}, 1)$ with uniformity condition. This remark leads to an immediate generalization of almost convergence if we replace $(\mathbf{C}, 1)$ by some other matrix method and impose a uniformity condition of the same type as in Definition 4.2.

DEFINITION 4.3. Let $\mathbf{A} = (a_{nk})$ be a given matrix method. A sequence (s_n) of real numbers is called **A**-*almost convergent to the value* s if

$$\lim_{n\to\infty} \sum_{k=1}^{\infty} a_{nk} s_{k+h} = s \qquad \text{uniformly in } h = 0, 1, 2, \ldots. \tag{4.27}$$

We denote the resulting summation method by $F_{\mathbf{A}}$. Corollary 4.3 suggests then the following generalization of the notion of μ-well distributivity.

DEFINITION 4.4. A sequence (x_n) in X is called *$(\mathbf{A}, \mu)$-well distributed in X* if for every $f \in \mathscr{R}(X)$, the sequence $(f(x_n))$ is summable by $F_{\mathbf{A}}$ to the value $\int_X f\, d\mu$.

THEOREM 4.6. Every $(\mathbf{A}, \mu)$-well distributed sequence in X is also μ-well distributed in X.

PROOF. It will suffice to show that every bounded sequence $\sigma = (s_n)$ of real numbers summable by $F_{\mathbf{A}}$ to the value s is also summable by F to the same value. By the definition of $\mathbf{A}$-almost convergence, there will exist, for given $\varepsilon > 0$, a positive integer N such that

$$\left| \sum_{k=1}^{\infty} a_{nk} s_{k+h} - s \right| < \varepsilon \qquad \text{for all } n > N \text{ and all } h = 0, 1, 2, \ldots. \tag{4.28}$$

With $t_{nh} = \sum_{k=1}^{\infty} a_{nk} s_{k+h}$ for $n \geq 1$ and $h \geq 0$, let τ_n be the sequence $\tau_n = (t_{n0}, t_{n1}, t_{n2}, \ldots)$, which is again in Σ, since $|t_{nh}| \leq \|\mathbf{A}\|\ \|\sigma\|$. If η denotes the constant sequence $\eta = (1, 1, 1, \ldots)$, then, by (4.28),

$$\|\tau_n - s\eta\| = \sup_{h=0,1,2,\ldots} |t_{nh} - s| \leq \varepsilon \qquad \text{for all } n > N. \tag{4.29}$$

So $\tau_n \to s\eta$ as $n \to \infty$ in the Banach space Σ. Any Banach limit L on Σ is continuous, and therefore,

$$\lim_{n\to\infty} L(\tau_n) = L(s\eta) = sL(\eta) = s. \tag{4.30}$$

Let us now look, for fixed $n \geq 1$, at the series $\sum_{k=1}^{\infty} a_{nk}(T^{k-1}\sigma)$ of elements in Σ. This series converges in norm to some element $\rho_n \in \Sigma$, since

$$\sum_{k=1}^{\infty} |a_{nk}|\ \|T^{k-1}\sigma\| \leq \sum_{k=1}^{\infty} |a_{nk}|\ \|\sigma\| \leq \|\mathbf{A}\|\ \|\sigma\|.$$

Let $\rho_n = (r_{n1}, r_{n2}, \ldots)$. Now norm convergence in Σ implies coordinatewise convergence; thus, $r_{ni} = \sum_{k=1}^{\infty} a_{nk} s_{k+i-1} = t_{n,i-1}$ for all $i \geq 1$. Hence, $\rho_n = \tau_n$, and

$$L(\tau_n) = L(\rho_n) = \sum_{k=1}^{\infty} a_{nk} L(T^{k-1}\sigma) = \sum_{k=1}^{\infty} a_{nk} L(\sigma), \tag{4.31}$$

by the shift-invariance of L. Letting $n \to \infty$ in (4.31) and using (4.30), we obtain $L(\sigma) \lim_{n\to\infty} \sum_{k=1}^{\infty} a_{nk} = s$, or $L(\sigma) = s$ for any Banach limit L in Σ. By Theorem 4.5, the proof is then complete. ∎

Theorem 4.6 is remarkable in the light of the fact that the corresponding statement for μ-u.d. sequences is not true. One can even construct regular matrices $\mathbf{A}$ such that $(\mathbf{A}, \mu)$-u.d. does not imply μ-u.d. (see Exercises 4.13 and 4.14; see below for a strongly regular matrix with the same property).

If **A** is strongly regular, then F and $F_{\mathbf{A}}$ are even equivalent (Lorentz [1]). Thus, for strongly regular matrix methods **A**, the following result holds: The sequence (x_n) in X is $(\mathbf{A}, \mu)$-well distributed in X if and only if (x_n) is μ-well distributed in X. Again, the corresponding result for μ-u.d. is false. We know from Chapter 1, Theorem 7.16, and Chapter 1, Example 2.4, that the sequence $(\log n)$ is $(\mathbf{R}, 1/n)$-u.d. mod 1 but not u.d. mod 1; furthermore, $(\mathbf{R}, 1/n)$ is strongly regular by Exercise 4.9.

Notes

A detailed treatment of the theory of summability is given in Hardy [2], Cooke [1], Knopp [1], G. M. Petersen [2], Zeller and Beekmann [1] (with extensive bibliography), and Peyerimhoff [1].

The consideration of summation methods other than (C, 1) in conjunction with u.d. goes back to Tsuji [2], who considered $(\mathbf{R}, p_n)$ in the mod 1 case (compare with the notes in Section 7 of Chapter 1). The concept of $(\mathbf{A}, \mu)$-u.d. for compact groups (with a positive matrix **A**) was introduced by Hlawka [1] and was extended by the same author ([3, 6]) to compact spaces with countable base. In fact, most of the theory of $(\mathbf{A}, \mu)$-u.d. was developed in these two papers.

The Hill condition (4.4) was first given by Hill [1] for the special case of the discrete space $X = \{0, 1\}$ and was further studied for special summation methods in Hill [2]. The counterexample in Example 4.4 is also due to Hill [1]. Our proof of Theorem 4.2 follows ideas of Kemperman [2] and is less involved than the original proof of Hlawka [3]. Another sufficient condition for the Borel property, the so-called Lorentz condition (Lorentz [2]), was given by Müller [1]: If **A** is regular and $\lim_{n\to\infty} \sum_{k=1}^{\infty} |a_{nk} - a_{n,k+1}| \log k = 0$, then **A** has the Borel property with respect to all measures μ on X (again, X is supposed to have a countable base). An analysis of the relations between the Borel property and properties akin to it was carried out by Fleischer [1]. In connection with the Borel property, the following analogue of Theorem 2.3 given by Hlawka [3] is of interest: If X has a countable base and contains more than one point and if **A** is regular, then the set of all $(\mathbf{A}, \mu)$-u.d. sequences in X is of the first category in X^{∞}. A theorem of K. Schmidt [1] also touches upon the Borel property. A brief discussion of $(\mathbf{A}, \mu)$-u.d. can also be found in Helmberg [5] and Cigler [10].

An existence theorem for $(\mathbf{A}, \mu)$-u.d. was shown by Descovich [1] for a restricted class of positive regular summation matrices **A**. The concept of a "strongly $(\mathbf{A}, \mu)$-u.d. sequence" in a compact Hausdorff space X with countable base was introduced and studied by Philipp [2]. A generalization of the theorem of de Bruijn and Post (see the notes in Chapter 1, Section 1) and a rearrangement theorem for $(\mathbf{A}, \mu)$-u.d. (compare with Theorem 2.5 and the notes in Section 2) were shown by Binder [1].

The summation method F (almost convergence) was introduced and studied by Lorentz [1]. The usefulness of this method for the theory of well-distributed sequences was pointed out by G. M. Petersen [1]. A discussion of the method F from the point of view of summability theory can also be found in G. M. Petersen [2, Chapter 3] and Zeller and Beekmann [1, Section 6].

Exercises

4.1. Let $\mathbf{A} = (a_{nk})$ be a matrix method with $a_{nk} \geq 0$ for all $n, k = 1, 2, \ldots$ (a so-called *positive* matrix method). Prove that the sequence (x_n) is

(**A**, μ)-u.d. in X if and only if for every μ-continuity set M in X, we have $\lim_{n\to\infty} \sum_{k=1}^{\infty} a_{nk} c_M(x_k) = \mu(M)$, where c_M denotes the characteristic function of M. *Hint:* Compare with the proof of Theorem 1.2.

4.2. Consider the simple Riesz means $(\mathbf{R}, p_n)$ with $p_n = n^\sigma$, $\sigma \geq -1$. Show that $(\mathbf{R}, p_n)$ has the Borel property. *Hint:* See Example 4.5.

4.3. Show that for $\sigma > -1$, the simple Riesz means $(\mathbf{R}, p_n)$ with $p_n = n^\sigma$ and (**C**, 1) are even equivalent. *Hint:* See Example 4.5.

4.4. If X has at least two points, then there is no matrix method **A** such that every sequence in X is (**A**, μ)-u.d. in X.

4.5. Prove that if **A** is strongly regular, then the sequence $(n\alpha)$, α irrational, is **A**-u.d. mod 1.

4.6. Prove that Theorem 4.4 remains true if in both the hypothesis and the conclusion, "(**A**, μ)-u.d." is replaced by "(**A**, μ)-well distributed."

4.7. Show that every almost-convergent sequence of real numbers is bounded.

4.8. Verify that the sequence $((-1)^n)$ is almost convergent to the value zero.

4.9. Show that $(\mathbf{R}, 1/n)$ is strongly regular.

4.10. Show, more generally, that a regular $(\mathbf{R}, p_n)$ with nonincreasing p_n is strongly regular.

4.11. Similarly, show that $(\mathbf{R}, p_n)$ with nondecreasing p_n is strongly regular if and only if $\lim_{n\to\infty} p_n/P_n = 0$.

4.12. Consider the discrete space $X = \{0, 1\}$ with measure μ defined by $\mu(\{0\}) = \mu(\{1\}) = \frac{1}{2}$. Prove that a sequence (x_n) in X is (**A**, μ)-u.d. in X if and only if (x_n) is summable by **A** to the value $\frac{1}{2}$.

4.13. Show that the following matrix $\mathbf{A} = (a_{nk})$ defines a regular matrix method: $a_{nk} = 1/2n$ if $1 \leq k \leq 3n$ and $k \not\equiv 1 \pmod 3$, and $a_{nk} = 0$ otherwise. With X and μ as in Exercise 4.12, prove that the following sequence (x_n) is (**A**, μ)-u.d. but no μ-u.d. in X: $x_n = 1$ if $n \equiv 0 \pmod 3$, and $x_n = 0$ otherwise.

4.14. With X and μ as in Exercise 4.12, the sequence (x_n) with $x_n = 1$ for $n = 2^s$, $s \geq 1$, and $x_n = 0$ otherwise, is given. This sequence is not μ-u.d. in X (proof!). Construct a regular matrix method **A** such that (x_n) is (**A**, μ)-u.d. in X. *Hint:* Compare with Exercise 4.13.

4.15. Let X be a compact metric space with metric d and with a nonnegative normed Borel measure μ, let **A** be a regular matrix method, and suppose that (x_n) is (**A**, μ)-u.d. in X. Prove that every sequence (y_n) in X satisfying $\lim_{n\to\infty} d(x_n, y_n) = 0$ is also (**A**, μ)-u.d. in X. *Hint:* Use Theorem 4.4.

4.16. Use the previous exercise, with X, μ, and **A** having the same meaning as there, to prove the following result: If X admits (**A**, μ)-u.d. sequences and H is an arbitrary dense set in X, then there exist (**A**, μ)-u.d. sequences consisting entirely of elements from H.

4

UNIFORM DISTRIBUTION IN TOPOLOGICAL GROUPS

Quite a number of results from the classical theory of u.d. mod 1 cannot conceivably have an analogue in the general theory of u.d. in an arbitrary compact Hausdorff space. In particular, all the results using, or being stated in terms of, the algebraic structure of the reals belong to this category.

To generalize this part of the classical theory, we are going to develop a theory of u.d. in compact topological groups. Evidently, all the results established in the preceding chapter will carry over to this case. It will turn out that in addition to those results, we can reveal a rather far-reaching analogy to the theory of u.d. mod 1. In the last section of this chapter, we shall also take a glance at locally compact topological groups and introduce a meaningful notion of u.d. for this case as well.

The reader is supposed to be familiar with the general theory of topological groups. We present only a brief sketch of the important concepts and facts that will be used throughout the entire chapter.

1. GENERALITIES

Haar Measure

Let G be a compact (topological) group with identity element $e \in G$. The topological groups that we consider will always satisfy the Hausdorff separation axiom. The following notation is standard: For $a \in G$ and a subset M of

G, we define $aM = \{ax: x \in M\}$ and $Ma = \{xa: x \in M\}$; furthermore, $M^{-1} = \{x^{-1}: x \in M\}$.

Among the nonnegative regular normed Borel measures on G, there is one singled out by a remarkable property. Namely, there exists a unique nonnegative regular normed Borel measure μ on G that is left translation invariant; that is, $\mu(xB) = \mu(B)$ for all $x \in G$ and all Borel sets B in G. This measure μ is called the (normed) *Haar measure* on G. Because of the compactness of G, the Haar measure is also right translation invariant; that is, $\mu(Bx) = \mu(B)$ for all $x \in G$ and all Borel sets B in G; also, $\mu(B^{-1}) = \mu(B)$ for all Borel sets B in G. Apart from a brief interlude in Section 3, we shall only study u.d. with respect to the Haar measure. Therefore, we may adopt a more convenient way of speaking.

DEFINITION 1.1. A sequence (x_n) in G is called *u.d. in* G if (x_n) is u.d. with respect to the Haar measure on G. In a similar way, we define the terms *well distributed in* G and **A**-*u.d. in* G for a matrix method **A**.

It follows easily from the translation invariance of the Haar measure μ that the support of μ is G (see Exercise 1.1). In other words, every nonvoid open set in G has positive μ-measure. In particular, a u.d. sequence in G will necessarily be everywhere dense (compare with Exercise 1.8 of Chapter 3). We recall the following simple fact from Section 1 of Chapter 3.

LEMMA 1.1. The sequence (x_n) is u.d. in G if and only if

$$\lim_{N\to\infty} \frac{1}{N} \sum_{n=1}^{N} f(x_n) = \int_G f \, d\mu \tag{1.1}$$

holds for all $f \in \mathscr{C}(G)$.

As we already remarked in Section 1 of Chapter 3, we can show in the same way as in Theorem 1.1 of Chapter 3 that the validity of (1.1) for certain restricted classes of functions $f \in \mathscr{C}(G)$ guarantees u.d. Let $\mathscr{V}$ be a class of functions from $\mathscr{C}(G)$. For the purposes of the present chapter, sp $(\mathscr{V})$ shall denote the linear subspace of $\mathscr{C}(G)$ generated by $\mathscr{V}$. Thus, sp $(\mathscr{V})$ consists now of all finite linear combinations of elements from $\mathscr{V}$ with complex coefficients.

LEMMA 1.2. Let $\mathscr{V}$ be a class of functions from $\mathscr{C}(G)$ with $\overline{\text{sp}\,(\mathscr{V})} = \mathscr{C}(G)$. Then (x_n) is u.d. in G if and only if (1.1) holds for all $f \in \mathscr{V}$.

Furthermore, by using the version of the Stone-Weierstrass theorem for $\mathscr{C}(G)$, we arrive at the following consequence, which we also enunciated in Corollary 1.2 of Chapter 3.

COROLLARY 1.1. If sp $(\mathscr{V})$ is a subalgebra of $\mathscr{C}(G)$ that separates points, contains the constant functions, and is closed under complex conjugation, then $\overline{\text{sp}(\mathscr{V})} = \mathscr{C}(G)$, and so, $\mathscr{V}$ can serve as a class of functions for the criterion given in Lemma 1.2.

Representations and Linear Groups

By using representation theory, a very important special case of Lemma 1.2 can be exhibited. A (linear) *representation* of a (not necessarily compact) topological group is a continuous homomorphism (i.e., a group homomorphism that is continuous) from the given topological group into a multiplicative topological group of nonsingular complex $k \times k$ matrices with some fixed k; the positive integer k is then called the *degree* of the representation. By a topology on a set of complex matrices of fixed order, we agree to mean the topology of entrywise convergence.

Important examples of topological groups of matrices are the general linear group $\mathbf{GL}(k)$ of all nonsingular complex matrices of order k and the unitary group $\mathbf{U}(k)$ of all unitary matrices of order k. We recall that a matrix $\mathbf{U}$ is called *unitary* if $\bar{\mathbf{U}}^T = \mathbf{U}^{-1}$, where $\bar{\mathbf{U}}^T$ is the transpose of the conjugate of $\mathbf{U}$. The topological group $\mathbf{GL}(k)$ is locally compact and has a countable base, and $\mathbf{U}(k)$ is compact with countable base. The topology in both $\mathbf{GL}(k)$ and $\mathbf{U}(k)$ (and in the set of all matrices of order k) could also be introduced by means of the following matrix norm. For an arbitrary complex square matrix $\mathbf{A} = (a_{ij})$ of order k, define

$$\|\mathbf{A}\| = \left(\sum_{i,j=1}^{k} |a_{ij}|^2 \right)^{1/2}. \tag{1.2}$$

Note that this notion of norm is, of course, completely different from the norm of a summation matrix defined in Chapter 3, Section 4. In this chapter, only the norm in (1.2) will be used. We have the usual properties of a norm:

i. $\|\mathbf{A}\| = 0$ if and only if $\mathbf{A}$ is the zero matrix
ii. $\|\alpha\mathbf{A}\| = |\alpha| \, \|\mathbf{A}\|$ for complex α
iii. $\|\mathbf{A} + \mathbf{B}\| \le \|\mathbf{A}\| + \|\mathbf{B}\|$.

Furthermore, we have
iv. $\|\mathbf{AB}\| \le \|\mathbf{A}\| \, \|\mathbf{B}\|$
v. If $\mathbf{U} \in \mathbf{U}(k)$, then $\|\mathbf{U}\| = \sqrt{k}$ and $\|\mathbf{AU}\| = \|\mathbf{UA}\| = \|\mathbf{A}\|$ for all $\mathbf{A}$.
The proofs are straightforward.

EXAMPLE 1.1. As a sample result, we prove that $\|\mathbf{AB}\| \le \|\mathbf{A}\| \, \|\mathbf{B}\|$ for complex square matrices of the same order k. Let $\mathbf{A} = (a_{ij})$ and $\mathbf{B} = (b_{ij})$ with $1 \le i, j \le k$. Then $\mathbf{AB} = (c_{ij})$ with $c_{ij} = \sum_{r=1}^{k} a_{ir} b_{rj}$. Therefore, by the

Cauchy-Schwarz inequality,

$$\|\mathbf{AB}\|^2 = \sum_{i,j=1}^{k} |c_{ij}|^2 = \sum_{i,j=1}^{k} \left| \sum_{r=1}^{k} a_{ir} b_{rj} \right|^2$$
$$\leq \sum_{i,j=1}^{k} \left(\sum_{r=1}^{k} |a_{ir}|^2 \right) \left(\sum_{r=1}^{k} |b_{rj}|^2 \right) = \|\mathbf{A}\|^2 \, \|\mathbf{B}\|^2.$$

Taking square roots completes the proof. ■

EXAMPLE 1.2. An application of the Cauchy-Schwarz inequality also yields an estimate for the norm of the sum of matrices that is often more useful than the triangle inequality. Let $\mathbf{A}_1, \ldots, \mathbf{A}_N$ be complex square matrices of the same order k, say $\mathbf{A}_r = (a_{ij}^{(r)})$ for $1 \leq r \leq N$, $1 \leq i, j \leq k$. Then, we show that

$$\left\| \sum_{r=1}^{N} \mathbf{A}_r \right\|^2 \leq N \sum_{r=1}^{N} \|\mathbf{A}_r\|^2. \tag{1.3}$$

We have

$$\left\| \sum_{r=1}^{N} \mathbf{A}_r \right\|^2 = \sum_{i,j=1}^{k} \left| \sum_{r=1}^{N} a_{ij}^{(r)} \right|^2. \tag{1.4}$$

We apply the Cauchy-Schwarz inequality to $|\sum_{r=1}^{N} 1.a_{ij}^{(r)}|^2$. Consequently,

$$\left\| \sum_{r=1}^{N} \mathbf{A}_r \right\|^2 \leq \sum_{i,j=1}^{k} N \sum_{r=1}^{N} |a_{ij}^{(r)}|^2 = N \sum_{r=1}^{N} \sum_{i,j=1}^{k} |a_{ij}^{(r)}|^2 = N \sum_{r=1}^{N} \|\mathbf{A}_r\|^2. \quad ■ \tag{1.5}$$

The matrix norm that we introduced above can also be defined by means of an inner product. For a complex square matrix $\mathbf{A}$, let the trace tr $(\mathbf{A})$ be the sum of the diagonal elements of $\mathbf{A}$. If $\mathbf{A}$ and $\mathbf{B}$ are complex square matrices of the same order, we define $(\mathbf{A} \mid \mathbf{B}) = \operatorname{tr}(\bar{\mathbf{B}}^T \mathbf{A})$. It is easy to see that we have the following rules:

i. $(\alpha\mathbf{A} \mid \mathbf{B}) = \alpha(\mathbf{A} \mid \mathbf{B})$ and $(\mathbf{A} \mid \alpha\mathbf{B}) = \bar{\alpha}(\mathbf{A} \mid \mathbf{B})$ for complex α
ii. $(\mathbf{A}_1 + \mathbf{A}_2 \mid \mathbf{B}) = (\mathbf{A}_1 \mid \mathbf{B}) + (\mathbf{A}_2 \mid \mathbf{B})$ and $(\mathbf{A} \mid \mathbf{B}_1 + \mathbf{B}_2) = (\mathbf{A} \mid \mathbf{B}_1) + (\mathbf{A} \mid \mathbf{B}_2)$
iii. $(\mathbf{B} \mid \mathbf{A}) = \overline{(\mathbf{A} \mid \mathbf{B})}$
iv. If $\mathbf{U}$ is unitary, then $(\mathbf{UA} \mid \mathbf{UB}) = (\mathbf{AU} \mid \mathbf{BU}) = (\mathbf{A} \mid \mathbf{B})$.

Furthermore, the relation to the matrix norm is exhibited by the identity

$$\|\mathbf{A}\|^2 = (\mathbf{A} \mid \mathbf{A}). \tag{1.6}$$

Many of the properties of the matrix norm can therefore be shown by using a corresponding property of the inner product.

Two representations $\mathbf{D}_1$ and $\mathbf{D}_2$ of G of the same degree k are called *equivalent* if there exists a nonsingular $k \times k$ matrix $\mathbf{S}$ such that

$$\mathbf{D}_2(x) = \mathbf{S}^{-1}\mathbf{D}_1(x)\mathbf{S} \qquad \text{for all} \quad x \in G. \tag{1.7}$$

For the compact group G, every representation is equivalent to a *unitary representation*, that is, a representation $\mathbf{D}$ of G such that $\mathbf{D}(x)$ is a unitary matrix for all $x \in G$. A representation $\mathbf{D}$ of G of degree k is called *reducible* if there exists a linear subspace V of the k-dimensional vector space $\mathbb{C}^k$ over the complex numbers such that $0 < \dim V < k$ and $\mathbf{D}(x)V \subseteq V$ for all $x \in G$, where we think of the matrix $\mathbf{D}(x)$ as a linear operator on $\mathbb{C}^k$. A representation that is not reducible is called *irreducible*.

If $\mathbf{D}$ is a representation of G, then the *character* χ of $\mathbf{D}$ is defined by $\chi(x) = \operatorname{tr}(\mathbf{D}(x))$ for all $x \in G$. The character χ is a complex-valued continuous function on G. Two equivalent representations have the same character. The character χ is invariant on the classes of conjugate elements in G; that is, $\chi(y^{-1}xy) = \chi(x)$ for all $x, y \in G$. If the representation $\mathbf{D}$ has degree 1, then $\mathbf{D}$ is identical with its own character. In case G is abelian, any irreducible representation is of degree 1; therefore, we shall usually speak of characters instead of representations for abelian G. To emphasize this important case, let us repeat that a character χ of a compact abelian group is a continuous mapping from G into the multiplicative topological group $\boldsymbol{T}$ of complex numbers of unit modulus, the so-called (one-dimensional) circle group, such that $\chi(xy) = \chi(x)\chi(y)$ for all $x, y \in G$.

EXAMPLE 1.3. Let φ be the following mapping from the additive group $\mathbb{R}$ of the reals in the usual topology into the circle group $\boldsymbol{T}$: $\varphi(x) = e^{2\pi i x}$ for $x \in \mathbb{R}$. The mapping φ is a continuous homomorphism onto $\boldsymbol{T}$ with kernel $\mathbb{Z}$, the additive group of integers. Therefore, $\boldsymbol{T}$ and the quotient group $\mathbb{R}/\mathbb{Z}$ are isomorphic as topological groups (see Theorems 1.8 and 1.9). The reals mod 1 (see introduction to Chapter 3) can be identified in a canonical fashion with $\mathbb{R}/\mathbb{Z}$. The measure on the reals mod 1 induced by ordinary Lebesgue measure becomes then the Haar measure of $\mathbb{R}/\mathbb{Z}$. The characters of the compact abelian group $\mathbb{R}/\mathbb{Z}$ are exactly the mappings χ_m defined by $\chi_m(x\mathbb{Z}) = e^{2\pi i m x}$ for $x \in \mathbb{R}$, where m attains every integral value. ■

EXAMPLE 1.4. For $m \geq 1$, let C_m be a finite cyclic group of order m generated by $a \in C_m$. In the discrete topology, C_m is a compact abelian group. The characters of C_m are exactly the mappings χ_h, $h = 0, 1, \ldots, m-1$, defined by $\chi_h(a^k) = e^{2\pi i (h/m)k}$, $k = 0, 1, \ldots, m-1$. ■

We shall use the following theorem, which is a special case of the famous Gel'fond-Raikov theorem for locally compact groups. Let us remark at this point that we do not provide formal proofs for the basic results from the general theory of topological groups. The reader is referred to the literature mentioned in the notes.

THEOREM 1.1: Gel'fond-Raikov Theorem. For any element $x \neq e$ of the compact group G, there exists an irreducible unitary representation $\mathbf{D}$ of G such that $\mathbf{D}(x)$ is not the identity matrix.

Thus, we have what is sometimes described as an "adequate system" of representations. Let us now look at the family of all irreducible unitary representations of G. By virtue of the notion of equivalence between representations that we introduced in (1.7), this family is divided into equivalence classes. From each equivalence class, we choose one representation. We thereby arrive at a system $\{\mathbf{D}^{(\lambda)}: \lambda \in \Lambda\}$ of nonequivalent irreducible unitary representations, where Λ denotes a suitable index set. Let $\mathbf{D}^{(\lambda)}(x)$ be the matrix $\mathbf{D}^{(\lambda)}(x) = (d_{ij}^{(\lambda)}(x))$ with $x \in G$. For varying x, the entry $d_{ij}^{(\lambda)}$ can be viewed as a complex-valued continuous function on G. By collecting those entry functions for all $\mathbf{D}^{(\lambda)}$, $\lambda \in \Lambda$, we obtain a class $\mathscr{D}$ of functions from $\mathscr{C}(G)$. We want to verify that sp $(\mathscr{D})$ satisfies all the conditions of Corollary 1.1.

Let us first show that $\mathscr{D}$ already separates points. Take two distinct points x and y from G. By Theorem 1.1, there exists an irreducible unitary representation $\mathbf{D}$ of G such that $\mathbf{D}(xy^{-1})$ is not the identity matrix. The representation $\mathbf{D}$ is equivalent to some $\mathbf{D}^{(\lambda)}$, $\lambda \in \Lambda$. Then $\mathbf{D}^{(\lambda)}(xy^{-1})$ is not the identity matrix; hence, $\mathbf{D}^{(\lambda)}(x) \neq \mathbf{D}^{(\lambda)}(y)$. Consequently, for one of the entries $d_{ij}^{(\lambda)}$ of $\mathbf{D}^{(\lambda)}$, we have $d_{ij}^{(\lambda)}(x) \neq d_{ij}^{(\lambda)}(y)$.

Suppose we have arranged matters so that the integer 0 is an element of the index set Λ. There is one trivial representation among the $\mathbf{D}^{(\lambda)}$ that we shall call $\mathbf{D}^{(0)}$. The representation $\mathbf{D}^{(0)}$ has degree 1 and is defined by $\mathbf{D}^{(0)}(x) = 1$ for all $x \in G$. In particular, we have the function $d^{(0)} \equiv 1$ in the class $\mathscr{D}$, and so, sp $(\mathscr{D})$ contains the constant functions.

Furthermore, we must verify that with every function in sp $(\mathscr{D})$, we find its conjugate in sp $(\mathscr{D})$. It will suffice to prove that for every function $d_{ij}^{(\lambda)}$ we have $\overline{d_{ij}^{(\lambda)}} \in$ sp $(\mathscr{D})$. We use that with every $\mathbf{D}^{(\lambda)}$ the mapping $\mathbf{E}^{(\lambda)}(x) = \overline{\mathbf{D}^{(\lambda)}(x)}$, $x \in G$, is also an irreducible unitary representation of G. Therefore, $\mathbf{E}^{(\lambda)}$ is equivalent to a representation $\mathbf{D}^{(\nu)}$ for some $\nu \in \Lambda$. By equating corresponding entries in the matrix equation $\mathbf{E}^{(\lambda)}(x) = \mathbf{S}^{-1}\mathbf{D}^{(\nu)}(x)\mathbf{S}$, we infer that $\overline{d_{ij}^{(\lambda)}}$ is a complex linear combination of functions from $\mathscr{D}$.

There is one more point that we have to settle, namely, that sp $(\mathscr{D})$ is closed under multiplication of functions. Evidently, it will be sufficient to show that a product of the form $d_{ij}^{(\lambda)} d_{mn}^{(\nu)}$ is again in sp $(\mathscr{D})$. To this end, we form the Kronecker product $\mathbf{F} = \mathbf{D}^{(\lambda)} \oplus \mathbf{D}^{(\nu)}$ of the two representations $\mathbf{D}^{(\lambda)}$ and $\mathbf{D}^{(\nu)}$; that is, we set $\mathbf{F}(x) = \mathbf{D}^{(\lambda)}(x) \oplus \mathbf{D}^{(\nu)}(x)$ for all $x \in G$, where the Kronecker product $\mathbf{A} \oplus \mathbf{B}$ of two matrices $\mathbf{A} = (a_{pq})$ of order c and $\mathbf{B} = (b_{rs})$ of order d is defined as follows. The matrix $\mathbf{A} \oplus \mathbf{B}$ is a matrix of order cd that is built up from d^2 blocks of $c \times c$ submatrices. The $c \times c$ submatrix appearing in the rth row of blocks and sth column of blocks is $b_{rs}\mathbf{A}$. In other words, we get the matrix $\mathbf{A} \oplus \mathbf{B}$ by taking the matrix $\mathbf{B}$ and replacing each entry b_{rs} by the $c \times c$ matrix $b_{rs}\mathbf{A}$. It is easily seen that $\mathbf{F} = \mathbf{D}^{(\lambda)} \oplus \mathbf{D}^{(\nu)}$ is again a unitary representation of G but that $\mathbf{F}$ need not be irreducible. We apply the important fact that any representation of a compact group is equivalent to a so-called *completely reducible* representation, that is, to a representation $\mathbf{E}$ of the form

$$\mathbf{E}(x) = \begin{pmatrix} \mathbf{E}_1(x) & \mathbf{0} & \cdots & \mathbf{0} \\ \mathbf{0} & \mathbf{E}_2(x) & \cdots & \mathbf{0} \\ \cdot & \cdot & \cdot & \cdot \\ \cdot & \cdot & \cdot & \cdot \\ \cdot & \cdot & \cdot & \cdot \\ \mathbf{0} & \mathbf{0} & \cdots & \mathbf{E}_t(x) \end{pmatrix},$$

where the blocks $\mathbf{E}_1(x), \ldots, \mathbf{E}_t(x)$ are irreducible unitary representations and the zeros denote zero matrices of appropriate order. By choosing, if necessary, a representation equivalent to $\mathbf{E}$, we may assume that $\mathbf{E}_1, \ldots, \mathbf{E}_t$ belong to the system $\{\mathbf{D}^{(\lambda)}: \lambda \in \Lambda\}$. Thus, we have $\mathbf{F}(x) = \mathbf{S}^{-1}\mathbf{E}(x)\mathbf{S}$ with the nonzero entry functions in $\mathbf{E}(x)$ belonging to the class $\mathscr{D}$. In particular, the entry function $d_{ij}^{(\lambda)}\, d_{mn}^{(\nu)}$ in $\mathbf{F}$ will be a complex linear combination of functions from $\mathscr{D}$.

Combining all the above considerations, we have thus sketched the proof of an important result, namely, the so-called Peter-Weyl theorem.

THEOREM 1.2: Peter-Weyl Theorem. The subalgebra sp $(\mathscr{D})$ is dense in $\mathscr{C}(G)$.

Weyl Criterion

The Peter-Weyl theorem leads us back to Lemma 1.2. A sequence (x_n) in G will be u.d. in G if and only if $\lim_{N\to\infty} (1/N) \sum_{n=1}^{N} d_{ij}^{(\lambda)}(x_n) = \int_G d_{ij}^{(\lambda)}\, d\mu$ holds for all functions $d_{ij}^{(\lambda)} \in \mathscr{D}$. If $\lambda = 0$ (i.e., if we consider the trivial representation $\mathbf{D}^{(0)}$), then this limit relation is clearly satisfied for any sequence in G. Thus, we may confine our attention to $\lambda \in \Lambda$ with $\lambda \neq 0$. By the well-known orthogonality relations for compact groups, $\int_G d_{ij}^{(\lambda)}\, d\mu = 0$ holds for all $d_{ij}^{(\lambda)}$ with $\lambda \neq 0$. Since we defined convergence of matrices entrywise, we may collect the limit relations for all the entry functions of a single representation $\mathbf{D}^{(\lambda)}$ into a limit relation of the type (1.8) below. We arrive at the following fundamental criterion.

THEOREM 1.3: Weyl Criterion. Let $\{\mathbf{D}^{(\lambda)}: \lambda \in \Lambda\}$ be a system of representations of G that is obtained by choosing exactly one representation from each equivalence class of irreducible unitary representations of G. Let $\mathbf{D}^{(0)}$ be the trivial representation. Then the sequence (x_n) in G is u.d. in G if and only if

$$\lim_{N\to\infty} \frac{1}{N} \sum_{n=1}^{N} \mathbf{D}^{(\lambda)}(x_n) = \mathbf{0} \tag{1.8}$$

holds for all $\lambda \in \Lambda$ with $\lambda \neq 0$, where $\mathbf{0}$ denotes a zero matrix of appropriate order. By the properties of the matrix norm, the condition (1.8) is equivalent to

$$\lim_{N\to\infty} \left\| \frac{1}{N} \sum_{n=1}^{N} \mathbf{D}^{(\lambda)}(x_n) \right\| = 0 \qquad \text{for all } \lambda \in \Lambda \text{ with } \lambda \neq 0. \tag{1.9}$$

Because of Example 1.3, the classical Weyl criterion for u.d. mod 1 (see Chapter 1, Theorem 2.1) is a special case of Theorem 1.3. Because of its importance, we shall state Theorem 1.3 once again for the special case of an abelian group G. Then the representations become the characters, the assumptions of irreducibility and unitarity are redundant, and the equivalence classes from Theorem 1.3 are singletons. Obviously, the trivial character shall then be defined by $\chi_0(x) = 1$ for all $x \in G$.

COROLLARY 1.2. Let G be a compact abelian group. Then the sequence (x_n) in G is u.d. in G if and only if

$$\lim_{N\to\infty} \frac{1}{N} \sum_{n=1}^{N} \chi(x_n) = 0 \tag{1.10}$$

holds for all nontrivial characters χ of G.

Using Corollary 3.1 of Chapter 3 and the Peter-Weyl theorem, we are also led to the following convenient criterion for well-distributivity in a compact group G.

COROLLARY 1.3. Let $\{\mathbf{D}^{(\lambda)}: \lambda \in \Lambda\}$ be as in Theorem 1.3. Then the sequence (x_n) in G is well distributed in G if and only if for all $\lambda \in \Lambda$ with $\lambda \neq 0$, we have

$$\lim_{N\to\infty} \frac{1}{N} \sum_{n=1+h}^{N+h} \mathbf{D}^{(\lambda)}(x_n) = \mathbf{0} \qquad \text{uniformly in } h = 0, 1, 2, \ldots, \tag{1.11}$$

or, equivalently,

$$\lim_{N\to\infty} \left\| \frac{1}{N} \sum_{n=1+h}^{N+h} \mathbf{D}^{(\lambda)}(x_n) \right\| = 0 \qquad \text{uniformly in } h = 0, 1, 2, \ldots. \tag{1.12}$$

It is certainly of interest to know "how many" conditions we have to check when applying Theorem 1.3. This question is answered by a well-known result from representation theory that says that the cardinality of the index set Λ is equal to the weight of G. By the *weight* of a topological space, we mean the minimal cardinality of a base for the open sets in the space. In particular, if G satisfies the second axiom of countability, then there is a countable system of irreducible unitary representations of G that determines the u.d. of a given sequence in G. In the light of Theorem 2.1 of Chapter 3, this should not come as a surprise.

Some Consequences of Earlier Results

The compact group G can be viewed as a uniform space in a natural sense. For each neighborhood V of e, define $V_L = \{(x, y) \in G \times G: x^{-1}y \in V\}$. The family of all sets V_L forms a base for the so-called *left uniformity* of G. In the same vein, we define $V_R = \{(x, y) \in G \times G: xy^{-1} \in V\}$, and we let the *right uniformity* of G be the uniformity that has the family of all sets V_R as a base. The topology of both the left and the right uniformity is identical with the original topology of G. We shall now apply to the present situation two results from Chapter 3 that we found for compact Hausdorff uniform spaces. For the first application, we look at a very special case of Theorem 4.4 of Chapter 3. For the matrix method **A** occurring there, we choose the Cesàro means (**C**, 1), and for the measure μ we take, of course, the Haar measure on G.

THEOREM 1.4. Let (x_n) be a u.d. sequence in G, and suppose that (c_n) is a sequence in G such that $\lim_{n\to\infty} c_n$ exists. Then the sequences $(c_n x_n)$ and $(x_n c_n)$ are u.d. in G.

PROOF. We confine our attention to the sequence $(c_n x_n)$. The proof for the second sequence is analogous. We set $c = \lim_{n\to\infty} c_n$. For $n \geq 1$, let T_n be the transformation on G defined by $T_n x = c_n x$ for all $x \in G$. In addition, we consider the transformation T on G given by $Tx = cx$ for all $x \in G$. The transformation T is both continuous and measure-preserving. By Theorem 4.4 of Chapter 3, it remains to show that the transformations T_n converge uniformly to T. For this purpose, it will suffice to look at the sets V_R in the right uniformity of G, where V is again some neighborhood of e. But $(T_n x, Tx) \in V_R$ precisely if $(T_n x)(Tx)^{-1} = c_n x(cx)^{-1} = c_n c^{-1} \in V$. Therefore, $(T_n x, Tx) \in V_R$ holds for sufficiently large n and every $x \in G$ because the sequence $(c_n c^{-1})$ converges to e. ∎

By using Exercise 4.6 of Chapter 3, one can prove a corresponding statement for well-distributivity. That is, if (x_n) is well distributed in G and (c_n) converges, then the sequences $(c_n x_n)$ and $(x_n c_n)$ are both well distributed in G. Alternative proofs for the above results can be based on the Weyl criterion as given in Theorem 1.3 and Corollary 1.3 (see Exercises 1.2 and 1.5).

Employing in a consistent way the convention that originates in Definition 1.1, we call a family of sequences in G a *family of equi-u.d. sequences in G* if it is a family of equi-μ-u.d. sequences with respect to the Haar measure μ. We consider now a special case of Example 3.2 of Chapter 3. Let us first note the following immediate consequence of Theorem 1.4: If (x_n) is u.d. in G, then the sequences (cx_n) and $(x_n c)$ are u.d. in G for fixed $c \in G$. It turns out that the sequences of this type are tied together very closely with respect to their distribution behavior.

THEOREM 1.5. Let (x_n) be a given u.d. sequence in G. Then $\{(cx_n): c \in G\}$ and $\{(x_nc): c \in G\}$ are families of equi-u.d. sequences in G.

PROOF. We already mentioned that we shall apply the result from Example 3.2 of Chapter 3. Evidently, we consider the transformations P_c, $c \in G$, defined by $P_cx = cx$ for $x \in G$. Certainly, all the transformations P_c are measure-preserving. Thus, we only have to show that the family $\{P_c: c \in G\}$ is equicontinuous at every point $x \in G$. Choose a neighborhood V of e and consider $V_L = \{(x, y) \in G \times G: x^{-1}y \in V\}$ from the left uniformity of G. Then $(P_cx, P_cy) \in V_L$ if and only if $(P_cx)^{-1}(P_cy)$, or $x^{-1}y$, is an element of V. Hence, $(P_cx, P_cy) \in V_L$ for all y in the neighborhood xV of x and for all $c \in G$. So $\{(cx_n): c \in G\}$ is a family of equi-u.d. sequences in G. A similar argument holds for the family $\{(x_nc): c \in G\}$. ∎

Again, the above theorem could also be proved by means of a Weyl criterion for equi-u.d. (see Exercise 1.4). There is a simple, but interesting, consequence of Theorem 1.5.

COROLLARY 1.4. If, for some $a \in G$, the sequence (a^n) is u.d. in G, then the sequence (a^n) is even well distributed in G.

PROOF. By Theorem 1.5, the family $\{(a^ha^n): h = 0, 1, 2, \ldots\}$ is a family of equi-u.d. sequences in G. In other words, $\{(a^{n+h}): h = 0, 1, 2, \ldots\}$ is a family of equi-u.d. sequences in G, and so, by definition, (a^n) is well distributed in G. ∎

In Section 4, we will study sequences of the form (a^n) in detail. For the present time, let us just point out that there are important classes of compact groups G for which the hypothesis of the above corollary can be satisfied.

Applying Homomorphisms

A very natural thing happens when a surjective continuous homomorphism is applied to a u.d. sequence; namely, the property of u.d. is preserved.

THEOREM 1.6. Let φ be a continuous homomorphism from the compact group G onto the compact group G_1. If (x_n) is u.d. in G, then the sequence $(\varphi(x_n))$ is u.d. in G_1.

PROOF. We proceed by Theorem 1.3. Let $\mathbf{D}_1$ be a nontrivial irreducible unitary representation of G_1. Then the composite mapping $\mathbf{D}$, defined by $\mathbf{D}(x) = \mathbf{D}_1(\varphi(x))$ for $x \in G$, is a nontrivial irreducible unitary representation of G. Therefore,

$$\lim_{N\to\infty}\frac{1}{N}\sum_{n=1}^{N}\mathbf{D}_1(\varphi(x_n)) = \lim_{N\to\infty}\frac{1}{N}\sum_{n=1}^{N}\mathbf{D}(x_n) = \mathbf{0}, \tag{1.13}$$

and the proof is complete. ■

The above argument is characteristic of the simple way in which the group structure and, in particular, the Weyl criterion may be exploited to arrive at short proofs. A purely measure-theoretic proof of Theorem 1.6 can also be given (see Exercise 1.6). We arrive at an important special case if we take for the second group G_1 a quotient group of G. Let H be a closed normal subgroup of the topological group G, that is, a normal subgroup of the group G that is closed in the topology of G. The quotient group G/H becomes a topological group under the following topology: As the open sets in G/H take all sets of the form $\{bH\colon\ b \in B\}$, where B is open in G. The canonical mapping $\varphi_H\colon\ x \in G \mapsto xH \in G/H$ is then a continuous homomorphism from G onto G/H, and G/H is compact whenever G is compact. The following result is now an immediate consequence of Theorem 1.6.

COROLLARY 1.5. Let H be a closed normal subgroup of the compact group G, and let (x_n) be u.d. in G. Then the sequence (x_nH) is u.d. in the quotient group G/H.

EXAMPLE 1.5. We determine all closed subgroups of the circle group $\boldsymbol{T}$. For $z \in \boldsymbol{T}$, let $\arg z$ be the value of the argument of z with $0 \leq \arg z < 2\pi$. Let $H \neq \{1\}$ be a closed subgroup of $\boldsymbol{T}$, and put $\alpha = \inf\{\arg z\colon\ z \in H,\ z \neq 1\}$. We distinguish two cases. If $\alpha = 0$, then we consider an arbitrary open interval $(\beta, \gamma) \subseteq [0, 2\pi)$. There exists $z \in H$ with $0 < \arg z < \gamma - \beta$; thus, there is a power z^n of z with $\arg z^n \in (\beta, \gamma)$. Since $z^n \in H$, we have shown that H is dense in $\boldsymbol{T}$. But H is closed, and so, $H = \boldsymbol{T}$. In the remaining case, we have $\alpha > 0$. Since H is closed, we have $z_0 = e^{i\alpha} \in H$. The subgroup H is not dense in this case, therefore $\alpha/2\pi$ is rational. Let now z be any element in H. Then $\arg z = n\alpha + \delta$ for some nonnegative integer n and $0 \leq \delta < \alpha$. We get $\arg(zz_0^{-n}) = \delta$ with $zz_0^{-n} \in H$. By the construction of α, we must have $\delta = 0$; thus, $z = z_0^n$. Therefore, H is cyclic and discrete because $\alpha/2\pi$ is rational. Consequently, our final result is as follows: The closed subgroups of $\boldsymbol{T}$ are precisely the discrete cyclic groups, generated by some root of unity, and $\boldsymbol{T}$ itself. ■

The preceding example has an important consequence for the characters of a compact abelian group G: namely, the image $\chi(G)$ of G under the character χ is a closed subgroup of $\boldsymbol{T}$ and therefore falls into one of the categories listed above. This leads to the following distinction: If $\chi(G)$ is a discrete cyclic group, then χ is said to be a *discrete* character. If $\chi(G) = \boldsymbol{T}$, then χ is called *nondiscrete*.

There is another interesting characterization of u.d. that may be partially inferred from Theorem 1.6. Theoretically, the subsequent theorem enables us to reduce the study of u.d. in the compact group G to the study of sequences of matrices.

THEOREM 1.7. If (x_n) is u.d. in G, then, for every representation $\mathbf{D}$ of G, the sequence $(\mathbf{D}(x_n))$ is u.d. in the image of $\mathbf{D}$. Conversely, if $\{\mathbf{D}^{(\lambda)}: \lambda \in \Lambda\}$ is a system of representations of G as in Theorem 1.3 and if for every $\lambda \in \Lambda$ with $\lambda \neq 0$, the sequence $(\mathbf{D}^{(\lambda)}(x_n))$ is u.d. in the image of $\mathbf{D}^{(\lambda)}$, then (x_n) is u.d. in G.

PROOF. The first part is an immediate consequence of Theorem 1.6. On the other hand, suppose the condition in the second part is satisfied. For a fixed $\lambda \in \Lambda$ with $\lambda \neq 0$, let $\mathscr{M}^{(\lambda)}$ be the image of $\mathbf{D}^{(\lambda)}$. Then the identity mapping on $\mathscr{M}^{(\lambda)}$ gives a nontrivial irreducible unitary representation of $\mathscr{M}^{(\lambda)}$. Since $(\mathbf{D}^{(\lambda)}(x_n))$ is u.d. in $\mathscr{M}^{(\lambda)}$, the Weyl criterion implies $\lim_{N\to\infty} (1/N)\sum_{n=1}^{N}\mathbf{D}^{(\lambda)}(x_n) = \mathbf{0}$. This holds for all λ under consideration, and so, (x_n) is u.d. in G. ■

Let G_1 and G_2 be topological groups; a mapping from G_1 onto G_2 that is both a group isomorphism and a homeomorphism is called a *topological isomorphism*. If such a mapping exists, the groups G_1 and G_2 are called *topologically isomorphic*. To point out the difference, a (not necessarily continuous) homomorphism resp. isomorphism will sometimes be referred to as an *algebraic homomorphism* resp. *algebraic isomorphism*. In most of the cases that we consider, a continuous homomorphism will automatically be an open mapping. Recall that a topological space is called *σ-compact* if it can be written as the union of at most countably many compact subsets.

THEOREM 1.8. A continuous homomorphism from a locally compact σ-compact group onto a locally compact group is an open mapping.

COROLLARY 1.6. Every character of a compact abelian group is an open mapping.

It is often important to know that a continuous homomorphism is also open. For instance, for such homomorphisms the well-known first isomorphism theorem for discrete groups carries over to topological groups as well.

THEOREM 1.9: Isomorphism Theorem. Let $\varphi\colon G \mapsto G_1$ be an open continuous homomorphism from the topological group G onto the topological group G_1 with kernel H. Then G_1 is topologically isomorphic to the quotient group G/H.

Duality Theory

Let now G be a locally compact abelian group. By a *character* of G, we mean again a continuous homomorphism from G into the circle group $\boldsymbol{T}$. The set of all characters of G can be made into a group by taking as the product $\chi_1\chi_2$ of two characters χ_1 and χ_2 of G the character defined by $(\chi_1\chi_2)(x) = \chi_1(x)\chi_2(x)$ for all $x \in G$. The identity of this group is the trivial character of G. With the so-called compact-open topology, the group of characters becomes a topological group, the *character group* or *dual group* $\hat{G}$ of G. As a basis for the neighborhoods of the identity in $\hat{G}$, we choose all subsets $U(K; \varepsilon)$ of $\hat{G}$ of the form $U(K; \varepsilon) = \{\chi \in \hat{G}: |\chi(x) - 1| < \varepsilon \text{ for all } x \in K\}$, where K is an arbitrary compact subset of G and ε is an arbitrary positive number. Then $\hat{G}$ is again a locally compact abelian group. If G is topologically isomorphic to the locally compact abelian group G_1, then $\hat{G}$ is topologically isomorphic to $\hat{G}_1$.

THEOREM 1.10. If G is compact, then $\hat{G}$ is discrete. If G is discrete, then $\hat{G}$ is compact.

Since $\hat{G}$ is a locally compact abelian group, we may ask for the character group of $\hat{G}$. Let us first note that the original group G can be (algebraically) embedded into the dual group of $\hat{G}$. For if we take a fixed $x \in G$, then the mapping $\hat{x}: \hat{G} \mapsto T$, defined by $\hat{x}(\chi) = \chi(x)$ for $\chi \in \hat{G}$, is easily seen to be a character of $\hat{G}$. The set of all those characters forms a subgroup of the dual group of $\hat{G}$ that is isomorphic to G. The following fundamental theorem tells us that far more is true.

THEOREM 1.11: Duality Theorem of Pontryagin-van Kampen. If G is a locally compact abelian group, then the character group $\hat{\hat{G}}$ of $\hat{G}$ is topologically isomorphic to G, and the topological isomorphism is given by the mapping $x \mapsto \hat{x}$.

In particular, every character of $\hat{G}$ is a mapping of the form $\hat{x}$ for some $x \in G$. Furthermore, another case of the Gel'fond-Raikov theorem follows: For every $x \neq e$ in G, there exists $\chi \in \hat{G}$ with $\chi(x) \neq 1$. In the sequel, we will often identify G and $\hat{\hat{G}}$. To discuss the character groups of closed subgroups and quotient groups of G, we introduce the notion of an annihilator. For an arbitrary nonvoid subset H of G, let the *annihilator* $A(\hat{G}, H)$ of H in $\hat{G}$ be defined as the set

$$A(\hat{G}, H) = \{\chi \in \hat{G}: \chi(x) = 1 \quad \text{for all} \quad x \in H\}. \tag{1.14}$$

The annihilator $A(\hat{G}, H)$ is a closed subgroup of $\hat{G}$. The importance of this concept is revealed in the following theorem.

THEOREM 1.12. Let H be a closed subgroup of the locally compact abelian group G. Then the character group of G/H is topologically isomorphic to $A(\hat{G}, H)$. Furthermore, the character group of H is topologically isomorphic to $\hat{G}/A(\hat{G}, H)$. We also have $H = A(G, A(\hat{G}, H))$.

When discussing representation theory, we pointed out the intimate relation between the weight of the compact group G and the cardinality of the set of equivalence classes of irreducible unitary representations. This situation has a close analogue in the present setting.

THEOREM 1.13. For a locally compact abelian group G, the weight $\mathfrak{w}(G)$ of G is identical with the weight $\mathfrak{w}(\hat{G})$ of $\hat{G}$.

Since the weight of a discrete space is identical with the cardinality of the space, we obtain the following immediate consequence of Theorems 1.10 and 1.13, which may also be inferred from representation theory.

COROLLARY 1.7. For a compact abelian group G, we have $\mathfrak{w}(G) = \operatorname{card} \hat{G}$.

The duality theorem also gives rise to important structure theorems for locally compact abelian groups. One of them that will turn out to be useful in Section 4 is given below. For $n \geq 1$, $\mathbb{R}^n$ denotes the n-dimensional euclidean space, considered as an additive group in the usual topology. We define $\mathbb{R}^0 = \{e\}$.

THEOREM 1.14. Every locally compact abelian group G is topologically isomorphic to a direct product of the form $\mathbb{R}^n \times H$, where H is a locally compact abelian group containing a compact open subgroup. The nonnegative integer n is uniquely determined by G.

Another result on the duality theory of locally compact abelian groups is of a more specific nature. We first recall the notion of the weak direct product of a family of groups. Let J be an arbitrary nonvoid index set, and for each $j \in J$, we are given a discrete group G_j with identity e_j. Then the *weak direct product* $\prod_{j \in J}^* G_j$ is the group consisting of all tuples $(x_j)_{j \in J}$ with $x_j \in G_j$ and $x_j \neq e_j$ for at most finitely many j and with multiplication defined coordinatewise.

THEOREM 1.15. For each j from the nonvoid index set J, let G_j be a discrete abelian group. If the weak direct product $\prod_{j \in J}^* G_j$ is given the discrete topology, then its dual is topologically isomorphic to the direct product $\prod_{j \in J} \hat{G}_j$.

Let G be a compact abelian group, and let C be the (connected) component of e. Then C is a closed subgroup of G, and the quotient group G/C is totally disconnected. There is a strong interrelation between the connectivity character of G and the torsion character of $\hat{G}$. Recall that in a (discrete) abelian group A the elements of finite order form a subgroup of A, the so-called *torsion subgroup* F. If $F = \{e\}$, then A is called *torsion-free;* if $F = A$, then A is called a *torsion group*. It is evident that a character $\chi \in \hat{G}$ is of finite order in $\hat{G}$ if and only if χ is discrete. Thus, the torsion subgroup of $\hat{G}$ is simply the collection of all discrete characters of G. It is easy to verify that a connected G cannot possess a nontrivial discrete character. For if $\chi \in \hat{G}$ is nontrivial and discrete, then the inverse image of $1 \in T$ under χ (which is then also the inverse image of some small open neighborhood of 1 under χ) would be a nontrivial subset of G that is both open and closed. For a generalization, see Theorem 1.16.

An element x in the locally compact abelian group G is called *compact* if the closed subgroup generated by x is compact. The set of all compact elements of G is a closed subgroup of G but not necessarily compact. We collect some relevant information in the following theorem.

THEOREM 1.16. Let G be a locally compact abelian group, and let C be its component of the identity. If G is compact, then the annihilator $A(\hat{G}, C)$ is exactly the torsion subgroup of $\hat{G}$. In the general case, let B be the closed subgroup of $\hat{G}$ consisting of all compact elements of $\hat{G}$. Then, $B = A(\hat{G}, C)$, and dually, $C = A(G, B)$.

COROLLARY 1.8. The compact abelian group G is connected if and only if it admits no nontrivial discrete character. Moreover, G is totally disconnected if and only if every character of G is discrete.

Notes

Various parts or all of the structure theory and representation theory for topological groups presented here can be found in the standard treatises on the subject, for example, Weil [1], Pontryagin [1], Hewitt and Ross [1], and Rudin [1]. A detailed survey on duality theory is given in Heyer [1]. Theorem 1.15, which is maybe not so widely known, is shown in Hewitt and Ross [1, Theorem 23.22].

U.d. in compact groups was first studied by Eckmann [1]. Unfortunately, this paper contains a serious error in that the condition $\lim_{N\to\infty} A(M; N)/N = \mu(M)$ is required to hold for all closed sets M instead of permitting only (closed) μ-continuity sets. The theory was brought to its present state by Hlawka and other authors, notably Cigler, Hartman, Helmberg, and Kemperman. The basic criteria, namely, Theorem 1.3 and Corollary 1.2, were already established by Eckmann [1]. Zaretti [2] gives a slight improvement of the necessary condition. Theorems 1.4 and 1.5, and Corollary 1.4, are from Hlawka [1].

We shall now mention various aspects of the theory that could not be included in our treatment. A completely satisfactory quantitative theory in compact groups has not yet been developed. The first step in this direction was undertaken by Hlawka and Niederreiter [1] and Niederreiter [1]. Another notion of discrepancy is in K. Schmidt [2, 3], who defined it for sequences of measures in locally compact abelian groups with countable base. If sequences of points in a compact group are considered, then Schmidt's definition reduces to a special case of the notion of maximal deviation introduced in Definition 2.2 of Chapter 3.

A remarkable result was shown by Veech [3]. He calls a sequence $r_1, r_2, \ldots$ of positive integers a "u.d.-sequence generator" if, for every compact group G and every sequence (y_n) in G that is not contained in any proper closed subgroup of G, the sequence (x_n) in G defined by $x_n = y_{r_1} y_{r_2} \cdots y_{r_n}$ is u.d. in G. The author not only shows that such u.d.-sequence generators exist but gives also explicit constructions, one of them based on normal numbers.

Normal numbers, or rather a generalization thereof, also occur in a paper of Cigler [3]. Let G be a compact abelian group with countable base, and let T be a continuous endomorphism of G that is ergodic with respect to the Haar measure μ. Then the element $x \in G$ is called *normal* with respect to T if the sequence $(T^n x)$ is u.d. in G. By Exercise 2.17 of Chapter 3, this definition generalizes the classical notion of normality. By the individual ergodic theorem and the fact that $\hat{G}$ is countable, one obtains that μ-almost all $x \in G$ are normal with respect to T (for a more general result, see Philipp [2]). Other properties of normal numbers extend also to this case. For instance, if x is normal with respect to T, then x is normal with respect to T^k for all $k = 1, 2, \ldots$; conversely, if x is normal with respect to T^k for some $k \geq 2$, then x is normal with respect to T. It was shown by Rohlin [1] that the ergodicity of the continuous endomorphism T may also be characterized algebraically: T is ergodic with respect to μ if and only if, for each nontrivial character χ of G, the functions $\chi \circ T^n$, $n = 1, 2, \ldots$, on G are all distinct. For another application of ergodic theory, see Couot [4].

Cigler [11] introduces the notion of a "strongly u.d. sequence" (*stark gleichverteilte Folge*) in a compact group. Every well-distributed sequence and every completely u.d. sequence (see Chapter 3, Section 3) is strongly u.d. but not every u.d. sequence is necessarily strongly u.d. (in fact, as soon as $\lim_{n\to\infty} x_{n+1} x_n^{-1} = e$ and G contains more than one element, then (x_n) cannot be strongly u.d. in G).

An interesting application of u.d. in compact groups occurs in connection with Artin's conjecture on primitive roots. See Serre [1, Chapter 1] and Goldstein [1].

Maak [2, 3] studies analogues of Kronecker's theorem for abstract groups, using a notion of independence for unitary representations of the group. See also Helmberg [3]. For the special case of abelian groups, similar investigations were carried out by Bundgaard [1, 2], who also introduced a notion of u.d. for functions on the group with values in a finite-dimensional circle group.

R. C. Baker [1, 2, 3] investigates sequences (χ_n) of characters of a locally compact abelian group G such that $(\chi_n(x))$ is u.d. in the circle group for almost all $x \in G$ in the sense of Haar measure. A basic role is played by sequences of characters that satisfy conditions generalizing the growth condition of Weyl (see Chapter 1, Exercise 4.6).

Starting from certain sequences of finite abelian groups, Dennis [1] studies sequences in these groups that are independently distributed in a certain technical sense. A Weyl criterion is shown in the case where all groups are elementary abelian.

For compact groups, the existence problem for u.d. sequences can be settled in a very satisfactory manner: namely, the compact group G admits u.d. sequences if and only if G is separable (see Corollary 5.4 and the notes in Section 5). This follows also from the results of Veech [3] quoted above.

Sequences of measures on compact groups were studied by Cigler [6]. This viewpoint was pursued further by K. Schmidt [2, 3] and Sigmund [1].

Exercises

1.1. Prove that the compact group G is the support of its Haar measure μ.

1.2. Prove Theorem 1.4 by means of the Weyl criterion.

1.3. Prove that $\{(x_{n,\sigma}): \sigma \in J\}$ is a family of equi-u.d. sequences in G if and only if, for each $\mathbf{D}^{(\lambda)}$ with $\lambda \neq 0$ from the system in Theorem 1.3 and for each $\varepsilon > 0$, there exists a positive integer $N_0(\lambda, \varepsilon)$ such that $\|(1/N)\sum_{n=1}^{N}\mathbf{D}^{(\lambda)}(x_{n,\sigma})\| < \varepsilon$ holds for all $N \geq N_0(\lambda, \varepsilon)$ and for all $\sigma \in J$.

1.4. Use the criterion in the previous exercise to give an alternative proof for Theorem 1.5.

1.5. Use the criterion in Corollary 1.3 to prove that if (x_n) is well distributed in G and if (c_n) converges, then the sequences $(c_n x_n)$ and $(x_n c_n)$ are both well distributed in G.

1.6. Prove Theorem 1.6 by going back to Chapter 3, Exercise 1.10.

1.7. For a complex square matrix $\mathbf{A}$, show that $\|\mathbf{A}\|^2 = \operatorname{tr}(\bar{\mathbf{A}}^T\mathbf{A})$.

1.8. Deduce from the previous exercise that $\|\mathbf{U}\| = \sqrt{k}$ and $\|\mathbf{UA}\| = \|\mathbf{AU}\| = \|\mathbf{A}\|$, where $\mathbf{U}$ is unitary of order k and $\mathbf{A}$ is an arbitrary complex square matrix of the same order.

1.9. Show that $\|\mathbf{A} + \mathbf{B}\| \leq \|\mathbf{A}\| + \|\mathbf{B}\|$ and $\|\mathbf{A} + \mathbf{B}\| \geq \big|\, \|\mathbf{A}\| - \|\mathbf{B}\| \,\big|$ for complex square matrices $\mathbf{A}$ and $\mathbf{B}$ of the same order.

1.10. Consult a book on topological groups and work out a detailed proof of the fact that any representation of a compact group is equivalent to a unitary one.

1.11. Prove that $\operatorname{tr}(\mathbf{AB}) = \operatorname{tr}(\mathbf{BA})$ for complex square matrices $\mathbf{A}$ and $\mathbf{B}$ of the same order.

1.12. Prove that two equivalent representations have the same character.

1.13. Give a detailed proof of the fact that the Kronecker product of two unitary matrices is again unitary.

1.14. Prove the following generalization of Theorem 1.2 of Chapter 1: If (x_n) is u.d. in G and if $\lim_{n\to\infty} x_n y_n^{-1}$ exists, then (y_n) is u.d. in G.

1.15. We have seen in this section how the Gel'fond–Raikov theorem for compact groups implies the Peter-Weyl theorem. Show, conversely, that the latter theorem also implies the former.

2. THE GENERALIZED DIFFERENCE THEOREM

Proof via Fundamental Inequality

We have seen in Chapter 1 that one of the fundamental results in the theory of u.d. is the so-called difference theorem of van der Corput (see Chapter 1, Theorem 3.1). Since this theorem relies on the presence of an algebraic structure in the underlying space, its generalization to a more abstract setting had to be deferred in Chapter 3. The time has now come to return to this topic.

By virtue of representation theory, a complete analogue of the difference theorem can in fact be shown to hold in every compact topological group. For the proof, we try to proceed along the same lines as in the classical case.

This means, first of all, that we have to generalize van der Corput's fundamental inequality (see Lemma 3.1 of Chapter 1). This is done in the following lemma.

LEMMA 2.1. Let $\mathbf{D}_1, \ldots, \mathbf{D}_N$ be complex square matrices of the same order k, and let H be an integer with $1 \leq H \leq N$. Then we have

$$H^2 \left\| \sum_{i=1}^{N} \mathbf{D}_i \right\|^2 \leq H(N+H-1) \sum_{i=1}^{N} \|\mathbf{D}_i\|^2 + 2(N+H-1) \sum_{h=1}^{H-1} (H-h) \left| \sum_{j=1}^{N-h} (\mathbf{D}_j \mid \mathbf{D}_{j+h}) \right|. \quad (2.1)$$

PROOF. We extend the definition of the $\mathbf{D}_i$ by putting $\mathbf{D}_i = \mathbf{0}$, the zero matrix of order k, for all $i \leq 0$ and all $i > N$. Consider the matrix $H \sum_{i=1}^{N} \mathbf{D}_i$. We may write this matrix as follows: $H \sum_{i=1}^{N} \mathbf{D}_i = \sum_{i=1}^{N} \sum_{h=0}^{H-1} \mathbf{D}_i$. Now we put $p = i + h$; then $1 \leq p \leq N + H - 1$, and so,

$$H \sum_{i=1}^{N} \mathbf{D}_i = \sum_{p=1}^{N+H-1} \sum_{h=0}^{H-1} \mathbf{D}_{p-h}$$

because of the extended definition of the $\mathbf{D}_i$. It follows that

$$H^2 \left\| \sum_{i=1}^{N} \mathbf{D}_i \right\|^2 = \left\| \sum_{p=1}^{N+H-1} \sum_{h=0}^{H-1} \mathbf{D}_{p-h} \right\|^2. \quad (2.2)$$

Using Example 1.2 and (1.6), we conclude that

$$\begin{aligned} H^2 \left\| \sum_{i=1}^{N} \mathbf{D}_i \right\|^2 &\leq (N+H-1) \sum_{p=1}^{N+H-1} \left\| \sum_{h=0}^{H-1} \mathbf{D}_{p-h} \right\|^2 \\ &= (N+H-1) \sum_{p=1}^{N+H-1} \left(\sum_{h=0}^{H-1} \mathbf{D}_{p-h} \,\middle|\, \sum_{h=0}^{H-1} \mathbf{D}_{p-h} \right) \\ &= (N+H-1) \sum_{p=1}^{N+H-1} \sum_{r,s=0}^{H-1} (\mathbf{D}_{p-r} \mid \mathbf{D}_{p-s}). \end{aligned} \quad (2.3)$$

For fixed i with $1 \leq i \leq N$, the number of terms $(\mathbf{D}_i \mid \mathbf{D}_i)$ occurring in the sum $\sum_{p=1}^{N+H-1} \sum_{r,s=0}^{H-1} (\mathbf{D}_{p-r} \mid \mathbf{D}_{p-s})$ is equal to H. We arrive at

$$\begin{aligned} H^2 \left\| \sum_{i=1}^{N} \mathbf{D}_i \right\|^2 &\leq H(N+H-1) \sum_{i=1}^{N} \|\mathbf{D}_i\|^2 \\ &\quad + (N+H-1) \sum_{p=1}^{N+H-1} \sum_{\substack{r,s=0 \\ r \neq s}}^{H-1} (\mathbf{D}_{p-r} \mid \mathbf{D}_{p-s}) \\ &= H(N+H-1) \sum_{i=1}^{N} \|\mathbf{D}_i\|^2 \\ &\quad + (N+H-1) \sum_{p=1}^{N+H-1} \sum_{\substack{r,s=0 \\ s < r}}^{H-1} ((\mathbf{D}_{p-r} \mid \mathbf{D}_{p-s}) + (\mathbf{D}_{p-s} \mid \mathbf{D}_{p-r})). \end{aligned} \quad (2.4)$$

Now $(\mathbf{D}_{p-r} \mid \mathbf{D}_{p-s}) + (\mathbf{D}_{p-s} \mid \mathbf{D}_{p-r}) = (\mathbf{D}_{p-r} \mid \mathbf{D}_{p-s}) + \overline{(\mathbf{D}_{p-r} \mid \mathbf{D}_{p-s})} = 2 \operatorname{Re} (\mathbf{D}_{p-r} \mid \mathbf{D}_{p-s})$, where $\operatorname{Re} z$ denotes the real part of the complex number z. We get

$$H^2 \left\| \sum_{i=1}^{N} \mathbf{D}_i \right\|^2 \leq H(N + H - 1) \sum_{i=1}^{N} \|\mathbf{D}_i\|^2$$

$$+ 2(N + H - 1) \operatorname{Re} \sum_{p=1}^{N+H-1} \sum_{\substack{r,s=0\\ s<r}}^{H-1} (\mathbf{D}_{p-r} \mid \mathbf{D}_{p-s}). \quad (2.5)$$

For fixed h with $1 \leq h \leq H - 1$ and j with $1 \leq j \leq N - h$, the number of terms $(\mathbf{D}_j \mid \mathbf{D}_{j+h})$ occurring in the sum $\sum_{p=1}^{N+H-1} \sum_{r,s=0,\, s<r}^{H-1} (\mathbf{D}_{p-r} \mid \mathbf{D}_{p-s})$ is equal to $H - h$. Consequently, we can write

$$\operatorname{Re} \sum_{p=1}^{N+H-1} \sum_{\substack{r,s=0\\ s<r}}^{H-1} (\mathbf{D}_{p-r} \mid \mathbf{D}_{p-s}) = \operatorname{Re} \sum_{h=1}^{H-1} (H - h) \sum_{j=1}^{N-h} (\mathbf{D}_j \mid \mathbf{D}_{j+h})$$

$$= \sum_{h=1}^{H-1} (H - h) \operatorname{Re} \sum_{j=1}^{N-h} (\mathbf{D}_j \mid \mathbf{D}_{j+h}). \quad (2.6)$$

But $\operatorname{Re} z \leq |z|$ for any complex number z, and so, together with (2.5), we arrive at (2.1). ■

We are now ready to prove the desired generalization of the difference theorem. Let G be a compact group, and let (x_n) be a sequence of elements in G. We assume that for each $h = 1, 2, \ldots$, the sequence $(x_{n+h}x_n^{-1})$ is u.d. in G. We even want to show a bit more than in the mod 1 case. Namely, not only is the sequence (x_n) itself u.d. but every subsequence of (x_n) where the subscripts run through an arithmetic progression is also u.d. in G.

THEOREM 2.1. Let (x_n) be a sequence in the compact group G such that for each $h = 1, 2, \ldots$, the sequence $(x_{n+h}x_n^{-1})$ is u.d. in G. Then, for every positive integer q and nonnegative integer r, the sequence (x_{qn+r}) is u.d. in G. In particular, the sequence (x_n) itself is u.d. in G.

PROOF. We suppose that q and r are fixed throughout the proof. We show the u.d. of (x_{qn+r}) by using the Weyl criterion as given in Theorem 1.3. In fact, we shall verify the desired limit relation (1.9) for any nontrivial irreducible unitary representation $\mathbf{D}$ of G. Such a representation being chosen, we first note that

$$\sum_{n=1}^{N} \mathbf{D}(x_{qn+r}) = \frac{1}{q} \sum_{j=1}^{q} \sum_{s=1}^{qN} \exp\left(\frac{js}{q}\right) \mathbf{D}(x_{s+r}), \quad (2.7)$$

where $\exp(\alpha) = e^{2\pi i\alpha}$ for $\alpha \in \mathbb{R}$. The above identity stems, of course, from the simple fact that $(1/q) \sum_{j=1}^{q} \exp(js/q) = 1$ for $s \equiv 0 \pmod{q}$ and $= 0$ for

$s \not\equiv 0 \pmod q$. It follows that

$$\left\| \sum_{n=1}^{N} \mathbf{D}(x_{qn+r}) \right\| \leq \frac{1}{q} \sum_{j=1}^{q} \left\| \sum_{s=1}^{qN} \exp\left(\frac{js}{q}\right) \mathbf{D}(x_{s+r}) \right\|$$

$$\leq \max_{j=1,\ldots,q} \left\| \sum_{s=1}^{qN} \exp\left(\frac{js}{q}\right) \mathbf{D}(x_{s+r}) \right\|. \tag{2.8}$$

We now choose an integer H with $1 \leq H \leq qN$. An application of Lemma 2.1 yields for each fixed j,

$$H^2 \left\| \sum_{s=1}^{qN} \exp\left(\frac{js}{q}\right) \mathbf{D}(x_{s+r}) \right\|^2 \leq H(qN + H - 1) \sum_{s=1}^{qN} \|\mathbf{D}(x_{s+r})\|^2$$

$$+ 2(qN + H - 1) \sum_{h=1}^{H-1} (H - h) \left|\textstyle\sum_h\right|, \tag{2.9}$$

with

$$\textstyle\sum_h = \displaystyle\sum_{p=1}^{qN-h} \left(\exp\left(\frac{jp}{q}\right) \mathbf{D}(x_{p+r}) \,\middle|\, \exp\left(\frac{j(p+h)}{q}\right) \mathbf{D}(x_{p+r+h}) \right). \tag{2.10}$$

Suppose that the degree of the representation $\mathbf{D}$ is m. Since $\mathbf{D}$ is unitary, we have $\|\mathbf{D}(x)\| = \sqrt{m}$ for every $x \in G$ by property (v) of the matrix norm. We can simplify $\sum_h$ by using, first of all, property (i) of the inner product:

$$\textstyle\sum_h = \exp\left(-\dfrac{jh}{q}\right) \displaystyle\sum_{p=1}^{qN-h} (\mathbf{D}(x_{p+r}) \mid \mathbf{D}(x_{p+r+h})). \tag{2.11}$$

Then, using properties (iv) and (ii) of the inner product and the fact that $\mathbf{D}$ is a homomorphism:

$$\textstyle\sum_h = \exp\left(-\dfrac{jh}{q}\right) \displaystyle\sum_{p=1}^{qN-h} (\mathbf{E} \mid \mathbf{D}(x_{p+r+h} x_{p+r}^{-1}))$$

$$= \exp\left(-\frac{jh}{q}\right) \left(\mathbf{E} \,\middle|\, \sum_{p=1}^{qN-h} \mathbf{D}(x_{p+r+h} x_{p+r}^{-1}) \right), \tag{2.12}$$

where $\mathbf{E}$ denotes the identity matrix of order m. Now, for any complex square matrix $\mathbf{A} = (a_{ik})$ of order m, we have $(\mathbf{E} \mid \mathbf{A}) = \operatorname{tr}(\bar{\mathbf{A}}^T) = \sum_{i=1}^m \bar{a}_{ii}$. By the Cauchy-Schwarz inequality, we get $|(\mathbf{E} \mid \mathbf{A})|^2 = |\sum_{i=1}^m \bar{a}_{ii}|^2 \leq m \|\mathbf{A}\|^2$, or $|(\mathbf{E} \mid \mathbf{A})| \leq \sqrt{m} \|\mathbf{A}\|$. In particular, we obtain

$$\left|\textstyle\sum_h\right| \leq \sqrt{m} \left\| \displaystyle\sum_{p=1}^{qN-h} \mathbf{D}(x_{p+r+h} x_{p+r}^{-1}) \right\| \qquad \text{for all } h = 1, 2, \ldots, H-1. \tag{2.13}$$

Combining all these facts, the inequality (2.9) leads to

$$H^2 \left\| \sum_{s=1}^{qN} \exp\left(\frac{js}{q}\right) \mathbf{D}(x_{s+r}) \right\|^2$$
$$\leq HNqm(qN + H - 1)$$
$$+ 2\sqrt{m}\,(qN + H - 1) \sum_{h=1}^{H-1} (H - h) \left\| \sum_{p=1}^{qN-h} \mathbf{D}(x_{p+r+h}x_{p+r}^{-1}) \right\|. \quad (2.14)$$

Dividing by H^2N^2, we arrive at

$$\left\| \frac{1}{N} \sum_{s=1}^{qN} \exp\left(\frac{js}{q}\right) \mathbf{D}(x_{s+r}) \right\|^2$$
$$\leq qm \frac{qN + H - 1}{HN} + \frac{2\sqrt{m}\,(qN + H - 1)}{H^2N}$$
$$\cdot \sum_{h=1}^{H-1} (H - h) \frac{qN - h}{N} \left\| \frac{1}{qN - h} \sum_{p=1}^{qN-h} \mathbf{D}(x_{p+r+h}x_{p+r}^{-1}) \right\|. \quad (2.15)$$

For each fixed h, the sequence $(x_{p+r+h}x_{p+r}^{-1})$, $p = 1, 2, \ldots$, is u.d. by hypothesis. Hence, $\|(1/(qN - h)) \sum_{p=1}^{qN-h} \mathbf{D}(x_{p+r+h}x_{p+r}^{-1})\|$ tends to zero as $N \to \infty$, and this for each $h = 1, 2, \ldots, H - 1$. Altogether, the second term on the right-hand side of (2.15) tends to zero as $N \to \infty$. The first term tends to q^2m/H as $N \to \infty$. Letting H attain arbitrarily large values, we see that

$$\lim_{N\to\infty} \left\| \frac{1}{N} \sum_{s=1}^{qN} \exp\left(\frac{js}{q}\right) \mathbf{D}(x_{s+r}) \right\| = 0 \qquad \text{for every } j = 1, 2, \ldots, q. \quad (2.16)$$

The inequality (2.8) implies then that

$$\lim_{N\to\infty} \left\| \frac{1}{N} \sum_{n=1}^{N} \mathbf{D}(x_{qn+r}) \right\| = 0, \quad (2.17)$$

which is exactly what we wanted to show. ■

Difference Theorems for Well-Distributed Sequences

THEOREM 2.2. Let (x_n) be a sequence in the compact group G such that for each $h = 1, 2, \ldots,$ the sequence $(x_{n+h}x_n^{-1})$ is well distributed in G. Then, for every positive integer q and nonnegative integer r, the sequence (x_{qn+r}) is well distributed in G. In particular, the sequence (x_n) itself is well distributed in G.

PROOF. In order to show that (x_{qn+r}) is well distributed in G, it suffices to verify that

$$\lim_{N\to\infty}\left\|\frac{1}{N}\sum_{n=1}^{N}\mathbf{D}(x_{q(n+b)+r})\right\|=0 \tag{2.18}$$

holds uniformly in $b = 0, 1, 2, \ldots$, for every nontrivial irreducible unitary representation $\mathbf{D}$ of G (see Corollary 1.3). We will show slightly more, namely, that

$$\lim_{N\to\infty}\left\|\frac{1}{N}\sum_{n=1}^{N}\mathbf{D}(x_{qn+r})\right\|=0 \tag{2.19}$$

holds uniformly in $r = 0, 1, 2, \ldots$, for every such representation $\mathbf{D}$ of G. For a given $\varepsilon > 0$, we choose an integer $H = H(\varepsilon) > 1$ with $H > 4q^2m/\varepsilon^2$. For each h with $1 \le h \le H - 1$, the sequence $(x_{n+h}x_n^{-1})$ is well distributed by assumption. This means, in particular, that there exists a positive integer $M_h(\varepsilon)$ such that

$$\left\|\frac{1}{M}\sum_{n=1}^{M}\mathbf{D}(x_{n+r+h}x_{n+r}^{-1})\right\|<\frac{\varepsilon^2}{4\sqrt{m}\,q^2} \tag{2.20}$$

holds for all $M > M_h(\varepsilon)$ and for all $r = 0, 1, 2, \ldots$. Put $M_0 = M_0(\varepsilon) = \max_{1\le h\le H-1} M_h(\varepsilon)$; then, for $M > M_0$, the inequality (2.20) holds simultaneously for all $h = 1, 2, \ldots, H - 1$.

We choose now an integer $N(\varepsilon)$ with $qN(\varepsilon) \ge M_0 + H$. For all $N > N(\varepsilon)$, we have then $1 \le H \le qN$, and so, the inequality (2.15) is applicable. The first term on the right-hand side of (2.15) is dominated by $qm(2qN/HN)$, and so, by $\varepsilon^2/2$. As to the second term, we clearly have $2\sqrt{m}(qN + H - 1)/H^2N \le 4q\sqrt{m}/H^2$ and $(qN - h)/N \le q$ for all $h = 1, 2, \ldots, H - 1$. To estimate the matrix norm occurring in the second term, we observe that $qN - h > qN(\varepsilon) - h \ge M_0 + H - h > M_0$ holds for all $h = 1, 2, \ldots, H - 1$, and so, the inequality (2.20) is available. Altogether, we obtain the following estimate from (2.15):

$$\begin{aligned}\left\|\frac{1}{N}\sum_{s=1}^{qN}\exp\left(\frac{js}{q}\right)\mathbf{D}(x_{s+r})\right\|^2 &\le \frac{\varepsilon^2}{2}+\frac{4q\sqrt{m}}{H^2}\sum_{h=1}^{H-1}(H-h)q\,\frac{\varepsilon^2}{4\sqrt{m}q^2}\\ &=\frac{\varepsilon^2}{2}+\frac{\varepsilon^2}{H^2}\sum_{h=1}^{H-1}(H-h)<\varepsilon^2\end{aligned} \tag{2.21}$$

for each $j = 1, 2, \ldots, q$, for all $N > N(\varepsilon)$ and for all $r = 0, 1, 2, \ldots$. By (2.8), we get then

$$\left\|\frac{1}{N}\sum_{n=1}^{N}\mathbf{D}(x_{qn+r})\right\|<\varepsilon \tag{2.22}$$

for all $N > N(\varepsilon)$ and for all $r = 0, 1, 2, \ldots$, and so, (2.19) holds. ■

Another remarkable property of well-distributed sequences that depends on an algebraic structure in the underlying space can be shown. We first need an auxiliary result on unitary matrices.

LEMMA 2.2. Let $\mathbf{U}_1, \mathbf{U}_2, \ldots, \mathbf{U}_r$ be unitary matrices of the same order, and let $\mathbf{E}$ be the identity matrix of this order. Then

$$\|\mathbf{U}_1\mathbf{U}_2 \cdots \mathbf{U}_r - \mathbf{E}\| \leq \sum_{i=1}^{r} \|\mathbf{U}_i - \mathbf{E}\|. \tag{2.23}$$

PROOF. We proceed by induction on r. The inequality is certainly correct for $r = 1$. Suppose we have already shown the inequality for some $r \geq 1$. Then $\|\mathbf{U}_1\mathbf{U}_2 \cdots \mathbf{U}_r\mathbf{U}_{r+1} - \mathbf{E}\| = \|(\mathbf{U}_1\mathbf{U}_2 \cdots \mathbf{U}_r - \mathbf{E})\mathbf{U}_{r+1} + \mathbf{U}_{r+1} - \mathbf{E}\| \leq \|(\mathbf{U}_1\mathbf{U}_2 \cdots \mathbf{U}_r - \mathbf{E})\mathbf{U}_{r+1}\| + \|\mathbf{U}_{r+1} - \mathbf{E}\|$. Using property (v) of the matrix norm, we conclude that $\|\mathbf{U}_1\mathbf{U}_2 \cdots \mathbf{U}_r\mathbf{U}_{r+1} - \mathbf{E}\| \leq \|\mathbf{U}_1\mathbf{U}_2 \cdots \mathbf{U}_r - \mathbf{E}\| + \|\mathbf{U}_{r+1} - \mathbf{E}\| \leq \sum_{i=1}^{r+1} \|\mathbf{U}_i - \mathbf{E}\|$. ■

THEOREM 2.3. Let (y_n) be a well-distributed sequence in the compact group G. If (x_n) is a sequence in G with $\lim_{n\to\infty} y_{n+1}^{-1}x_{n+1}x_n^{-1}y_n = e$, then (x_n) is again well distributed in G.

PROOF. Let $\mathbf{D}$ be a nontrivial irreducible unitary representation of G that will be fixed throughout the subsequent consideration. For a given $\varepsilon > 0$, there exists an $N_0 = N_0(\varepsilon)$ such that $\|(1/N) \sum_{n=1+h}^{N+h} \mathbf{D}(y_n)\| < \varepsilon/4$ holds for all $N > N_0$ and for all $h = 0, 1, 2, \ldots$. We have to show that a similar statement holds true for the sequence (x_n). To establish a link between the two sequences, we put $u_n = y_{n+1}^{-1}x_{n+1}x_n^{-1}y_n$ for all $n \geq 1$ and observe that $x_j = y_ju_{j-1}u_{j-2} \cdots u_iy_i^{-1}x_i$ whenever $j > i$. Using this and property (v) of the matrix norm, we get for all $N > N_0$ and $h \geq 1$:

$$\begin{aligned}
\left\| \sum_{n=1+h}^{N+h} \mathbf{D}(x_n) \right\| &= \left\| \left(\sum_{n=1+h}^{N+h} \mathbf{D}(y_nu_{n-1}u_{n-2} \cdots u_h) \right) \mathbf{D}(y_h^{-1}x_h) \right\| \\
&= \left\| \sum_{n=1+h}^{N+h} \mathbf{D}(y_nu_{n-1}u_{n-2} \cdots u_h) \right\| \\
&\leq \left\| \sum_{n=1+h}^{N+h} (\mathbf{D}(y_nu_{n-1}u_{n-2} \cdots u_h) - \mathbf{D}(y_n)) \right\| \\
&\quad + \left\| \sum_{n=1+h}^{N+h} \mathbf{D}(y_n) \right\| \\
&< \sum_{n=1+h}^{N+h} \|\mathbf{D}(u_{n-1}u_{n-2} \cdots u_h) - \mathbf{E}\| + N\frac{\varepsilon}{4}.
\end{aligned} \tag{2.24}$$

We now choose an integer K with $K > N_0$ and $K > 4/\varepsilon$. Since $\lim_{n\to\infty} u_n = e$, we have $\lim_{n\to\infty}\mathbf{D}(u_n) = \mathbf{E}$; therefore, there exists a positive integer H such that

$$\|\mathbf{D}(u_j) - \mathbf{E}\| < \frac{1}{K^2} \qquad \text{for all } j \geq H. \tag{2.25}$$

For $n > h \geq H$, we have then, by Lemma 2.2,

$$\|\mathbf{D}(u_{n-1}u_{n-2}\cdots u_h) - \mathbf{E}\| = \|\mathbf{D}(u_{n-1})\cdots\mathbf{D}(u_h) - \mathbf{E}\| < \frac{n-h}{K^2}. \tag{2.26}$$

Combining (2.24) and (2.26), the following inequality holds for all $h \geq H$:

$$\left\|\sum_{n=1+h}^{K+h} \mathbf{D}(x_n)\right\| < \sum_{n=1+h}^{K+h} \frac{n-h}{K^2} + K\frac{\varepsilon}{4} \leq 1 + K\frac{\varepsilon}{4}. \tag{2.27}$$

Let m denote the degree of the representation $\mathbf{D}$. We put $N_1 = N_1(\varepsilon) = \max(4K\sqrt{m}/\varepsilon, 8H\sqrt{m}/\varepsilon)$. We consider an integer $N > N_1$. Using the division algorithm, we write $N = qK + r$ with $0 \leq r < K$. Then, by (2.27) we obtain the following estimate for all $h \geq H$:

$$\begin{aligned}\left\|\frac{1}{N}\sum_{n=1+h}^{N+h} \mathbf{D}(x_n)\right\| &\leq \left\|\frac{1}{N}\sum_{j=0}^{q-1}\sum_{n=1+jK+h}^{K+jK+h} \mathbf{D}(x_n)\right\| + \left\|\frac{1}{N}\sum_{n=1+qK+h}^{N+h} \mathbf{D}(x_n)\right\| \\ &\leq \frac{q}{N}\left(1 + K\frac{\varepsilon}{4}\right) + \frac{r}{N}\sqrt{m}.\end{aligned} \tag{2.28}$$

Hence, for all $N > N_1$ and all $h \geq H$, we have

$$\left\|\frac{1}{N}\sum_{n=1+h}^{N+h} \mathbf{D}(x_n)\right\| < \frac{1}{K}\left(1 + K\frac{\varepsilon}{4}\right) + \frac{K\sqrt{m}}{N} < \frac{\varepsilon}{4} + \frac{\varepsilon}{4} + \frac{\varepsilon}{4} = \frac{3\varepsilon}{4}. \tag{2.29}$$

It remains to consider those h with $0 \leq h < H$. Using (2.29) we get, for all $N > N_1$ and for all h in the indicated range,

$$\begin{aligned}&\left\|\frac{1}{N}\sum_{n=1+h}^{N+h} \mathbf{D}(x_n)\right\| \\ &\quad = \left\|\frac{1}{N}\sum_{n=1+h}^{H} \mathbf{D}(x_n) + \frac{1}{N}\sum_{n=1+H}^{N+H} \mathbf{D}(x_n) - \frac{1}{N}\sum_{n=N+h+1}^{N+H} \mathbf{D}(x_n)\right\| \\ &\quad < \frac{(H-h)\sqrt{m}}{N} + \frac{3\varepsilon}{4} + \frac{(H-h)\sqrt{m}}{N} < \varepsilon.\end{aligned} \tag{2.30}$$

Altogether, we have shown that the inequality $\|(1/N)\sum_{n=1+h}^{N+h} \mathbf{D}(x_n)\| < \varepsilon$ holds for all $N > N_1(\varepsilon)$ and for all $h = 0, 1, 2, \ldots$. Consequently, the sequence (x_n) is well distributed. ■

It should be noted that Theorem 2.3 enunciates a special feature of well-distributed sequences. If (y_n) is only supposed to be u.d. in G and

$$\lim_{n\to\infty} y_{n+1}^{-1} x_{n+1} x_n^{-1} y_n = e$$

holds, then (x_n) need not be u.d. in G. Simple counterexamples can already be constructed in the classical case $G = \mathbb{R}/\mathbb{Z}$. For (y_n) we take the sequence $(\sqrt{n})$, viewed as a sequence in $\mathbb{R}/\mathbb{Z}$. We know from Example 2.7 of Chapter 1 that (y_n) is u.d. in $\mathbb{R}/\mathbb{Z}$. Now every constant sequence (x_n) satisfies

$$\lim_{n\to\infty} y_{n+1}^{-1} x_{n+1} x_n^{-1} y_n = e$$

in $\mathbb{R}/\mathbb{Z}$ but is evidently not u.d. Other examples of this type can be found easily.

The Method of Correlation Functions

We offer now a completely different approach to the difference theorem by using so-called correlation functions. This approach is, first of all, of great theoretical interest; second, it avoids the fundamental inequality; and last, it leads to more general versions of the difference theorem. We will develop as much of the theory of correlation functions as we need. Since it does not require a greater effort, we carry out the investigation in the more general context of **A**-u.d.

Let $\mathbf{A} = (a_{nk})$ be a positive strongly regular matrix method. By what we know from Section 4 of Chapter 3, this means that the following conditions are satisfied: $a_{nk} \geq 0$ for all $n, k = 1, 2, \ldots$; $\lim_{n\to\infty} \sum_{k=1}^{\infty} a_{nk} = 1$; $\lim_{n\to\infty} a_{nk} = 0$ for all $k = 1, 2, \ldots$; $\lim_{n\to\infty} \sum_{k=1}^{\infty} |a_{nk} - a_{n,k+1}| = 0$. Furthermore, we are given a fixed sequence ω of complex square matrices $\mathbf{M}(1), \mathbf{M}(2), \ldots, \mathbf{M}(k), \ldots$ that are of the same order and are uniformly bounded in norm; that is, $\|\mathbf{M}(k)\| \leq c$ for some positive constant c and all $k \geq 1$.

DEFINITION 2.1. Let $n_1 < n_2 < \cdots < n_s < \cdots$ by an increasing sequence of positive integers such that the limit

$$\gamma(h) = \lim_{s\to\infty} \sum_{k=1}^{\infty} a_{n_s k}(\mathbf{M}(k+h) \mid \mathbf{M}(k)) \tag{2.31}$$

exists for $h = 0, 1, 2, \ldots$. If we extend the definition of γ by putting $\gamma(-h) = \overline{\gamma(h)}$ for $h = 1, 2, \ldots$, then the resulting function γ on $\mathbb{Z}$ is called an **A**-*correlation function* of the sequence ω.

LEMMA 2.3. The sequence ω has at least one **A**-correlation function.

PROOF. Let $\mathbb{C}$ be the set of complex numbers in the usual topology, and let $\mathbb{C}^\infty$ denote the Cartesian product of denumerably many copies of $\mathbb{C}$ equipped with the product topology. For $n \geq 1$, let $\xi_n \in \mathbb{C}^\infty$ be given by $\xi_n = (\xi_n^{(0)}, \xi_n^{(1)}, \ldots)$ with $\xi_n^{(h)} = \sum_{k=1}^{\infty} a_{nk}(\mathbf{M}(k+h) \mid \mathbf{M}(k))$ for $h \geq 0$. Since the inner products $(\mathbf{M}(i) \mid \mathbf{M}(j))$, $i, j = 1, 2, \ldots$, are uniformly bounded, the sequence (ξ_n) in $\mathbb{C}^\infty$ is contained in a compact subset of $\mathbb{C}^\infty$. Therefore, (ξ_n) has a convergent subsequence (ξ_{n_s}) in $\mathbb{C}^\infty$. Since convergence in $\mathbb{C}^\infty$ implies coordinatewise convergence, we are done. ■

On the other hand, it may well be that the sequence ω has more than one **A**-correlation function. We want to establish an important property satisfied by any **A**-correlation function of ω. We recall the following well-known definition.

DEFINITION 2.2. A complex-valued function p on an (abstract) group G is called *positive-definite* if the inequality

$$\sum_{n,m=1}^{N} c_n \bar{c}_m p(x_n x_m^{-1}) \geq 0 \tag{2.32}$$

holds for every choice of finitely many elements $x_1, \ldots, x_N$ in G and for every choice of complex numbers $c_1, \ldots, c_N$.

It is a fundamental fact for our approach that the **A**-correlation functions of ω are positive-definite on the additive group of integers. Before we can show this, we need an auxiliary result that follows easily from the strong regularity of the summation matrix **A**.

LEMMA 2.4. Let $n_1 < n_2 < \cdots < n_s < \cdots$ be an increasing sequence of positive integers, and let $\ldots, b_{-2}, b_{-1}, b_0, b_1, b_2, \ldots$ be a doubly infinite sequence of complex numbers such that the b_k with positive k are uniformly bounded. Then, for every integer q, we have

$$\lim_{s\to\infty} \sum_{k=1}^{\infty} a_{n_s k} b_k = \lim_{s\to\infty} \sum_{k=1}^{\infty} a_{n_s k} b_{k+q}, \tag{2.33}$$

whenever one of the two limits exists.

PROOF. It suffices to prove the assertion for $q = 1$. For then we can show the result for every nonnegative integer q by induction. If q is negative, say

$q = -m$, then we put $b'_k = b_{k-m}$, and the result follows from the corresponding result for m. For fixed $s \geq 1$, we get

$$\left|\sum_{k=1}^{\infty} a_{n_s k} b_{k+1} - \sum_{k=1}^{\infty} a_{n_s k} b_k\right| = \left|\sum_{k=1}^{\infty} a_{n_s k} b_{k+1} - \sum_{k=1}^{\infty} a_{n_s, k+1} b_{k+1} - a_{n_s 1} b_1\right|$$

$$\leq \sum_{k=1}^{\infty} |a_{n_s k} - a_{n_s, k+1}|\, |b_{k+1}| + a_{n_s 1}\, |b_1|$$

$$\leq \left(\sup_{k \geq 2} |b_k|\right) \sum_{k=1}^{\infty} |a_{n_s k} - a_{n_s, k+1}| + a_{n_s 1}\, |b_1|. \quad (2.34)$$

If we now let s tend to infinity, then both terms in the last sum tend to zero because of the strong regularity of **A**. Our result follows immediately. ■

LEMMA 2.5. Every **A**-correlation function of the sequence ω is a positive-definite function on the additive group $\mathbb{Z}$ of integers.

PROOF. Let γ be an **A**-correlation function of ω, say

$$\gamma(h) = \lim_{s \to \infty} \sum_{k=1}^{\infty} a_{n_s k} (\mathbf{M}(k+h) \mid \mathbf{M}(k)) \quad (2.35)$$

for $h \geq 0$ and $\gamma(-h) = \overline{\gamma(h)}$ for $h \geq 1$ (actually, $h = 0$ might be included here as well, since $\gamma(0)$ is real). We first want to show that the formula (2.35) for $\gamma(h)$ with $h \geq 0$ also holds for negative h if we just agree to define the matrices $\mathbf{M}(k)$ with $k \leq 0$ in an arbitrary way but of the same order as the $\mathbf{M}(k)$ with $k \geq 1$. For $h \geq 1$, we get

$$\gamma(-h) = \overline{\gamma(h)} = \lim_{s \to \infty} \sum_{k=1}^{\infty} a_{n_s k} (\mathbf{M}(k) \mid \mathbf{M}(k+h)).$$

Using Lemma 2.4 with $b_k = (\mathbf{M}(k) \mid \mathbf{M}(k+h))$ for all $k \in \mathbb{Z}$ and $q = -h$, it follows that $\gamma(-h) = \lim_{s\to\infty} \sum_{k=1}^{\infty} a_{n_s k} (\mathbf{M}(k-h) \mid \mathbf{M}(k))$, and our assertion is justified.

To show that γ is positive-definite on $\mathbb{Z}$, we choose a natural number N, integers $x_1, \ldots, x_N$, and complex numbers $c_1, \ldots, c_N$. Then,

$$\sum_{n,m=1}^{N} c_n \bar{c}_m \gamma(x_n - x_m)$$

$$= \sum_{n,m=1}^{N} c_n \bar{c}_m \lim_{s \to \infty} \sum_{k=1}^{\infty} a_{n_s k} (\mathbf{M}(k + x_n - x_m) \mid \mathbf{M}(k)) \quad (2.36)$$

by the extended formula (2.35) for γ. For each fixed n and m, we apply Lemma 2.4 with $b_k = (\mathbf{M}(k + x_n - x_m) \mid \mathbf{M}(k))$ and $q = x_m$. Therefore,

$$\begin{aligned}\sum_{n,m=1}^{N} c_n\bar{c}_m\gamma(x_n - x_m) &= \sum_{n,m=1}^{N} c_n\bar{c}_m \lim_{s\to\infty}\sum_{k=1}^{\infty} a_{n_sk}(\mathbf{M}(k + x_n) \mid \mathbf{M}(k + x_m))\\ &= \lim_{s\to\infty}\sum_{k=1}^{\infty} a_{n_sk}\sum_{n,m=1}^{N} c_n\bar{c}_m(\mathbf{M}(k + x_n) \mid \mathbf{M}(k + x_m))\\ &= \lim_{s\to\infty}\sum_{k=1}^{\infty} a_{n_sk}\left(\sum_{n=1}^{N} c_n\mathbf{M}(k + x_n)\,\middle|\,\sum_{m=1}^{N} c_m\mathbf{M}(k + x_m)\right)\\ &= \lim_{s\to\infty}\sum_{k=1}^{\infty} a_{n_sk}\left\|\sum_{n=1}^{N} c_n\mathbf{M}(k + x_n)\right\|^2 \geq 0. \quad \blacksquare\end{aligned}$$

The crucial step in our argument is the application of the classical theorem of Bochner-Herglotz: For every continuous positive-definite function p on a locally compact abelian group G there exists a uniquely determined non-negative bounded regular Borel measure σ in the dual group $\hat{G}$ of G such that

$$p(x) = \int_{\hat{G}} \hat{x}(\chi)\, d\sigma(\chi) \qquad \text{for all } x \in G. \tag{2.37}$$

As in Section 1, the symbol $\hat{x}$ denotes the character of $\hat{G}$ defined by $\hat{x}(\chi) = \chi(x)$ for all $\chi \in \hat{G}$.

We consider now a positive-definite function γ on $\mathbb{Z}$. Since $\mathbb{Z}$ is discrete, γ is continuous. The dual group of $\mathbb{Z}$ is T. For $h \in \mathbb{Z}$, the character $\hat{h}$ of T has the form $\hat{h}(z) = z^h$ for all $z \in T$. The following representation of γ is thus obtained by the Bochner-Herglotz theorem:

$$\gamma(h) = \int_{T} z^h\, d\sigma(z) \qquad \text{for all } h \in \mathbb{Z}, \tag{2.38}$$

where σ is a uniquely determined nonnegative bounded regular Borel measure in T. The surprising usefulness of the present approach stems from the fact that rather weak conditions on those measures σ suffice to allow conclusions about the distribution of the original sequence $(\mathbf{M}(k))$. In particular, it is enough to know the σ-measure of the singleton $\{1\}$ to draw such conclusions. A good deal of the mystery will be resolved by the following identity.

LEMMA 2.6. Let γ be an **A**-correlation function of ω. If σ is the measure in T corresponding to the positive-definite function γ on $\mathbb{Z}$, and if $\mathbf{B} = (b_{nk})$ is a positive strongly regular matrix method (not necessarily the same as **A**), then

$$\lim_{n\to\infty}\sum_{k=1}^{\infty} b_{nk}\gamma(k) = \sigma(\{1\}). \tag{2.39}$$

PROOF. Using the representation of γ given in (2.38), we arrive at

$$\sum_{k=1}^{\infty} b_{nk}\gamma(k) = \sum_{k=1}^{\infty} b_{nk}\int_T z^k \, d\sigma(z)$$
$$= \int_T \left(\sum_{k=1}^{\infty} b_{nk}z^k\right) d\sigma(z) \qquad \text{for every } n \geq 1. \qquad (2.40)$$

Let us look at the function $g(z) = \lim_{n\to\infty} \sum_{k=1}^{\infty} b_{nk}z^k$ on T (the limit will turn out to exist for every $z \in T$). We clearly have $g(1) = 1$. For $z \neq 1$, it is an easy matter to prove that the sequence (z^k) is almost convergent to the value 0 (using the complex analogue of Definition 4.2 of Chapter 3). Now **B**, being strongly regular, includes the summation method F of almost convergence, and so, $g(z) = 0$ for $z \neq 1$ (strictly speaking, we have to apply the results of Section 4 of Chapter 3 to the real and imaginary part of (z^k)). Thus, $g(z)$ is nothing else but the characteristic function of the singleton $\{1\}$.

Returning to (2.40), we use the dominated convergence theorem to get

$$\lim_{n\to\infty} \sum_{k=1}^{\infty} b_{nk}\gamma(k) = \int_T \left(\lim_{n\to\infty} \sum_{k=1}^{\infty} b_{nk}z^k\right) d\sigma(z) = \int_T g(z)\, d\sigma(z) = \sigma(\{1\}). \quad \blacksquare$$

We are now going to exhibit how some information about $\sigma(\{1\})$ leads to results for the original sequence $\omega = (\mathbf{M}(k))$.

LEMMA 2.7. Suppose that for every **A**-correlation function of ω the corresponding measure σ in T satisfies $\sigma(\{1\}) = 0$. Then,

$$\lim_{n\to\infty} \sum_{k=1}^{\infty} a_{nk}\mathbf{M}(k) = \mathbf{0},$$

the zero matrix of appropriate order.

PROOF. Let $\mathbf{C}_n$ be the matrix $\mathbf{C}_n = \sum_{k=1}^{\infty} a_{nk}\mathbf{M}(k)$ for $n \geq 1$. We shall show that the only limit point of the norm-bounded sequence $(\mathbf{C}_n)$ is the zero matrix $\mathbf{0}$, thereby proving the lemma. Let the matrix $\mathbf{C}$ be a limit point of $(\mathbf{C}_n)$; thus, $\mathbf{C} = \lim_{s\to\infty} \mathbf{C}_{n_s}$ for some sequence $n_1 < n_2 < \cdots < n_s < \cdots$ of positive integers. With $\mathbf{G}(k) = \mathbf{M}(k) - \mathbf{C}$ for $k \geq 1$ we arrive then at

$$\lim_{s\to\infty} \sum_{k=1}^{\infty} a_{n_s k}\mathbf{G}(k) = \mathbf{0}. \qquad (2.41)$$

Let us introduce the summation method $\mathbf{R} = (r_{sk})$ defined by $r_{sk} = a_{n_s k}$. It follows readily that $\mathbf{R}$ is positive and strongly regular. By Lemma 2.3, the norm-bounded sequence $(\mathbf{G}(k))$ has an $\mathbf{R}$-correlation function δ. Thus, there exists a sequence $s_1 < s_2 < \cdots < s_i < \cdots$ of positive integers such that

$$\delta(h) = \lim_{i\to\infty} \sum_{k=1}^{\infty} r_{s_i k}(\mathbf{G}(k+h) \mid \mathbf{G}(k)) = \lim_{i\to\infty} \sum_{k=1}^{\infty} a_{n_{s_i} k}(\mathbf{G}(k+h) \mid \mathbf{G}(k)) \quad (2.42)$$

for $h \geq 0$. Evidently, the function δ is also an **A**-correlation function of $(\mathbf{G}(k))$. To simplify the notation, we write $m_i = n_{s_i}$ for $i \geq 1$. Then, $\delta(h) = \lim_{i\to\infty} \sum_{k=1}^{\infty} a_{m_ik}(\mathbf{G}(k+h) \mid \mathbf{G}(k))$, and also $\lim_{i\to\infty} \sum_{k=1}^{\infty} a_{m_ik}\mathbf{G}(k) = \mathbf{0}$ by (2.41). We note that $(\mathbf{M}(k+h) \mid \mathbf{M}(k)) = (\mathbf{G}(k+h) + \mathbf{C} \mid \mathbf{G}(k) + \mathbf{C}) = (\mathbf{G}(k+h) \mid \mathbf{G}(k)) + (\mathbf{C} \mid \mathbf{G}(k)) + (\mathbf{G}(k+h) \mid \mathbf{C}) + \|\mathbf{C}\|^2$ for all $k \geq 1$ and all $h \geq 0$. Using Lemma 2.4, it follows that

$$\lim_{i\to\infty} \sum_{k=1}^{\infty} a_{m_ik}(\mathbf{M}(k+h) \mid \mathbf{M}(k)) = \lim_{i\to\infty} \sum_{k=1}^{\infty} a_{m_ik}(\mathbf{G}(k+h) \mid \mathbf{G}(k)) + \|\mathbf{C}\|^2 = \delta(h) + \|\mathbf{C}\|^2 \quad (2.43)$$

for all $h \geq 0$. In other words, the function $\gamma(h) = \delta(h) + \|\mathbf{C}\|^2$ is an **A**-correlation function of $\omega = (\mathbf{M}(k))$.

By the given hypothesis and Lemma 2.6 with $\mathbf{B} = \mathbf{A}$, we have

$$\lim_{n\to\infty} \sum_{k=1}^{\infty} a_{nk}\gamma(k) = 0.$$

This implies $\lim_{n\to\infty} \sum_{k=1}^{\infty} a_{nk}\, \delta(k) = -\|\mathbf{C}\|^2$. On the other hand, δ is an **A**-correlation function of the sequence $(\mathbf{G}(k))$, and so Lemma 2.6 yields $\lim_{n\to\infty} \sum_{k=1}^{\infty} a_{nk}\, \delta(k) \geq 0$, since the measure τ in $\boldsymbol{T}$ corresponding to δ is nonnegative; hence, $\|\mathbf{C}\|^2 \leq 0$. But this is only possible if $\mathbf{C} = \mathbf{0}$. ■

The generalization of van der Corput's theorem which we are heading for is now a simple consequence of Lemmas 2.6 and 2.7.

THEOREM 2.4. Let G be a compact group, let (x_k) be a sequence in G, and let $\mathbf{A} = (a_{nk})$ and $\mathbf{B} = (b_{nk})$ be two positive strongly regular matrix methods. Suppose that for every nontrivial irreducible unitary representation $\mathbf{D}$ of G, all **A**-correlation functions γ of the sequence $(\mathbf{D}(x_k))$ satisfy

$$\lim_{n\to\infty} \sum_{k=1}^{\infty} b_{nk}\gamma(k) = 0.$$

Then the sequence (x_k) is **A**-u.d. in G.

PROOF. For the sequence $\omega = (\mathbf{M}(k))$ in the previous considerations, we take now the sequence $(\mathbf{D}(x_k))$ for some nontrivial irreducible unitary representation $\mathbf{D}$ of G. Together with Lemmas 2.6 and 2.7, our assumption implies that $\lim_{n\to\infty} \sum_{k=1}^{\infty} a_{nk}\mathbf{D}(x_k) = \mathbf{0}$ holds for every nontrivial irreducible unitary representation $\mathbf{D}$ of G. Then from Theorem 4.1 of Chapter 3 (or, more exactly, from its obvious analogue for complex-valued functions) and from the Peter-Weyl theorem, we infer that (x_k) is **A**-u.d. in G. ■

EXAMPLE 2.1. Let us verify that Theorem 2.4 really includes the difference theorem for **A**-u.d. sequences as a special case. So suppose that for each $h = 1, 2, \ldots$, the sequence $(x_{k+h}x_k^{-1})$ is **A**-u.d. in G. For a given nontrivial irreducible unitary representation $\mathbf{D}$ of G of degree m, we get then

$$\lim_{n\to\infty} \sum_{k=1}^{\infty} a_{nk}(\mathbf{D}(x_{k+h}) \mid \mathbf{D}(x_k)) = \lim_{n\to\infty} \sum_{k=1}^{\infty} a_{nk}(\mathbf{D}(x_{k+h}x_k^{-1}) \mid \mathbf{E})$$

$$= \left(\lim_{n\to\infty} \sum_{k=1}^{\infty} a_{nk}\mathbf{D}(x_{k+h}x_k^{-1}) \mid \mathbf{E}\right) = 0 \qquad \text{for all } h \geq 1.$$

Consequently, the sequence $(\mathbf{D}(x_k))$ has only one $\mathbf{A}$-correlation function, namely, that one given by $\gamma(0) = m$ and $\gamma(h) = 0$ for $h \neq 0$. Therefore, the condition of Theorem 2.4 is trivially satisfied, and (x_k) is $\mathbf{A}$-u.d. in G. ■

Several other new results are contained as special cases in Theorem 2.4. Let us mention one of them that is highly interesting because it allows us to restrict the values of h for which we have to require the sequence $(x_{k+h}x_k^{-1})$ to be $\mathbf{A}$-u.d. in G.

THEOREM 2.5. Let G be a compact group, let (x_k) be a sequence in G, and let $\mathbf{A} = (a_{nk})$ and $\mathbf{B} = (b_{nk})$ be two positive strongly regular matrix methods. Suppose that P is a set of positive integers such that

$$\lim_{n\to\infty} \sum_{k\in P} b_{nk} = 1.$$

If the sequence $(x_{k+h}x_k^{-1})$ is $\mathbf{A}$-u.d. in G for every $h \in P$, then the sequence (x_k) itself is $\mathbf{A}$-u.d. in G.

PROOF. Let $\mathbf{D}$ be a nontrivial irreducible unitary representation of G of degree m. Suppose

$$\gamma(h) = \lim_{s\to\infty} \sum_{k=1}^{\infty} a_{n_s k}(\mathbf{D}(x_{k+h}) \mid \mathbf{D}(x_k)) = \left(\lim_{s\to\infty} \sum_{k=1}^{\infty} a_{n_s k}\mathbf{D}(x_{k+h}x_k^{-1}) \,\middle|\, \mathbf{E}\right)$$

for $h \geq 0$ is an $\mathbf{A}$-correlation function of the sequence $(\mathbf{D}(x_k))$. The hypothesis implies that $\gamma(h) = 0$ for $h \in P$. For all $h \geq 0$, we obtain

$$|\gamma(h)| \leq m \lim_{s\to\infty} \sum_{k=1}^{\infty} a_{n_s k} = m,$$

since $|\mathrm{tr}\,(\mathbf{U})| \leq m$ for every unitary matrix $\mathbf{U}$ of order m. Let $\mathbb{Z}^+$ denote the set of positive integers. Then, for $n \geq 1$, we get

$$\left|\sum_{k=1}^{\infty} b_{nk}\gamma(k)\right| = \left|\sum_{k\in\mathbb{Z}^+\setminus P}^{\infty} b_{nk}\gamma(k)\right| \leq m \sum_{k\in\mathbb{Z}^+\setminus P}^{\infty} b_{nk}.$$

But

$$\lim_{n\to\infty} \sum_{k\in\mathbb{Z}^+\setminus P}^{\infty} b_{nk} = \lim_{n\to\infty} \left(\sum_{k=1}^{\infty} b_{nk} - \sum_{k\in P} b_{nk}\right) = 0,$$

and so, $\lim_{n\to\infty} \sum_{k=1}^{\infty} b_{nk}\gamma(k) = 0$. The rest follows from Theorem 2.4. ■

The above theorem attains a particularly simple form if we take for $\mathbf{B}$ the summation method of arithmetic means. We introduce the following notion, which is fundamental in additive number theory. Let P be a set of positive integers. For $n \geq 1$, let $C(P; n)$ denote the number of elements from P that are less than, or equal to, n. If $\lim_{n\to\infty} C(P; n)/n$ exists, then its value is called the *natural density* of the set P.

COROLLARY 2.1. Let (x_k) be a sequence in a compact group G. Suppose that the sequence $(x_{k+h}x_k^{-1})$ is $\mathbf{A}$-u.d. in G for every value of h from a set of natural density 1. Then, (x_k) is $\mathbf{A}$-u.d. in G.

PROOF. We consider the positive strongly regular matrix method $\mathbf{B} = (b_{nk})$ defined by $b_{nk} = 1/n$ for $1 \leq k \leq n$, and $b_{nk} = 0$ for $k > n$. Let P be a set of natural density 1. Then we simply note that

$$\lim_{n\to\infty} \sum_{k\in P} b_{nk} = \lim_{n\to\infty} C(P; n)/n = 1,$$

and Theorem 2.5 implies what we want. ∎

Weakening the Hypothesis

It suffices to impose conditions on the sequence $(x_{k+h}x_k^{-1})$ for rather sparsely scattered values of h only. In this direction, we present the following result.

THEOREM 2.6. Let (x_k) be a sequence in the compact group G, let $\mathbf{A} = (a_{nk})$ be a positive strongly regular matrix method, and let r be a fixed positive integer. Suppose that the sequence $(x_{k+h}x_k^{-1})$ is $\mathbf{A}$-u.d. in G for every value of h that is a positive multiple of r. Then the sequence (x_k) itself is $\mathbf{A}$-u.d. in G.

PROOF. Take a nontrivial irreducible unitary representation $\mathbf{D}$ of G of degree m, and let γ be any $\mathbf{A}$-correlation function of the sequence $(\mathbf{D}(x_k))$. By Lemma 2.5, the function γ on $\mathbb{Z}$ is positive-definite. It is then clear that the function δ on $\mathbb{Z}$, defined by $\delta(h) = \gamma(rh)$ for $h \in \mathbb{Z}$, is also positive-definite. By the Bochner-Herglotz theorem, there exist uniquely determined non-negative bounded regular Borel measures σ and λ on $\boldsymbol{T}$ such that $\gamma(h) = \int_{\boldsymbol{T}} z^h\, d\sigma(z)$ and $\delta(h) = \int_{\boldsymbol{T}} z^h\, d\lambda(z)$ for all $h \in \mathbb{Z}$. The measures σ and λ can be identified in an obvious fashion with measures in $[0, 1)$; thus, we can write $\gamma(h) = \int_{[0,1)} e^{2\pi i h x}\, d\sigma(x)$ and $\delta(h) = \int_{[0,1)} e^{2\pi i h x}\, d\lambda(x)$. On the other hand, $\delta(h) = \gamma(rh) = \int_{[0,1)} e^{2\pi i r h x}\, d\sigma(x)$ for all $h \in \mathbb{Z}$.

Let us find out in what way the measures σ and λ are related. Let V be the transformation $V: x \in [0, 1) \mapsto rx \in [0, r)$. Then, $\delta(h) = \int_{[0,r)} e^{2\pi i h t}\, d\tau(t)$, where τ is the measure in $[0, r)$ defined by $\tau(B) = \sigma(V^{-1}B)$ for every Borel set B in $[0, r)$. It follows that $\delta(h) = \sum_{j=0}^{r-1} \int_{[j,j+1)} e^{2\pi i h t}\, d\tau_j(t)$, where τ_j is the measure in $[j, j + 1)$ induced by τ. For fixed j with $0 \leq j \leq r - 1$, let W_j be the transformation $W_j\colon t \in [j, j+1) \mapsto t - j \in [0, 1)$. Then

$$\delta(h) = \sum_{j=0}^{r-1} \int_{[0,1)} e^{2\pi i h t}\, d\nu_j(t),$$

where $\nu_j(B) = \tau_j(W_j^{-1}B)$ for a Borel set B in $[0, 1)$. In other words, we arrive at

$$\delta(h) = \int_{[0,1)} e^{2\pi i h t}\, d\left(\sum_{j=0}^{r-1} \nu_j\right)(t) \qquad \text{for all } h \in \mathbb{Z}. \tag{2.44}$$

Hence, the uniqueness of the measure λ implies that $\lambda = \sum_{j=0}^{r-1} \nu_j$. Note that ν_j is defined in terms of σ, so this is the desired relation between λ and σ.

After those general considerations, we shall now use our hypothesis. According to this, we have $\delta(h) = 0$ for all $h \neq 0$ and $\delta(0) = m$ at any rate. The uniqueness of the measure λ implies that λ has to be the Lebesgue measure in $[0, 1)$ multiplied by the constant m. Then, $0 = \lambda(\{0\}) = \sum_{j=0}^{r-1} \nu_j(\{0\})$ yields $\nu_0(\{0\}) = 0$, since all the ν_j are nonnegative measures. It follows that $\tau(\{0\}) = \tau_0(\{0\}) = 0$, and so, $\sigma(\{0\}) = 0$. Thus, the corresponding measure σ in $\boldsymbol{T}$ satisfies $\sigma(\{1\}) = 0$, and an application of Lemma 2.7 completes the proof. ■

One might wonder whether Corollary 2.1 and Theorem 2.6 also hold with the stronger conclusion that we obtained in Theorem 2.1, namely, that all sequences of the form (x_{qk+r}), $k = 1, 2, \ldots$, have to be **A**-u.d. In fact, this can be shown at least in the case where **A** is the summation method of arithmetic means. To do this, we first need a generalized version of our pivotal Lemma 2.7. The following result will still hold for arbitrary positive strongly regular **A**. We return to our basic sequence $\omega = (\mathbf{M}(k))$.

LEMMA 2.8. Let q be a given positive integer. Suppose that for every **A**-correlation function of ω the corresponding measure σ in $\boldsymbol{T}$ satisfies $\sigma(\{\xi\}) = 0$ for all qth roots of unity ξ in $\boldsymbol{T}$. Then, $\lim_{n\to\infty} \sum_{k=1}^{\infty} a_{n,qk+r}\mathbf{M}(qk + r) = \mathbf{0}$ holds for every $r = 0, 1, \ldots, q-1$.

PROOF. For $0 \leq r \leq q-1$, we introduce an auxiliary matrix $\mathbf{B}^{(r)} = (b_{nk}^{(r)})$ with $b_{nk}^{(r)} = qa_{n,qk+r}$. The $\mathbf{B}^{(r)}$ are again positive strongly regular matrix methods. The only property that does not follow readily is $\lim_{n\to\infty} \sum_{k=1}^{\infty} b_{nk}^{(r)} = 1$. Define a sequence (c_k) by $c_k = q$ if $k \geq q$ and $k \equiv r \pmod q$, and $c_k = 0$ otherwise. Then (c_k) is almost convergent to the value 1, and so, the strong regularity of **A** implies $1 = \lim_{n\to\infty} \sum_{k=1}^{\infty} a_{nk}c_k = \lim_{n\to\infty} \sum_{k=1}^{\infty} qa_{n,qk+r} = \lim_{n\to\infty} \sum_{k=1}^{\infty} b_{nk}^{(r)}$.

Let us now look at the sequences $\omega^{(r)} = (\mathbf{M}(qk + r))$ with $0 \leq r \leq q-1$. For fixed r, suppose that

$$\gamma^{(r)}(h) = \lim_{s\to\infty} \sum_{k=1}^{\infty} b_{n_s k}^{(r)}(\mathbf{M}(qk + qh + r) \mid \mathbf{M}(qk + r))$$

is a $\mathbf{B}^{(r)}$-correlation function of $\omega^{(r)}$. By passing to a suitable subsequence of (n_s), which we will also denote by (n_s) for simplicity, we can assume that

$$\gamma^{(j)}(h) = \lim_{s\to\infty} \sum_{k=1}^{\infty} b^{(j)}_{n_s k}(\mathbf{M}(qk + qh + j) \mid \mathbf{M}(qk + j)) \tag{2.45}$$

is an $\mathbf{B}^{(j)}$-correlation function of $\omega^{(j)}$ for $0 \le j \le q - 1$ and that

$$\gamma(h) = \lim_{s\to\infty} \sum_{k=1}^{\infty} a_{n_s k}(\mathbf{M}(k + h) \mid \mathbf{M}(k)) \tag{2.46}$$

is an $\mathbf{A}$-correlation function of ω (use the argument in the proof of Lemma 2.3). We deduce the following important identity:

$$\begin{aligned}\sum_{j=0}^{q-1} \gamma^{(j)}(h) &= \lim_{s\to\infty} \sum_{k=1}^{\infty} \sum_{j=0}^{q-1} b^{(j)}_{n_s k}(\mathbf{M}(qk + qh + j) \mid \mathbf{M}(qk + j)) \\ &= \lim_{s\to\infty} \sum_{k=1}^{\infty} \sum_{j=0}^{q-1} q a_{n_s, qk+j}(\mathbf{M}(qk + j + qh) \mid \mathbf{M}(qk + j)) \\ &= \lim_{s\to\infty} \sum_{k=q}^{\infty} q a_{n_s k}(\mathbf{M}(k + qh) \mid \mathbf{M}(k)) \\ &= \lim_{s\to\infty} \sum_{k=1}^{\infty} q a_{n_s k}(\mathbf{M}(k + qh) \mid \mathbf{M}(k)),\end{aligned}$$

since $\lim_{s\to\infty} a_{n_s k} = 0$ for $1 \le k \le q - 1$. Therefore,

$$\sum_{j=0}^{q-1} \gamma^{(j)}(h) = q\gamma(qh) \qquad \text{for all } h \in \mathbb{Z}. \tag{2.47}$$

We put $\delta(h) = \gamma(qh)$ for $h \in \mathbb{Z}$. The function δ is positive-definite on $\mathbb{Z}$ and therefore corresponds to a certain measure in $\boldsymbol{T}$. As we already did in the proof of Theorem 2.6, measures in $\boldsymbol{T}$ will be identified with measures in $[0, 1)$. Let σ be the measure in $[0, 1)$ corresponding to γ. Our hypothesis implies that $\sigma(\{0\}) = \sigma(\{1/q\}) = \cdots = \sigma(\{(q - 1)/q\}) = 0$. Let λ be the measure in $[0, 1)$ corresponding to δ. In the first part of the proof of Theorem 2.6, we studied the relation between λ and σ. It turned out that $\lambda = \sum_{j=0}^{q-1} \nu_j$ with the notation introduced there. But then

$$\lambda(\{0\}) = \sum_{j=0}^{q-1} \nu_j(\{0\}) = \sum_{j=0}^{q-1} \tau_j(\{j\}) = \sum_{j=0}^{q-1} \tau(\{j\}) = \sum_{j=0}^{q-1} \sigma\left(\left\{\frac{j}{q}\right\}\right) = 0. \tag{2.48}$$

For $0 \le j \le q - 1$, let $\lambda^{(j)}$ be the measure in $[0, 1)$ corresponding to $\gamma^{(j)}$. By (2.47), we have

$$\delta(h) = (1/q) \textstyle\sum_{j=0}^{q-1} \gamma^{(j)}(h) = \int_{[0,1)} e^{2\pi i h x}\, d((1/q) \sum_{j=0}^{q-1} \lambda^{(j)})(x)$$

for all $h \in \mathbb{Z}$. The uniqueness of the measure λ implies $\lambda = (1/q)\sum_{j=0}^{q-1}\lambda^{(j)}$. Since all the measures $\lambda^{(j)}$ are nonnegative, we deduce from (2.48) that $\lambda^{(r)}(\{0\}) = 0$. In other words, the measure $\lambda^{(r)}$ in $\boldsymbol{T}$ corresponding to the $\mathbf{B}^{(r)}$-correlation function $\gamma^{(r)}$ of the sequence $\omega^{(r)}$ satisfies $\lambda^{(r)}(\{1\}) = 0$. Note that $\gamma^{(r)}$ was an arbitrary $\mathbf{B}^{(r)}$-correlation function of $\omega^{(r)}$. Therefore, an application of Lemma 2.7 yields

$$\lim_{n\to\infty}\sum_{k=1}^{\infty} b_{nk}^{(r)}\mathbf{M}(qk + r) = \lim_{n\to\infty}\sum_{k=1}^{\infty} q a_{n,qk+r}\mathbf{M}(qk + r) = \mathbf{0},$$

and the desired result follows. ■

EXAMPLE 2.2. The condition of Lemma 2.8 is certainly satisfied if the measures σ even vanish for all singletons in $\boldsymbol{T}$, that is, $\sigma(\{z\}) = 0$ for all $z \in \boldsymbol{T}$. A measure of this type is called a *continuous measure* in $\boldsymbol{T}$. We present a useful criterion for the continuity of a measure that is in fact just a reformulation of the criterion given in Theorem 7.5 of Chapter 1. Suppose that σ corresponds to the positive-definite function γ on $\mathbb{Z}$, that is, $\gamma(h) = \int_{\boldsymbol{T}} z^h\, d\sigma(z)$ for all $h \in \mathbb{Z}$. Then σ is continuous if and only if

$$\lim_{H\to\infty}\frac{1}{H}\sum_{h=1}^{H}|\gamma(h)|^2 = 0.$$

To show this, let τ denote the product measure in the Cartesian product $\boldsymbol{T} \times \boldsymbol{T}$ induced by σ and let x and y be variables ranging over $\boldsymbol{T}$. Then, using Fubini's theorem and the fact that $\lim_{H\to\infty}(1/H)\sum_{h=1}^{H} z^h = 0$ for $z \in \boldsymbol{T}$, $z \neq 1$, and $\lim_{H\to\infty}(1/H)\sum_{h=1}^{H} z^h = 1$ for $z = 1$, we have

$$\begin{aligned}
\lim_{H\to\infty}\frac{1}{H}\sum_{h=1}^{H}|\gamma(h)|^2 &= \lim_{H\to\infty}\frac{1}{H}\sum_{h=1}^{H}\gamma(h)\overline{\gamma(h)}\\
&= \lim_{H\to\infty}\frac{1}{H}\sum_{h=1}^{H}\int_{\boldsymbol{T}} x^h\, d\sigma(x)\int_{\boldsymbol{T}} y^{-h}\, d\sigma(y)\\
&= \lim_{H\to\infty}\frac{1}{H}\sum_{h=1}^{H}\int_{\boldsymbol{T}\times\boldsymbol{T}} x^h y^{-h}\, d\tau(x, y)\\
&= \int_{\boldsymbol{T}\times\boldsymbol{T}}\left(\lim_{H\to\infty}\frac{1}{H}\sum_{h=1}^{H}(xy^{-1})^h\right) d\tau(x, y)\\
&= \tau(\{(x, y) \in \boldsymbol{T}\times\boldsymbol{T}\colon x = y\}) = \int_{\boldsymbol{T}}\sigma(\{x\})\, d\sigma(x) = 0
\end{aligned}$$

if and only if the measure σ is continuous. ■

Combining the above lemma with what we have already seen in the proof of Theorem 2.6, we obtain the following stronger version of this theorem in the case of $\mathbf{A}$ being the summation method of arithmetic means.

COROLLARY 2.2. Suppose that (x_k) is a sequence in the compact group G such that for some positive integer r, every sequence of the form $(x_{k+h}x_k^{-1})$ with h a positive multiple of r is u.d. in G. Then the sequence (x_{qk+s}) is u.d. in G for every positive integer q and for every nonnegative integer s.

PROOF. Let us first work with an arbitrary summation method **A** (of course, positive and strongly regular) to see exactly where we need the specific form of **A**. For a nontrivial irreducible unitary representation **D** of G, let γ be an **A**-correlation function of $(\mathbf{D}(x_k))$. It turned out in the proof of Theorem 2.6 that the measure λ in $[0, 1)$ corresponding to the positive-definite function $\delta(h) = \gamma(rh)$ on $\mathbb{Z}$ is just the Lebesgue measure multiplied by some positive constant. If σ denotes the measure in $[0, 1)$ corresponding to γ, then we find the relation $\lambda = \sum_{j=0}^{r-1} \nu_j$. Now suppose that $\sigma(\{p/q\}) > 0$ for some rational p/q with $0 \le p \le q - 1$. Then, for $i = [rp/q]$, the measure ν_i is positive at the point $(rp/q) - [rp/q]$. Hence, λ is positive at that point, an obvious impossibility. Thus, the condition of Lemma 2.8 is satisfied for all q, and so, we get

$$\lim_{n\to\infty} \sum_{k=1}^{\infty} a_{n,qk+s}\mathbf{D}(x_{qk+s}) = \mathbf{0} \qquad \text{for all } q \text{ and } s. \tag{2.49}$$

Thus, for general **A**, this method does not allow us to conclude that the sequences (x_{qk+s}) are **A**-u.d.; for that purpose, we would need

$$\lim_{n\to\infty} \sum_{k=1}^{\infty} a_{nk}\mathbf{D}(x_{qk+s}) = \mathbf{0}.$$

But if **A** is the method of arithmetic means, then (2.49) yields

$$\begin{aligned}\mathbf{0} = q \lim_{n\to\infty} \frac{1}{n} \sum_{\substack{k\\ qk+s\le n}} \mathbf{D}(x_{qk+s}) &= \lim_{n\to\infty} \frac{q}{n} \sum_{k=1}^{[(n-s)/q]} \mathbf{D}(x_{qk+s}) \\ &= \lim_{n\to\infty} \frac{q[(n-s)/q]}{n} \cdot \frac{1}{[(n-s)/q]} \sum_{k=1}^{[(n-s)/q]} \mathbf{D}(x_{qk+s}) \\ &= \lim_{N\to\infty} \frac{1}{N} \sum_{k=1}^{N} \mathbf{D}(x_{qk+s}),\end{aligned}$$

and we are done. ■

Using a slightly different method, the conclusion in Corollary 2.1 can be strengthened if we agree to take for **A** again the summation method of arithmetic means.

COROLLARY 2.3. Let (x_k) be a sequence in a compact group G. Suppose that the sequence $(x_{k+h}x_k^{-1})$ is u.d. in G for every value of h from a set of natural density 1. Then the sequence (x_{qk+s}) is u.d. in G for every positive integer q and for every nonnegative integer s.

PROOF. As usual, we choose a nontrivial irreducible unitary representation $\mathbf{D}$ of G of degree m, say, and consider an $\mathbf{A}$-correlation function γ of $(\mathbf{D}(x_k))$, where $\mathbf{A}$ is now the matrix of arithmetic means. If P is the set of positive integers of natural density 1 from which the values of h are taken, then we have $\gamma(h) = 0$ for all $h \in P$. We have seen in the proof of Theorem 2.5 that $|\gamma(h)| \leq m$ holds for all $h \geq 1$. For $H \geq 1$, we get

$$\frac{1}{H}\sum_{h=1}^{H}|\gamma(h)|^2 \leq \frac{m^2}{H}(H - C(P; H)) = m^2\left(1 - \frac{C(P; H)}{H}\right). \qquad (2.50)$$

For the definition of $C(P; H)$, see the discussion preceding Corollary 2.1. The condition on P implies that $\lim_{H\to\infty} (1/H) \sum_{h=1}^{H} |\gamma(h)|^2 = 0$. By the criterion given in Example 2.2, the measure σ in T corresponding to γ is continuous. Thus, Lemma 2.8 is applicable for every q, and we obtain

$$\lim_{n\to\infty} \sum_{k=1}^{\infty} a_{n,qk+s}\mathbf{D}(x_{qk+s}) = \mathbf{0}$$

for all q and s. The proof is completed as in Corollary 2.2. ■

Notes

Theorems 2.1, 2.2, and 2.3 are all from Hlawka [1]. For the mod 1 case, the strengthened version of van der Corput's theorem, as given in Theorem 2.1, was established earlier by Korobov and Postnikov [1]. Theorems 2.2 and 2.3 remain true if "well distributed" is replaced by "weakly well distributed" (Hlawka [2, 8]). For the definition of "weakly well-distributed sequences", see Chapter 3, Exercise 3.13. A detailed study of "hereditary properties" (*erbliche Eigenschaften*), that is, properties that hold for all sequences (x_{qn+r}) whenever they hold for all sequences $(x_{n+h}x_n^{-1})$, $h = 1, 2, \ldots$, was carried out by Hlawka [8]. Other variants of van der Corput's theorem for compact groups can be found in Hlawka [5], Kemperman [1], and Cigler [11].

The possibility of proving van der Corput's theorem in the mod 1 case by using correlation functions was first realized by Bass and Bertrandias [1]. A detailed account of their method is given in Bass [1]. J.-P. Bertrandias [3] shows most of the results of this section but again only for the mod 1 case. Corollary 2.2 is attributed to Delange. For further applications of the method to u.d. mod 1, see Bass [3], J.-P. Bertrandias [2], Bésineau [1, 2], Lesca and Mendès France [1], and Mendès France [1, 4]. We refer also to Donoghue [1, p. 199] for an interesting approach.

In its full generality, the method was developed by Cigler [7, 8]. In a different direction, Cigler [6] shows that van der Corput's theorem extends to sequences of measures on a compact group. For the latter aspect, see also K. Schmidt [3] and Sigmund [1].

Proofs of the fundamental Bochner-Herglotz theorem can be found in Loomis [1], Rudin [1], and Weil [1]. For very general versions of the difference theorem, see Kemperman [2, 3].

Exercises

2.1. Let x be an element in a compact group G such that the sequence $((x^k)^n)$ is u.d. in G for every nonzero integer k. Show that the sequence $(x^{f(n)})$ is u.d. in G for every polynomial f of positive degree with integral coefficients.

2.2. Prove that every character of a locally compact abelian group G is a positive-definite function. What is the measure in $\hat{G}$ corresponding to a character according to the Bochner-Herglotz theorem?

2.3. Prove the uniqueness of the measure σ in the Bochner-Herglotz theorem.

2.4. Prove the following converse of the Bochner-Herglotz theorem: The function $p(x) = \int_{\hat{G}} \hat{x}(\chi)\, d\sigma(\chi)$ on G is positive-definite.

2.5. Generalize Theorem 2.6 by showing that the hypothesis can be weakened to the following one: The sequence $(x_{k+h}x_k^{-1})$ is **A**-u.d. in G for every h of the form $h = pr$, where p is taken from a set of natural density 1.

2.6. Generalize Corollary 2.2 by showing that the hypothesis can be weakened to the following one: The sequence $(x_{k+h}x_k^{-1})$ is u.d. in G for every h of the form $h = pr$, where p is taken from a set of natural density 1.

2.7. Explain why Theorem 2.6 is not contained as a special case in Theorem 2.5 for $r > 1$.

2.8. Prove that if (y_n) is well distributed in the compact group G and (x_n) satisfies $\lim_{n\to\infty} y_n^{-1}x_n = e$, then (x_n) is well distributed in G.

2.9. Let G be a compact group that has at least two elements. Prove that a sequence (x_n) for which $\lim_{n\to\infty} x_{n+1}x_n^{-1} = e$ cannot be well distributed in G. *Hint:* Use Lemma 3.3 of Chapter 3.

2.10. Show that Theorem 2.3 generalizes Theorem 3.3 of Chapter 1.

2.11. Prove Lemma 2.3 using the method in Exercise 7.7 of Chapter 1.

3. CONVOLUTION OF SEQUENCES

Convolution of Measures

In this section, we want to reveal a remarkable analogy between the well-known operation of convolution for measures in a compact group G and a certain binary operation in the set of all sequences in G that, because of this analogy, will also be referred to as convolution. Let us first collect some of the pertinent data about convolution of measures. In our treatment, we are only interested in nonnegative normed regular Borel measures in G. But one should be aware of the fact that convolution may be defined in exactly the same way for bounded complex regular Borel measures in locally compact groups (as is done in abstract harmonic analysis). The following simple lemma and the Riesz representation theorem provide the basis for our definition.

LEMMA 3.1. Let λ and ν be two nonnegative normed regular Borel measures in the compact group G. The functional L on $\mathscr{R}(G)$ defined by $L(f) = \int_G \int_G f(xy)\, d\lambda(x)\, d\nu(y)$ is a nonnegative normed linear functional.

PROOF. All the properties are easily checked. ■

LEMMA 3.2: Riesz Representation Theorem. Let X be a compact Hausdorff space, and let L be a nonnegative linear functional on $\mathscr{R}(X)$. Then there exists a unique bounded nonnegative regular Borel measure σ in X such that $L(f) = \int_X f\, d\sigma$ for all $f \in \mathscr{R}(X)$. If L is normed, then σ is also normed.

DEFINITION 3.1. The unique nonnegative normed regular Borel measure in G corresponding to the functional L in Lemma 3.1 by virtue of the Riesz representation theorem is called the *convolution* of λ and ν, and denoted by $\lambda * \nu$.

More explicitly, we thus have

$$\int_G f\, d(\lambda * \nu) = \int_G \int_G f(xy)\, d\lambda(x)\, d\nu(y) \qquad \text{for all } f \in \mathscr{R}(G), \tag{3.1}$$

and in the usual manner this identity can be seen to hold as well for $f \in \mathscr{C}(G)$. To simplify our notation, we write $\mathscr{M}^+(G)$ for the set of all nonnegative normed regular Borel measures in G. For $a \in G$, let ε_a denote the normed point measure at a (compare with Exercise 1.1 of Chapter 3).

LEMMA 3.3. The set $\mathscr{M}^+(G)$ is a semigroup under convolution.

PROOF. Convolution is clearly a binary operation in $\mathscr{M}^+(G)$. To show associativity, choose three measures $\lambda, \nu, \sigma \in \mathscr{M}^+(G)$. For every $f \in \mathscr{R}(G)$, we obtain $\int_G f\, d((\lambda*\nu)*\sigma) = \int_G \int_G f(xz)\, d(\lambda*\nu)(x)\, d\sigma(z)$, and with $g_z(x) = f(xz)$ for fixed $z \in G$ we arrive at $\int_G f(xz)\, d(\lambda*\nu)(x) = \int_G g_z(x)\, d(\lambda*\nu)(x) = \int_G \int_G g_z(xy)\, d\lambda(x)\, d\nu(y) = \int_G \int_G f(xyz)\, d\lambda(x)\, d\nu(y)$. Consequently, we have

$$\int_G f\, d((\lambda*\nu)*\sigma) = \int_G \int_G \int_G f(xyz)\, d\lambda(x)\, d\nu(y)\, d\sigma(z).$$

On the other hand, putting $g(y) = \int_G f(xy)\, d\lambda(x)$, we get

$$\int_G f\, d(\lambda*(\nu*\sigma)) = \int_G \int_G f(xy)\, d\lambda(x)\, d(\nu*\sigma)(y) = \int_G g(y)\, d(\nu*\sigma)(y)$$

$$= \int_G \int_G g(yz)\, d\nu(y)\, d\sigma(z) = \int_G \int_G \int_G f(xyz)\, d\lambda(x)\, d\nu(y)\, d\sigma(z).$$

Therefore, $(\lambda*\nu)*\sigma = \lambda*(\nu*\sigma)$. ■

LEMMA 3.4. $\varepsilon_a * \varepsilon_b = \varepsilon_{ab}$ for all $a, b \in G$.

PROOF. By an easy computation, we get

$$\int_G f\, d(\varepsilon_a * \varepsilon_b) = \int_G \int_G f(xy)\, d\varepsilon_a(x)\, d\varepsilon_b(y) = \int_G f(ay)\, d\varepsilon_b(y) = f(ab) = \int_G f\, d\varepsilon_{ab}$$

for all $f \in \mathscr{R}(G)$. ■

LEMMA 3.5. The semigroup $\mathscr{M}^+(G)$ is commutative if and only if G is commutative.

PROOF. If G is commutative, and $\lambda, \nu \in \mathscr{M}^+(G)$, then

$$\int_G f\, d(\lambda * \nu) = \int_G \int_G f(xy)\, d\lambda(x)\, d\nu(y) = \int_G \int_G f(yx)\, d\nu(y)\, d\lambda(x) = \int_G f\, d(\nu * \lambda)$$

for all $f \in \mathscr{R}(G)$, and so, $\lambda * \nu = \nu * \lambda$. But if G is not commutative (i.e., $ab \neq ba$ for some $a, b \in G$), then Lemma 3.4 implies $\varepsilon_a * \varepsilon_b \neq \varepsilon_b * \varepsilon_a$, and so, $\mathscr{M}^+(G)$ is not commutative. ■

LEMMA 3.6. A measure $\lambda \in \mathscr{M}^+(G)$ is the Haar measure in G if and only if $\lambda * \nu = \nu * \lambda = \lambda$ holds for all $\nu \in \mathscr{M}^+(G)$.

PROOF. It is clear that there can only be one such measure λ, for if $\lambda_1 \in \mathscr{M}^+(G)$ has the same property, then $\lambda * \lambda_1 = \lambda$ and $\lambda * \lambda_1 = \lambda_1$. Now let us show that the Haar measure μ in G enjoys this property. For $\nu \in \mathscr{M}^+(G)$ and for any $f \in \mathscr{R}(G)$, we get

$$\int_G f\, d(\mu * \nu) = \int_G \int_G f(xy)\, d\mu(x)\, d\nu(y) = \int_G f(x)\, d\mu(x) \int_G d\nu = \int_G f\, d\mu$$

because of the translation invariance of μ. Thus, $\mu * \nu = \mu$, and in the same way, one shows $\nu * \mu = \mu$. ■

The operation of convolution has also a remarkably nice behavior with respect to representations of the group G. Let $\mathbf{D} = (d_{ij})$ be a representation of G of degree r, with entry functions $d_{ij} \in \mathscr{C}(G)$. We introduce the following convenient abbreviation. For a measure $\nu \in \mathscr{M}^+(G)$, let $\nu(\mathbf{D})$ be the $r \times r$ matrix $\nu(\mathbf{D}) = (\alpha_{ij})$ with $\alpha_{ij} = \int_G d_{ij}\, d\nu$.

LEMMA 3.7. For $\lambda, \nu \in \mathscr{M}^+(G)$ and any representation $\mathbf{D}$ of G, the matrix identity $(\lambda * \nu)(\mathbf{D}) = \lambda(\mathbf{D})\nu(\mathbf{D})$ holds.

PROOF. Let $\mathbf{D} = (d_{ij})$ be a representation of degree r, and put $\lambda(\mathbf{D}) = (\alpha_{ij})$, $\nu(\mathbf{D}) = (\beta_{ij})$, and $\lambda(\mathbf{D})\nu(\mathbf{D}) = (\gamma_{ij})$. Then

$$\gamma_{ij} = \sum_{k=1}^{r} \alpha_{ik}\beta_{kj} = \int_G \int_G \left(\sum_{k=1}^{r} d_{ik}(x)\, d_{kj}(y) \right) d\lambda(x)\, d\nu(y). \tag{3.2}$$

But $\sum_{k=1}^{r} d_{ik}(x)\, d_{kj}(y)$ is just the entry in the ith row and jth column of the matrix product $\mathbf{D}(x)\mathbf{D}(y) = \mathbf{D}(xy)$ and is therefore equal to $d_{ij}(xy)$. Consequently, $\gamma_{ij} = \int_G\int_G d_{ij}(xy)\, d\lambda(x)\, d\nu(y) = \int_G d_{ij}\, d(\lambda*\nu)$, and the proof is complete. ■

Convolution and Uniform Distribution

We shall be interested in u.d. of sequences with respect to measures in $\mathscr{M}^+(G)$. By exactly the same argument as in Section 1 (namely, application of the Peter-Weyl theorem), the Weyl criterion as given in Theorem 1.3 can be modified to yield the following criterion: For $\nu \in \mathscr{M}^+(G)$, a sequence (x_n) is ν-u.d. in G if and only if

$$\lim_{N\to\infty} \frac{1}{N} \sum_{n=1}^{N} \mathbf{D}(x_n) = \nu(\mathbf{D}) \tag{3.3}$$

holds for all irreducible unitary representations $\mathbf{D}$ of G.

In the Cartesian product G^∞ of denumerably many copies of G, which we identify with the set of all sequences in G, we define now a binary operation that we shall also call convolution. Given two sequences (x_n) and (y_n) in G, we construct a sequence (z_n) in such a way that its first k^2 elements are just all possible products x_iy_j with $1 \le i \le k$ and $1 \le j \le k$. Specifically, we define z_n by taking the unique integer $k \ge 1$ with $(k-1)^2 < n \le k^2$, and setting $z_n = x_ky_i$ if $n = (k-1)^2 + 2i - 1$, and $z_n = x_iy_k$ if $n = (k-1)^2 + 2i$. Thus, the first terms of the sequence (z_n) are $x_1y_1, x_2y_1, x_1y_2, x_2y_2, x_3y_1, x_1y_3, x_3y_2, x_2y_3, x_3y_3, \ldots$.

DEFINITION 3.2. The sequence (z_n) defined above is called the *convolution* of the sequences (x_n) and (y_n), and is denoted by $(x_n)*(y_n)$.

The intimate connection between the two convolution operations is revealed by the following theorem, which also provides a good justification for using the same terminology for both operations.

THEOREM 3.1. Let $\lambda, \nu \in \mathscr{M}^+(G)$. If (x_n) is λ-u.d. and (y_n) is ν-u.d. in G, then $(x_n)*(y_n)$ is $\lambda*\nu$-u.d. in G.

PROOF. Let (z_n) be the sequence $(x_n)*(y_n)$, and choose an irreducible unitary representation $\mathbf{D}$ of G of degree r. By (3.3) we have to show that $\lim_{N\to\infty} (1/N) \sum_{n=1}^{N} \mathbf{D}(z_n) = (\lambda*\nu)(\mathbf{D}) = \lambda(\mathbf{D})\nu(\mathbf{D})$. For an integer $N > 1$, there exists a unique integer $k = k(N) \ge 1$ such that $k^2 < N \le (k+1)^2$.

Using the properties of the matrix norm as listed in Section 1, we obtain

$$\left\| \frac{1}{N} \sum_{n=1}^{N} \mathbf{D}(z_n) - \lambda(\mathbf{D})\nu(\mathbf{D}) \right\|$$

$$= \left\| \frac{1}{N} \sum_{n=1}^{k^2} \mathbf{D}(z_n) - \lambda(\mathbf{D})\nu(\mathbf{D}) + \frac{1}{N} \sum_{n=k^2+1}^{N} \mathbf{D}(z_n) \right\|$$

$$\leq \left\| \frac{k^2}{N} \cdot \frac{1}{k^2} \sum_{n=1}^{k^2} \mathbf{D}(z_n) - \lambda(\mathbf{D})\nu(\mathbf{D}) \right\| + \frac{2k+1}{N} \sqrt{r}.$$

By the construction of the sequence (z_n), its first k^2 elements are exactly all possible products $x_i y_j$ with $1 \leq i, j \leq k$ in some order. Therefore,

$$\sum_{n=1}^{k^2} \mathbf{D}(z_n) = \left(\sum_{n=1}^{k} \mathbf{D}(x_n) \right) \left(\sum_{n=1}^{k} \mathbf{D}(y_n) \right).$$

It follows that

$$\left\| \frac{1}{N} \sum_{n=1}^{N} \mathbf{D}(z_n) - \lambda(\mathbf{D})\nu(\mathbf{D}) \right\|$$

$$\leq \left\| \frac{k^2}{N} \left(\frac{1}{k} \sum_{n=1}^{k} \mathbf{D}(x_n) \right) \left(\frac{1}{k} \sum_{n=1}^{k} \mathbf{D}(y_n) \right) - \lambda(\mathbf{D})\nu(\mathbf{D}) \right\| + \frac{2k+1}{k^2} \sqrt{r}. \quad (3.4)$$

If we now let N tend to infinity (or, equivalently, $k \to \infty$), then

$$\lim_{N\to\infty} \frac{k(N)^2}{N} = 1, \qquad \lim_{k\to\infty} \frac{1}{k} \sum_{n=1}^{k} \mathbf{D}(x_n) = \lambda(\mathbf{D}),$$

and $\lim_{k\to\infty} (1/k) \sum_{n=1}^{k} \mathbf{D}(y_n) = \nu(\mathbf{D})$ imply that the right-hand side of (3.4) tends to zero. ■

An important special case occurs when $G = G_1 \times G_2$, that is, when G is the direct product of two other compact groups G_1 and G_2. The groups G_1 and G_2, together with their Haar measures μ_1 and μ_2, may be identified with the subgroups $H_1 = G_1 \times \{e_2\}$ and $H_2 = \{e_1\} \times G_2$ of G and their corresponding Haar measures. We define measures ν_1 and ν_2 on G by $\nu_1(B) = \mu_1(B \cap H_1)$ and $\nu_2(B) = \mu_2(B \cap H_2)$ for all Borel sets B in G. Let us compute $\nu_1 * \nu_2$. For a function $f \in \mathscr{R}(G)$, we get

$$\int_G f\, d(\nu_1 * \nu_2) = \int_G \int_G f(xy)\, d\nu_1(x)\, d\nu_2(y) = \int_{H_2} \int_{H_1} f(xy)\, d\mu_1(x)\, d\mu_2(y).$$

Since x ranges over H_1, we have $x = (x_1, e_2)$ with $x_1 \in G_1$, and similarly, $y = (e_1, x_2)$ with $x_2 \in G_2$. Thus,

$$\textstyle\int_G f\, d(\nu_1 * \nu_2) = \int_{G_2} \int_{G_1} f((x_1, x_2))\, d\mu_1(x_1)\, d\mu_2(x_2) = \int_G f\, d\mu.$$

Therefore, we conclude that $\nu_1 * \nu_2 = \mu$, the Haar measure in G. Now consider a u.d. sequence (x_n) in G_1, and a u.d. sequence (y_n) in G_2. Then, the sequence $((x_n, e_2))$ is ν_1-u.d. and the sequence $((e_1, y_n))$ is ν_2-u.d. in G. By Theorem 3.1, the sequence $((x_n, e_2)) * ((e_1, y_n))$ is u.d. in G. But this sequence can be written as $(x_1, y_1), (x_2, y_1), (x_1, y_2), (x_2, y_2), \ldots$, and so on in obvious analogy to the convolution of sequences. To be exact, the nth term w_n of the resulting sequence is $w_n = (x_k, y_i)$ if $n = (k-1)^2 + 2i - 1$, and $w_n = (x_i, y_k)$ if $n = (k-1)^2 + 2i$, where n and k are connected as in the definition of convolution. We have thus established a simple method of constructing u.d. sequences in direct products of groups from u.d. sequences in the factors.

COROLLARY 3.1. Let (x_n) be u.d. in G_1, and let (y_n) be u.d. in G_2. Then the sequence (w_n) as defined above is u.d. in $G_1 \times G_2$.

A Family of Criteria for Uniform Distribution

THEOREM 3.2. The following properties of a sequence (x_n) in a compact group G are equivalent:

1. The sequence (x_n) is u.d. in G.
2. For every sequence (y_n) in G, the sequence $(x_n) * (y_n)$ is ν-u.d. in G for some $\nu \in \mathscr{M}^+(G)$.
3. For every sequence (y_n) in G, the sequence $(y_n) * (x_n)$ is u.d. in G.
4. The sequence $(x_n) * (x_n^{-1})$ is u.d. in G.
5. The subsequence of $(x_n) * (x_n^{-1})$ consisting of all products $x_i x_j^{-1}$ with $i > j$ is u.d. in G.

PROOF. It is our aim to verify the following two chains of implications: $(1) \Rightarrow (3) \Rightarrow (2) \Rightarrow (1)$, and $(1) \Rightarrow (5) \Rightarrow (4) \Rightarrow (1)$. Thereby we will have shown the equivalence of all five properties.

$(1) \Rightarrow (3)$: This step bears a resemblance to the proof of Theorem 3.1. We may therefore abbreviate our arguments a bit. We take a nontrivial irreducible unitary representation $\mathbf{D}$ of G of degree r, and put $(z_n) = (x_n) * (y_n)$. We note that $\lim_{N \to \infty} \|(1/N) \sum_{n=1}^{N} \mathbf{D}(x_n)\| = 0$. Choosing N and k as in the proof of Theorem 3.1, we obtain

$$\left\| \frac{1}{N} \sum_{n=1}^{N} \mathbf{D}(z_n) \right\| \leq \frac{k^2}{N} \left\| \frac{1}{k} \sum_{n=1}^{k} \mathbf{D}(x_n) \right\| \left\| \frac{1}{k} \sum_{n=1}^{k} \mathbf{D}(y_n) \right\| + \frac{2k+1}{N} \sqrt{r}$$
$$\leq \frac{k^2}{N} \sqrt{r} \left\| \frac{1}{k} \sum_{n=1}^{k} \mathbf{D}(x_n) \right\| + \frac{2k+1}{k^2} \sqrt{r},$$

and it follows immediately that $\lim_{N \to \infty} \|(1/N) \sum_{n=1}^{N} \mathbf{D}(z_n)\| = 0$.

$(3) \Rightarrow (2)$: Trivial.

(2) ⇒ (1): For (y_n) we first select the constant sequence $e, e, \ldots$. Then $(z_n) = (x_n)*(e)$ is ν-u.d. in G for some $\nu \in \mathscr{M}^+(G)$. The integers N and k are again chosen as in the proof of Theorem 3.1. We note that among the first k^2 elements of (z_n) we find each x_j with $1 \leq j \leq k$ exactly k times. Therefore, $\sum_{n=1}^{N} \mathbf{D}(z_n) = k \sum_{n=1}^{k} \mathbf{D}(x_n) + \sum_{n=k^2+1}^{N} \mathbf{D}(z_n)$, where $\mathbf{D}$ is an irreducible unitary representation of G. It follows that

$$\frac{1}{k}\sum_{n=1}^{k}\mathbf{D}(x_n) = \frac{N}{k^2}\cdot\frac{1}{N}\sum_{n=1}^{N}\mathbf{D}(z_n) - \frac{1}{k^2}\sum_{n=k^2+1}^{N}\mathbf{D}(z_n),$$

and thus, $\lim_{k\to\infty} (1/k)\sum_{n=1}^{k} \mathbf{D}(x_n) = \nu(\mathbf{D})$. Consequently, the sequence (x_n) is ν-u.d in G.

It remains to show that $\nu = \mu$, the Haar measure in G. If $\nu \neq \mu$, then ν could not be translation invariant; hence, $\nu \neq \nu_c$ for some $c \in G$, where ν_c is defined by $\nu_c(B) = \nu(Bc)$ for all Borel sets B in G. Now consider the constant sequence $c, c, c, \ldots$. This sequence is clearly ε_c-u.d. in G. By Theorem 3.1, the sequence $(u_n) = (x_n)*(c)$ is then ν_c-u.d. in G, since it can be shown easily that $\nu * \varepsilon_c = \nu_c$. From $\nu \neq \nu_c$ and from the Peter-Weyl theorem, it follows that there exists a nontrivial irreducible unitary representation $\mathbf{D}$ of G with $\nu(\mathbf{D}) \neq \nu_c(\mathbf{D})$. We are going to construct a sequence (w_n) in G such that $\lim_{N\to\infty} (1/N)\sum_{n=1}^{N} \mathbf{D}(t_n)$ does not exist for the sequence $(t_n) = (x_n)*(w_n)$, thereby contradicting property (2).

The sequence (w_n) is defined as follows: The first 2^{2^0} terms are all equal to c, the next 2^{2^1} terms are all equal to e, the next 2^{2^2} terms are all equal to c, the next 2^{2^3} terms are all equal to e, and so on. For $M \geq 0$, we set $a(M) = \sum_{k=0}^{M} 2^{2^k}$. Let us first show that

$$\lim_{M\to\infty} \frac{1}{a^2(2M)} \sum_{n=1}^{a^2(2M)} \mathbf{D}(t_n) = \nu_c(\mathbf{D}). \tag{3.5}$$

The first $a^2(2M)$ elements of (t_n) just consist of all possible products $x_i w_j$ with $1 \leq i, j \leq a(2M)$. If we put $b(M) = \sum_{k=0}^{M} 2^{2^{2k}}$, then exactly $b^2(M)$ of those $a^2(2M)$ elements will be of the form $x_m c$ with $1 \leq m \leq b(M)$. Those $b^2(M)$ elements will also be the first $b^2(M)$ elements of the ν_c-u.d. sequence (u_n) in some order. Altogether, we get

$$\left\| \frac{1}{a^2(2M)} \sum_{n=1}^{a^2(2M)} \mathbf{D}(t_n) - \nu_c(\mathbf{D}) \right\|$$

$$\leq \frac{b^2(M)}{a^2(2M)} \left\| \frac{1}{b^2(M)} \sum_{n=1}^{b^2(M)} \mathbf{D}(u_n) - \nu_c(\mathbf{D}) \right\|$$

$$+ \frac{a^2(2M) - b^2(M)}{a^2(2M)} (\sqrt{r} + \|\nu_c(\mathbf{D})\|),$$

where r denotes the degree of $\mathbf{D}$. It can be shown by straightforward estimates that $\lim_{M\to\infty} b(M)/a(2M) = 1$. Our desired limit relation (3.5) follows then immediately. In exactly the same way, it can be verified that

$$\lim_{M\to\infty} \frac{1}{a^2(2M+1)} \sum_{n=1}^{a^2(2M+1)} \mathbf{D}(t_n) = \nu(\mathbf{D}), \tag{3.6}$$

by using the fact that $(x_n)*(e)$ is ν-u.d. in G. The relations (3.5) and (3.6) together imply the non-existence of

$$\lim_{N\to\infty} \frac{1}{N} \sum_{n=1}^{N} \mathbf{D}(t_n).$$

(1) $\Rightarrow$ (5): If (x_n) is u.d. in G, then (x_n^{-1}) is u.d. in G because of

$$\lim_{N\to\infty} (1/N) \textstyle\sum_{n=1}^{N} \mathbf{D}(x_n^{-1}) = \lim_{N\to\infty} (1/N) \sum_{n=1}^{N} \overline{\mathbf{D}(x_n)}^T = \mathbf{0}$$

for all nontrivial irreducible unitary representations $\mathbf{D}$ of G. Such a representation $\mathbf{D}$ of degree r being chosen, we thus have $\|(1/N) \sum_{n=1}^{N} \mathbf{D}(x_n^{-1})\| < \varepsilon$ for all $N \geq N_0 = N_0(\varepsilon)$. Let now (z_n) be the subsequence of $(x_n)*(x_n^{-1})$ described in (5). For given $N \geq 1$, there exists a unique integer $m = m(N)$ such that $m(m-1)/2 \leq N < (m+1)m/2$. Since

$$\left\| \frac{1}{N} \sum_{n=1}^{N} \mathbf{D}(z_n) \right\| \leq \left\| \frac{2}{m(m-1)} \sum_{n=1}^{m(m-1)/2} \mathbf{D}(z_n) \right\| + \frac{m}{N}\sqrt{r},$$

it will suffice to show

$$\lim_{m\to\infty} \left\| \frac{2}{m(m-1)} \sum_{n=1}^{m(m-1)/2} \mathbf{D}(z_n) \right\| = 0.$$

We choose $m > N_0$. By the construction of the sequence (z_n), its first $m(m-1)/2$ terms just consist of all the products $x_i x_j^{-1}$ with $1 \leq j < i \leq m$. Therefore,

$$\begin{aligned}
&\left\| \frac{2}{m(m-1)} \sum_{n=1}^{m(m-1)/2} \mathbf{D}(z_n) \right\| \\
&\quad = \left\| \frac{2}{m(m-1)} \sum_{i=2}^{m} \sum_{j=1}^{i-1} \mathbf{D}(x_i x_j^{-1}) \right\| \\
&\quad \leq \left\| \frac{2}{m(m-1)} \sum_{i=2}^{N_0} \sum_{j=1}^{i-1} \mathbf{D}(x_i x_j^{-1}) \right\| \\
&\qquad + \left\| \frac{2}{m(m-1)} \sum_{i=N_0+1}^{m} (i-1)\mathbf{D}(x_i)\left(\frac{1}{i-1} \sum_{j=1}^{i-1} \mathbf{D}(x_j^{-1})\right) \right\| \\
&\quad \leq \frac{N_0(N_0-1)}{m(m-1)} \sqrt{r} + \frac{2\varepsilon\sqrt{r}}{m(m-1)} \sum_{i=N_0+1}^{m} (i-1) < 2\sqrt{r}\,\varepsilon
\end{aligned}$$

for sufficiently large m.

(5) $\Rightarrow$ (4): Let the sequence (z_n) from above be u.d. in G. Then (z_n^{-1}) is also u.d. in G and so is the sequence (y_n) resulting from those two by superposition in the following way: For $m \geq 1$, we set $y_{2m-1} = z_m$ and $y_{2m} = z_m^{-1}$. The sequence (y_n) contains all the products $x_i x_j^{-1}$ with $i \neq j$ in exactly the same order in which they occur in $(x_n)*(x_n^{-1})$. Thus, we obtain the sequence $(x_n)*(x_n^{-1})$ from (y_n) by inserting the terms $x_i x_i^{-1} = e$ at appropriate places. Namely, there will be a term e preceding y_1, and between any two terms of the form y_{n^2-n} and y_{n^2-n+1} with $n \geq 2$ we have to insert a term e. These new terms will not affect the u.d. of the sequence, since their asymptotic relative frequency is zero.

(4) $\Rightarrow$ (1): Let $(w_n) = (x_n)*(x_n^{-1})$ be u.d. in G, and choose a nontrivial irreducible unitary representation $\mathbf{D}$ of G. Using properties of the matrix norm and of the inner product for matrices as listed in Section 1, we obtain

$$\begin{aligned}\lim_{N\to\infty} \left\| \frac{1}{N} \sum_{n=1}^{N} \mathbf{D}(x_n) \right\|^2 &= \lim_{N\to\infty} \left(\frac{1}{N} \sum_{n=1}^{N} \mathbf{D}(x_n) \,\middle|\, \frac{1}{N} \sum_{n=1}^{N} \mathbf{D}(x_n) \right) \\ &= \lim_{N\to\infty} \frac{1}{N^2} \sum_{i=1}^{N} \sum_{j=1}^{N} (\mathbf{D}(x_i) \mid \mathbf{D}(x_j)) \\ &= \lim_{N\to\infty} \frac{1}{N^2} \sum_{i=1}^{N} \sum_{j=1}^{N} (\mathbf{D}(x_i x_j^{-1}) \mid \mathbf{E}) \\ &= \left(\lim_{N\to\infty} \frac{1}{N^2} \sum_{i=1}^{N} \sum_{j=1}^{N} \mathbf{D}(x_i x_j^{-1}) \,\middle|\, \mathbf{E} \right) \\ &= \left(\lim_{N\to\infty} \frac{1}{N^2} \sum_{n=1}^{N^2} \mathbf{D}(w_n) \,\middle|\, \mathbf{E} \right) = 0.\end{aligned}$$

Therefore, (x_n) is u.d. in G. ∎

The equivalence of properties (1) and (5) allows the following illustration, which also reveals a relation to van der Corput's difference theorem for compact groups (see Section 2). The difference theorem, in its simplest form, may be stated as follows: Let (x_n) be a sequence in the compact group G, and consider the infinite array of elements

$$\begin{array}{cccc} x_2x_1^{-1} & x_3x_2^{-1} & x_4x_3^{-1} & \cdots \\ x_3x_1^{-1} & x_4x_2^{-1} & x_5x_3^{-1} & \cdots \\ x_4x_1^{-1} & x_5x_2^{-1} & x_6x_3^{-1} & \cdots \\ \cdot & \cdot & \cdot & \\ \cdot & \cdot & \cdot & \\ \cdot & \cdot & \cdot & \end{array}$$

If every row in this array is a u.d. sequence, then so is the first (and therefore, every) column and the sequence (x_n) itself. The equivalence of (1) and (5) may be interpreted as follows: The u.d. of the first (and so, of every) column is equivalent to the u.d. of the sequence that is obtained by enumerating the elements in the array according to the well-known diagonal counting procedure: $x_2x_1^{-1}$, $x_3x_1^{-1}$, $x_3x_2^{-1}$, $x_4x_1^{-1}$, $x_4x_2^{-1}$, $x_4x_3^{-1}$, ...

Notes

The definition of convolution for sequences and all the main results in this section are from Helmberg [4, 6]. A detailed treatment of convolution of measures can be found in books on harmonic analysis (Rudin [1], Hewitt and Ross [1], Weil [1]). Some interesting results concerning the structure of the semigroup $\mathcal{M}^+(G)$ are contained in a paper of Wendel [1]. See also Stromberg [1, 2]. For the Riesz representation theorem, we refer to Hewitt and Stromberg [1] and books on functional analysis.

For related investigations on sequences of products, see Helmberg [1, 2]. The idea of arranging sum sets in specific abelian groups in a way that resembles convolution was applied to density theory by Volkmann [3, 5]. A special case of Corollary 3.1 using a slightly different arrangement was proved by Kuipers and Scheelbeek [2]. For a related result, see Kuipers and Scheelbeek [1]. Helmberg [8] shows results analogous to Theorems 3.1 and 3.2 for the convolution $(x_n)*'(y_n)$ defined by the diagonal arrangement e, x_1, y_1, x_2, x_1y_1, y_2, x_3, x_2y_1, x_1y_2, y_3, ...

In a somewhat different context, certain types of convoluted sequences were studied by Arnol'd and Krylov [1] and Každan [1].

Exercises

3.1. Show that $\mathcal{M}^+(G)$ is a semigroup with identity.

3.2. For $\nu \in \mathcal{M}^+(G)$ and $c \in G$, define the translated measures ν_c and ${}_c\nu$ by $\nu_c(B) = \nu(Bc)$ and ${}_c\nu(B) = \nu(cB)$ for all Borel sets B in G. Prove $\nu*\varepsilon_c = \nu_c$ and $\varepsilon_c*\nu = {}_c\nu$.

3.3. Let τ: $G \times G \mapsto G$ be the mapping $\tau((x, y)) = xy$. For $\lambda, \nu \in \mathcal{M}^+(G)$, show that $(\lambda*\nu)(B) = (\lambda \times \nu)(\tau^{-1}(B))$ holds for every Borel set B in G, where $\lambda \times \nu$ denotes the product measure of λ and ν in $G \times G$.

3.4. State and prove an analogue of Lemma 3.4 for convolution of sequences.

3.5. Let G have at least two elements. Show that convolution of sequences in G is not an associative operation.

3.6. Let G have at least two elements. Show that there is no identity with respect to convolution of sequences, that is, that there exists no sequence (x_n) in G such that $(x_n)*(y_n) = (y_n)*(x_n) = (y_n)$ holds for all sequences (y_n) in G.

3.7. Give a detailed argument for the fact stated in the proof of Theorem 3.2, part (2) $\Rightarrow$ (1): $\lim_{M\to\infty} b(M)/a(2M) = 1$.

3.8. Referring to the proof of Theorem 3.2, part (2) $\Rightarrow$ (1), show in detail that (3.6) holds.

3.9. Let (x_n) be u.d. in G, and let (y_n) be an arbitrary sequence in G. Prove that the sequence of products $x_i y_j$ in the diagonal arrangement x_1y_1, x_1y_2, x_2y_1, x_1y_3, x_2y_2, x_3y_1, x_1y_4, x_2y_3, x_3y_2, $x_4y_1, \ldots$ is u.d. in G.

4. MONOTHETIC GROUPS

Definition

In the theory of u.d. mod 1, the sequences $(n\alpha)$ with irrational α constitute a very important class of u.d. sequences. As a natural generalization to the compact group G, we consider now sequences of the form (a^n) where a is a fixed element in G. In particular, we will be interested in conditions on the group G that guarantee the existence of u.d. sequences of this type.

We have noted in Section 1 that a u.d. sequence in G is necessarily everywhere dense. Thus, if a sequence (a^n) is u.d. in G for some $a \in G$, then G has to contain a dense cyclic subgroup, namely, the subgroup generated by a. This observation leads to the following definition.

DEFINITION 4.1. A topological group H is called *monothetic* if it contains a dense cyclic subgroup. A generator of a dense cyclic subgroup of H is called a *generator of H*.

Evidently, we are mainly interested in compact monothetic groups, although we will take a brief look at the locally compact case as well (see Theorem 4.8). Let us first show an important necessary condition for a group to be monothetic.

THEOREM 4.1. Every monothetic group is abelian.

PROOF. Let C be the dense cyclic subgroup contained in the monothetic group H. We want to show that the set L of all ordered pairs (x, y) in the direct product $H \times H$, for which $xyx^{-1}y^{-1} = e$, is equal to $H \times H$. To this end, let us note that the mapping $(x, y) \mapsto xyx^{-1}y^{-1}$ from $H \times H$ into H is continuous. Therefore, the set L is closed. We certainly have $C \times C \subseteq L$. But since C is dense in H, the set $C \times C$ will be dense in $H \times H$. Therefore, $L = H \times H$. ■

It is a crucial consequence of this theorem that the duality theory for locally compact abelian groups can be applied to the monothetic groups that are of interest to us. Before we proceed, we provide some simple examples of compact monothetic groups. A detailed classification of such groups will be given later on (see Theorem 4.7, Corollary 4.5, and the notes).

EXAMPLE 4.1. Clearly, every finite cyclic group in the discrete topology is a compact monothetic group. To have a less trivial example, let us prove that the k-dimensional circle group T^k (i.e., the direct product of k copies of the one-dimensional circle group T) is monothetic. To see this, choose real numbers $\alpha_1, \ldots, \alpha_k$ such that $1, \alpha_1, \ldots, \alpha_k$ are linearly independent over the rationals. Then $(e^{2\pi i\alpha_1}, \ldots, e^{2\pi i\alpha_k})$ is a generator of T^k, because the u.d. mod 1 of $((n\alpha_1, \ldots, n\alpha_k))$, $n = 1, 2, \ldots$, in $\mathbb{R}^k$ implies the density of the sequence $((e^{2\pi in\alpha_1}, \ldots, e^{2\pi in\alpha_k}))$, $n = 1, 2, \ldots$, in T^k. ∎

Sequences of the form (a^n)

THEOREM 4.2. If the sequence (a^n) is dense in the compact group G, then (a^n) is u.d. in G.

PROOF. By Theorem 4.1, G is abelian. Thus, we will proceed by Corollary 1.2. If we can show $\chi(a) \neq 1$ for each nontrivial character χ of G, then we may use essentially the same argument as in the proof for the u.d. mod 1 of $(n\alpha)$ (see Example 2.1 of Chapter 1). In fact, we get then for each nontrivial character χ of G and for $N \geq 1$,

$$\frac{1}{N}\sum_{n=1}^{N}\chi(a^n) = \frac{1}{N}\sum_{n=1}^{N}(\chi(a))^n = \frac{(\chi(a))^{N+1} - \chi(a)}{N(\chi(a) - 1)}, \tag{4.1}$$

and since the numerator of the above fraction is bounded, we eventually obtain $\lim_{N\to\infty} (1/N)\sum_{n=1}^{N} \chi(a^n) = 0$. To complete the proof, let us assume that $\chi(a) = 1$ for some nontrivial character χ of G. Then, $\chi(x) = 1$ on a dense subset of G, namely, for all elements of the sequence (a^n), and so, $\chi(x) = 1$ for all $x \in G$. In other words, χ is the trivial character, a contradiction. ∎

A remark concerning the assumption in Theorem 4.2 is in order. Namely, we want to point out that (a^n) is dense in G if and only if a is a generator of G. One implication is clear: If the sequence (a^n) is dense in G, then, *a fortiori*, the cyclic subgroup generated by a is dense in G. For the converse, we must realize that in the cyclic subgroup generated by a we find all powers of a, whereas in the sequence (a^n) we only consider the positive powers of a. The case where G is discrete is easily dealt with, because G is then finite (by compactness), and the positive powers of a generator a already exhaust all elements of G. For nondiscrete G, let U be an open set that contains a power a^k of a with $k < 0$. Then $a^{-k}U$, as a neighborhood of e, contains a symmetric open neighborhood V of e, that is, an open neighborhood with $V^{-1} = V$. Without loss of generality, we may assume that V contains none of the elements $a, a^2, \ldots, a^{-k}$. It follows from the properties of V that V contains a positive power a^m of a with $m > -k$. Then a^{m+k} is a positive power of a lying in U.

Combining Corollary 1.4 with Theorem 4.2, we arrive at the following generalization of Exercise 5.6 of Chapter 1.

COROLLARY 4.1. If the sequence (a^n) is dense in the compact group G, then (a^n) is well distributed in G.

Characterizations of Generators

In the course of the proof of Theorem 4.2 we have seen that if a is a generator of the compact monothetic group G, then $\chi(a) \neq 1$ for each nontrivial character χ of G. The converse of this statement is also true. In addition, a characterization of generators in terms of the ergodicity of a certain transformation on G can be given.

THEOREM 4.3. For a compact abelian group G and an element $a \in G$, the following properties are equivalent:

(1) The transformation T_a on G defined by $T_a x = ax$ for $x \in G$ is ergodic (with respect to Haar measure).

(2) $\chi(a) \neq 1$ for each nontrivial character χ of G.

(3) a is a generator of G.

PROOF. (1) $\Rightarrow$ (2): Let χ be a nontrivial character of G. From the individual ergodic theorem (see Chapter 3, Lemma 2.2) we get

$$\lim_{N\to\infty} \frac{1}{N} \sum_{n=0}^{N-1} \chi(a^n x) = \lim_{N\to\infty} \frac{1}{N} \sum_{n=0}^{N-1} \chi(T_a{}^n x) = \int_G \chi \, d\mu = 0$$

for μ-almost all $x \in G$. In particular, there exists $x_0 \in G$ with

$$\lim_{N\to\infty} \frac{1}{N} \sum_{n=0}^{N-1} (\chi(a))^n \chi(x_0) = 0. \tag{4.2}$$

Now suppose that $\chi(a) = 1$. Then, (4.2) implies $\chi(x_0) = 0$, which is absurd.

(2) $\Rightarrow$ (3): Let H be the closure in G of the cyclic subgroup generated by a. Then H is a closed subgroup of G. Assume $H \neq G$. Then the quotient group G/H contains elements other than the identity. By the Gel'fond-Raikov theorem (see Theorem 1.1), there exists a nontrivial character χ_H of G/H. We obtain a nontrivial character χ of G by defining $\chi(x) = \chi_H(xH)$ for $x \in G$. Now $a \in H$ and H is the identity in G/H; therefore, $\chi(a) = \chi_H(aH) = \chi_H(H) = 1$. This is a contradiction. We note for later use that the same argument applies to a locally compact abelian group G.

(3) $\Rightarrow$ (1): By Theorem 4.2 and the subsequent remarks, the sequence (a^n) is u.d. in G. We infer from Theorem 1.4 that the sequences $(a^n x)$ are u.d. in G for each $x \in G$. Thus, for every real-valued continuous function f on G

and every $x \in G$, we have

$$\lim_{N\to\infty} \frac{1}{N} \sum_{n=1}^{N} f(a^n x) = \int_G f\,d\mu.$$

We conclude that

$$\lim_{N\to\infty} \int_G \left| \frac{1}{N} \sum_{n=1}^{N} f(a^n x) - \int_G f\,d\mu \right| d\mu(x) = 0. \tag{4.3}$$

We now want to show that (4.3) remains true if f is replaced by an arbitrary Haar-integrable function g. Take $f \in \mathscr{R}(G)$ and $g \in L^1(\mu)$; then,

$$\begin{aligned}
\int_G \left| \frac{1}{N} \sum_{n=1}^{N} g(a^n x) - \int_G g\,d\mu \right| d\mu(x) \\
&\leq \frac{1}{N} \sum_{n=1}^{N} \int_G |g(a^n x) - f(a^n x)|\,d\mu(x) \\
&\quad + \int_G |g - f|\,d\mu + \int_G \left| \frac{1}{N} \sum_{n=1}^{N} f(a^n x) - \int_G f\,d\mu \right| d\mu(x) \\
&= 2\int_G |g - f|\,d\mu + \int_G \left| \frac{1}{N} \sum_{n=1}^{N} f(a^n x) - \int_G f\,d\mu \right| d\mu(x).
\end{aligned}$$

In the last step, we used the translation invariance of μ. The expression we obtained as an upper bound can be made arbitrarily small because of (4.3) and the fact that $\mathscr{R}(G)$ is dense in $L^1(\mu)$.

To show that T_a is ergodic, we consider a Borel set E in X that is left invariant by T_a (i.e., $E = aE$). For the integrable function g in the above consideration, we take now the characteristic function c_E of E. Thus,

$$\lim_{N\to\infty} \int_G \left| \frac{1}{N} \sum_{n=1}^{N} c_E(a^n x) - \mu(E) \right| d\mu(x) = 0.$$

But $c_E(a^n x) = c_E(x)$ for all $x \in G$ and all $n \geq 1$; therefore,

$$\int_G |c_E(x) - \mu(E)|\,d\mu(x) = 0.$$

This implies $c_E(x) = \mu(E)$ μ-a.e., and since c_E only attains the values 0 or 1, we conclude $\mu(E) = 0$ or 1. ■

Sequences of Powers of Generators

If G is also connected, then we can prove results concerning the u.d. of more general sequences of powers of a generator a of G. The hypothesis of connectedness stems from the fact that in a compact connected abelian group there is no nontrivial discrete character (see Corollary 1.8).

THEOREM 4.4. Let G be a compact connected monothetic group with generator a. If (r_n) is a sequence of integers such that $(r_n\alpha)$ is u.d. mod 1 for every irrational α, then the sequence (a^{r_n}) is u.d. in G.

PROOF. From the assumption on G, we infer that $\chi(G) = \boldsymbol{T}$ for every nontrivial character χ of G. For nontrivial χ, we get then $\chi(a) = e^{2\pi i\alpha}$ with irrational α, for otherwise, χ would be discrete. The Weyl criterion for u.d. mod 1 (Theorem 2.1 of Chapter 1) implies

$$\lim_{N\to\infty} \frac{1}{N} \sum_{n=1}^{N} \chi(a^{r_n}) = \lim_{N\to\infty} \frac{1}{N} \sum_{n=1}^{N} e^{2\pi i r_n\alpha} = 0.$$

By the Weyl criterion for compact abelian groups (see Corollary 1.2), the sequence (a^{r_n}) is u.d. in G. ■

COROLLARY 4.2. Let G and $a \in G$ be as in Theorem 4.4, and let $f(x)$ be a nonconstant polynomial that attains integral values on the set of positive integers. Then the sequence $(a^{f(n)})$ is u.d. in G.

PROOF. By the Lagrange interpolation formula, all the coefficients of $f(x)$ are rational. For a given irrational α, the polynomial $\alpha f(x)$ has then an irrational leading coefficient, and so, the sequence $(\alpha f(n))$ is u.d. mod 1 by Theorem 3.2 of Chapter 1. An application of Theorem 4.4 completes the proof. ■

COROLLARY 4.3. Let G and $a \in G$ be as in Theorem 4.4, and let k be a nonzero integer. Then the sequence $((a^k)^n)$ is u.d. in G. In particular, a^k is again a generator of G.

If G is not connected, then Theorem 4.4 need not hold. It suffices to disprove Corollary 4.3 in this case. The duality theory tells us that for a compact abelian G that is not connected, there exists a nontrivial discrete character χ (in particular, see Corollary 1.8). Then $\chi^m \equiv 1$ for some positive integer m. Now $\chi(a^m) = \chi^m(a) = 1$; thus, a^m cannot be a generator by Theorem 4.3.

The Measure of the Set of Generators

In connection with the notion of a generator, it is natural to ask how large the set of generators can be. In the special case $G = \boldsymbol{T}$, we know that the set of generators is the complement of a countable subset of G. A measure-theoretic analogue of this result can be proved for an important class of compact abelian groups.

THEOREM 4.5. Let G be a compact connected abelian group satisfying the second axiom of countability. Then the set of generators of G has Haar measure 1.

PROOF. Since G has a countable base, there are only countably many characters of G (by Corollary 1.7). Let $\chi_1, \chi_2, \ldots$ be the nontrivial characters of G. They are all nondiscrete, since G is connected. For $i \geq 1$, let H_i be the kernel of χ_i. Then G/H_i has infinitely many elements. Since G is the disjoint union of the distinct cosets of H_i and all the cosets of H_i have equal μ-measure (by the translation invariance of μ), we must have $\mu(H_i) = 0$. By Theorem 4.3, the set S of nongenerators of G can be written as $S = \bigcup_{i=1}^{\infty} H_i$. Therefore, $\mu(S) = 0$, and the desired result follows. ■

COROLLARY 4.4. Every compact connected abelian group satisfying the second axiom of countability is monothetic.

Another proof of the above corollary based entirely on duality theory may be given (see Exercise 4.23). We shall now prove in a simple way that if G is not connected, then the set of generators can never have measure 1. To formulate this result in a precise manner, we introduce the outer Haar measure $\bar{\mu}$, which is defined for any subset A of G by

$$\bar{\mu}(A) = \inf \{\mu(B)\colon\ B \text{ is a Borel set in } G \text{ with } B \supseteq A\}. \tag{4.4}$$

The set function $\bar{\mu}$ is nonnegative, monotone, and countably subadditive. Naturally, for a Borel set B in G the identity $\bar{\mu}(B) = \mu(B)$ holds.

THEOREM 4.6. Let G be a compact abelian group that is not connected. Then the set of generators of G has outer Haar measure less than 1.

PROOF. By duality theory, G has a nontrivial discrete character χ. The kernel H of χ is an open subgroup of G; therefore, $\mu(H) > 0$. By Theorem 4.3, the set E of generators of G satisfies $E \subseteq H'$, and thus, $\bar{\mu}(E) \leq \bar{\mu}(H') = \mu(H') < 1$. ■

Structure Theory for Locally Compact Monothetic Groups

THEOREM 4.7. The compact abelian group G is monothetic if and only if its character group $\hat{G}$ is algebraically isomorphic to a subgroup of $\boldsymbol{T}$.

PROOF. Let G be monothetic, and let a be a generator of G. We consider the mapping $\psi\colon\ \hat{G} \mapsto \boldsymbol{T}$ defined by $\psi(\chi) = \chi(a)$ for $\chi \in \hat{G}$. The mapping ψ is certainly an algebraic homomorphism from $\hat{G}$ into $\boldsymbol{T}$ (it is even a character of $\hat{G}$). But ψ is also injective by Theorem 4.3. Therefore, $\hat{G}$ is algebraically isomorphic to a subgroup of $\boldsymbol{T}$.

Conversely, suppose that there exists an algebraic isomorphism ψ from $\hat{G}$ onto a subgroup of $\boldsymbol{T}$. Since $\hat{G}$ is discrete, the mapping ψ is continuous and thus a character of $\hat{G}$. By the duality theorem, there exists an element $a \in G$ such that $\psi(\chi) = \chi(a)$ for all $\chi \in \hat{G}$. Now ψ is injective; therefore, $\psi(\chi) \neq 1$

for nontrivial $\chi \in \hat{G}$, and so, $\chi(a) \neq 1$ for all nontrivial $\chi \in \hat{G}$. Hence, a is a generator of G by Theorem 4.3. ■

It is important to note that the mapping ψ considered in the above proof will, in general, not yield a topological isomorphism from $\hat{G}$ onto a subgroup of $\boldsymbol{T}$, since the inverse mapping of ψ need not be continuous. But we obtain a topological isomorphism if we change the topology in $\boldsymbol{T}$ from the ordinary one into the discrete topology, thereby obtaining a locally compact abelian group that we shall denote by $\boldsymbol{T}_d$. Theorem 4.7 may then be restated as follows: The compact abelian group G is monothetic if and only if its character group $\hat{G}$ is topologically isomorphic to a subgroup of $\boldsymbol{T}_d$.

The duality theory allows for still another interpretation of Theorem 4.7. Let G_0 be the character group of $\boldsymbol{T}_d$. As the dual of a discrete group, G_0 is a compact abelian group. Furthermore, the character group of G_0 is $\boldsymbol{T}_d$, and hence, Theorem 4.7 implies that G_0 is monothetic. The group G_0 is the largest compact monothetic group in the following sense.

COROLLARY 4.5. A topological group G is compact monothetic if and only if G is a continuous homomorphic image of G_0.

PROOF. It is clear that every continuous homomorphic image of the compact monothetic group G_0 is again compact monothetic (note that the image of a generator under a continuous surjective homomorphism is again a generator; see also Exercises 4.3 and 4.4). Now suppose that G is a compact monothetic group. By Theorem 4.7, its character group $\hat{G}$ is topologically isomorphic to a subgroup H of $\boldsymbol{T}_d$. Hence, G, as the dual of $\hat{G}$, is topologically isomorphic to the dual of H. By Theorem 1.12, the character group of the closed subgroup H of $\boldsymbol{T}_d$ is topologically isomorphic to the quotient group $G_0/A(G_0, H)$, where $A(G_0, H)$ is the annihilator of H in G_0. In other words, G is topologically isomorphic to a quotient group of G_0, and the proof is complete. ■

EXAMPLE 4.2. Let us give a more explicit description of the group G_0. This can be done by describing all characters of $\boldsymbol{T}_d$. For this purpose, it is convenient to view $\boldsymbol{T}_d$ as the quotient group $\mathbb{R}_d/\mathbb{Z}$, where $\mathbb{R}_d$ is the additive group of reals in the discrete topology. Then, G_0 consists of all characters χ of $\mathbb{R}_d$ for which $\chi(m) = 1$ for every $m \in \mathbb{Z}$. To find all such characters, we proceed in the following way. Let $\mathbb{Q}$ denote the additive group of rationals, and let B be a fixed Hamel basis of $\mathbb{R}$ over $\mathbb{Q}$ with $1 \in B$. Thus, every real number α has a unique representation as a finite sum of the form $\alpha = \sum_{i=1}^{k} r_i b_i$ with $r_i \in \mathbb{Q}\backslash\{0\}$ and $b_i \in B$ (for $\alpha = 0$, we take $k = 0$).

We start to define a character χ of $\mathbb{R}_d/\mathbb{Z}$ by setting $\chi(1) = 1$ (as it should be) and choosing an arbitrary element $\chi(b)$ of $\boldsymbol{T}$ for each $b \in B$ with $b \neq 1$. It suffices to extend the definition of χ to numbers of the form rb with $r \in \mathbb{Q}$

and $b \in B$, since for arbitrary $\alpha = \sum_{i=1}^{k} r_i b_i$ we have then necessarily $\chi(\alpha) = \prod_{i=1}^{k} \chi(r_i b_i)$. We note that every $r \in \mathbb{Q}$ can be written in the form $r = m/n!$ with $m \in \mathbb{Z}$ and n a positive integer. Again, it will suffice to define $\chi((1/n!)b)$, for then necessarily $\chi((m/n!)b) = (\chi((1/n!)b))^m$. For fixed $b \in B$, we define $\chi((1/n!)b)$ recursively. For $n = 1$, the expression is already defined. Suppose we have already defined $\chi((1/s!)b)$ for some $s \geq 1$. Then, necessarily, we must have

$$\left(\chi\left(\frac{1}{(s+1)!}\, b\right)\right)^{s+1} = \chi\left(\frac{1}{s!}\, b\right).$$

Consequently, we will take for $\chi((1/(s+1)!)b)$ one of the $s + 1$ roots of the equation $z^{s+1} = \chi((1/s!)b)$. Let us check whether $\chi(rb)$ is then well defined. If $r = m/n! = p/q!$ with $q > n$, say, then $\chi((m/n!)b) = (\chi((1/n!)b))^m$ and $\chi((p/q!)b)) = (\chi((1/q!)b))^p$. But $p = m(n+1)(n+2)\cdots q$. Thus,

$$\begin{aligned}\left(\chi\left(\frac{1}{q!}\, b\right)\right)^p &= \left(\left(\chi\left(\frac{1}{q!}\, b\right)\right)^q\right)^{m(n+1)\cdots(q-1)} \\ &= \left(\chi\left(\frac{1}{(q-1)!}\, b\right)\right)^{m(n+1)\cdots(q-1)} = \cdots = \left(\chi\left(\frac{1}{n!}\, b\right)\right)^m.\end{aligned}$$

From the construction of the mapping χ, it is evident that χ is a character of $\mathbb{R}_d$ that maps $\mathbb{Z}$ into $\{1\}$.

We used necessary conditions for characters in each of the steps of the construction, so clearly every character of $\mathbb{R}_d/\mathbb{Z}$ has to be of the above form. In summary, a character χ of $\mathbb{R}_d/\mathbb{Z}$ is uniquely determined by the following prescribed data: (i) a collection $\{\eta^{(b)}\colon b \in B\}$ of arbitrary elements from $\boldsymbol{T}$ that serve as the $\chi(b)$, the only restriction being $\eta^{(1)} = 1$; (ii) for each $b \in B$, a sequence of elements $\xi_1^{(b)}, \xi_2^{(b)}, \ldots, \xi_n^{(b)}, \ldots$ from $\boldsymbol{T}$ (which serve as the $\chi((1/n!)b)$) with $\xi_1^{(b)} = \eta^{(b)}$ and $(\xi_{n+1}^{(b)})^{n+1} = \xi_n^{(b)}$. We get an important special case if we write $\eta^{(b)} = e^{2\pi i \delta^{(b)}}$ with $\delta^{(b)} \in \mathbb{R}$, and define $\chi(rb) = e^{2\pi i r \delta^{(b)}}$ for $r \in \mathbb{Q}$. ■

THEOREM 4.8. A locally compact monothetic group is either compact or topologically isomorphic to the discrete additive group $\mathbb{Z}$ of integers.

PROOF. Let G be a locally compact monothetic group, and let a be a generator of G. By duality theory, the mapping $\psi\colon \hat{G} \mapsto \boldsymbol{T}$, defined by $\psi(\chi) = \chi(a)$ for all $\chi \in \hat{G}$, is a character of $\hat{G}$. Furthermore, ψ is injective. For suppose $\chi(a) = 1$ for a nontrivial character χ of G; then, $\chi(a^n) = 1$ for any integer n, and so, $\chi(x) = 1$ for any $x \in G$, a contradiction. We note that $\hat{G}$ is locally compact. Thus, by the structure theorem for locally compact abelian groups (see Theorem 1.14), $\hat{G}$ may be identified with the group $\mathbb{R}^n \times H$ where H is a locally compact abelian group containing a compact open subgroup K. Evidently, the mapping ψ restricted to $\mathbb{R}^n \times \{e\}$ is an injective character. But then, $\mathbb{R}^n$ is topologically isomorphic to a closed subgroup of $\boldsymbol{T}$ (apply Theorem 1.8), which is only possible for $n = 0$. Therefore, we can identify $\hat{G}$ with H. We consider ψ on the compact group K. The image $\psi(K)$ is a closed subgroup of $\boldsymbol{T}$. By Example 1.5, we have to distinguish two cases. In the first case, $\psi(K)$ is a finite cyclic group. Since ψ is injective, K itself is a finite cyclic group. But K is open in $\hat{G}$; therefore, $\hat{G}$ has to be discrete. Then, G is compact as the dual of a discrete group. In the remaining case, we have $\psi(K) = \boldsymbol{T}$. Since ψ is injective, K is already all of $\hat{G}$. In particular, $\hat{G}$ is topologically isomorphic to $\boldsymbol{T}$. Then G, as the dual of $\hat{G}$, is topologically isomorphic to the dual of $\boldsymbol{T}$, which is $\mathbb{Z}$ (see Example 1.3). ∎

The following lemma will be of technical importance later on, but it deserves some interest of its own. In a very convincing way, it constitutes another indication that there is an abundance of generators in a reasonable monothetic group (more precisely, in a compact monothetic group that is not totally disconnected).

LEMMA 4.1. Let G be a compact monothetic group, and let χ be a nondiscrete character of G. Then, for every irrational α, there exists a generator a of G with $\chi(a) = e^{2\pi i\alpha}$.

PROOF. It suffices to prove the assertion for the group G_0 from Corollary 4.5. For suppose the result is true for G_0. Now, if G is any compact monothetic group, then G can be identified with a quotient group G_0/H. Furthermore, if χ is a nondiscrete character of G_0/H, then $\varphi(x) = \chi(xH)$ for $x \in G_0$ defines a nondiscrete character of G_0. Thus, an irrational α being given, we can find a generator x_0 of G_0 with $\varphi(x_0) = e^{2\pi i\alpha}$. But then x_0H is a generator of G_0/H with $\chi(x_0H) = e^{2\pi i\alpha}$.

To prove the lemma for G_0, let us recall that G_0 was defined as the dual of $\boldsymbol{T}_d$. By Theorem 4.3 and the duality theorem, a character $\psi \in G_0$ will be a generator of G_0 if and only if $\psi(\gamma) \neq 1$ for all $\gamma \in \boldsymbol{T}_d$ with $\gamma \neq 1$. It will again be convenient to view $\boldsymbol{T}_d$ as the quotient group $\mathbb{R}_d/\mathbb{Z}$. The given nondiscrete character of G_0 corresponds to an element of infinite order in the dual group $\boldsymbol{T}_d$ of G_0, hence, to a coset $\beta + \mathbb{Z}$ with irrational β. The assertion that we have to prove reads then as follows: For a given irrational α, there exists a character ψ of $\mathbb{R}_d$ that is equal to 1 on $\mathbb{Z}$ (in other words, an element of G_0) such that $\psi(\beta) = e^{2\pi i\alpha}$ and $\psi(\gamma) \neq 1$ for all reals γ with $\gamma \notin \mathbb{Z}$. We extend the system $\{1, \beta\}$ to a Hamel basis $B = \{\beta_j\colon j \in J\}$ of $\mathbb{R}$ over $\mathbb{Q}$ with $\beta_0 = 1$ and $\beta_1 = \beta$; we extend $\{1, \alpha\}$ to a Hamel basis $A = \{\alpha_j\colon j \in J\}$ of $\mathbb{R}$ over $\mathbb{Q}$ with $\alpha_0 = 1$ and $\alpha_1 = \alpha$ (J denotes a suitable index set). For $\gamma \in \mathbb{R}$, we have a unique representation $\gamma = r_1\beta_{j_1} + \cdots + r_k\beta_{j_k}$ with $r_i \in \mathbb{Q}\backslash\{0\}$ and distinct subscripts $j_1, \ldots, j_k$ (for $\gamma = 0$, we take $k = 0$). We define

$$\psi(\gamma) = \exp\,(r_1\alpha_{j_1} + \cdots + r_k\alpha_{j_k}).$$

Then ψ is a character of $\mathbb{R}_d$ that is equal to 1 on $\mathbb{Z}$, and also $\psi(\beta) = e^{2\pi i\alpha}$. Moreover, if $\psi(\gamma) = 1$ and $\gamma = r_0\beta_0 + r_1\beta_{j_1} + \cdots + r_k\beta_{j_k}$ with the rationals r_s possibly being zero and the j_t distinct and different from zero, then $r_0\alpha_0 + r_1\alpha_{j_1} + \cdots + r_k\alpha_{j_k} = m$ for some integer m, and so, $r_0 = m$ and $r_1 = \cdots = r_k = 0$. This implies $\gamma \in \mathbb{Z}$. ∎

Equi-uniform Distribution

The general theory of families of equi-u.d. sequences was developed in Section 3 of Chapter 3. In accordance with the mainstream of investigation in the present section, we shall consider families of sequences (a^n), where a ranges over some subset of G. Since every individual sequence in a family of equi-u.d. sequences is u.d., we may confine our attention to families of the form $\{(a^n)\colon a \in A\}$, where A is a set of generators of G. As a matter of convenience, we shall denote by E the set of all generators of the compact monothetic group G.

THEOREM 4.9. Let A be a subset of E such that $\{(a^n)\colon a \in A\}$ is a family of equi-u.d. sequences in the compact monothetic group G. Then $\{(a^n)\colon a \in \bar{A}\}$ is also a family of equi-u.d. sequences in G.

PROOF. Again we consider families of sequences in G as subsets of the product space G^∞. By Theorem 3.4 of Chapter 3, the closure $\overline{\{(a^n)\colon a \in A\}}$ in G^∞ represents a family of equi-u.d. sequences in G. The mapping $g\colon G \mapsto G^\infty$, defined by $g(x) = (x, x^2, x^3, \ldots)$ for $x \in G$, is continuous, since each of the coordinate functions is continuous. Therefore, $g(\bar{A}) \subseteq \overline{g(A)}$, or $\{(a^n)\colon a \in \bar{A}\} \subseteq \overline{\{(a^n)\colon a \in A\}}$, and the result follows. ∎

COROLLARY 4.6. Suppose G has at least two elements, and let a be a generator of G. Then $\{((a^m)^n)\colon m = 1, 2, \ldots\}$ is not a family of equi-u.d. sequences in G.

PROOF. Assume the contrary. Since the sequence $a, a^2, a^3, \ldots$ is everywhere dense in G, the preceding theorem implies that $\{(b^n)\colon b \in G\}$ is a family of equi-u.d. sequences in G. In particular, the identity e would be a generator of G, an obvious absurdity. ∎

COROLLARY 4.7. Suppose G has at least two elements, and let A be a subset of E such that $\{(a^n)\colon a \in A\}$ is a family of equi-u.d. sequences in G. Then $\mu(\bar{A}) < 1$.

PROOF. By Theorem 4.9, $\bar{A}$ is again a subset of E. If $\mu(\bar{A}) = 1$, we would infer $\bar{A} = G$, since the support of μ is G. But then $E = G$, a contradiction. ∎

If G is not connected, then Corollary 4.7 may of course also be deduced by combining Theorem 4.6 with Theorem 4.9.

In Section 3 of Chapter 3 we saw that families of equi-u.d. sequences cannot be too large. Nevertheless, it can happen that the family consisting of all the sequences (a^n) that can possibly be u.d., namely, the family $\{(a^n)\colon a \in E\}$, is a family of equi-u.d. sequences. In our next theorem, we classify those compact monothetic groups for which this phenomenon occurs. First, we need an auxiliary result.

LEMMA 4.2. Let G be a compact monothetic group. Then the set E of generators of G is closed in G if and only if G is totally disconnected.

PROOF. If G is totally disconnected, then every character of G is discrete (see Corollary 1.8). By Theorem 4.3, E is the intersection of all sets of the form $\{x \in G\colon \chi(x) \neq 1\}$ with nontrivial characters χ. But, for discrete χ, every set of this form is closed, and so, E itself is closed.

On the other hand, if G is not totally disconnected, then there exists a nondiscrete character χ of G. Assume that E is closed. Then, $\chi(E)$ is closed in $\boldsymbol{T}$. By Lemma 4.1, all the elements $e^{2\pi i\alpha}$ in $\boldsymbol{T}$ with irrational α lie in $\chi(E)$. Consequently, we must have $\chi(E) = \boldsymbol{T}$. Hence, there exists $a \in E$ with $\chi(a) = 1$, which contradicts Theorem 4.3. ■

THEOREM 4.10. Let G be a compact monothetic group. The family $\{(a^n)\colon a \in E\}$ is a family of equi-u.d. sequences in G if and only if G is totally disconnected.

PROOF. We can easily deal with the case where G is not totally disconnected. If $\{(a^n)\colon a \in E\}$ were a family of equi-u.d. sequences in G, then, by Theorem 4.9, $\{(a^n)\colon a \in \bar{E}\}$ would also be a family of equi-u.d. sequences in G. But the only sequences (b^n) in G that are u.d. are those with $b \in E$. Therefore, we infer $\bar{E} = E$, which is incompatible with Lemma 4.2.

Now, suppose that G is totally disconnected. By the complex analogue of Theorem 3.1 in Chapter 3 and the Peter-Weyl theorem (see Theorem 1.2), it suffices to show that for each nontrivial character χ of G and for each given $\varepsilon > 0$, there exists an $N_0(\varepsilon)$ independent of $a \in E$ such that

$$|(1/N)\sum_{n=1}^{N} \chi(a^n)| \leq \varepsilon$$

for all $N \geq N_0(\varepsilon)$ and for all $a \in E$. Since χ is discrete and $\chi(a) \neq 1$ for $a \in E$, there exists a positive constant $c(\chi)$ independent of $a \in E$ such that $|\chi(a) - 1| \geq c(\chi)$ for all $a \in E$. Thus, we obtain

$$\left|\frac{1}{N}\sum_{n=1}^{N}\chi(a^n)\right| = \frac{1}{N}\left|\frac{\chi(a^{N+1}) - \chi(a)}{\chi(a) - 1}\right| \leq \frac{2}{Nc(\chi)} \leq \varepsilon \qquad \text{for all } N \geq \frac{2}{\varepsilon c(\chi)}$$

and all $a \in E$. ■

For totally disconnected G, we have settled the problem of equi-u.d. of sequences (a^n) in the optimal sense. Therefore, we shall now look at the case of not totally disconnected groups. We prove a theorem that bears some resemblance to results from Section 3 of Chapter 3.

THEOREM 4.11. Let G be a compact monothetic group that is not totally disconnected, and let A be a subset of E such that $\{(a^n)\colon a \in A\}$ is a family of equi-u.d. sequences in G. Then A is nowhere dense in G.

PROOF. By Theorem 4.9, $\bar{A}$ is again a subset of E. Suppose $\bar{A}$ contains a nonempty open set in G. We take a nondiscrete character χ of G, and note that χ is an open mapping from G onto $\boldsymbol{T}$ by Corollary 1.6. Consequently, $\chi(\bar{A})$ contains a nonempty open subset of $\boldsymbol{T}$. But the continuous homomorphism χ maps generators of G into generators of $\boldsymbol{T}$; therefore, $\chi(\bar{A})$ is a subset of the set $\{\zeta \in \boldsymbol{T}\colon \zeta = e^{2\pi i \alpha}$ with irrational $\alpha\}$, which does not contain a nonempty open subset of $\boldsymbol{T}$. ■

It should be noted that the above theorem need not hold for totally disconnected groups. A trivial counterexample is provided by a finite cyclic group with the discrete topology. Such a group has an open set of generators. Less trivial examples may also be constructed (see the notes).

Notes

Monothetic groups were introduced in a footnote of a paper by van Dantzig [1] and were also mentioned by the same author in [3]. The first systematic treatment of monothetic groups was given by Halmos and Samelson [1]. Another important contribution to the subject matter is contained in Anzai and Kakutani [1]. Apparently unaware of the work of Halmos and Samelson, Eckmann [1] proved some results on monothetic groups that were already known at that time. For a fairly up-to-date account of the general theory of monothetic groups, see Hewitt and Ross [1, Sections 9, 24, and 25].

In Theorem 4.3, the equivalence of (1) and (3) was established by Halmos and von Neumann [1]. A discussion of this point may also be found in Halmos [2, pp. 27–28]. Our proof follows the argument of Hartman and Ryll-Nardzewski [1, Satz 2]. The equivalence of (2) and (3) was first observed by Halmos and Samelson [1, Section 2(c)]. It should be noted that apart from the trivial case $G = \{e\}$, the transformation T_a is not mixing (see Exercise 4.12). Several other conditions characterizing generators were given by Hartman and Ryll-Nardzewski [1, Satz 2].

The basic result concerning the structure theory of compact monothetic groups, namely, Theorem 4.7, is from Halmos and Samelson [1]. The special role of the group G_0 was pointed out by Anzai and Kakutani [1]. Because of Corollary 4.5, the group G_0 is often referred to as the universal compact monothetic group. An explicit description of G_0 similar to Example 4.2 may also be found in Hewitt and Ross [1, Section 25] and Maak [1, Section 23]. Based on a detailed group-theoretic study of $\boldsymbol{T}_d$, the following important results can be proved (see Halmos and Samelson [1] and Hewitt and Ross [1, Section 25]). The compact abelian group G with (connected) component of the identity C is monothetic if and only if its weight $\mathfrak{w}(G)$ satisfies $\mathfrak{w}(G) \leq \mathfrak{c}$ and the totally disconnected quotient group G/C is topologically isomorphic to the direct product $\prod_p A_p$, where p is running through the prime numbers and each A_p is either the trivial group $\{e\}$ or cyclic of order p^n for some positive integer n or the additive group of p-adic integers. In particular, a compact connected abelian group G is monothetic if and only if $\mathfrak{w}(G) \leq \mathfrak{c}$. The direct product $T^{\mathfrak{m}}$ of $\mathfrak{m}$ copies of T, where $\mathfrak{m}$ is some cardinal number, is monothetic if and only if $\mathfrak{m} \leq \mathfrak{c}$. Anzai and Kakutani [1] have an interesting result involving the cardinality of G: A compact connected abelian group G is monothetic if and only if its cardinality is at most $2^{\mathfrak{c}}$ (compare with Exercise 4.20).

The characterization of locally compact monothetic groups given in Theorem 4.8 was already known to Weil [1, p. 97]. An interesting example of a topological group that is monothetic but not locally compact is constructed in Anzai and Kakutani [1]. A general structure theory for not locally compact monothetic groups was developed by Nienhuys [1]. For further remarks on this subject, see Rolewicz [1].

A very general viewpoint is taken in a paper by Hartman and Hulanicki [1], who propose to determine dense sets of minimal cardinality in a topological group. Among their results, the following one is pertinent: Under the assumption of the generalized continuum hypothesis, an infinite compact abelian group of cardinality at most $2^{2^{\mathfrak{m}}}$ contains a dense subset of cardinality at most $\mathfrak{m}$. This leads to an alternative proof of the fact that a compact connected abelian group is monothetic if and only if its cardinality is at most $2^{\mathfrak{c}}$.

Theorem 4.5 is from Halmos and Samelson [1]. Earlier results in this direction were given by Schreier [1] and Schreier and Ulam [1], and by Auerbach [1] for linear groups. Halmos and Samelson [1] showed also that for the compact connected monothetic group $T^{\mathfrak{c}}$ the set of generators is not Haar measurable. In fact, the set of generators of this group has outer Haar measure equal to 1 and inner Haar measure equal to 0 (see also Hewitt and Ross [1, Section 25]). For a detailed account of outer and inner measures, we refer to Halmos [1, Chapters 2 and 3]. As to totally disconnected compact monothetic groups, we note that in a cyclic group of order p^n, p prime, $n \geq 1$, the set of generators has measure $1 - (1/p)$. It can then be easily seen that for a group of the form $\prod_p A_p$, with $A_p = \{e\}$ or cyclic of order p^n and p running through the primes, the set of generators has a measure equal to $\prod_p (1 - (1/p))$, where now p runs through all the primes with $A_p \neq \{e\}$. But this product may attain any value from the interval [0, 1]. For details, see Halmos and Samelson [1] and Hewitt and Ross [1, Section 25].

The fact that for a generator a of a compact monothetic group the sequence (a^n) is u.d. was first discovered by Eckmann [1]. For a completely different proof, see Hewitt and Ross [1, p. 437]. An interesting generalization was given by Helmberg [1], who considered compact groups with a finitely generated dense subgroup. Even more general formulations are given in Helmberg [2]. For other generalizations of Eckmann's results, see Helmberg [8] and Kuipers and Scheelbeek [1].

In conjunction with Corollary 4.1, let us note here that the characterization of a strongly regular matrix method A as a matrix method including almost convergence (see Section 4 of Chapter 3) implies the A-u.d. of (a^n) for every generator $a \in G$. This was observed by Cigler [10] but not proved in this direct manner. Theorem 4.4 is essentially from Hartman and Ryll-Nardzewski [1]. A partial converse of Theorem 4.4 reads as follows: Let G be a compact monothetic group that is not totally disconnected; if (r_n) is a sequence of integers such that (a^{r_n}) is u.d. in G for every generator a of G, then the sequence $(r_n\alpha)$ is u.d. mod 1 for every irrational α. The proof is an immediate application of Lemma 4.1 (see also Exercise 4.14). Hartman and Ryll-Nardzewski [1] prove the following generalization of Theorem 4.1 of Chapter 1: If G is a compact connected abelian group with countable base and (r_n) is an increasing sequence of integers, then the sequence (a^{r_n}) is u.d. in G for μ-almost all $a \in G$. A refinement was achieved by Stapleton [1], using the growth condition in Exercise 4.5 of Chapter 1. For a slight improvement, see Philipp [2]. Another metric result for sequences of the form (a^{r_n}) was given by Zame [2].

The results on equi-u.d., along with Lemma 4.1, are from Baayen and Helmberg [1]. The same authors also construct an example of a compact totally disconnected monothetic group with infinitely many elements and an open set of generators. A simpler example is the group $\mathbb{Z}_p$ of p-adic integers. This proves again that Theorem 4.11 cannot hold for all totally disconnected groups (simple counterexamples were already given earlier). Armacost [1], apparently unaware of the work of Baayen and Helmberg, proves some of their results on the set of generators but also has original results.

Exercises

4.1. Prove in detail that if a is a generator of the compact monothetic group G and if χ is a nondiscrete character of G, then $\chi(a) = e^{2\pi i\alpha}$ with irrational α.

4.2. Let a and G be as in the preceding exercise, but let χ now be a discrete character of G. Show that $\chi(a)$ is a generator of the finite cyclic group $\chi(G)$.

4.3. Prove more generally that if $f\colon G \mapsto G'$ is a continuous homomorphism of a compact monothetic group G onto the topological group G', then the image of a generator of G under f is a generator of G'. In particular, the continuous homomorphic image of a compact monothetic group is again compact monothetic.

4.4. Prove the last part of Exercise 4.3 by using Theorem 4.7.

4.5. Use Theorem 1.7 and Exercises 4.1 and 4.2 to give an alternative proof of Theorem 4.2.

4.6. Let G be a compact group (not necessarily abelian), and let $a \in G$. Prove that the sequence (a^n) is u.d. in G if and only if for all nontrivial irreducible representations $\mathbf{D}$ of G, the matrix $\mathbf{D}(a) - \mathbf{E}$ is nonsingular (where $\mathbf{E}$ is an identity matrix of appropriate order).

4.7. Give an alternative proof of Corollary 4.3 by going back to Theorem 4.3.

4.8. Let G be a compact abelian group with a discrete character χ of order n in $\hat{G}$. Find $\mu(H)$, where H is the kernel of χ.

4.9. Let χ be a nondiscrete character of the compact abelian group G. Let $\arg z$ denote the unique value of the argument of the complex number z lying in the interval $[0, 2\pi)$. Show that the set $A = \{x \in G\colon 0 \leq (\arg \chi(x))/2\pi < \alpha\}$, with $0 < \alpha \leq 1$, has Haar measure $\mu(A) = \alpha$. *Hint:* Show the result first for $\alpha = 1/k$ with a positive integer k; then, use an approximation argument.

4.10. Generalize Exercise 4.9 to the following theorem: If B is a Borel set in $\boldsymbol{T}$ and $A = \{x \in G\colon \chi(x) \in B\}$, then $\mu(A) = \lambda(B)$, where λ is the Haar measure in $\boldsymbol{T}$. *Hint:* Show that the set function λ_1 on the Borel sets of $\boldsymbol{T}$ defined by $\lambda_1(B) = \mu(A)$ satisfies all the properties of Haar measure in $\boldsymbol{T}$.

4.11. Let $(Y, \mathscr{F}, \nu)$ be a measure space with ν being a nonnegative normed measure. A measure-preserving transformation T of Y is called *mixing* (with respect to ν) if $\lim_{n\to\infty} \nu(A \cap T^{-n}B) = \nu(A)\nu(B)$ holds for any two sets $A, B \in \mathscr{F}$. Prove that every mixing transformation is also ergodic with respect to the same measure.

4.12. Let G be a compact abelian group having at least two elements, and let a be an arbitrary element of G. Show that the transformation T_a: $x \mapsto ax$ is not mixing with respect to Haar measure. *Hint:* If G has a nontrivial discrete character χ, take the set H from Exercise 4.8 and look at $\lim_{n\to\infty} \mu(H \cap T_a^{-n}H)$; for a nondiscrete character χ, take the set A from Exercise 4.9 with $\alpha = \frac{1}{2}$ and look at $\lim_{n\to\infty} \mu(A \cap T_a^{-n}A)$. Thus, if G is monothetic and a is a generator of G, then T_a is an ergodic transformation that is not mixing.

4.13. What is the Haar measure of the set of generators in a finite cyclic group?

4.14. Prove that if G is a compact monothetic group that is not totally disconnected and if (r_n) is a sequence of integers such that (a^{r_n}) is u.d. in G for every generator a of G, then the sequence $(r_n\alpha)$ is u.d. mod 1 for every irrational α.

4.15. Let G be a compact monothetic group, and let A be a set of generators of G. Show that $\{(a^n)\colon a \in A\}$ is a family of equi-u.d. sequences in G if and only if $\{(\chi(a^n))\colon a \in A\}$ is a family of equi-u.d. sequences in $\boldsymbol{T}$ for every fixed nondiscrete character χ of G.

4.16. Let G be a compact monothetic group. For each nondiscrete character χ of G, choose a subset $B(\chi)$ of $\boldsymbol{T}$ such that $\{(b^n)\colon b \in B(\chi)\}$ is a family of equi-u.d. sequences in $\boldsymbol{T}$. Put $A = \bigcap_\chi \{x \in G\colon \chi(x) \in B(\chi)\}$, $E =$ set of all generators of G, and show that $\{(a^n)\colon a \in A \cap E\}$ is a family of equi-u.d. sequences in G. *Hint:* Use Exercise 4.15.

4.17. Let G be a compact monothetic group satisfying the second axiom of countability. Then, the set E of all generators of G is a Borel set in G.

4.18. Combine Exercises 4.10, 4.16, and 4.17 to verify the following result: Let G be a compact monothetic group satisfying the second axiom of countability, and let $\varepsilon > 0$ be given. Then, there exists a subset A of the set E of all generators of G such that $\mu(A) > \mu(E) - \varepsilon$ and $\{(a^n)\colon a \in A\}$ is a family of equi-u.d. sequences in G. We can even find a closed set A with these properties. *Hint:* Prove the result first for $G = \mathbb{R}/\mathbb{Z}$.

4.19. Why is $\mathfrak{w}(G) \leq \mathfrak{c}$ a necessary condition for a locally compact group G to be monothetic?

4.20. Prove that for a compact abelian group G the condition $\mathfrak{w}(G) \leq \mathfrak{c}$ is equivalent to card $G \leq 2^{\mathfrak{c}}$. *Hint:* Use Corollary 1.7.

4.21. Let a be a generator of the compact monothetic group G, and let (x_n) be a sequence in G with $\lim_{n\to\infty} x_{n+1}x_n^{-1} = a$. Prove that the sequence (x_n) is well distributed in G.

4.22. Prove that the direct product of two compact monothetic groups need not be monothetic.

4.23. Give a purely group-theoretic proof, based on Theorem 4.7, of the fact that a compact connected abelian group with countable base is monothetic, by constructing an injective homomorphism from a discrete countable torsion-free abelian group H into $\boldsymbol{T}$. *Hint:* Let $h_0 = e, h_1, h_2, \ldots$ be the elements of H, and let H_j be the subgroup of H generated by $h_0, h_1, \ldots, h_j$; construct the homomorphism by extension along the chain of subgroups $H_0 \subseteq H_1 \subseteq \cdots$.

5. LOCALLY COMPACT GROUPS

Definition and Some Examples

The natural measures on a locally compact noncompact group, the Haar measures, are not finite measures anymore. Therefore, it is not feasible to define u.d. in such groups in the spirit of Theorem 1.2 of Chapter 3. New concepts have to be found that lend themselves to a meaningful treatment in the more general situation. An important restriction on such a concept will be its compatibility with the earlier notion of u.d. in compact groups; that is, the new definition should include, as a special case, the familiar notion for compact groups.

There are essentially three ideas that we shall pursue. The first one amounts to considering, instead of the given locally compact group, its various compact quotient groups in which a notion of u.d. is already available (see Definition 5.2). Secondly, we may take the Weyl criterion for compact groups as a definition of u.d. (see Definition 5.6). Thirdly, one could transfer the problem of u.d. to a natural compactification of the given group in a canonical fashion. However, in the light of certain special cases that we are going to investigate in the subsequent chapter, the first alternative seems to deserve a preferential treatment. For this very reason, the notion corresponding to the first possibility will be called u.d. *per se*, whereas the other notions will have certain prefixes attached.

Let G be an arbitrary locally compact group. We introduce an important class of subgroups of G that are the topological analogues of subgroups of finite index.

DEFINITION 5.1. A closed normal subgroup H of G is called a *subgroup of compact index* if the quotient group G/H is compact.

COROLLARY 5.1. If G is discrete, then the subgroups of compact index of G are exactly the normal subgroups of finite index.

EXAMPLE 5.1. Let $\mathbb{Z}$ be the additive group of integers in the discrete topology. Since $\mathbb{Z}$ is cyclic, its subgroups are exactly the trivial subgroup $\{0\}$ and the cyclic subgroups $n\mathbb{Z}$, where $n = 1, 2, \ldots$. Apart from $\{0\}$, all subgroups are of compact index. ■

EXAMPLE 5.2. Let $\mathbb{R}$ be the additive group of real numbers in the usual topology. We determine, first of all, the closed subgroups of $\mathbb{R}$. The argument is almost identical with the one employed in Example 1.5. To avoid repetition, we give only a brief indication. Let H be a closed subgroup of $\mathbb{R}$, and let $\mathbb{R}^+$ denote the set of positive real numbers. If $\inf(H \cap \mathbb{R}^+) = 0$, then H is a dense subgroup and thus identical with $\mathbb{R}$ itself. If $\inf(H \cap \mathbb{R}^+) = \alpha > 0$, then H is the discrete cyclic group $\alpha\mathbb{Z}$ generated by α. If $H \cap \mathbb{R}^+$ is void, then $H = \{0\}$. Among those closed subgroups, the subgroups of compact index are exactly the groups $\alpha\mathbb{Z}$ with $\alpha > 0$ and $\mathbb{R}$ itself. Note that all the compact quotient groups $\mathbb{R} / \alpha\mathbb{Z}$ are topologically isomorphic to $\mathbb{R} / \mathbb{Z}$ and hence to $\boldsymbol{T}$. ■

EXAMPLE 5.3. Recall that a group D is called *divisible* if for every $x \in D$ and every positive integer n, there exists $y \in D$, so that $y^n = x$. We claim that a divisible group D in the discrete topology does not possess a subgroup of compact index apart from D itself. For let H be a subgroup of compact (i.e., finite) index, and let m be the order of D / H. Let $x \in D$ be arbitrary; then, we can find $y \in D$ with $y^m = x$. We have $(yH)^m = H$; thus, $x \in H$, and so, $H = D$. As important special cases, we mention the additive group $\mathbb{R}_d$ of real numbers in the discrete topology and the additive group $\mathbb{Q}_d$ of rational numbers in the discrete topology. ■

DEFINITION 5.2. A sequence (x_n) in the locally compact group G is called *u.d. in* G if for every subgroup H of G of compact index, the sequence (x_nH) is u.d. in G / H.

Our first task will be to show that in the case of a compact group G, the above definition coincides with the standard one. Indeed, if (x_n) is u.d. in G in the sense of Definition 5.2, then we observe that $H = \{e\}$ is a subgroup of compact index, and so, (x_n) is u.d. in the usual sense. Conversely, if (x_n) is u.d. in G in the usual sense, then Corollary 1.5 implies that (x_nH) is u.d. in G/H for every closed normal subgroup H of G, and so, (x_n) is u.d. in G in the sense of Definition 5.2.

EXAMPLE 5.4. Let $\mathbb{R}$ be the additive group of real numbers in the usual topology. We refer to Example 5.2 for a complete description of all subgroups of $\mathbb{R}$ of compact index. We may leave out the trivial subgroup $\mathbb{R}$ from the subsequent discussion, since every sequence in $\mathbb{R}/\mathbb{R}$ is u.d. Let now (x_n) be a given sequence of real numbers. We pointed out in Example 5.2 that the quotient group $\mathbb{R}/\alpha\mathbb{Z}$ with $\alpha > 0$ is topologically isomorphic to $\mathbb{R}/\mathbb{Z}$. More explicitly, a topological isomorphism is given by the mapping ψ: $\mathbb{R}/\alpha\mathbb{Z} \mapsto \mathbb{R}/\mathbb{Z}$, defined by $\psi(\beta + \alpha\mathbb{Z}) = (\beta/\alpha) + \mathbb{Z}$ for $\beta \in \mathbb{R}$. Thus, the sequence $(x_n + \alpha\mathbb{Z})$ will be u.d. in $\mathbb{R}/\alpha\mathbb{Z}$ if and only if the sequence $((x_n/\alpha) + \mathbb{Z})$ is u.d. in $\mathbb{R}/\mathbb{Z}$ (or, in other words, if and only if (x_n/α) is u.d. mod 1).

We arrive at the following useful criterion: The sequence (x_n) is u.d. in $\mathbb{R}$ if and only if the sequence (tx_n) is u.d. mod 1 for every real number $t \neq 0$. Our standard example in the mod 1 theory, namely, the sequence $(x_n) = (n\xi)$ for irrational ξ, is not u.d. in $\mathbb{R}$, because $((1/\xi)x_n)$ is not u.d. mod 1. Nevertheless, we can construct some simple examples of u.d. sequences in $\mathbb{R}$. We know from Example 2.7 of Chapter 1 that whatever the nonzero coefficient $\beta \in \mathbb{R}$, the sequence (βn^τ) with $\tau > 0$ and $\tau \notin \mathbb{Z}$ is u.d. mod 1. It follows immediately that the sequence (βn^τ) is even u.d. in $\mathbb{R}$.

To obtain another class of u.d. sequences in $\mathbb{R}$, let us determine the polynomials $f(x) = \alpha_k x^k + \alpha_{k-1}x^{k-1} + \cdots + \alpha_0$ with real coefficients and positive degree for which the sequence $(f(n))$ is u.d. in $\mathbb{R}$. We claim that $(f(n))$ is u.d. in $\mathbb{R}$ if and only if the system $\{\alpha_k, \alpha_{k-1}, \ldots, \alpha_1\}$ has rank at least two over the rationals. For suppose this condition is satisfied; then there exist coefficients α_i and α_j of $f(x)$ with $1 \leq i < j \leq k$ that are linearly independent over the rationals. For $\alpha \neq 0$, at least one of $\alpha\alpha_i$ and $\alpha\alpha_j$ has to be irrational. Thus, the polynomial $\alpha f(x)$ satisfies the condition of Theorem 3.2 of Chapter 1, and so, $(\alpha f(n))$ is u.d. mod 1. Since α was arbitrary, the sequence $(f(n))$ is u.d. in $\mathbb{R}$. On the other hand, if the rank of $\{\alpha_k, \alpha_{k-1}, \ldots, \alpha_1\}$ over the rationals is 1, then $\alpha_i = r_i\alpha$, $1 \leq i \leq k$, for some real number $\alpha \neq 0$ and rationals r_i. Consequently, the sequence $((1/\alpha)f(n))$ is not u.d. mod 1. ■

EXAMPLE 5.5. If D is a divisible group in the discrete topology, then it follows from Example 5.3 that the only compact quotient group of D is the one-element group. Therefore, every sequence in D is u.d. In particular, every sequence in $\mathbb{R}_d$ and every sequence in $\mathbb{Q}_d$ is u.d. More generally, we may introduce the following notion: A locally compact group G is called *topologically divisible* if its only subgroup of compact index is G itself. An example of a topologically divisible nondiscrete group is the additive group of p-adic numbers. Evidently, every sequence in a topologically divisible group is u.d. ■

General Properties

We ask first whether an analogue of the useful Theorem 1.6 holds for locally compact groups as well. If we pose the question in this generality, the answer has to be negative. Consider the continuous homomorphism φ: $\mathbb{R}_d \mapsto \mathbb{R}$, defined by $\varphi(x) = x$ for every $x \in \mathbb{R}_d$. Take (x_n) as a constant sequence (or as one of the many sequences that is not u.d. in $\mathbb{R}$, for that matter); then (x_n) is u.d. in $\mathbb{R}_d$ (by Example 5.5), but $(\varphi(x_n))$ is not u.d. in $\mathbb{R}$. To remedy this situation, we have to add the hypothesis that the continuous homomorphism be open. In fact, this condition is implicit in Theorem 1.6 as well, since a continuous homomorphism between compact groups is automatically open (see Theorem 1.8).

THEOREM 5.1. Let φ be a continuous open homomorphism from the locally compact group G onto the locally compact group G_1. If (x_n) is u.d. in G, then $(\varphi(x_n))$ is u.d. in G_1.

PROOF. Consider an arbitrary subgroup H_1 of G_1 of compact index. Let H be the inverse image of H_1 under φ; then H is a closed normal subgroup of G. The mapping ψ: $G/H \mapsto G_1/H_1$, given by $\psi(xH) = \varphi(x)H_1$, is easily seen to be well defined and is certainly a surjective homomorphism. It is also immediate that ψ is injective. Let us verify that ψ is an open mapping. We recall that an open set in G/H is of the form $\{xH\colon x \in U\}$ for some open set U in G. But $\psi(\{xH\colon x \in U\}) = \{yH_1\colon y \in \varphi(U)\}$, which is open in G_1/H_1, since $\varphi(U)$ is open in G_1. Now ψ being open implies that the inverse mapping ψ^{-1}: $G_1/H_1 \mapsto G/H$ is continuous. In particular, G/H is compact. But ψ^{-1}, as a continuous homomorphism between compact groups, is open, and so, ψ is continuous. We have (x_nH) u.d. in G/H; thus, $(\psi(x_nH))$—that is, $(\varphi(x_n)H_1)$—is u.d. in G_1/H_1 (by Theorem 1.6), and we are done. The attentive reader will have observed that we essentially worked with the second isomorphism theorem for topological groups. ■

Since u.d. in locally compact groups was defined in terms of u.d. in certain compact groups, many of the general properties listed in Sections 1 and 2 will carry over to the present case. We indicate two of them. The proofs are immediate.

THEOREM 5.2. Let (x_n) be a u.d. sequence in the locally compact group G, and suppose that (c_n) is a sequence in G such that $\lim_{n\to\infty} c_n$ exists. Then the sequences (c_nx_n) and (x_nc_n) are u.d. in G.

PROOF. For a closed normal subgroup H of G, the canonical mapping from G onto G/H is continuous. Therefore, $\lim_{n\to\infty} c_nH$ exists in G/H. The assertion follows then from Definition 5.2 and Theorem 1.4. ■

THEOREM 5.3. Let (x_n) be a sequence in the locally compact group G. If $(x_{n+h}x_n{}^{-1})$ is u.d. in G for every $h = 1, 2, \ldots$, then (x_n) itself is u.d. in G.

PROOF. This follows from Definition 5.2 and Theorem 2.1. ■

Periodic Functions and Periodic Representations

We recall that u.d. in a compact group may be characterized as follows: If M is a compact group and ν is the Haar measure in M, then a sequence (y_n) in M is u.d. in M if and only if $\lim_{N\to\infty} (1/N)\sum_{n=1}^{N} f(y_n) = \int_M f\,d\nu$ holds for all $f \in \mathscr{C}(M)$. If we apply this criterion to the locally compact group G, we have to consider continuous complex-valued functions on compact quotient groups G/H. Thus, let H be a subgroup of G of compact index, and let f_H be a continuous complex-valued function on G/H. The function f_H can be viewed as a continuous complex-valued function on G in a canonical manner. Namely, we define a function f on G by $f(x) = f_H(xH)$ for $x \in G$. Then f has the above mentioned properties. The characteristic feature of such a function f is the fact that f is constant on the left cosets of H. This observation suggests the following definition.

DEFINITION 5.3. A complex-valued function f on the locally compact group G is called *periodic* if f is constant on the left cosets of some subgroup of G of compact index. More generally, a mapping ψ from G into a set S is called *periodic* if there exists a subgroup H of G of compact index such that $\psi(x) = \psi(y)$ whenever $x^{-1}y \in H$ (or, equivalently, ψ is constant on the left cosets of H).

EXAMPLE 5.6. This notion of periodicity is a natural generalization of the standard notion of periodicity for functions defined on $\mathbb{R}$. For if f is a function on $\mathbb{R}$ with period $\alpha > 0$, then f is constant on the cosets of the subgroup $\alpha\mathbb{Z}$ of compact index. Conversely, if f is periodic in the sense of Definition 5.3, then, by Example 5.2, f is either a constant function or constant on the cosets of $\alpha\mathbb{Z}$ for some $\alpha > 0$. In other words, f is periodic in the standard sense.

In the case where G is topologically divisible, the periodic functions on G are precisely the constant functions. ■

We remark that if f is a continuous periodic function on G—say, f is constant on the left cosets of the subgroup H of G of compact index—then f may be viewed as a continuous function on G/H by considering the function f_H defined by $f_H(xH) = f(x)$. We now define an integral $\int f\,d\mu$ of f by setting $\int f\,d\mu = \int_{G/H} f_H\,d\mu_H$, where μ_H is the Haar measure on G/H. Of course, we must show that $\int f\,d\mu$ is well defined in this manner.

LEMMA 5.1. The integral $\int f\,d\mu$ is well defined.

PROOF. Suppose that K is another subgroup of G of compact index so that f is constant on the left cosets of K. We wish to show, first of all, that f is then constant on the left cosets of the normal subgroup HK of G. We note that for all $x \in G$, we have $f(xh) = f(x)$ for each $h \in H$ and $f(xk) = f(x)$ for each $k \in K$. If the elements y and z of G lie in the same left coset of HK, then $y = zhk$ for some elements $h \in H$ and $k \in K$. But then $f(y) = f(zhk) = f(zh) = f(z)$, and out first assertion is proved.

The continuity of f implies that f is also constant on the left cosets of the closed normal subgroup $L = \overline{HK}$. Now L is again a subgroup of compact index. To see this, we consider the mapping ψ: $G/H \mapsto G/L$ defined by $\psi(xH) = xL$ for $x \in G$. This mapping is well defined, since H is contained in L. An open set in G/L is of the form $\{xL\colon x \in U\}$ for some open set U in G. But $\psi^{-1}(\{xL\colon x \in U\}) = \{xH\colon x \in UL\}$, which is an open set in G/H. Therefore, G/L is compact as the continuous image of a compact space. Our goal will be achieved once we verify that

$$\int_{G/H} f_H \, d\mu_H = \int_{G/L} f_L \, d\mu_L \qquad \text{and} \qquad \int_{G/K} f_K \, d\mu_K = \int_{G/L} f_L \, d\mu_L.$$

It suffices to show the first identity, the proof of the second one is analogous. It is important to observe that the set function ν_L, defined for the Borel sets B of G/L by $\nu_L(B) = \mu_H(\psi^{-1}(B))$, has exactly the same properties as the Haar measure μ_L of G/L, and therefore, the uniqueness of the Haar measure implies $\mu_L(B) = \mu_H(\psi^{-1}(B))$ for all Borel sets B of G/L. By using the fact that the function f_H is identical with the composite function $f_L \circ \psi$, we obtain

$$\int_{G/H} f_H \, d\mu_H = \int_{G/H} (f_L \circ \psi) \, d\mu_H = \int_{G/L} f_L \, d\mu_L,$$

which is the desired result. ■

THEOREM 5.4. The sequence (x_n) in the locally compact group G is u.d. in G if and only if

$$\lim_{N\to\infty} \frac{1}{N} \sum_{n=1}^{N} f(x_n) = \int f \, d\mu \tag{5.1}$$

holds for all continuous periodic functions f on G.

PROOF. Let (x_n) be u.d. in G, and let f be a continuous complex-valued function that is constant on the left cosets of the subgroup H of G of compact index. As remarked earlier, the function f_H on G/H, defined by $f_H(xH) = f(x)$ for $x \in G$, is continuous. Since the sequence (x_nH) is u.d. in G/H, we conclude that

$$\lim_{N\to\infty} \frac{1}{N} \sum_{n=1}^{N} f(x_n) = \lim_{N\to\infty} \frac{1}{N} \sum_{n=1}^{N} f_H(x_nH) = \int_{G/H} f_H \, d\mu_H = \int f \, d\mu.$$

The converse is shown by similar arguments. ■

DEFINITION 5.4. A representation of a locally compact group G that is a periodic mapping is called a *periodic representation*. Similarly, a character of a locally compact abelian group that is a periodic function is called a *periodic character*.

LEMMA 5.2. A representation $\mathbf{D}$ of a locally compact group G is periodic if and only if the kernel of $\mathbf{D}$ is a subgroup of G of compact index.

PROOF. Since $\mathbf{D}$ is constant on the cosets of its kernel, the condition is certainly sufficient. On the other hand, if $\mathbf{D}$ is constant on the left cosets of the subgroup H of G of compact index, then, in particular, we have $\mathbf{D}(h) = \mathbf{D}(e)$ for every $h \in H$. Thus, H is contained in the kernel of $\mathbf{D}$, and by the reasoning in the proof of Lemma 5.1 we see that the kernel of $\mathbf{D}$ is a subgroup of compact index. ■

THEOREM 5.5. The sequence (x_n) in the locally compact group G is u.d. in G if and only if

$$\lim_{N\to\infty} \frac{1}{N} \sum_{n=1}^{N} \mathbf{D}(x_n) = \mathbf{0} \tag{5.2}$$

holds for every nontrivial periodic irreducible unitary representation $\mathbf{D}$ of G.

PROOF. If (x_n) is u.d. in G and $\mathbf{D}$ is a representation of the above form with kernel H, then, $\mathbf{D}_H(xH) = \mathbf{D}(x)$ for $x \in G$ defines a nontrivial irreducible unitary representation $\mathbf{D}_H$ of the compact group G/H. Since (x_nH) is u.d. in G/H, an application of Theorem 1.3 yields

$$\lim_{N\to\infty} \frac{1}{N} \sum_{n=1}^{N} \mathbf{D}(x_n) = \lim_{N\to\infty} \frac{1}{N} \sum_{n=1}^{N} \mathbf{D}_H(x_nH) = \mathbf{0}.$$

The converse is shown by similar arguments. ■

COROLLARY 5.2. The sequence (x_n) in the locally compact abelian group G is u.d. in G if and only if $\lim_{N\to\infty} (1/N) \sum_{n=1}^{N} \chi(x_n) = 0$ holds for all nontrivial periodic characters χ of G.

Compactifications

It will be a standing hypothesis for the remainder of this section that G is a locally compact abelian group.

We shall establish a connection between u.d. in G and u.d. in certain compact abelian groups. We consider the character group $\hat{G}$ of G, and denote by $\hat{G}^p$ the set of periodic characters of G. It is important to note that, in general, $\hat{G}^p$ is not a subgroup of $\hat{G}$. An example is given below.

EXAMPLE 5.7. Let G be the direct product $G = \mathbb{R} \times \mathbb{Z}$. Since $G = \bigcup_{m=1}^{\infty} ([-m, m] \times \{-m, -m+1, \ldots, m\})$, the space G is σ-compact. For a given irrational α, consider the characters χ_1 and χ_2 of G given by $\chi_1((r, z)) = e^{2\pi i(\alpha z - r)}$ and $\chi_2((r, z)) = e^{2\pi i r}$ for $(r, z) \in \mathbb{R} \times \mathbb{Z}$. Both χ_1 and χ_2 are clearly onto $\boldsymbol{T}$, and so, the isomorphism theorem (see Theorems 1.8 and 1.9) implies that $G/(\text{kernel } \chi_1)$ and $G/(\text{kernel } \chi_2)$ are both topologically isomorphic to $\boldsymbol{T}$. Then, χ_1 and χ_2 are periodic by Lemma 5.2. Now

$$(\chi_1 \chi_2)((r, z)) = e^{2\pi i \alpha z},$$

and the kernel of $\chi_1\chi_2$ is $\mathbb{R} \times \{0\}$. But $G/(\mathbb{R} \times \{0\})$ is topologically isomorphic to the noncompact group $\mathbb{Z}$. Therefore, $\chi_1\chi_2$ is not a periodic character of G. ∎

Having learned to be careful, we will rather look at the (algebraic) subgroup of $\hat{G}$ generated by $\hat{G}^p$. If we furnish this subgroup of $\hat{G}$ with the discrete topology (which, in general, will not be the same as the relative topology), then its dual is a compact abelian group, the so-called *periodic compactification $\bar{G}^p$* of G. More generally, if Γ is an arbitrary subgroup of $\hat{G}$ with the discrete topology, then its compact dual is called a *compactification* of the original group G. If we take for Γ the group $\hat{G}$ itself, then we arrive at the well-known *Bohr compactification $\bar{G}$* of G. By Theorem 1.12, any compactification is a quotient group of the Bohr compactification. The following lemma provides a justification for the use of the term *compactification.*

LEMMA 5.3. Let $K = \hat{\Gamma}$ be an arbitrary compactification of the locally compact abelian group G. Then there exists a natural continuous homomorphism φ from G into K such that $\varphi(G)$ is dense in K. Furthermore, φ is injective if and only if the subgroup Γ of $\hat{G}$ separates points; that is, if for any two distinct points x and y in G, there exists $\chi \in \Gamma$ with $\chi(x) \neq \chi(y)$.

PROOF. The mapping φ: $G \mapsto K$ is constructed in the following way. For a given $x \in G$, the function $\hat{x}$ on $\hat{G}$ defined by $\hat{x}(\chi) = \chi(x)$ for $\chi \in \hat{G}$, is a character of $\hat{G}$. The restriction of $\hat{x}$ to Γ, which we shall denote by $\tilde{x}$, is then clearly a character of Γ and hence an element of K; we define now $\varphi(x) = \tilde{x}$. Evidently, the mapping φ is a homomorphism. Therefore, it suffices to show the continuity of φ at the identity element $e \in G$. We note that the topology in K is the compact-open topology, and that the compact subsets of Γ are precisely the finite subsets. Thus, an open neighborhood from the base of open neighborhoods of the identity element in K has the form $A = \{k \in K$: $|k(\chi_i) - 1| < \varepsilon$ for $1 \leq i \leq n\}$ with $\varepsilon > 0$ and $\chi_1, \ldots, \chi_n \in \Gamma$. These χ_i are characters of G; therefore, the set $U = \{x \in G$: $|\chi_i(x) - 1| < \varepsilon$ for $1 \leq i \leq n\}$ is an open neighborhood of $e \in G$. Once we have shown $\varphi(U) \subseteq A$, the continuity of φ is established. But for any $x \in U$ we get $|\hat{x}(\chi_i) - 1| < \varepsilon$

for $1 \leq i \leq n$, and so, $|\tilde{x}(\chi_i) - 1| < \varepsilon$ for $1 \leq i \leq n$. In other words, for any $x \in U$ we have $\tilde{x} \in A$.

Suppose that $\varphi(G)$ is not dense in K. Then $\overline{\varphi(G)}$ is a proper closed subgroup of K. Using the fact that $K/\overline{\varphi(G)}$ admits a nontrivial character and that every character of a quotient group of K can be viewed as a character of K, we arrive at the following situation: There exists a nontrivial character τ of K with $\tau(\overline{\varphi(G)}) = \{1\}$. Since $K = \hat{\Gamma}$, an application of the duality theorem tells us that there exists a nontrivial character $\chi \in \Gamma$ such that $\tau(k) = k(\chi)$ for all $k \in K$. But then, for all $x \in G$, we have $1 = \tau(\tilde{x}) = \tilde{x}(\chi) = \hat{x}(\chi) = \chi(x)$, an obvious contradiction to χ being nontrivial.

To prove the last assertion, let H be the kernel of φ and let η denote the identity in K. The homomorphism φ is injective if and only if $H = \{e\}$. We have the following chain of equivalences: $x \in H \Leftrightarrow \varphi(x) = \eta \Leftrightarrow \hat{x}(\chi) = 1$ for all $\chi \in \Gamma \Leftrightarrow \chi(x) = 1$ for all $\chi \in \Gamma$. If H contains an element $x \neq e$, then Γ would not separate x from e. On the other hand, if Γ does not separate the distinct elements x and y, then $\chi(x^{-1}y) = 1$ for all $\chi \in \Gamma$, and $x^{-1}y$ would be an element of H distinct from e. ■

It should be remarked that a converse of the above lemma also holds true. Namely, whenever K is a compact abelian group and φ is a continuous homomorphism from G into K with dense image, then K is topologically isomorphic to a compactification of G. Let $\hat{K}$ be the discrete character group of K. Consider the mapping σ: $\psi \in \hat{K} \mapsto \psi^* \in \hat{G}$, where ψ^* is the character of G given by $\psi^*(x) = \psi(\varphi(x))$ for $x \in G$. The mapping σ is not only a continuous homomorphism, but σ is also injective. For if $\psi_1^* = \psi_2^*$ for $\psi_1, \psi_2 \in \hat{K}$, then the characters ψ_1 and ψ_2 are identical on the dense subgroup $\varphi(G)$, and so, $\psi_1 = \psi_2$. If we furnish the image Γ of σ with the discrete topology, then $\hat{K}$ is even topologically isomorphic to Γ. But then K is topologically isomorphic to $\hat{\Gamma}$, which is a compactification of G.

LEMMA 5.4. Let Γ be a subgroup of $\hat{G}$ with the discrete topology, and let $K = \hat{\Gamma}$ be the corresponding compactification of G. Let φ: $G \mapsto K$ be the mapping constructed in Lemma 5.3, and suppose that (x_n) is a sequence in G. Then $(\varphi(x_n))$ is u.d. in K if and only if $\lim_{N\to\infty} (1/N) \sum_{n=1}^{N} \chi(x_n) = 0$ holds for all nontrivial characters $\chi \in \Gamma$.

PROOF. By Corollary 1.2, the sequence $(\varphi(x_n))$ is u.d. in K if and only if $\lim_{N\to\infty} (1/N) \sum_{n=1}^{N} \psi(\varphi(x_n)) = 0$ holds for every nontrivial character ψ of K. By duality theory, the nontrivial characters of K are precisely the functions $\hat{\chi}$ for some nontrivial character $\chi \in \Gamma$, where as usual $\hat{\chi}(\tau) = \tau(\chi)$ for all $\tau \in K$. With the notation employed in the proof of Lemma 5.3, we write $\varphi(x_n) = \tilde{x}_n$. Then $\hat{\chi}(\varphi(x_n)) = \hat{\chi}(\tilde{x}_n) = \tilde{x}_n(\chi) = \hat{x}_n(\chi) = \chi(x_n)$, and the proof is complete. ■

THEOREM 5.6. Let $\bar{G}^p$ be the periodic compactification of the locally compact abelian group G, and let $\varphi\colon G \mapsto \bar{G}^p$ be the natural mapping constructed in Lemma 5.3. Suppose that (x_n) is a sequence in G such that $(\varphi(x_n))$ is u.d. in $\bar{G}^p$. Then (x_n) is u.d. in G.

PROOF. We apply Lemma 5.4 with Γ being the subgroup of $\hat{G}$ generated by $\hat{G}^p$, the set of periodic characters. The assumption that $(\varphi(x_n))$ is u.d. in $\bar{G}^p = \hat{\Gamma}$ implies, in particular, that $\lim_{N\to\infty} (1/N) \sum_{n=1}^N \chi(x_n) = 0$ holds for all nontrivial periodic characters χ of G. By Corollary 5.2, the sequence (x_n) is then u.d. in G. ∎

Under what circumstances is the converse of this theorem true? Obviously, if G admits no u.d. sequence at all, then the converse holds trivially. In the interesting case where G admits u.d. sequences, it turns out that the validity of the converse of Theorem 5.6 is equivalent to $\hat{G}^p$ being a subgroup of $\hat{G}$. In particular, if $\hat{G}^p$ is a subgroup of $\hat{G}$, then we arrive at the following criterion: (x_n) is u.d. in G if and only if $(\varphi(x_n))$ is u.d. in $\bar{G}^p$.

THEOREM 5.7. Let G be a locally compact abelian group that admits a u.d. sequence. Then the following two conditions are equivalent:

i. The set $\hat{G}^p$ of all periodic characters of G forms a subgroup of $\hat{G}$.

ii. The sequence $(\varphi(x_n))$ is u.d. in the periodic compactification $\bar{G}^p$ of G, whenever (x_n) is u.d. in G.

PROOF. The implication (i) $\Rightarrow$ (ii) follows immediately from Corollary 5.2 and Lemma 5.4. The implication (ii) $\Rightarrow$ (i) requires more work. We assume that $\hat{G}^p$ is not a subgroup of $\hat{G}$. Let Γ be the subgroup of $\hat{G}$ generated by $\hat{G}^p$; then $\hat{G}^p$ is a proper subset of Γ. We have to construct a u.d. sequence (x_n) in G such that $(\varphi(x_n))$ is not u.d. in $\bar{G}^p$. In terms of characters, we search for a sequence (x_n) in G such that $\lim_{N\to\infty} (1/N) \sum_{n=1}^N \chi(x_n) = 0$ for all nontrivial $\chi \in \hat{G}^p$ but such that $\lim_{N\to\infty} (1/N) \sum_{n=1}^N \chi_1(x_n) = 0$ does not hold for some $\chi_1 \in \Gamma \backslash \hat{G}^p$.

By assumption, there exists a sequence (y_n) in G such that

$$\lim_{N\to\infty} \frac{1}{N} \sum_{n=1}^{N} \chi(y_n) = 0$$

holds for all nontrivial $\chi \in \hat{G}^p$. It might well be that (y_n) is already the sequence we are looking for. If this is not so, then $\lim_{N\to\infty} (1/N) \sum_{n=1}^N \chi(y_n) = 0$ has to hold for all nontrivial $\chi \in \Gamma$. We shall now prove a bit more than we originally intended to. Namely, given a character $\chi_1 \in \Gamma\backslash\hat{G}^p$, we can rearrange and repeat the elements y_n so as to produce a sequence (x_n) in G satisfying $\lim_{N\to\infty} (1/N) \sum_{n=1}^N \chi(x_n) = 0$ for all nontrivial characters $\chi \in \hat{G}^p$, and

$$\lim_{N\to\infty} \frac{1}{N} \sum_{n=1}^{N} \chi_1(x_n) \neq 0.$$

This is done in the following way: Let f be the real-valued function on G defined by $f(x) = 1 + \frac{1}{2}(\chi_1(x) + \overline{\chi_1(x)})$ for $x \in G$. Then $f(x) \geq 0$ for all $x \in G$. For $j \geq 1$ and $k \geq 1$, we define the integer $p_{jk} = [kf(y_j)]$. Then $|f(y_j)/k - p_{jk}/k^2| < 1/k^2$, and so, $\sum_{j=1}^{k} |f(y_j)/k - p_{jk}/k^2| < 1/k$. Put $A_k = \sum_{j=1}^{k} p_{jk}$. We describe the sequence (x_n) blockwise, the first A_1 terms constituting the first block, the next A_2 terms constituting the second block, and so on. The kth block will be the finite sequence consisting of the A_k terms $y_1, \ldots, y_1, y_2, \ldots, y_2, \ldots, y_k, \ldots, y_k$, where y_j occurs p_{jk} times for $j = 1, 2, \ldots, k$. To simplify the notation, let us label the terms of the kth block by $z_1^{(k)}, z_2^{(k)}, \ldots, z_{A_k}^{(k)}$. Now choose an arbitrary character $\chi \in \Gamma$. We have

$$\left|\frac{1}{k^2}\sum_{\nu=1}^{A_k}\chi(z_\nu^{(k)}) - \frac{1}{k}\sum_{j=1}^{k} f(y_j)\chi(y_j)\right| = \left|\frac{1}{k^2}\sum_{j=1}^{k} p_{jk}\chi(y_j) - \frac{1}{k}\sum_{j=1}^{k} f(y_j)\chi(y_j)\right|$$

$$= \left|\sum_{j=1}^{k}\chi(y_j)\left(\frac{p_{jk}}{k^2} - \frac{f(y_j)}{k}\right)\right| < \frac{1}{k}.$$

Therefore,

$$\lim_{k\to\infty}\frac{1}{k^2}\sum_{\nu=1}^{A_k}\chi(z_\nu^{(k)}) = \lim_{k\to\infty}\frac{1}{k}\sum_{j=1}^{k} f(y_j)\chi(y_j)$$

$$= \lim_{k\to\infty}\frac{1}{k}\sum_{j=1}^{k}\chi(y_j) + \frac{1}{2}\lim_{k\to\infty}\frac{1}{k}\sum_{j=1}^{k}(\chi_1\chi)(y_j)$$

$$+ \frac{1}{2}\lim_{k\to\infty}\frac{1}{k}\sum_{j=1}^{k}(\bar{\chi}_1\chi)(y_j). \tag{5.3}$$

Let us abbreviate the latter sum by $\alpha(\chi)$. Taking for χ the trivial character, we deduce that $\lim_{k\to\infty} A_k/k^2 = 1$ (in particular, $A_k > 0$ for sufficiently large k). This limit relation, together with (5.3), implies that

$$\lim_{k\to\infty}\frac{1}{A_k}\sum_{\nu=1}^{A_k}\chi(z_\nu^{(k)}) = \alpha(\chi) \qquad \text{for all } \chi \in \Gamma.$$

Furthermore, since A_k is of the order of magnitude k^2, we have

$$\lim_{k\to\infty} A_{k+1}/(A_1 + \cdots + A_k) = 0.$$

By Lemma 4.1 of Chapter 2, we obtain

$$\lim_{N\to\infty}\frac{1}{N}\sum_{n=1}^{N}\chi(x_n) = \alpha(\chi) \qquad \text{for all } \chi \in \Gamma.$$

It remains to compute $\alpha(\chi)$. For a nontrivial character $\chi \in \hat{G}^p$, we note that both $\chi_1\chi$ and $\bar{\chi}_1\chi$ are nontrivial, for otherwise, χ_1 would be periodic. Thus,

$\alpha(\chi) = 0$ for such χ, as it should be. On the other hand, we get $\alpha(\chi_1) = \frac{1}{2}\lim_{k\to\infty}(1/k)\sum_{j=1}^{k}\chi_1^2(y_j) + \frac{1}{2}$, which is either 1 or $\frac{1}{2}$, depending on whether χ_1^2 is trivial or nontrivial (in fact, it is not hard to see that χ_1^2 has to be nontrivial). At any rate, we have shown the desired result. ∎

EXAMPLE 5.8. Let $\mathbb{R}$, as usual, be the additive group of real numbers in the ordinary topology. The group $\mathbb{R}$ is self-dual; that is, the dual group of $\mathbb{R}$ is topologically isomorphic to $\mathbb{R}$ itself. The characters of $\mathbb{R}$ are exactly the functions χ_α with $\alpha \in \mathbb{R}$, where $\chi_\alpha(x) = e^{2\pi i\alpha x}$ for $x \in \mathbb{R}$. We observe that χ_0 is the trivial character and that for $\alpha \neq 0$, the kernel of χ_α is the subgroup $(1/\alpha)\mathbb{Z}$ of $\mathbb{R}$ of compact index. Thus, every character of $\mathbb{R}$ is periodic. In particular, the periodic characters form a group, and the periodic compactification of $\mathbb{R}$ is identical with the Bohr compactification $\overline{\mathbb{R}}$ of $\mathbb{R}$. From the previous theorems, we conclude that (x_n) is u.d. in $\mathbb{R}$ if and only if $(\varphi(x_n))$ is u.d. in $\overline{\mathbb{R}}$. How can we describe the Bohr compactification in this case? By definition, $\overline{\mathbb{R}}$ is the compact dual of $\mathbb{R}_d$. Thus, $\overline{\mathbb{R}}$ consists of all homomorphisms (continuous or not) of $\mathbb{R}$ into T. As to the explicit form of $\overline{\mathbb{R}}$, we may refer to an earlier example. Namely, the elements of $\overline{\mathbb{R}}$ are constructed in exactly the same way as in Example 4.2 if we just abolish the condition $\chi(1) = 1$, which was required there. It follows from Theorem 1.12 that $\overline{\mathbb{R}}$ contains the universal compact monothetic group G_0 as a closed subgroup. The mapping φ: $\mathbb{R} \mapsto \overline{\mathbb{R}}$ has the explicit form $\varphi(\alpha) = \chi_\alpha$ for $\alpha \in \mathbb{R}$. ∎

EXAMPLE 5.9. For $G = \mathbb{Z}$, the characters are the functions χ_t with $t \in \mathbb{R}$, where $\chi_t(m) = e^{2\pi itm}$ for $m \in \mathbb{Z}$ (the number t is, of course, only relevant mod 1). If t is irrational, then the kernel of χ_t is just $\{0\}$. For rational t, the kernel of χ_t is a subgroup of compact index. Thus, the periodic characters of $\mathbb{Z}$ are exactly those corresponding to rational values of t. Consequently, the periodic characters form a group in this case. More conveniently, the dual of $\mathbb{Z}$ can be viewed as $\mathbb{R}/\mathbb{Z}$, which is topologically isomorphic to T. Therefore, the group of periodic characters of $\mathbb{Z}$ can be regarded as the subgroup U of T consisting of all roots of unity. The Bohr compactification $\overline{\mathbb{Z}}$ of $\mathbb{Z}$ is a group that we already know, for $\overline{\mathbb{Z}}$ is the dual of T_d, which is nothing else but the universal compact monothetic group G_0. The periodic compactification $\overline{\mathbb{Z}}^p$ of $\mathbb{Z}$, being a quotient group of the Bohr compactification, will thus be monothetic by Corollary 4.5. The group $\overline{\mathbb{Z}}^p$ is commonly referred to as the universal monothetic Cantor group G_1. To give some idea of what G_1 looks like, we first offer an alternative description of the group U. To be sure, let us emphasize once again that if we furnish U with the discrete topology, then its dual will be G_1.

For a fixed prime p, we define the so-called quasi-cyclic group $\mathbb{Z}(p^\infty)$ as the subgroup of T consisting of all the numbers of the form $e^{2\pi ikp^{-n}}$, with n a nonnegative and k an arbitrary integer. We claim that U is isomorphic to the weak direct product $\prod_p^* \mathbb{Z}(p^\infty)$, where p runs through all the primes. For typographic convenience, we write exp (α) instead of $e^{2\pi i\alpha}$ with $\alpha \in \mathbb{R}$. Let ξ be an arbitrary element of U, say, $\xi = \exp(r/m)$ with $m > 0$ but r/m not necessarily in reduced form. Furthermore, let $p_1, p_2, \ldots$ be the primes in ascending order. Then we have $m = \prod_{i=1}^{\infty} p_i^{e_i}$ with $e_i \geq 0$, and at most finitely many e_i are positive. For each $i \geq 1$, let a_i be a solution of the linear congruence $x \prod_{\substack{j=1 \\ j \neq i}}^{\infty} p_j^{e_j} \equiv r \pmod{p_i^{e_i}}$. We define a mapping $\tau\colon U \mapsto \prod_p^* \mathbb{Z}(p^\infty)$ by $\tau(\xi) = (\exp(a_1/p_1^{e_1}), \ldots, \exp(a_i/p_i^{e_i}), \ldots)$. It is a straightforward exercise in elementary number theory to show that $\tau(\xi)$ is well defined and that τ is an isomorphism of U onto $\prod_p^* \mathbb{Z}(p^\infty)$.

If both groups are considered in the discrete topology, then τ is even a topological isomorphism. The compact group $\mathbb{Z}_p$ of p-adic integers has the discrete group $\mathbb{Z}(p^\infty)$ as its dual (see Exercise 2.3 of Chapter 5). By the duality theorem, the dual of $\mathbb{Z}(p^\infty)$ is then just $\mathbb{Z}_p$. An application of Theorem 1.15 yields that G_1 is topologically isomorphic to the direct product $\prod_p \mathbb{Z}_p$. ■

Monogenic Groups

We found out that for every generator a of a compact monothetic group, the sequence (a^n) is u.d. (see Theorem 4.2). This clearly holds as well for $\mathbb{Z}$, the only locally compact noncompact monothetic group. These results suggest the following definition.

DEFINITION 5.5. A locally compact abelian group G is called *monogenic* if there exists an element $a \in G$ such that the sequence (a^n) is u.d. in G. The element a is called a *monogenic generator of G*.

EXAMPLE 5.10. By the brief discussion preceding Definition 5.5, every locally compact monothetic group is monogenic. For compact abelian groups, "monothetic" and "monogenic" are equivalent concepts. There are many monogenic groups that are not monothetic. For instance, every topologically divisible abelian group G (see Example 5.5) is evidently monogenic, whereas if G has more than one element, then G cannot be monothetic (for otherwise, G would have to be either compact or topologically isomorphic to $\mathbb{Z}$, and both alternatives are incompatible with G being topologically divisible). A word of warning is in order. A continuous homomorphic image of a monogenic group need not be monogenic again (compare with Exercise 4.3). Just take the group $\mathbb{R}_d$, which is monogenic by Example 5.5. The mapping $\tau\colon \mathbb{R}_d \mapsto \mathbb{R}$ with $\tau(x) = x$ for $x \in \mathbb{R}_d$ is certainly a continuous homomorphism. But $\mathbb{R}$ is not monogenic, since no sequence of the form $(n\alpha)$, $\alpha \in \mathbb{R}$, is u.d. in $\mathbb{R}$ (see Example 5.4). However, it follows from Theorem 5.1 that if

φ: $G \mapsto G_1$ is a continuous open homomorphism from the monogenic group G onto the locally compact group G_1, then G_1 is also monogenic. Also, the direct product of two monogenic groups need not be monogenic. This is certainly clear by Exercise 4.22. But we can also give an example with noncompact groups. We know that $\mathbb{Z}$ is monogenic, but $\mathbb{Z}^2$ is not monogenic, since it is easily seen that no sequence of the form $((na, nb))$ with $a, b \in \mathbb{Z}$ can be u.d. in $\mathbb{Z}^2$. ∎

The structure of monogenic groups is completely known (see the notes). As can be expected from Example 5.10, their classification is much harder than for locally compact monothetic groups. For properties that may be proved readily, we refer to the Exercises 5.13–5.18. The next theorem exhibits another relation between monogenic and monothetic groups.

THEOREM 5.8. If G is monogenic, then every compact quotient group of G is monothetic and every discrete subgroup of $\hat{G}$ is algebraically isomorphic to a subgroup of $\boldsymbol{T}$.

PROOF. Let a be a monogenic generator of G. Then (a^n) is u.d. in G and so for every compact quotient group G/H, the sequence $(a^n H)$ is u.d. in G/H. Consequently, the coset aH is a generator of G/H. For the second assertion, we note that every discrete subgroup of $\hat{G}$ is the dual of some compact quotient group G/H. The rest follows from Theorem 4.7. ∎

Unfortunately, the converse of the above result does not hold true. For $G = \mathbb{R}$, the nontrivial compact quotient groups are all topologically isomorphic to $\boldsymbol{T}$ (see Example 5.2), and so, monothetic. But $\mathbb{R}$ itself is not monogenic by Example 5.10.

Hartman-Uniform Distribution

DEFINITION 5.6. A sequence (x_n) in a locally compact abelian group G is called *Hartman-u.d. in* G if $\lim_{N\to\infty} (1/N) \sum_{n=1}^{N} \chi(x_n) = 0$ holds for every nontrivial character χ of G.

COROLLARY 5.3. Every Hartman-u.d. sequence in G is u.d. in G.

PROOF. This is an immediate consequence of Definition 5.6 and Corollary 5.2. ∎

One might ask under what conditions on G the notions of Hartman-u.d. and u.d. *per se* are completely equivalent. This is not difficult to answer.

THEOREM 5.9. The notions of Hartman-u.d. and u.d. in the locally compact abelian group G coincide if and only if either G admits no u.d. sequence or every character of G is periodic.

PROOF. The sufficiency of the condition is clear. If G admits no u.d. sequence, then both notions are vacuous by Corollary 5.3, and thus identical. So, suppose that G admits a u.d. sequence (y_n), and assume that G has a nonperiodic character χ_1. Then $\lim_{N\to\infty} (1/N) \sum_{n=1}^{N} \chi(y_n) = 0$ for every nontrivial $\chi \in \hat{G}^p$. By just repeating the argument in the proof of Theorem 5.7, we can then construct a sequence (x_n) in G such that

$$\lim_{N\to\infty} (1/N) \sum_{n=1}^{N} \chi(x_n) = 0$$

for every nontrivial $\chi \in \hat{G}^p$, but $\lim_{N\to\infty} (1/N) \sum_{n=1}^{N} \chi_1(x_n) = 0$ does not hold. In other words, the sequence (x_n) is u.d. but not Hartman-u.d. in G. ■

EXAMPLE 5.11. By Example 5.8, every character of $\mathbb{R}$ is periodic. Thus, in the group $\mathbb{R}$, Hartman-u.d. and u.d. are equivalent. The picture changes if we consider $G = \mathbb{Z}$. In Example 5.9 we saw that not every character of $\mathbb{Z}$ is periodic. Then, Theorem 5.9 implies that there have to exist u.d. sequences in $\mathbb{Z}$ that are not Hartman-u.d. We shall exhibit such sequences explicitly. We first give another characterization of Hartman-u.d. sequences in $\mathbb{Z}$. A sequence (x_n) in $\mathbb{Z}$ is Hartman-u.d. in $\mathbb{Z}$ if and only if (x_n) is u.d. in $\mathbb{Z}$ and (αx_n) is u.d. mod 1 for every irrational $\alpha \in \mathbb{R}$. To prove our assertion, we note that a sequence (x_n) in $\mathbb{Z}$ is Hartman-u.d. precisely if

$$\lim_{N\to\infty} (1/N) \sum_{n=1}^{N} e^{2\pi i t x_n} = 0$$

holds for all $t \in \mathbb{R}\backslash\mathbb{Z}$. Thus, a Hartman-u.d. sequence (x_n) and an irrational α being given, we have $\lim_{N\to\infty} (1/N) \sum_{n=1}^{N} e^{2\pi i m\alpha x_n} = 0$ for all nonzero integers m. The classical Weyl criterion (see Theorem 2.1 of Chapter 1) implies then that (αx_n) is u.d. mod 1. Conversely, from the u.d. in $\mathbb{Z}$ of (x_n) we infer that $\lim_{N\to\infty} (1/N) \sum_{n=1}^{N} e^{2\pi i t x_n} = 0$ holds for all $t \in \mathbb{Q}\backslash\mathbb{Z}$. The corresponding limit relation for irrational t follows from the fact that (tx_n) is u.d. mod 1.

We will show in Theorem 1.5 of Chapter 5 that for irrational α, the sequence $([n\alpha])$ is u.d. in $\mathbb{Z}$. By Theorem 1.8 of Chapter 5, this sequence is not Hartman-u.d. in $\mathbb{Z}$. However, examples of Hartman-u.d. sequences in $\mathbb{Z}$ can easily be given (see Exercise 5.28). ■

When discussing u.d. in G, we realized that in many cases the theory could be reduced to the distribution theory in one single compact group (namely, the periodic compactification). This is even more so for Hartman-u.d. sequences, as the following simple result shows.

THEOREM 5.10. Let G be a locally compact abelian group, let $\bar{G}$ be its Bohr compactification, and let φ: $G \mapsto \bar{G}$ be the natural homomorphism (see Lemma 5.3). Then (x_n) is Hartman-u.d. in G if and only if $(\varphi(x_n))$ is u.d. in $\bar{G}$.

PROOF. This is, in fact, a special case of Lemma 5.4 with $\Gamma = \hat{G}$. ■

Almost-Periodic Functions

Although Theorem 5.10 is trivial once we have Lemma 5.4, it gives rise to an interesting relation between Hartman-u.d. and almost-periodic functions on G. There are many equivalent characterizations of almost-periodic functions, we shall choose one that is suitable for our purposes.

We adopt the same notation as in Theorem 5.10, and we note that $\varphi: G \mapsto \bar{G}$ is injective by the second part of Lemma 5.3. Now let f be a given complex-valued function on G. Then the composite function $f \circ \varphi^{-1}$ is a complex-valued function on $\varphi(G)$. If $f \circ \varphi^{-1}$ can be extended to a continuous function on $\bar{G}$, then f itself is called an *almost-periodic function* on G. Since $\varphi(G)$ is dense in the compact group $\bar{G}$, the function $f \circ \varphi^{-1}$ can have at most one continuous extension to $\bar{G}$. Alternatively, the function f is almost-periodic if $f \circ \varphi^{-1}$ is the restriction to $\varphi(G)$ of a continuous function on $\bar{G}$. Clearly, every almost-periodic function on G is bounded. Trivial examples of almost-periodic functions on G are the constant functions. For further examples, see Exercises 5.19 and 5.22. With addition, multiplication, and scalar multiplication defined pointwise, and equipped with the norm $\|f\| = \sup_{x \in G} |f(x)|$, the set $\mathscr{A}(G)$ of all almost-periodic functions on G forms a Banach algebra. This Banach algebra is closed under still another operation. For fixed $a \in G$, define the translate ${}_af$ of a function f on G by ${}_af(x) = f(ax)$ for all $x \in G$. Then we have ${}_af \in \mathscr{A}(G)$ whenever $f \in \mathscr{A}(G)$.

We introduce the *mean value* $M(f)$ of the almost-periodic function f on G in the following way: Let g be the unique continuous extension of $f \circ \varphi^{-1}$ to $\bar{G}$, and let ν be the Haar measure on $\bar{G}$. Then we define $M(f) = \int_{\bar{G}} g\, d\nu$. It is not hard to see that M is a complex linear functional on $\mathscr{A}(G)$ that is normed in the sense that if $f \equiv 1$, then $M(f) = 1$. Furthermore, M satisfies $M({}_af) = M(f)$ for all $f \in \mathscr{A}(G)$ and all $a \in G$, and M is strictly positive—that is, $M(f) > 0$ for all $f \in \mathscr{A}(G)$ with $f \geq 0$ and $f \not\equiv 0$. Actually, all the properties of M listed here follow easily from the corresponding properties of the Haar integral on $\bar{G}$.

As is true for the concept of almost periodicity itself, the mean value $M(f)$ allows an internal description; that is, we need not go "outside" of G. This characterization becomes particularly simple if G is σ-compact. In this case the following result holds (see Hewitt and Ross [1, Theorem 18.14]). There exists an increasing sequence $H_1 \subseteq H_2 \subseteq \cdots \subseteq H_n \subseteq \cdots$ of relatively compact open sets that exhausts G such that

$$M(f) = \lim_{n \to \infty} \frac{1}{\lambda(H_n)} \int_{H_n} f\, d\lambda \qquad \text{for every continuous } f \in \mathscr{A}(G),$$

where λ is a Haar measure on G. Two special cases are worth mentioning. For $G = \mathbb{R}$, we have the identity

$$M(f) = \lim_{T\to\infty} \frac{1}{2T} \int_{-T}^{T} f(x)\, dx = \lim_{T\to\infty} \frac{1}{T} \int_{0}^{T} f(x)\, dx$$

for every continuous $f \in \mathscr{A}(\mathbb{R})$. For $G = \mathbb{Z}$, the identity

$$M(f) = \lim_{N\to\infty} \frac{1}{2N+1} \sum_{n=-N}^{N} f(n) = \lim_{N\to\infty} \frac{1}{N} \sum_{n=1}^{N} f(n)$$

holds for every $f \in \mathscr{A}(\mathbb{Z})$.

The pertinence of the above concepts in the theory of Hartman-u.d. is revealed by the following theorem, which is a counterpart to Theorem 5.4.

THEOREM 5.11. The sequence (x_n) in the locally compact abelian group G is Hartman-u.d. in G if and only if

$$\lim_{N\to\infty} \frac{1}{N} \sum_{n=1}^{N} f(x_n) = M(f) \tag{5.4}$$

holds for every almost-periodic function f on G.

PROOF. If (x_n) is Hartman-u.d. in G, then $(\varphi(x_n))$ is u.d. in $\bar{G}$ by Theorem 5.10. Thus, $\lim_{N\to\infty} (1/N) \sum_{n=1}^{N} g(\varphi(x_n)) = \int_{\bar{G}} g\, d\nu$ holds for every $g \in \mathscr{C}(\bar{G})$. Now let $f \in \mathscr{A}(G)$; then, on $\varphi(G)$, the function $f \circ \varphi^{-1}$ is identical with its continuous extension g to $\bar{G}$. Thus, we get $g(\varphi(x_n)) = (f \circ \varphi^{-1})(\varphi(x_n)) = f(x_n)$ for all $n \geq 1$. Consequently, $\lim_{N\to\infty} (1/N) \sum_{n=1}^{N} f(x_n) = \int_{\bar{G}} g\, d\nu = M(f)$. For the converse, we use exactly the same ideas. We start out from an arbitrary $g \in \mathscr{C}(\bar{G})$. Then $f = g \circ \varphi$ defines an almost-periodic function on G. By hypothesis, we get

$$\lim_{N\to\infty} \frac{1}{N} \sum_{n=1}^{N} g(\varphi(x_n)) = \lim_{N\to\infty} \frac{1}{N} \sum_{n=1}^{N} f(x_n) = M(f) = \int_{\bar{G}} g\, d\nu.$$

Hence, $(\varphi(x_n))$ is u.d. in $\bar{G}$, and an application of Theorem 5.10 completes the proof. ■

Existence of Hartman-Uniformly Distributed Sequences

THEOREM 5.12. The locally compact abelian group G admits a Hartman-u.d. sequence if and only if card $\hat{G} \leq \mathfrak{c}$.

PROOF. It is rather easy to see that the condition is necessary. For if (x_n) is Hartman-u.d. in G, then Theorem 5.10 implies that $(\varphi(x_n))$ is u.d. in the compact group $\bar{G}$. In particular, $\bar{G}$ is separable. Since a character χ of $\bar{G}$ is determined by its values on a countable dense set in $\bar{G}$ and since χ takes values in the set T of cardinality $\mathfrak{c}$, we have at most $\mathfrak{c}^{\mathfrak{a}} = \mathfrak{c}$ characters of $\bar{G}$. By the construction of the Bohr compactification $\bar{G}$, its dual is algebraically isomorphic to $\hat{G}$. Thus, we arrive at card $\hat{G} \leq \mathfrak{c}$.

The converse of our result is much harder to prove. It will be advantageous to have the following special cases of two theorems of Kakutani [1, Theorems 1 and 3] available:

i. If card $H \leq \mathfrak{c}$ for a compact abelian group H, then card $\hat{H} \leq \mathfrak{a}$ (see also Hewitt and Ross [1, (24.47)]).

ii. A compact abelian group H is separable if and only if card $\hat{H} \leq \mathfrak{c}$ (incidentally, we showed the "only if" in the first part of the proof).

We first want to show that the hypothesis card $\hat{G} \leq \mathfrak{c}$ implies that G is separable. By Theorem 1.14, it suffices to consider a group G of the form $\mathbb{R}^n \times H$ where the locally compact abelian group H contains a compact open subgroup K. The quotient group $Q = G/(\mathbb{R}^n \times K)$ is discrete. Its compact dual $\hat{Q}$ is topologically isomorphic to a subgroup of $\hat{G}$; thus, card $\hat{Q} \leq \mathfrak{c}$. It follows from (i) that Q itself is countable. Moreover, the dual of the compact group K is topologically isomorphic to a quotient group of $\hat{H}$, and so, card $\hat{K} \leq \mathfrak{c}$. But then (ii) implies that K is separable. Evidently, $\mathbb{R}^n$ is separable, and so, the direct product $\mathbb{R}^n \times K$ is separable. Combining all those results, we see that G can be written as the countable union of the pairwise disjoint cosets of $\mathbb{R}^n \times K$, all of which are separable. Hence, G itself is separable.

Let now (y_n) be a dense sequence in G. We shall use the elements of this sequence to construct a Hartman-u.d. sequence in G. Consider the following sequence (x_n), whose law of construction will be described in detail:

$$y_1, y_1y_2, y_1^2y_2, y_1y_2^2, y_1^2y_2^2, y_1y_2^3, y_1^2y_2^3, y_1y_2^4, y_1^2y_2^4, y_1y_2y_3, y_1^2y_2y_3, y_1^3y_2y_3, \ldots, y_1^3y_2^9y_3^{27}, \ldots, y_1y_2\cdots y_m, \ldots, y_1^m y_2^{m^2}\cdots y_m^{m^m}, \ldots.$$

The sequence is built up from certain blocks that begin with $y_1y_2\cdots y_m$ and end with $y_1^m y_2^{m^2}\cdots y_m^{m^m}$. Let us describe the way in which such a block is constructed for given $m \geq 1$. The terms of the mth block B_m are elements of the form $y_1^{a_1}y_2^{a_2}\cdots y_m^{a_m}$ with the exponents a_j satisfying $1 \leq a_j \leq m^j$ for $1 \leq j \leq m$. The exponent of y_1 runs in ascending order through the integers from 1 to m and then cycles. The exponent of y_2 runs through the integers from 1 to m^2 in blocks of m and then cycles. In general, for $1 \leq j \leq m$, the exponent of y_j runs through the integers from 1 to m^j in blocks of $m^{j(j-1)/2}$ and then cycles. The mth block B_m has $m^{m(m+1)/2}$ terms.

Let χ be a given nontrivial character of G. We have to show that

$$\lim_{N\to\infty} \frac{1}{N} \sum_{n=1}^{N} \chi(x_n) = 0.$$

Since (y_n) is dense in G, the continuous function χ cannot be identical to 1 for all the y_n. Let $r \geq 1$ be the smallest subscript such that $\chi(y_r) \neq 1$. We consider one of the above blocks B_m with $m > r$. Each element of B_m is of the form $y_1^{a_1} \cdots y_{r-1}^{a_{r-1}} y_r^{b} y_{r+1}^{a_{r+1}} \cdots y_m^{a_m}$, where b runs from 1 to m^r in blocks of $m^{r(r-1)/2}$ and then cycles. One such cycle for b has then a length of $m^{r(r+1)/2}$. We divide B_m into subblocks according to the cycles of b. The first cycle of b defines a subblock C_{m1} of B_m (which is nothing else but the first $m^{r(r+1)/2}$ elements of B_m), the second cycle of b defines a subblock C_{m2} of B_m (comprising the next $m^{r(r+1)/2}$ elements of B_m), and so on to the last subblock C_{ms_m} with $s_m = m^{m(m+1)/2-r(r+1)/2}$. Now let us look at the behavior of the exponents a_j with $r < j \leq m$ in such a subblock. We notice that such an a_j attains its values in blocks of length $m^{j(j-1)/2}$. In other words, each value of a_j is first repeated $m^{j(j-1)/2}$ times before a_j proceeds on to the next value. But note that $j(j-1)/2 \geq r(r+1)/2$ for $j > r$, and so the length $m^{r(r+1)/2}$ of one of our subblocks is a divisor of $m^{j(j-1)/2}$; this is an important motivation for the particular choice of the sequence (x_n). Consequently, the exponents a_j with $j > r$ are simply constant on a fixed subblock C_{mt} with $1 \leq t \leq s_m$. This makes it easy to evaluate χ at one of the terms $x = y_1^{a_1} \cdots y_{r-1}^{a_{r-1}} y_r^{b} y_{r+1}^{a_{r+1}} \cdots y_m^{a_m}$ of a given subblock C_{mt}. We use that $\chi(y_j) = 1$ for $1 \leq j \leq r-1$, and that $y_{r+1}^{a_{r+1}} \cdots y_m^{a_m}$ is a fixed element z depending only on the subblock C_{mt}. Thus, $\chi(x) = \chi(y_r^{b})\chi(z)$ for all terms x of C_{mt}. We shall mean by $\sum_{x \in C_{mt}}$ a sum over all the terms of C_{mt} (thus an element x occurs in the sum with the same multiplicity with which it occurs in the block C_{mt}), and by $|C_{mt}|$, the number of terms in the block C_{mt}. Similar conventions will be employed for other blocks. Using the behavior of the exponents b in a fixed block C_{mt}, we arrive at

$$\left|\frac{1}{|C_{mt}|}\sum_{x \in C_{mt}} \chi(x)\right| = \left|\frac{1}{|C_{mt}|}\chi(z)\sum_{b=1}^{m^r} m^{r(r-1)/2}\chi(y_r^{b})\right|$$

$$= \frac{m^{r(r-1)/2}}{m^{r(r+1)/2}}\left|\sum_{b=0}^{m^r-1} (\chi(y_r))^b\right| \leq \frac{2m^{-r}}{|\chi(y_r) - 1|}. \tag{5.5}$$

Now we enumerate all the blocks C_{mt} in the same order as they occur in the sequence (x_n) from the block B_{r+1} onward: $D_1, D_2, D_3, \ldots$. Since m eventually becomes arbitrarily large, (5.5) implies

$$\lim_{p \to \infty} \frac{1}{|D_p|} \sum_{x \in D_p} \chi(x) = 0.$$

To establish $\lim_{N \to \infty}(1/N)\sum_{n=1}^{N} \chi(x_n) = 0$, we use Lemma 4.1 of Chapter 2. It suffices then to show that

$$\lim_{p \to \infty} \frac{|D_{p+1}|}{|D_1| + \cdots + |D_p|} = 0. \tag{5.6}$$

We consider sufficiently large p. Let $D_{p+1} = C_{mt}$ for some $m \geq r + 2$ and $1 \leq t \leq s_m$. Then,

$$\frac{|D_{p+1}|}{|D_1| + \cdots + |D_p|} \leq \frac{|C_{mt}|}{|B_{m-1}|} = \frac{m^{r(r+1)/2}}{(m-1)^{m(m-1)/2}}. \tag{5.7}$$

Letting $p \to \infty$ (or, equivalently, $m \to \infty$) in (5.7), we arrive at (5.6). ∎

COROLLARY 5.4. The compact abelian group G admits a u.d. sequence if and only if G is separable.

PROOF. In the compact abelian group G, the notions of Hartman-u.d. and u.d. are equivalent. Thus, Theorem 5.12, together with the auxiliary result (ii) in the proof of this theorem, implies our assertion. ∎

Many More Notions of Uniform Distribution

Let K be a compactification of the locally compact abelian group G with natural homomorphism φ: $G \mapsto K$. We may call a sequence (x_n) *K-u.d. in G* if $(\varphi(x_n))$ is u.d. in K. If Γ is the subgroup of $\hat{G}$, equipped with the discrete topology, for which $K = \hat{\Gamma}$, then Lemma 5.4 implies that (x_n) is K-u.d. in G if and only if $\lim_{N\to\infty} (1/N) \sum_{n=1}^{N} \chi(x_n) = 0$ holds for all nontrivial characters $\chi \in \Gamma$. Clearly, Hartman-u.d. is equivalent to $\bar{G}$-u.d. Moreover, if $\hat{G}^p$ is a subgroup of $\hat{G}$, then u.d. in G is equivalent to $\bar{G}^p$-u.d. If Γ separates the points of G, then φ is injective by Lemma 5.3; in this case, one can define a complex-valued function f on G to be *K-almost periodic* if the function $f \circ \varphi^{-1}$ on $\varphi(G)$ has a continuous extension to K. It is easy to see that every K-almost periodic function is almost-periodic (see Exercise 5.20) and that (x_n) is K-u.d. if and only if $\lim_{N\to\infty} (1/N) \sum_{n=1}^{N} f(x_n) = M(f)$ holds for all K-almost periodic functions f on G (see Exercise 5.21).

Notes

The general definition for u.d. in locally compact groups is from Rubel [1]. For two special noncompact groups—namely, for $G = \mathbb{Z}$ and $G = \mathbb{R}$—the notion was already studied earlier by Niven [2] and Cigler [6], respectively. In Cigler's paper, our Theorem 5.11 was taken as the definition (note that by Example 5.11 u.d. and Hartman-u.d. are equivalent for $G = \mathbb{R}$).

For locally compact abelian groups, a satisfactory theory of u.d. was developed in a joint paper by Berg, Rajagopalan, and Rubel [1]. Many results on periodic characters that are closely related to the material in this section can be found in Berg and Rubel [2]. Let us report on some of the theorems that could not be accommodated here.

Most important, the following existence theorem holds for u.d. sequences in a locally compact abelian group G: If card $\hat{G}^p \leq \mathfrak{c}$, then G admits a u.d. sequence; if $\hat{G}^p$ is a subgroup of $\hat{G}$ and G admits a u.d. sequence, then card $\hat{G}^p \leq \mathfrak{c}$; for any cardinal number $\mathfrak{m}$, there exists a locally compact abelian group G with card $\hat{G}^p \geq \mathfrak{m}$ that admits a u.d. sequence.

A necessary and sufficient condition can also be given, but it is not so satisfactory. The group G is called *K-separable* if there exists a sequence (x_n) in G such that, for every subgroup H of G of compact index, the sequence (x_nH) is dense in G/H. Then G admits a u.d. sequence if and only if G is K-separable.

The following notion is useful. A *D-compactification* of G is a compactification K of G, with natural homomorphism $\varphi: G \mapsto K$, such that (x_n) is u.d. in G if and only if $(\varphi(x_n))$ is u.d. in K. Theorems 5.6 and 5.7 together show that whenever $\hat{G}^p$ is a subgroup of $\hat{G}$, then $\bar{G}^p$ is a D-compactification of G. In this case, $\bar{G}^p$ is the only D-compactification of G. If $\hat{G}^p$ is not a subgroup of $\hat{G}$, then there exists no D-compactification. Moreover, $\hat{G}^p$ is a subgroup of $\hat{G}$ if and only if either G is totally disconnected or every discrete quotient group of G is a group of bounded order (some interesting relations between topological groups and the algebraic property of bounded order in a group are presented in Rudin [1, Section 2.5]). All those results are from Berg, Rajagopalan, and Rubel [1].

Some of the above results were generalized to arbitrary locally compact groups by Benzinger [1]. A construction similar to that in the proof of Theorem 5.12 can be carried out in the nonabelian case as well. In particular, Corollary 5.4 holds for arbitrary compact groups. Using the theory of commutative Banach algebras, the author also introduces and proves results about D-compactifications.

The notion of monogenic group was introduced by Rubel [1]. The discrete monogenic groups were completely characterized by Rajagopalan and Rotman [1]. They showed that a discrete monogenic group is the direct product of a divisible group with certain pure subgroups of products of the form $\prod_{p \in P} C_p$, where P is a set of distinct primes and C_p is either a cyclic p-group or the group of p-adic integers. Moreover, a discrete abelian group is monogenic if and only if it is monothetic in the Prüfer (or $\mathfrak{n}$-adic) topology. A complete characterization of all monogenic groups, and also of all topologically divisible abelian groups, was given by Rajagopalan [1]. The results are too complicated to be restated here.

The notion of Hartman-u.d. was first discussed by Hartman [4]. The intimate relation to almost-periodic functions was already pointed out in this paper. The existence criterion given in Theorem 5.12, together with the ingenious construction of a Hartman-u.d. sequence reproduced in our proof, is again from Berg, Rajagopalan, and Rubel [1]. Zame [6] uses a somewhat similar method in a related construction problem.

The universal monothetic Cantor group G_1, which emerged in Example 5.9 as the periodic compactification of $\mathbb{Z}$, was studied in detail by van Dantzig [2]. For this matter, see also Hewitt and Ross [1, (25.7)]. We note that results that are in the same spirit as (i) and (ii) in the proof of Theorem 5.12 can be found in Hartman and Hulanicki [1].

There is also a very fruitful measure-theoretic aspect in the theory of u.d. on locally compact abelian groups. This viewpoint was explored by Berg and Rubel [1, 2] and Rubel [2] and led to many interesting problems.

For a general survey of the theory of almost-periodic functions, see Maak [1]. A presentation whose spirit is closer to ours can be found in Loomis [1, Section 41] and Weil [1, Sections 33–35].

Exercises

The symbol G always denotes a locally compact abelian group.

5.1. Give an example of a group G and subgroups H_1 and H_2 of G of compact index such that $H_1 \cap H_2$ is not of compact index.

5.2. Show that a bounded sequence cannot be u.d. in $\mathbb{R}$.

5.3. Prove the following criterion: G is topologically divisible if and only if G admits no nontrivial periodic character.

5.4. Prove that G is topologically divisible if and only if $\hat{G}$ is torsion-free and contains only compact elements.

5.5. If G is topologically divisible, then every quotient group G/H is topologically divisible, where H is a closed subgroup of G.

5.6. Prove that if the closed subgroup H of G contains the connected component C of the identity in G, then G/H is totally disconnected.

5.7. Use Exercise 5.6 to prove the following: If the kernel H of a character χ of G contains C, then G/H is discrete, and so, χ is periodic if and only if G/H is finite. Prove also that if H does not contain C, then G/H is topologically isomorphic to $\boldsymbol{T}$, and so, χ is periodic.

5.8. Prove that every topologically divisible abelian group is totally disconnected.

5.9. Show the following characterization for periodic characters: The character χ of G is periodic if and only if the closed subgroup of $\hat{G}$ generated by χ is discrete.

5.10. If G is not totally disconnected, then $\hat{G}^p$ generates $\hat{G}$ algebraically. Thus, in this case, the periodic compactification $\bar{G}^p$ is identical with the Bohr compactification $\bar{G}$.

5.11. Whenever G is a discrete group of bounded order (i.e., a torsion group with a uniform bound on the order of its elements), then $\bar{G}$ is totally disconnected.

5.12. Prove more generally that $\bar{G}$ is totally disconnected if and only if G is totally disconnected and every discrete quotient group of G is of bounded order.

5.13. The element $x \in G$ is a monogenic generator of G if and only if $\chi(x) \neq 1$ for every nontrivial periodic character χ of G.

5.14. If x is a monogenic generator of G, then xH is a monogenic generator of G/H, where H is a closed subgroup of G.

5.15. Let H be a topologically divisible closed subgroup of G such that G/H is monogenic. Prove that G itself is monogenic.

5.16. Suppose G is monogenic and $\hat{G}^p$ is a subgroup of $\hat{G}$. Then $\hat{G}^p$ is isomorphic to a subgroup of $\boldsymbol{T}$.

5.17. Let $x \in G$ be a monogenic generator of G. Then all powers x^k, $k = 1, 2, \ldots$, are monogenic generators as well if and only if every compact quotient group G/H is connected.

5.18. In the same situation as in Exercise 5.17, all powers x^k, $k = 1, 2, \ldots$, are monogenic generators of G if and only if every discrete subgroup of $\hat{G}$ is torsion-free.

5.19. Prove that every character of G is an almost-periodic function on G.

5.20. Let $K = \hat{\Gamma}$ be a compactification of G with Γ separating the points of G. Prove that every K-almost periodic function on G is almost-periodic.

5.21. Let K be as in Exercise 5.20. Show that (x_n) is K-u.d. in G if and only if $\lim_{N\to\infty}(1/N)\sum_{n=1}^{N} f(x_n) = M(f)$ holds for every K-almost periodic function f on G.

5.22. Show that every continuous periodic function on G is almost-periodic.

5.23. Verify the following identity for continuous periodic functions f on G: $M(f) = \int f\, d\mu$.

5.24. If the compactification K of G is a quotient group of the compactification L of G, then every L-u.d. sequence in G is K-u.d. in G.

5.25. Let $K = \hat{\Gamma}$ be a compactification of G with a closed subgroup Γ of $\hat{G}$. Prove the following existence theorem for K-u.d.: If card $\Gamma \leq \mathfrak{c}$, then G admits K-u.d. sequences.

5.26. Let $K = \hat{\Gamma}$ be a compactification of G. We say that G is *K-monogenic* if there exists a sequence of the form (x^n) in G that is K-u.d. in G. Prove that if G is K-monogenic, then Γ is isomorphic to a subgroup of $\boldsymbol{T}$.

5.27. Consider the same situation as in Exercise 5.26. Give additional conditions on Γ that guarantee that G is K-monogenic if and only if Γ (in the relative topology of $\hat{G}$) is topologically isomorphic to a subgroup of $\boldsymbol{T}$.

5.28. Prove that $\mathbb{Z}$ is $\bar{\mathbb{Z}}$-monogenic by showing that the sequence of positive integers is Hartman-u.d. in $\mathbb{Z}$.

5.29. Call a sequence (x_n) in G *well distributed in* G if (x_nH) is well distributed in G/H for every subgroup H of G of compact index. Show that this is a natural generalization of the concept of well-distributivity in compact groups.

5.30. Exhibit a well-distributed sequence in $\mathbb{Z}$.

5.31. Exhibit some well-distributed sequences in $\mathbb{R}$.

5.32. Let x be a monogenic generator of G. Prove that the sequence (x^n) is well distributed in G.

5

SEQUENCES OF INTEGERS AND POLYNOMIALS

In this chapter, we are concerned with the distribution of sequences in special domains, such as the ring of rational integers, rings of p-adic integers, and polynomial rings over finite fields. The general theory developed in Chapter 4 applies to many of these cases, but for the most part our exposition will be independent of that chapter.

1. UNIFORM DISTRIBUTION OF INTEGERS

Basic Properties

Let (a_n), $n = 1, 2, \ldots$, be a sequence of rational integers. For integers $N \geq 1$, $m \geq 2$, and j, define $A(j, m, N)$ as the number of terms among $a_1, a_2, \ldots, a_N$ that satisfy the congruence $a_i \equiv j \pmod{m}$.

DEFINITION 1.1. The sequence (a_n) is said to be *uniformly distributed modulo m* (u.d. mod m) in case

$$\lim_{N\to\infty} \frac{A(j, m, N)}{N} = \frac{1}{m} \qquad \text{for } j = 1, 2, \ldots, m, \tag{1.1}$$

and (a_n) is said to be *uniformly distributed in* $\mathbb{Z}$ (u.d. in $\mathbb{Z}$) if (1.1) is satisfied for every integer $m \geq 2$.

This definition is, of course, a special case of Definition 5.2 of Chapter 4 (compare with Example 5.1 of Chapter 4). It is easily seen that a sequence that is u.d. mod m is also u.d. mod k whenever $k \geq 2$ is a divisor of m (see Exercise 1.1). On the other hand, the periodic sequence $0, 1, \ldots, m-1$, $0, 1, \ldots, m-1, \ldots$ is an example of a sequence that is u.d. mod m but not u.d. mod k if k does not divide m. We also have the following result.

THEOREM 1.1. There is a sequence (a_n) of integers that is not u.d. in $\mathbb{Z}$ but is u.d. mod p^α for every prime p and every integer $\alpha \geq 1$.

PROOF. We prove even more. For $n \geq 1$, define $a_n = n$ if $n \equiv 0, 1, 2$, or $5 \pmod 6$, $a_n = n - 2$ if $n \equiv 3 \pmod 6$, and $a_n = n + 2$ if $n \equiv 4 \pmod 6$. Then (a_n) is u.d. mod m for all $m \geq 2$ that are not divisible by 6. But (a_n) is not u.d. mod 6. ■

A Weyl criterion for u.d. mod m follows from Corollary 1.2 and Example 1.4 of Chapter 4. The criterion may also be verified directly by using elementary arguments (see Exercise 1.6).

THEOREM 1.2. Let (a_n) be a sequence of integers. A necessary and sufficient condition that (a_n) be u.d. mod m is that

$$\lim_{N\to\infty} \frac{1}{N} \sum_{n=1}^{N} e^{2\pi i h a_n/m} = 0 \qquad \text{for } h = 1, 2, \ldots, m-1. \tag{1.2}$$

COROLLARY 1.1. A necessary and sufficient condition that (a_n) be u.d. in $\mathbb{Z}$ is that

$$\lim_{N\to\infty} \frac{1}{N} \sum_{n=1}^{N} e^{2\pi i t a_n} = 0 \qquad \text{for all rational numbers } t \notin \mathbb{Z}. \tag{1.3}$$

Restriction to increasing sequences (a_n) of positive integers leads to an alternative definition of u.d. mod m. Let $A^*(j, m, N)$ be the number of terms of (a_n) that satisfy the conditions $a_k \leq N$ and $a_k \equiv j \pmod m$. Let $A(N)$ be the number of terms a_k satisfying $a_k \leq N$. Then the equivalent of (1.1) is

$$\lim_{N\to\infty} \frac{A^*(j, m, N)}{A(N)} = \frac{1}{m} \qquad \text{for } j = 1, 2, \ldots, m. \tag{1.4}$$

We remark that $\underline{\lim}_{N\to\infty} A(N)/N$ is called the *lower (asymptotic) density* of the sequence (a_n).

THEOREM 1.3. Let (a_n) be an increasing sequence of positive integers whose complement $\overline{(a_n)}$ with respect to the positive integers, arranged in increasing order, has positive lower density. Then if (a_n) is u.d. mod m, so is $\overline{(a_n)}$.

PROOF. First we have

$$A^*(j, m, N) + \bar{A}^*(j, m, N) = \frac{N}{m} + \alpha, \qquad |\alpha| < 1, \tag{1.5}$$

where bars refer to the complementary sequence $\overline{(a_n)}$. Since $A(N) + \bar{A}(N) = N$, we get from (1.5),

$$\frac{A^*(j, m, N)}{A(N)}\left(1 - \frac{\bar{A}(N)}{N}\right) + \frac{\bar{A}^*(j, m, N)}{\bar{A}(N)} \cdot \frac{\bar{A}(N)}{N} = \frac{1}{m} + \frac{\alpha}{N},$$

or

$$\frac{\bar{A}(N)}{N}\left(\frac{\bar{A}^*(j, m, N)}{\bar{A}(N)} - \frac{A^*(j, m, N)}{A(N)}\right) = \frac{1}{m} - \frac{A^*(j, m, N)}{A(N)} + \frac{\alpha}{N}. \tag{1.6}$$

Now let $N \to \infty$. Then the right-hand side of (1.6) goes to zero, and since $\underline{\lim}_{N\to\infty} \bar{A}(N)/N$ is positive, we see that

$$\lim_{N\to\infty} \frac{\bar{A}^*(j, m, N)}{\bar{A}(N)} = \frac{1}{m}. \quad \blacksquare$$

Uniform Distribution in $\mathbb{Z}$ and Uniform Distribution Mod 1

THEOREM 1.4. Let (x_n), $n = 1, 2, \ldots$, be a sequence of real numbers such that the sequence (x_n/m) is u.d. mod 1 for all integers $m \geq 2$. Then the sequence $([x_n])$ of integral parts is u.d. in $\mathbb{Z}$.

PROOF. For fixed $m \geq 2$ and for j with $0 \leq j \leq m - 1$, the relation $[x_n] \equiv j \pmod m$ is equivalent to $j/m \leq \{x_n/m\} < (j + 1)/m$. Hence, $A(j, m, N) = A([j/m, (j + 1)/m); N)$, where the second counting function refers to the sequence (x_n/m). Since (x_n/m) is u.d. mod 1, we get

$$\lim_{N\to\infty} \frac{A(j, m, N)}{N} = \lim_{N\to\infty} \frac{A([j/m, (j + 1)/m); N)}{N} = \frac{1}{m}$$

for all $j = 0, 1, \ldots, m - 1$. Thus, $([x_n])$ is u.d. mod m for all $m \geq 2$. $\blacksquare$

Theorem 1.4 is a powerful result. By means of this theorem we can find a great variety of u.d. sequences of integers. We mention the following applications.

EXAMPLE 1.1. The sequence $([f(n)])$ is u.d. in $\mathbb{Z}$ in each of the following cases:

i. $f(t) = \alpha_k t^k + \alpha_{k-1} t^{k-1} + \cdots + \alpha_1 t + \alpha_0$ is a polynomial over $\mathbb{R}$ with at least one of the coefficients α_i, $i \geq 1$, being irrational (see Chapter 1, Theorem 3.2).

ii. $f(t)$, $t \geq 1$, is a real-valued differentiable function with $f'(t) \downarrow 0$ and $tf'(t) \to \infty$ as $t \to \infty$ (see Chapter 1, Corollary 2.1).

iii. $f(t)$, $t \geq 0$, has a continuous derivative with $f'(t) \log t \to C$, a positive constant, as $t \to \infty$ (see Chapter 1, Theorem 9.8). ■

THEOREM 1.5. *For $\theta \in \mathbb{R}$, the sequence $([n\theta])$ is u.d. in $\mathbb{Z}$ if and only if θ is irrational or $\theta = 1/d$ for some nonzero integer d.*

PROOF. For irrational θ, Example 1.1, part (i), yields the desired conclusion. Now let $\theta \neq 0$ be rational, say $\theta = a/b$ with $(a, b) = 1$ and $b \geq 1$. We note that the sequence $([na/b])$ is periodic mod $|a|$ with period b. Thus, if $([na/b])$ is u.d. in $\mathbb{Z}$, then $|a|$ must divide b. Together with $(a, b) = 1$ we get $a = \pm 1$. On the other hand, it is easily seen that the sequence $([n/d])$ is u.d. in $\mathbb{Z}$ for every nonzero integer d. ■

COROLLARY 1.2. *If θ is irrational, $|\theta| < 1$, then the following sequence (a_n) is u.d. in $\mathbb{Z}$: Take all positive integers a_n such that there is an integer between θa_n and $\theta(a_n + 1)$, and arrange them in increasing order.*

PROOF. It suffices to prove this for $\theta > 0$. Let b_n be an integer with $\theta a_n < b_n < \theta(a_n + 1)$. Then $a_n < b_n\theta^{-1} < a_n + 1$, or $a_n = [b_n\theta^{-1}]$. Moreover, $\theta < 1$ implies that $b_n = n$, since there is always a multiple of θ between two consecutive nonnegative integers. The rest follows from Theorem 1.5. ■

Another theorem that reveals the close relation between u.d. mod 1 and u.d. of integers is the following.

THEOREM 1.6. *The sequence (x_n) in $\mathbb{R}$ is u.d. mod 1 if and only if the sequence $([mx_n])$ is u.d. mod m for all integers $m \geq 2$.*

PROOF. As in the proof of Theorem 1.4, the u.d. mod 1 of (x_n) implies the u.d. mod m of $([mx_n])$. Conversely, if $([mx_n])$ is u.d. mod m, we get by the same arguments that

$$\lim_{N\to\infty} \frac{A([j/m, (j+1)/m); N; (x_n))}{N} = \lim_{N\to\infty} \frac{A(j, m, N)}{N} = \frac{1}{m}$$

for all $j = 0, 1, \ldots, m - 1$. Letting m run through all the integers ≥ 2, it follows that

$$\lim_{N\to\infty} \frac{A([\alpha, \beta); N; (x_n))}{N} = \beta - \alpha$$

holds for all subintervals $[\alpha, \beta)$ of the unit interval with rational end points. By Exercise 1.3 of Chapter 1 the proof is complete. ■

The set Σ of u.d. sequences of positive integers contains as members all sequences $([n\theta])$ with θ irrational and > 1, so Σ has at least cardinality $\mathfrak{c}$ of the continuum. Since Σ is a subset of the set of all sequences of positive integers, Σ has cardinality $\mathfrak{c}$. The sequences that are not u.d. in $\mathbb{Z}$ have also cardinality $\mathfrak{c}$, for if (a_n) is u.d. in $\mathbb{Z}$, then (ma_n), $m \geq 2$, is not u.d. mod m, and distinct sequences (a_n) yield distinct sequences (ma_n).

EXAMPLE 1.2. Every increasing sequence (a_n), $n = 1, 2, \ldots$, of positive integers can be mapped onto an infinite decimal in the binary system, with a digit 0 in the position a_n and 1 elsewhere. In this way we obtain a one-to-one correspondence between increasing sequences of positive integers and the real numbers x with $0 \leq x < 1$, expressed in the binary system. But if x is normal to base 2, then the corresponding (a_n) is u.d. in $\mathbb{Z}$, and moreover, almost all numbers are normal to base 2. Hence, in the sense of the above mapping, almost all increasing sequences of positive integers are u.d. in $\mathbb{Z}$. ■

THEOREM 1.7. Let the sequence (a_n) of integers be u.d. in $\mathbb{Z}$. Then the sequence $(a_n\alpha)$ is almost u.d. mod 1 for almost all real numbers α.

PROOF. It suffices to consider $0 \leq \alpha \leq 1$. Let $B(k)$ be the number of terms a_n, $1 \leq n \leq N$, that are equal to k. Then

$$A(j, m, N) = \sum_{k \equiv j (\text{mod } m)} B(k).$$

For integers $h \neq 0$, we have

$$\int_0^1 \left| \frac{1}{N} \sum_{n=1}^{N} e^{2\pi i h a_n \alpha} \right|^2 d\alpha = \frac{1}{N^2} \int_0^1 \left| \sum_k B(k) e^{2\pi i h k \alpha} \right|^2 d\alpha = \frac{1}{N^2} \sum_k B^2(k). \quad (1.7)$$

Given any real number ε, $0 < \varepsilon < 1$, we choose an integer m such that $5\varepsilon^{-1} < m < 25(4\varepsilon)^{-1}$. For sufficiently large N we have

$$\left| \frac{A(j, m, N)}{N} - \frac{1}{m} \right| < \frac{\varepsilon}{5} \qquad \text{for } j = 1, 2, \ldots, m.$$

Then

$$\frac{1}{N^2} \sum_k B^2(k) = \frac{1}{N^2} \sum_{j=1}^{m} \sum_{k \equiv j (\text{mod } m)} B^2(k) \leq \sum_{j=1}^{m} \frac{1}{N^2} (A(j, m, N))^2$$

$$< \sum_{j=1}^{m} \left(\frac{1}{m} + \frac{\varepsilon}{5} \right)^2 < m \frac{4\varepsilon^2}{25} < \varepsilon. \quad (1.8)$$

Thus, according to (1.7) and (1.8), we have

$$\lim_{N \to \infty} \sum_{h \neq 0} 2^{-|h|} \int_0^1 \left| \frac{1}{N} \sum_{n=1}^{N} e^{2\pi i h a_n \alpha} \right|^2 d\alpha = 0.$$

Because of Fatou's lemma, we obtain

$$\int_0^1 \underline{\lim}_{N \to \infty} \left(\sum_{h \neq 0} 2^{-|h|} \left| \frac{1}{N} \sum_{n=1}^{N} e^{2\pi i h a_n \alpha} \right|^2 \right) d\alpha = 0,$$

and therefore, for almost all α, there exists a sequence of positive integers $N_1 < N_2 < \cdots$, which may depend on α, such that

$$\lim_{i\to\infty} \frac{1}{N_i} \sum_{n=1}^{N_i} e^{2\pi i h a_n \alpha} = 0 \qquad \text{for all } h \neq 0.$$

This proves the theorem. ■

For special sequences (a_n) of integers, one may characterize the real numbers α for which $(a_n\alpha)$ is u.d. mod 1, as is done in the following theorem.

THEOREM 1.8. For rational θ, the sequence $([n\theta]\alpha)$, $n = 1, 2, \ldots$, is u.d. mod 1 either for all irrationals α or for no real number α, depending on whether $\theta \neq 0$ or $\theta = 0$. If θ is irrational, then $([n\theta]\alpha)$, $n = 1, 2, \ldots$, is u.d. mod 1 if and only if $1, \theta, \theta\alpha$ are linearly independent over the rationals.

PROOF. The case $\theta = 0$ being trivial, we suppose $\theta = r/s$ where $s \geq 1$ and $r \neq 0$ are integers. Let α be any irrational number. Given $N \geq s$, put $M = [N/s]$. With $\exp(x) = e^{2\pi i x}$ for $x \in \mathbb{R}$, we get for each integer $h \neq 0$,

$$\begin{aligned}\sum_{n=1}^{N} \exp(h[n\theta]\alpha) &= \sum_{n=1}^{Ms} \exp(h[nr/s]\alpha) + O(1)\\ &= \sum_{k=0}^{M-1} \sum_{m=1}^{s} \exp(h[(ks+m)r/s]\alpha) + O(1)\\ &= \sum_{m=1}^{s} \exp(h[mr/s]\alpha) \sum_{k=0}^{M-1} \exp(hkr\alpha) + O(1).\end{aligned}$$

Now $hr\alpha$ is irrational, so

$$\sum_{k=0}^{M-1} \exp(khr\alpha) = \frac{\exp(Mhr\alpha) - 1}{\exp(hr\alpha) - 1} = O(1).$$

Therefore, $\sum_{n=1}^{N} \exp(h[n\theta]\alpha) = O(1)$ for all integers $h \neq 0$, and so, by the Weyl criterion for u.d. mod 1, we are done.

Now let θ be irrational, and suppose first that $1, \theta, \theta\alpha$ are linearly dependent over the rationals. There exist integers u, v, w, not all zero, such that $u + v\theta = w\theta\alpha$. Note that $w \neq 0$, since θ is irrational. In order to prove that $([n\theta]\alpha)$ is not u.d. mod 1, it suffices to show that

$$\lim_{N\to\infty} \frac{1}{N} \sum_{n=1}^{N} \exp(w[n\theta]\alpha) \neq 0. \tag{1.9}$$

Using the linear dependence relation, we get

$$\begin{aligned}\exp(w[n\theta]\alpha) &= \exp(n(w\theta\alpha) - w\{n\theta\}\alpha) = \exp(n(u + v\theta) - w\{n\theta\}\alpha)\\ &= \exp(v(n\theta) - w\{n\theta\}\alpha) = \exp((v - w\alpha)\{n\theta\}) = g(n\theta),\end{aligned}$$

where $g(x) = \exp((v - w\alpha)\{x\})$ for $x \in \mathbb{R}$. Clearly, g is periodic mod 1. Since $(n\theta)$, $n = 1, 2, \ldots$, is u.d. mod 1, we have

$$\lim_{N\to\infty} \frac{1}{N} \sum_{n=1}^{N} \exp(w[n\theta]\alpha) = \lim_{N\to\infty} \frac{1}{N} \sum_{n=1}^{N} g(n\theta) = \int_0^1 g(x)\,dx$$

by Corollary 1.1 of Chapter 1. But

$$\int_0^1 g(x)\,dx = \int_0^1 \exp((v - w\alpha)x)\,dx = \frac{\exp(v - w\alpha) - 1}{2\pi i(v - w\alpha)} \neq 0,$$

which proves (1.9).

Finally, suppose that $1, \theta, \theta\alpha$ are linearly independent over the rationals. For each nonzero integer h, we shall show that

$$\lim_{N\to\infty} \frac{1}{N} \sum_{n=1}^{N} \exp(h[n\theta]\alpha) = 0. \tag{1.10}$$

We have

$$\exp(h[n\theta]\alpha) = \exp(h(n\theta\alpha) - h\{n\theta\}\alpha) = f(n\theta, n\theta\alpha),$$

where $f(x, y) = \exp(hy - h\{x\}\alpha)$ for $(x, y) \in \mathbb{R}^2$. Note that f is periodic mod 1 in each variable. Since $((n\theta, n\theta\alpha))$, $n = 1, 2, \ldots$, is u.d. mod 1 in $\mathbb{R}^2$, we have

$$\lim_{N\to\infty} \frac{1}{N} \sum_{n=1}^{N} \exp(h[n\theta]\alpha) = \lim_{N\to\infty} \frac{1}{N} \sum_{=1}^{N} f(n\theta, n\theta\alpha) = \int_0^1\int_0^1 f(x, y)\,dx\,dy$$

by Exercise 6.3 of Chapter 1. Since the above double integral is zero, (1.10) is established. ■

Sequences of Polynomial Values

Let $f(x)$ be a polynomial with integral coefficients. A useful observation is that the question whether or not the sequence $(f(n))$, $n = 1, 2, \ldots$, is u.d. mod m is equivalent to the question whether or not the integers $f(1), f(2), \ldots, f(m)$ constitute a complete residue system mod m. Let us first consider monomials ax^k with $a \neq 0$ and $k \geq 1$. Obviously, if $k = 1$, the sequence (an), $n = 1, 2, \ldots$, is u.d. mod m if and only if $(a, m) = 1$.

THEOREM 1.9. Let p be a prime and let $k \geq 1$. Determine the integer K by the conditions $k = p^s K$ and $(p, K) = 1$. Then (an^k), $n = 1, 2, \ldots$, is u.d. mod p if and only if $(a, p) = (K, p - 1) = 1$.

PROOF. Let (an^k) be u.d. mod p. Then first $(a, p) = 1$. Moreover, for any integer x, we have $x^k = x^{p^s K}$ and $x^{p^s K} \equiv x^K \pmod{p}$ by Fermat's theorem. Let $(b, p) = 1$. Then $x^K \equiv b \pmod{p}$ has $(K, p - 1)$ incongruent solutions or no solution depending on whether

$$b^{(p-1)/(K,p-1)} \equiv 1 \pmod{p} \tag{1.11}$$

or not. Now (1.11) is satisfied by $b = 1$. Thus, the congruence $x^K \equiv 1 \pmod{p}$ has $(K, p - 1)$ incongruent solutions and therefore the congruence $ax^k \equiv a \pmod{p}$ has $(K, p - 1)$ incongruent solutions. But according to our assumption, $ax^k \equiv a \pmod{p}$ has only one solution; hence, $(K, p - 1) = 1$.

Now we shall show the sufficiency of the condition. Suppose $(a, p) = (K, p - 1) = 1$. Then, (1.11) is true for each $b = 1, 2, \ldots, p - 1$. To each such b there corresponds a unique x, $0 < x < p$, with $x^k \equiv b \pmod{p}$. It follows that the sequence (an^k) is u.d. mod p. ■

COROLLARY 1.3. For $k \geq 2$, there are infinitely many primes p such that (an^k), $n = 1, 2, \ldots$, is not u.d. mod p.

PROOF. Let q be any prime such that $q \mid k$. The arithmetic progression $1 + q, 1 + 2q, \ldots$ contains an infinite number of primes. Let p be any such prime with $p > k$. Then q is a divisor of $p - 1$. Thus, $(k, p - 1) > 1$ and according to Theorem 1.9 the sequence (an^k) is not u.d. mod p. Obviously there are infinitely many primes of the required type. ■

COROLLARY 1.4. If $k \geq 1$ and odd, then there exist infinitely many primes p such that (an^k), $n = 1, 2, \ldots$, is u.d. mod p.

PROOF. We have $(2, k) = 1$. Then the arithmetic progression $2 + k$, $2 + 2k, \ldots$ contains an infinite number of primes. Let $p = 2 + mk$ be any such prime with $p > |a|$. If d is a divisor of $p - 1 = 1 + mk$ and if d is also a divisor of k, then d must be a divisor of 1. Hence, $(k, p - 1) = 1$ and by Theorem 1.9 we see that (an^k) is u.d. mod p. Clearly there are infinitely many primes of that type. ■

THEOREM 1.10. For $k \geq 2$, the sequence (an^k), $n = 1, 2, \ldots$, is u.d. mod m if and only if m is square-free and (an^k) is u.d. mod p for each prime divisor p of m.

PROOF. Set $f(n) = an^k$. Let m be square-free, say, $m = p_1 \cdots p_r$ with distinct primes $p_1, \ldots, p_r$, and suppose $(f(n))$ is u.d. mod p_i for $1 \leq i \leq r$. If $f(x) \equiv f(y) \pmod{m}$ with $1 \leq x, y \leq m$, then $f(x) \equiv f(y) \pmod{p_i}$ for $1 \leq i \leq r$, and so, $x \equiv y \pmod{p_i}$ for $1 \leq i \leq r$. It follows that $x = y$; therefore, $(f(n))$ is u.d. mod m.

Conversely, suppose $(f(n))$ is u.d. mod m. Assume that there exists a prime p with $p^2 \mid m$. Since $(f(n))$ is u.d. mod p, we must have $(a, p) = 1$. But then the congruence $ax^k \equiv p \pmod{p^2}$ is not solvable for x, a contradiction to the fact that $(f(n))$ should be u.d. mod p^2. Thus, m must be square-free, and of course $(f(n))$ is u.d. modulo every prime divisor of m. ■

We mention the following result concerning arbitrary polynomials over $\mathbb{Z}$.

THEOREM 1.11. Let $f(x)$ be any polynomial with integer coefficients. Then:

i. The sequence $(f(n))$, $n = 1, 2, \ldots$, is u.d. in $\mathbb{Z}$ if and only if $f(x)$ is of the form $\pm x + c$.
ii. If $f(x)$ is not linear, then there are infinitely many primes p for which $(f(n))$ is not u.d. mod p.

PROOF. We first show (ii). The case of constant polynomials being trivial, assume that $f(x)$ has degree at least 2. Since $(f(n))$ and $(f(n) - f(0))$ behave the same way, it suffices to consider $F(x) = f(x) - f(0)$. If $F(x)$ is a monomial, then we are done by Corollary 1.3. If $F(x)$ is not a monomial, write $F(x) = x^j g(x)$ with $j \geq 1$ and

$$g(x) = a_0 + a_1 x + \cdots + a_k x^k, \qquad k \geq 1,\ a_0 \neq 0,\ a_k \neq 0.$$

Now for any such nonconstant polynomial $g(x)$ there are infinitely many primes p such that $g(x) \equiv 0 \pmod{p}$ is solvable (see Exercise 1.14). Choose such a prime $p > |a_0|$, and let r be an integer such that $g(r) \equiv 0 \pmod{p}$. Then $r \not\equiv 0 \pmod{p}$, since $g(0) = a_0 \not\equiv 0 \pmod{p}$. Hence, $F(x) \equiv 0 \pmod{p}$ has solutions $x \equiv r \pmod{p}$ and $x \equiv 0 \pmod{p}$, and therefore, $(F(n))$ is not u.d. mod p.

By (ii), only sequences $(f(n))$ with linear $f(x)$ can be u.d. in $\mathbb{Z}$. Now the sequence $(an + c)$, $n = 1, 2, \ldots$, is u.d. mod m if and only if $(a, m) = 1$. Hence, if $f(x)$ is of the form $ax + c$, then $(f(n))$ is u.d. in $\mathbb{Z}$ if and only if $a = \pm 1$. ■

Measure-Theoretic Approach

Making use of the so-called *measure* μ *of Banach-Buck* on the set $\mathbb{Z}^+$ of positive integers, the above theory can be extended in the following sense. Let $\mathscr{R}$ be the algebra generated by all finite subsets of $\mathbb{Z}^+$ and all subsets of $\mathbb{Z}^+$ that are arithmetic progressions. If $E \subseteq \mathbb{Z}^+$ and E is finite, then define $\mu(E) = 0$; if $E \subseteq \mathbb{Z}^+$ is an arithmetic progression with common difference d, then $\mu(E) = 1/d$, and if $A \in \mathscr{R}$, $B \in \mathscr{R}$, and $A \cap B$ is finite, then $\mu(A \cup B) = \mu(A) + \mu(B)$. Let $\mathscr{S}$ be the class of all subsets of $\mathbb{Z}^+$; then we define an outer measure μ^* on $\mathscr{S}$ by

$$\mu^*(E) = \inf \{\mu(H) \colon E \mathrel{\dot{\subseteq}} H \in \mathscr{R}\},$$

where $E \mathrel{\dot{\subseteq}} H$ means $E \backslash C \subseteq H \backslash C$ for some finite $C \subseteq \mathbb{Z}^+$. Let E' denote the complement of E with respect to $\mathbb{Z}^+$. Let $\mathscr{M}$ be the class of all sets $E \in \mathscr{S}$

such that

$$\mu^*(X) = \mu^*(X \cap E) + \mu^*(X \cap E') \qquad \text{for all } X \in \mathscr{S}.$$

Then μ^* is a finitely additive measure on $\mathscr{M}$, $\mathscr{R} \subseteq \mathscr{M}$, and $\mu^*(E) = \mu(E)$ for $E \in \mathscr{R}$. Call $\mathscr{M}$ the set of all measurable sets of $\mathbb{Z}^+$ and write $\mu(E)$ instead of $\mu^*(E)$ for $E \in \mathscr{M}$.

THEOREM 1.12. (i) Let $A = (a_n)$ be a sequence of positive integers u.d. in $\mathbb{Z}$. Then $\mu^*(A) = 1$. (ii) Conversely, let $A = (a_n)$ be an increasing sequence of positive integers. If A is measurable and $\mu(A) = 1$, then the sequence (a_n) is u.d. in $\mathbb{Z}$.

PROOF. (i) If the sequence (a_n) is u.d. in $\mathbb{Z}$, then

$$\lim_{N\to\infty} \frac{1}{N} \sum_{n=1}^{N} c_E(a_n) = \mu(E)$$

for all $E \in \mathscr{R}$, and conversely. (Here c_E denotes the characteristic function of the set E.) Because of the structure of $\mathscr{R}$ the equivalence of both definitions of u.d. is quite obvious. If $A \doteq H$ and $H \in \mathscr{R}$, then $A \subseteq H \cup C$ for some finite set C, $C \in \mathscr{R}$, $\mu(C) = 0$, and

$$1 = \lim_{N\to\infty} \frac{1}{N} \sum_{n=1}^{N} c_A(a_n) \leq \lim_{N\to\infty} \frac{1}{N} \sum_{n=1}^{N} c_H(a_n) + \lim_{N\to\infty} \frac{1}{N} \sum_{n=1}^{N} c_C(a_n)$$
$$= \mu(H) + \mu(C) = \mu(H).$$

Since $\mu(H) \leq 1$ it follows that $\mu(H) = 1$, and hence, A has outer measure 1. This establishes part (i).

(ii) Let m, j be integers such that $m \geq 2$, $0 \leq j < m$, and let $F = \{n \in \mathbb{Z}^+ : n \equiv j \pmod{m}\}$. Then

$$\frac{1}{N} \sum_{n=1}^{N} c_{F\cap A}(n) + \frac{1}{N} \sum_{n=1}^{N} c_{F\cap A'}(n) = \frac{1}{N} \sum_{n=1}^{N} c_F(n) = \frac{1}{m} + o(1) \tag{1.12}$$

as $N \to \infty$.

If $a_k \leq N < a_{k+1}$, then $A(N)$, the number of terms a_n with $a_n \leq N$, is equal to k, and hence,

$$\frac{1}{N} \sum_{n=1}^{N} c_{F\cap A}(n) = \frac{A(N) \cdot A(j, m, k)}{Nk}. \tag{1.13}$$

Moreover,

$$0 \leq \frac{1}{N} \sum_{n=1}^{N} c_{F\cap A'}(n) \leq \frac{1}{N} \sum_{n=1}^{N} c_{A'}(n) = 1 - \frac{A(N)}{N}. \tag{1.14}$$

Now it is known that for any set E with a density $D(E) = \lim_{n\to\infty} E(n)/n$, we have $D(E) \leq \mu^*(E) \leq 1$ and that in case E is measurable, we have the existence of $D(E)$ and the relation $D(E) = \mu(E)$ (see Buck [1, p. 572]). Hence, since A is measurable and $\mu(A) = 1$, we have $\lim_{N\to\infty} A(N)/N = 1$, and so, from (1.12), (1.13), and (1.14), we obtain the relation

$$\lim_{k\to\infty} \frac{A(j, m, k)}{k} = \frac{1}{m}. \quad \blacksquare$$

EXAMPLE 1.3. The condition in Theorem 1.12, part (ii), that the sequence (a_n) be increasing cannot be omitted as is shown by the following example: Set $a_{2n} = n$ and $a_{2n-1} = 1$, $n = 1, 2, \ldots$; then $A = \mathbb{Z}^+$ and $\mu(\mathbb{Z}^+) = 1$, but obviously the sequence is not u.d. in $\mathbb{Z}$. Furthermore, the condition in (ii) that A be measurable with $\mu(A) = 1$ cannot be weakened to $\mu^*(A) = 1$, as is shown by the following example: Let α be irrational and >1. Set $b_n = [n\alpha]$, then $B = (b_n)$ is u.d. in $\mathbb{Z}$ by Theorem 1.5, and as can be shown, B has outer measure 1 (see Buck [1, p. 570]). Let $A = B \cup \{2n\colon n \in \mathbb{Z}^+\}$, and let (a_n) be the increasing sequence consisting of all elements of A. Then evidently $\mu^*(A) = 1$, but the sequence (a_n) is not u.d. mod 2, for if $[N\alpha] = a_{k(N)}$, then $A(1, 2, k(N)) = \frac{1}{2}N + o(N)$ and $A(0, 2, k(N)) = \frac{1}{2}N\alpha + o(N)$ as $N \to \infty$. Furthermore $A(1, 2, k(N)) + A(0, 2, k(N)) = k(N)$, so that

$$\lim_{N\to\infty} A(1, 2, k(N))/k(N) = (1 + \alpha)^{-1}. \quad \blacksquare$$

Independence

Let m be an integer ≥ 2. Let $(f(n))$ be a sequence of integers all reduced mod m, or $0 \leq f(n) \leq m - 1$ for $n \geq 1$. Let $A(j, m, N)$ have the usual meaning. If $h(j) = \lim_{N\to\infty} A(j, m, N)/N$ exists for every $j = 0, 1, \ldots, m - 1$, then the sequence $(f(n))$ is said to be *relatively measurable*. It is easy to see that if the sequence $(f(n))$ is relatively measurable, then the mean value

$$M(f) = \lim_{N\to\infty} \frac{1}{N} \sum_{n=1}^{N} f(n)$$

of the sequence $(f(n))$ exists.

Now let $(f_1(n)), (f_2(n)), \ldots, (f_k(n))$ be k relatively measurable sequences all reduced mod m. Let $j_1, j_2, \ldots, j_k$ be k integers with $0 \leq j_r < m$ for $1 \leq r \leq k$. The sequences $(f_1(n)), (f_2(n)), \ldots, (f_k(n))$ are called *independent* if the limit

$$h(j_1, \ldots, j_k) = \lim_{N\to\infty} \frac{1}{N} \operatorname{card}\{n\colon f_1(n) = j_1, f_2(n) = j_2, \ldots, f_k(n) = j_k;\ 1 \leq n \leq N\}$$

exists and if, moreover,

$$h(j_1, \ldots, j_k) = h_1(j_1)h_2(j_2) \cdots h_k(j_k)$$

for all $(j_1, \ldots, j_k)$. Here $h_r(j_r)$ refers to the sequence $(f_r(n))$, $1 \leq r \leq k$. There is a criterion for the independence of k sequences of integers reduced mod m. First, the notion of mean value $M(f)$ can be extended to more general sequences. We have then the property that the relatively measurable sequences $(f_1(n)), (f_2(n)), \ldots, (f_k(n))$ are independent if and only if

$$M(e^{2\pi i(h_1 f_1 + h_2 f_2 + \cdots + h_k f_k)/m}) = \prod_{r=1}^{k} M(e^{2\pi i h_r f_r/m})$$

for all integers $h_1, h_2, \ldots, h_k$ with $0 \leq h_r < m$ for $1 \leq r \leq k$. In the following, we present some examples of how the notion of independence can be used in the theory of u.d. mod m.

EXAMPLE 1.4. Let the sequences $(f_1(n)), (f_2(n)), \ldots, (f_k(n))$, all reduced mod m, be independent. Suppose that the sequence $(f_1(n))$ is u.d. mod m. Then the sequence $(h_1 f_1(n) + h_2 f_2(n) + \cdots + h_k f_k(n))$ is u.d. mod m whenever $h_r \in \mathbb{Z}$ for $1 \leq r \leq k$ and $(h_1, m) = 1$. See Exercise 1.18. ■

EXAMPLE 1.5. Let (a_n) and (b_n) be two sequences of integers, both reduced mod m. Then the sequence $(a_n + mb_n)$ is u.d. mod m^2 if and only if simultaneously: (i) (a_n), (b_n) are u.d. mod m; (ii) (a_n) and (b_n) are independent. See Exercise 1.19. ■

Measurable Functions

Let $f(t)$ be a real-valued function with $[0, \infty)$ as its domain and Lebesgue measurable on each interval $[0, T]$, $T > 0$. Let $m \geq 2$ be an integer, let $[f(t)]_m$ denote the greatest multiple of m less than, or equal to, $f(t)$ and set $\{f(t)\}_m = f(t) - [f(t)]_m$. Let $\lambda(E)$ denote the Lebesgue measure of a Lebesgue-measurable set E. By $E(T)$ we denote the intersection $E \cap [0, T]$. In the following, we consider the m sets

$$E_j = \{t \in [0, \infty): j \leq \{f(t)\}_m < j + 1\}, \qquad j = 0, 1, \ldots, m - 1.$$

DEFINITION 1.2. The function $f(t)$ is said to be *c.u.d. mod m* if for every $j = 0, 1, \ldots, m - 1$,

$$\lim_{T \to \infty} \frac{1}{T} \lambda(E_j(T)) = \frac{1}{m}.$$

If these relations hold for every $m = 2, 3, \ldots$, then $f(t)$ is said to be *c.u.d. in* $\mathbb{Z}$.

THEOREM 1.13. Let the function $f(t)$ be c.u.d. in $\mathbb{Z}$. Let $w(u)$ be a Riemann-integrable function on $[0, 1]$. Then

$$\lim_{m\to\infty} \lim_{T\to\infty}{}^{*} \frac{1}{T}\int_0^T w(\{f(t)\}_m/m)\,dt = \int_0^1 w(u)\,du, \tag{1.15}$$

where lim* means $\underline{\lim}$ or $\overline{\lim}$.

PROOF. Let $W(t)$ denote the integrand of the integral on the left-hand side of (1.15). We have

$$\int_0^T W(t)\,dt = \sum_{j=0}^{m-1}\int_{E_j(T)} W(t)\,dt.$$

Let $W^*(j, T)$ denote the supremum of $W(t)$ in the domain $E_j(T)$, and $W_*(j, T)$ its infimum in $E_j(T)$. Then we have

$$\sum_{j=0}^{m-1}\frac{\lambda(E_j(T))}{T}W_*(j, T) \leq \frac{1}{T}\int_0^T W(t)\,dt \leq \sum_{j=0}^{m-1}\frac{\lambda(E_j(T))}{T}W^*(j, T).$$

Now let $T\to\infty$. Let $W^*(j)$ denote the supremum of $W(t)$ in E_j and let $w^*(j)$ denote the supremum of $w(u)$ in the interval $[j/m, (j+1)/m)$. Let $W_*(j)$ and $w_*(j)$ have similar meanings. Taking into account that $f(t)$ is c.u.d. in $\mathbb{Z}$, we see that the above inequalities lead to

$$\sum_{j=0}^{m-1} w_*(j)/m \leq \sum_{j=0}^{m-1} W_*(j)/m \leq \underline{\lim}_{T\to\infty}\frac{1}{T}\int_0^T W(t)\,dt$$

$$\leq \overline{\lim}_{T\to\infty}\frac{1}{T}\int_0^T W(t)\,dt \leq \sum_{j=0}^{m-1} W^*(j)/m \leq \sum_{j=0}^{m-1} w^*(j)/m.$$

According to the definition of the Riemann integral, the extreme members of the above inequalities tend to $\int_0^1 w(u)\,du$ as $m\to\infty$. ■

Notes

The notions of u.d. mod m and u.d. in $\mathbb{Z}$ are from Niven [2] (see also Niven [4]). Zame [6] has a generalization of Theorem 1.1. For Theorem 1.2 see Niven [2], who has a partial result, and S. Uchiyama [1] and Kuipers [12]. We refer also to Exercise 1.6. Various results relating u.d. in $\mathbb{Z}$ with density were established by Niven [2] (see Theorem 1.3), Vanden Eynden [1], Dijksma and Meijer [1] (see Exercise 1.17), Kelly [1], and Carlson [1]. The latter author proved the following interesting theorem: Suppose $0 \leq \alpha \leq \beta \leq 1$; then there is an increasing sequence of positive integers that is u.d. in $\mathbb{Z}$ and that has lower asymptotic density α and upper asymptotic density β. For Theorems 1.4 and 1.6 see Vanden Eynden [1]. Theorem 1.5 and Corollary 1.2 were shown by Niven [2]. See also Niven [3, pp. 27–28]. Theorem 1.7 is from Kuipers and Uchiyama [1], who corrected a statement of S. Uchiyama [1]. It should be noted that Meijer and Sattler [1] constructed a sequence (b_n) of integers such that (b_n) is u.d. in $\mathbb{Z}$ and $(b_n\alpha)$ is not u.d. mod 1 for all α in a set

$V \subseteq [0, 1]$ with Lebesgue measure $\lambda(V) = 1$. This improves an earlier result of Meijer [4]. Theorem 1.8 is from Carlson [1], who also studies sequences $([P(n)]\alpha)$ with a polynomial $P(x)$ over $\mathbb{R}$. Sequences of polynomial values were investigated by Niven [2], Zane [1], and Cavior [1]. Theorem 1.12 is from Dijksma and Meijer [1], who corrected an earlier version by M. and S. Uchiyama [1]. The measure μ on $\mathbb{Z}^+$ was introduced by Buck [1]. The sections on independence and on measurable functions are based on the work of Kuipers and Shiue [5] and Kuipers [8], respectively.

The Fibonacci sequence and its generalizations were investigated with respect to u.d. by Kuipers and Shiue [1, 2, 3, 4] and Niederreiter [7]. For sequences arising from q-adic expansions, see Kuipers and Shiue [5] and Carlson [1]. The distribution mod m of integer-valued additive functions was studied by Delange [1, 2, 5, 8, 11]. Veech [1, 2, 4] and Hlawka [27] discuss u.d. mod m of counting functions. For sequences of integral parts of exponential sequences, see Forman and Shapiro [1] and Shapiro and Sparer [1]. A notion of weak u.d. mod m was introduced by Narkiewicz [1]. See also Śliwa [1] and S. Uchiyama [3]. For "irregularities of distribution", see Hodges [4] (his results can be improved using Schmidt's lower bounds in Section 2 of Chapter 2). U.d. mod m with respect to summation methods is discussed in Schnabl [1]. Metric results for u.d. in $\mathbb{Z}$ were shown by Chauvineau [4, 6]. A notion of Hartman-u.d. in $\mathbb{Z}$ (see Chapter 4, Example 5.11) with uniformity condition was studied by Veech [4]. The theory of u.d. in $\mathbb{Z}^k$, the additive group of k-dimensional lattice points, was developed by Niederreiter [8, 12]. In particular, a multidimensional version of Theorem 1.5 is established in these papers.

Exercises

1.1. If a sequence of integers is u.d. mod m and if $k \mid m$ and $k \geq 2$, then the sequence is also u.d. mod k.

1.2. If a sequence of integers is u.d. mod m and u.d. mod k with $(m, k) = 1$, then the sequence is not necessarily u.d. mod mk.

1.3. If a sequence of integers is not u.d. in $\mathbb{Z}$, then there are infinitely many moduli m for which the sequence is not u.d. mod m.

1.4. Let both sequences (a_n) and (b_n) be u.d. mod m. Then the sequence (a_nb_n) is not u.d. mod m^2.

1.5. For a sequence (a_n) in $\mathbb{Z}$, prove that

$$\sum_{h=1}^{m-1} \left| \frac{1}{N} \sum_{n=1}^{N} e^{2\pi i h a_n/m} \right|^2 = m \sum_{j=0}^{m-1} \left(\frac{A(j, m, N)}{N} - \frac{1}{m} \right)^2 \qquad \text{for every } N \geq 1.$$

1.6. Deduce Theorem 1.2 from Exercise 1.5.

1.7. Prove the equivalence of (1.1) and (1.4) for increasing sequences of positive integers.

1.8. Prove statements (i)–(iii) in Example 1.1.

1.9. The sequence of the integral parts of the logarithms of the Fibonacci numbers is u.d. in $\mathbb{Z}$ (compare with Exercise 3.3 of Chapter 1).

1.10. For $\sigma > 0$, $\sigma \notin \mathbb{Z}$, prove that the sequence (a_n), $n = 1, 2, \ldots$, defined by $a_n = [n^\sigma x]$, is u.d. in $\mathbb{Z}$ for every real number $x \neq 0$.

1.11. Show that in the sufficiency part of Theorem 1.6, we need only assume that $([mx_n])$ is u.d. mod m for infinitely many $m \geq 2$.

1.12. See Example 1.2. Prove in detail that if x is normal to base 2, then the corresponding (a_n) is u.d. in $\mathbb{Z}$.

1.13. Does the converse of the statement in Exercise 1.12 hold?

1.14. Prove that every nonconstant polynomial $f(x)$ with integral coefficients has infinitely many prime divisors, that is, that there exist infinitely many primes p such that $f(x) \equiv 0 (\text{mod } p)$ is solvable. *Hint:* Proceed as in Euclid's proof of the infinitude of primes.

1.15. Let $f(x)$ be a polynomial with integral coefficients. If $(f(n))$ is u.d. mod m and u.d. mod k with $(m, k) = 1$, then $(f(n))$ is u.d. mod mk. Compare with Exercise 1.2.

1.16. Show that the measure μ on $\mathbb{Z}^+$ is not σ-additive.

1.17. Let $A = (a_n)$ be an increasing sequence of positive integers. Prove that if A has density $D(A) = 1$, then (a_n) is u.d. in $\mathbb{Z}$.

1.18. Prove the result enunciated in Example 1.4. *Hint:* Use the criterion for independence.

1.19. Prove the result enunciated in Example 1.5. *Hint:* Use elementary counting arguments.

1.20. If the sequence (x_n/m), $n = 1, 2, \ldots$, is u.d. mod 1, then the sequence $([x_n])$ is u.d. mod m.

1.21. For any $m \geq 2$, the function $\log(t + 1)$, $t \geq 0$, is not c.u.d. mod m.

1.22. Let the function $f(t)$, $t \geq 0$, be c.u.d. mod m and let $w(u)$ be defined at the points $u = j/m$, $j = 0, 1, \ldots, m - 1$. Then,

$$\lim_{T\to\infty} \frac{1}{T} \int_0^T w([\{f(t)\}_m]/m)\, dt = \frac{1}{m} \sum_{j=0}^{m-1} w(j/m).$$

1.23. The function $f(t)$, $t \geq 0$, is c.u.d. mod m if and only if

$$\lim_{T\to\infty} \frac{1}{T} \int_0^T e^{2\pi i h [\{f(t)\}_m]/m}\, dt = 0 \qquad \text{for } h = 1, 2, \ldots, m - 1.$$

1.24. Prove that Theorem 1.7 remains true if (a_n) is u.d. mod m for infinitely many $m \geq 2$.

2. ASYMPTOTIC DISTRIBUTION IN $\mathbb{Z}_p$

Definitions and Important Properties

Let p be a prime number. Let $\mathbb{Z}_p$ be the ring of p-adic integers. Let (x_n), $n = 1, 2, \ldots$, be a sequence of elements of $\mathbb{Z}_p$, and let E be a subset of $\mathbb{Z}_p$. Let $A(E; N)$ denote the number of elements among $x_1, \ldots, x_N$ that are contained in E. Let $\mathscr{A}$ be the algebra of finite unions of open spheres in $\mathbb{Z}_p$.

DEFINITION 2.1. If for every $E \in \mathscr{A}$ the limiting value

$$\chi(E) = \lim_{N\to\infty} \frac{A(E; N)}{N}$$

exists, then the sequence (x_n) is said to have the *distribution function* χ.

The function χ is a σ-additive set function on $\mathscr{A}$ with $\chi(\mathbb{Z}_p) = 1$ (compare with Exercise 2.10). Let now E be the sphere $\alpha + p^k\mathbb{Z}_p$, where $\alpha \in \mathbb{Z}_p$ and k is a positive integer. Write $E = S_k(\alpha)$, $A(E; N) = A(\alpha, k, N)$, $\chi(E) = \chi_k(\alpha)$. Evidently the sequence (x_n) in $\mathbb{Z}_p$ has the distribution function χ if and only if for every $\alpha \in \mathbb{Z}_p$ and every positive integer k,

$$\chi_k(\alpha) = \lim_{N\to\infty} \frac{A(\alpha, k, N)}{N}.$$

DEFINITION 2.2. For given $k \geq 1$, the sequence (x_n) in $\mathbb{Z}_p$ is said to be *uniformly distributed of order* k (abbreviated k-u.d.) in $\mathbb{Z}_p$ if for every $\alpha \in \mathbb{Z}_p$, $\chi_k(\alpha)$ exists and

$$\chi_k(\alpha) = \lim_{N\to\infty} \frac{A(\alpha, k, N)}{N} = p^{-k}. \tag{2.1}$$

The sequence (x_n) is said to be *u.d. in* $\mathbb{Z}_p$ if (x_n) is k-u.d. for every $k = 1, 2, \ldots$.

THEOREM 2.1. Let $a, b \in \mathbb{Z}_p$. The sequence $(na + b)$, $n = 1, 2, \ldots$, is u.d. in $\mathbb{Z}_p$ if and only if a is a unit.

PROOF. Let $E = S_k(\alpha)$ be the sphere $\alpha + p^k\mathbb{Z}_p$. Let a be a unit of $\mathbb{Z}_p$. Distinguish two cases. First, suppose that a and b are nonnegative rational integers. We have $(a, p) = 1$. The sequence $(na + b)$, $n = 1, 2, \ldots$, has the property that p^k consecutive terms form a complete residue system mod p^k. Now it is easily shown that (2.1) holds. For the condition $na + b \in \alpha + p^k\mathbb{Z}_p$ is equivalent to the first k coefficients in the canonical representations of $na + b$ and α being the same. Hence, $A(\alpha, k, N)/N \to p^{-k}$ as $N \to \infty$. Second, let a and b be arbitrary p-adic integers with a being a unit. Consider two nonnegative rational integers a^* and b^* whose p-adic expansions coincide with those of a and b, respectively, for indices $<k$ and whose coefficients for indices $\geq k$ vanish. We have then $a \in S_k(a^*)$ and $b \in S_k(b^*)$, and therefore, $na + b \in S_k(na^* + b^*)$ for $n = 1, 2, \ldots$, or the p-adic integers $na + b$ and $na^* + b^*$ have the same first k coefficients. According to the first part of the proof, the sequence $(na^* + b^*)$ is u.d. in $\mathbb{Z}_p$, and hence, $(na + b)$ is k-u.d. This holds for all $k = 1, 2, \ldots$.

If a is not a unit, then no element of $(na + b)$ lies in the sphere $b + 1 + p\mathbb{Z}_p$, and so, the sequence is not u.d. in $\mathbb{Z}_p$. ■

Theorem 2.1 implies that the sequence of natural numbers is u.d. in $\mathbb{Z}_p$.

Furnished with the norm given by the p-adic valuation, the quotient field $\mathbb{Q}_p$ of $\mathbb{Z}_p$ is a complete, separable and locally compact space. With respect to addition, $\mathbb{Q}_p$ is a locally compact abelian group containing $\mathbb{Z}_p$ as a compact subgroup. The notions of u.d. in $\mathbb{Z}_p$ given by Definition 1.1 of Chapter 4 and Definition 2.2 of this chapter coincide (compare with Theorem 2.3). We note that $\mathbb{Q}_p$ can be provided with a Haar measure μ such that $\mu(\mathbb{Z}_p) = 1$.

Consider a set $V \subseteq \mathbb{Z}_p$ coinciding with some residue class mod $p^v\mathbb{Z}_p$, $v \geq 1$, or a union of some classes mod $p^v\mathbb{Z}_p$. Let $\mu(V)$ denote the Haar measure of V.

DEFINITION 2.3. A sequence (x_n) of elements of V is said to be *u.d. in* V if for all $\alpha \in V$ and all integers $k \geq v$,

$$\lim_{N\to\infty} \frac{A(\alpha, k, N)}{N} = \frac{p^{-k}}{\mu(V)}.$$

THEOREM 2.2. Let p be an odd prime and let $b \in \mathbb{Z}_p$ be a unit. Then there exists a rational integer a (in fact, there exist infinitely many) such that the sequence $(a^n b)$, $n = 1, 2, \ldots$, is u.d. in the set U formed by the units of $\mathbb{Z}_p$.

PROOF. Note first that U is the union of $p - 1$ residue classes mod $p\mathbb{Z}_p$. Choose the rational integer a to be a primitive root mod p^2. For given $k \geq 1$, let $\bar{b}$ be the rational integer whose expansion coincides with that of b for terms of index $<k$ and whose terms of index $\geq k$ vanish. Note that a is a primitive root mod p^k. Hence, if n runs through $p^{k-1}(p - 1)$ consecutive positive integers, the product $a^n\bar{b}$ runs through a reduced residue system mod p^k. Since $a^n b$ and $a^n\bar{b}$ are identical mod $p^k\mathbb{Z}_p$, we have for all $\alpha \in U$,

$$A(\alpha, k, N) = \frac{N}{p^{k-1}(p-1)} + \theta, \qquad |\theta| < 1. \tag{2.2}$$

However, $\mu(U) = (p - 1)/p$; hence, from (2.2), it follows that

$$\lim_{N\to\infty} A(\alpha, k, N)/N = p^{-k}/\mu(U). \quad \blacksquare$$

Weyl Criteria

THEOREM 2.3. The sequence (x_n) of p-adic integers is u.d. in $\mathbb{Z}_p$ if and only if for every real-valued Riemann-integrable function f on $\mathbb{Z}_p$, we have

$$\lim_{N\to\infty} \frac{1}{N} \sum_{n=1}^{N} f(x_n) = \int_{\mathbb{Z}_p} f\, d\mu. \tag{2.3}$$

PROOF. The condition is sufficient. For let $c_{S_k(\alpha)}$ be the characteristic function of a sphere $S_k(\alpha)$; then (2.3) becomes

$$\lim_{N\to\infty} \frac{1}{N} \sum_{n=1}^{N} c_{S_k(\alpha)}(x_n) = \int_{\mathbb{Z}_p} c_{S_k(\alpha)}\, d\mu = p^{-k},$$

or the sequence (x_n) is u.d. in $\mathbb{Z}_p$.

Now, we show the necessity of (2.3). Without loss of generality, we suppose that f is a nonnegative function on $\mathbb{Z}_p$ with $K = \sup_{x\in\mathbb{Z}_p} f(x) > 0$. For $k \geq 1$, consider the subdivision of $\mathbb{Z}_p$ into pairwise disjoint spheres $S_k(\alpha)$, say, $S^{(1)}, \ldots, S^{(q)}$ with $q = p^k$. Let $r^{(i)}$, $R^{(i)}$ be the infimum resp. supremum of f on $S^{(i)}$. Set

$$m_k = \frac{1}{q} \sum_{i=1}^{q} r^{(i)}, \qquad M_k = \frac{1}{p} \sum_{i=1}^{q} R^{(i)}.$$

Suppose that f is Riemann-integrable. Then, given $\varepsilon > 0$, there exists a k_0 such that for all $k \geq k_0$,

$$\int_{\mathbb{Z}_p} f\, d\mu - \frac{\varepsilon}{2} \leq m_k \leq M_k \leq \int_{\mathbb{Z}_p} f\, d\mu + \frac{\varepsilon}{2}.$$

On the other hand, we have

$$\frac{1}{N} \sum_{n=1}^{N} f(x_n) = \sum_{i=1}^{q} \frac{1}{N} \sum_{\substack{n=1\\ x_n\in S^{(i)}}}^{N} f(x_n) \leq \sum_{i=1}^{q} R^{(i)} \left(\frac{1}{N} \sum_{n=1}^{N} c^{(i)}(x_n) \right), \tag{2.4}$$

where $c^{(i)}$ is the characteristic function of $S^{(i)}$. Since (x_n) is u.d. in $\mathbb{Z}_p$, we get for sufficiently large N,

$$\left| \frac{1}{N} \sum_{n=1}^{N} c^{(i)}(x_n) - p^{-k} \right| < \frac{\varepsilon p^{-k}}{2K} \qquad \text{for } i = 1, 2, \ldots, q. \tag{2.5}$$

Because of (2.4) and (2.5), we have

$$\frac{1}{N} \sum_{n=1}^{N} f(x_n) \leq \sum_{i=1}^{q} R^{(i)} \left(p^{-k} + \frac{\varepsilon p^{-k}}{2K} \right) \leq M_k + \frac{\varepsilon}{2} \leq \int_{\mathbb{Z}_p} f\, d\mu + \varepsilon$$

for sufficiently large N. Similarly, for N large enough,

$$\frac{1}{N} \sum_{n=1}^{N} f(x_n) \geq \int_{\mathbb{Z}_p} f\, d\mu - \varepsilon.$$

This proves the theorem completely. ■

In the following, $\mathbf{R}_p^{(h)}$ stands for the set of rationals in [0, 1) that in lowest terms have p^h as denominator, and we set

$$\mathbf{R}_p^k = \bigcup_{0\leq h\leq k} \mathbf{R}_p^{(h)} \quad \text{for } k \geq 1 \quad \text{and} \quad \mathbf{R}_p = \bigcup_{h\geq 0} \mathbf{R}_p^{(h)}.$$

THEOREM 2.4. The sequence (x_n) of p-adic integers has a distribution function χ in $\mathbb{Z}_p$ if and only if

$$c_r = \lim_{N\to\infty} \frac{1}{N} \sum_{n=1}^{N} e^{2\pi i r x_n} \tag{2.6}$$

exists for each $r \in \mathbf{R}_p$. Moreover, (2.6) implies

$$c_r = \sum_{j=0}^{p^k-1} e^{2\pi i r j} \chi_k(j) \tag{2.7}$$

for every $k \geq 1$ and every $r \in \mathbf{R}_p^k$, and

$$\chi_k(\alpha) = \frac{1}{p^k} \sum_{r\in\mathbf{R}_p^k} c_r e^{-2\pi i r \alpha} \tag{2.8}$$

for every $k \geq 1$ and every $\alpha \in \mathbb{Z}_p$.

PROOF. Let $r \in \mathbf{R}_p^k$ and $0 \leq j < p^k$. Then for every $x_n \in S_k(j)$ one has

$$e^{2\pi i r x_n} = e^{2\pi i r j}.$$

Hence, by partitioning,

$$\frac{1}{N} \sum_{n=1}^{N} e^{2\pi i r x_n} = \sum_{j=0}^{p^k-1} e^{2\pi i r j} \frac{A(j, k, N)}{N},$$

and if (x_n) has a distribution function χ, then c_r exists and (2.7) holds.

Conversely, suppose that c_r exists for every $r \in \mathbf{R}_p$. For $\alpha \in \mathbb{Z}_p$, consider the expression

$$A_N = \sum_{r\in\mathbf{R}_p^k} e^{-2\pi i r \alpha} \sum_{n=1}^{N} e^{2\pi i r x_n} = \sum_{n=1}^{N} \sum_{r\in\mathbf{R}_p^k} e^{2\pi i r (x_n-\alpha)}. \tag{2.9}$$

If $x_n \in S_k(\alpha)$, then the inner sum on the extreme right of (2.9) is equal to p^k, the number of r in $\mathbf{R}_p^k$, and if $x_n \notin S_k(\alpha)$, this sum is equal to

$$\sum_{j=0}^{p^k-1} e^{2\pi i j (x_n-\alpha)/p^k} = \frac{e^{2\pi i (x_n-\alpha)} - 1}{e^{2\pi i (x_n-\alpha)/p^k} - 1} = 0.$$

Hence,

$$A_N = \sum_{\substack{n=1 \\ x_n\in S_k(\alpha)}}^{N} p^k = p^k A(\alpha, k, N),$$

and so,

$$\frac{A(\alpha, k, N)}{N} \to \frac{1}{p^k} \sum_{r\in\mathbf{R}_p^k} e^{-2\pi i r \alpha} c_r \qquad \text{as } N \to \infty.$$

Therefore, (x_n) has a distribution function χ satisfying (2.8). ■

We recall that χ is a normed σ-additive set function on $\mathscr{A}$. Let $\mathscr{B}$ be the σ-algebra generated by $\mathscr{A}$, that is, the collection of all Borel sets of $\mathbf{Z}_p$. The unique extension $\bar{\chi}$ of χ to $\mathscr{B}$ is a probability measure on $\mathscr{B}$.

For fixed $\alpha \in \mathbf{Z}_p$, the sequence $(\chi_k(\alpha))$, $k = 1, 2, \ldots$, is nonincreasing. Moreover, $\chi_k(\alpha) \geq 0$ for $k \geq 1$; hence, $(\chi_k(\alpha))$ is convergent. So, one can define

$$j(\alpha) = \lim_{k\to\infty} \chi_k(\alpha)$$

for every $\alpha \in \mathbf{Z}_p$. We call $j(\alpha)$ the *jump function* of χ. The set function χ is said to be *continuous at* α if $j(\alpha) = 0$. Furthermore, χ is *continuous on* $\mathbf{Z}_p$ if $j(\alpha) = 0$ for all $\alpha \in \mathbf{Z}_p$.

THEOREM 2.5. If the sequence (x_n) has a distribution function χ in $\mathbf{Z}_p$, then

$$\lim_{k\to\infty} \frac{1}{p^k} \sum_{r\in\mathbf{R}_p^k} |c_r|^2 = \sum_{\alpha\in D} j^2(\alpha), \tag{2.10}$$

where $j(\alpha)$ denotes the jump function of χ on $\mathbf{Z}_p$ and D is the (countable) set of values $\alpha \in \mathbf{Z}_p$ where χ is discontinuous.

PROOF. According to (2.7) one has for every $r \in \mathbf{R}_p^k$, $k \geq 1$, the following identities:

$$|c_r|^2 = c_r\bar{c}_r = \sum_{h=0}^{p^k-1} e^{2\pi i r h}\chi_k(h) \sum_{m=0}^{p^k-1} e^{-2\pi i r m}\chi_k(m);$$

hence,

$$\sum_{r\in\mathbf{R}_p^k} |c_r|^2 = \sum_{h,m=0}^{p^k-1} \chi_k(h)\chi_k(m) \sum_{r\in\mathbf{R}_p^k} e^{2\pi i r(h-m)}.$$

On the right-hand side, the last sum is equal to p^k if $h = m$, and this sum vanishes if $h \neq m$. Hence, for every $k \geq 1$, one has

$$\frac{1}{p^k} \sum_{r\in\mathbf{R}_p^k} |c_r|^2 = \sum_{m=0}^{p^k-1} \chi_k^2(m). \tag{2.11}$$

Now $\chi_k(m)$ is the constant value of $\chi_k(\alpha)$ on the sphere $S_k(m)$ of $\bar{\chi}$-measure $\chi_k(m)$. Hence, the right-hand side of (2.11) can be expressed by means of an integral with respect to the measure $\bar{\chi}$ on $\mathscr{B}$, and so, we have

$$\frac{1}{p^k} \sum_{r\in\mathbf{R}_p^k} |c_r|^2 = \int_{\mathbf{Z}_p} \chi_k(\alpha)\, d\bar{\chi}(\alpha).$$

But $\chi_k(\alpha) \downarrow j(\alpha)$ as $k \to \infty$, and thus, according to the monotone convergence theorem,

$$\lim_{k\to\infty} \frac{1}{p^k} \sum_{r\in\mathbf{R}_p^k} |c_r|^2 = \int_{\mathbf{Z}_p} j(\alpha)\, d\bar{\chi}(\alpha) = \sum_{\alpha\in D} j^2(\alpha). \qquad \blacksquare$$

THEOREM 2.6. The sequence (x_n) has a continuous distribution function χ on $\mathbb{Z}_p$ if and only if c_r exists for every $r \in \mathbf{R}_p$ and

$$\lim_{k\to\infty} \frac{1}{p^k} \sum_{r\in\mathbf{R}_p^k} |c_r|^2 = 0. \tag{2.12}$$

PROOF. This follows immediately from Theorem 2.4, equation (2.10), and the fact that χ is continuous on $\mathbb{Z}_p$ if and only if $\sum_{\alpha\in D} j^2(\alpha) = 0$. ■

Notes

An introduction to p-adic number theory can be found in the books of Bachman [1], Borevich and Shafarevich [1], and Mahler [5]. U.d. in $\mathbb{Z}_p$ was first studied by Cugiani [1], who gave Definitions 2.2 and 2.3 and proved Theorems 2.1, 2.2, and 2.3. Definition 2.1 is from Chauvineau [6], where Theorems 2.4, 2.5, and 2.6 can also be found. Concerning these definitions and the Weyl criterion, see also Zaretti [1, 2] and Kuipers and Scheelbeek [2]. Chauvineau [3] considered some special sequences. Quantitative results were shown by Beer [1, 2, 3]. For metric theorems, see Chauvineau [4, 6] and F. Bertrandias [1]. P.V. numbers in $\mathbb{Q}_p$ were studied by Chabauty [1] and Schreiber [1, 2]. A notion of "very well distributed sequence" (*suite très bien répartie*) was introduced by Amice [1]. Chauvineau [5, 6] studied u.d. of functions in $\mathbb{Q}_p$.

The theory was extended to g-adic numbers (see Mahler [3]) by Meijer [2, 3]. See also Shiue [1]. Another extension is to rings of adeles. We refer to F. Bertrandias [1], Cantor [2], Decomps-Guilloux [1, 2], Grandet-Hugot [1], and Lesca [1].

Exercises

2.1. Let p be an odd prime, and let a and b be units of $\mathbb{Z}_p$ such that $a^x \notin 1 + p\mathbb{Z}_p$ for $1 \leq x \leq p - 2$ and $a^{p-1} \notin 1 + p^2\mathbb{Z}_p$. Prove that the sequence $(a^n b)$ is u.d. in U.

2.2. Prove the Weyl criterion for u.d. in $\mathbb{Z}_p$: The sequence (x_n) is u.d. in $\mathbb{Z}_p$ if and only if

$$\lim_{N\to\infty} \frac{1}{N} \sum_{n=1}^{N} e^{2\pi i r x_n} = 0 \qquad \text{for each } r \in \mathbf{R}_p,\ r \neq 0.$$

Hint: Use Theorem 2.4.

2.3. Prove that the characters of the compact additive group $\mathbb{Z}_p$ are exactly given by the functions ψ_r, $r \in \mathbf{R}_p$, defined by $\psi_r(\alpha) = e^{2\pi i r\alpha}$ for $\alpha \in \mathbb{Z}_p$. Note: In the light of Exercise 2.2, this shows again that the notions of u.d. in $\mathbb{Z}_p$ given by Definition 1.1 of Chapter 4 and Definition 2.2 of this chapter coincide.

2.4. For a prime p, let $\varphi_p\colon \mathbb{Z}_p \mapsto [0, 1]$ be the Monna map (see Monna [1]) defined as follows: For $\alpha = \sum_{i=0}^{\infty} a_i p^i$ in $\mathbb{Z}_p$, we set $\varphi_p(\alpha) = \sum_{i=0}^{\infty} a_i p^{-i-1}$. Prove that (x_n) is u.d. in $\mathbb{Z}_p$ if and only if $(\varphi_p(x_n))$ is u.d. mod 1.

2.5. The numbers c_r in Theorem 2.4 are all real if and only if $\chi_k(\alpha) = \chi_k(-\alpha)$ for every $k \geq 1$ and every $\alpha \in \mathbb{Z}_p$.
2.6. The set of points $\alpha \in \mathbb{Z}_p$ where χ is not continuous is countable.
2.7. Prove that (2.12) is equivalent to the condition

$$\lim_{k\to\infty} \frac{1}{p^k} \sum_{r\in\mathbf{R}_p^k} |c_r| = 0.$$

Hint: Use $|c_r| \leq 1$ and the Cauchy-Schwarz inequality.
2.8. Show the identity

$$\frac{1}{p^k} \sum_{r\in\mathbf{R}_p^{(k)}} c_r e^{-2\pi i r\alpha} = \chi_k(\alpha) - \frac{1}{p}\chi_{k-1}(\alpha)$$

for every $k \geq 1$ and every $\alpha \in \mathbb{Z}_p$, where we set $\chi_0(\alpha) = 1$ for all $\alpha \in \mathbb{Z}_p$.
2.9. Prove that the condition in Exercise 2.7 is equivalent to

$$\lim_{k\to\infty} \frac{1}{p^k} \sum_{r\in\mathbf{R}_p^{(k)}} |c_r| = 0.$$

Hint: Use Exercise 2.8.
2.10. Prove that every finitely additive set function on $\mathscr{A}$ is also σ-additive. *Hint:* Note that an open sphere is also closed in $\mathbb{Z}_p$.

3. UNIFORM DISTRIBUTION OF SEQUENCES IN $GF[q, x]$ AND $GF\{q, x\}$

Uniform Distribution in $GF[q, x]$

Let $\Phi = GF[q, x]$ denote the ring of polynomials in x over an arbitrary finite field $GF(q)$ of q elements, where $q = p^r$, p is a prime, and r is a positive integer. Let M be any element of Φ of degree $m > 0$. Then a complete residue system mod M contains q^m elements.

DEFINITION 3.1. Let $\theta = (A_n)$, $n = 1, 2, \ldots$, be an infinite sequence of elements of Φ. For any $B \in \Phi$ and any integer $N \geq 1$, define $A(B, M, N) = A(B, M, N, \theta)$ as the number of terms among $A_1, A_2, \ldots, A_N$ such that $A_n \equiv B(\text{mod } M)$. Then the sequence θ is said to be *u.d. mod M* if

$$\lim_{N\to\infty} \frac{A(B, M, N)}{N} = q^{-m} \qquad \text{for all } B \in \Phi. \tag{3.1}$$

Furthermore, we say that the sequence θ is *u.d.* if (3.1) holds for every $M \in \Phi$ of positive degree.

It is evident that in (3.1) it suffices to let B run through a complete residue system mod M. Moreover, we can restrict the investigation to the case where M is monic.

THEOREM 3.1. (i) If a sequence θ is u.d. mod M and if F divides M, then θ is u.d. mod F. (ii) If a sequence θ is not u.d., then there exist infinitely many moduli M for which θ is not u.d. mod M.

PROOF. (ii) follows from (i), for there is an F for which θ is not u.d. mod F and then this property is shared by all of the infinitely many distinct monic multiples M of F. The statement (i) is shown in the following manner. Assume that the sequence θ is u.d. mod M and that F divides M. Let B be any element of a complete residue system mod F and of degree $< f$, the degree of F. Suppose C is any element of a complete residue system mod M of degree $< m$, the degree of M, such that $C \equiv B(\text{mod } F)$, or $C = KF + B$, where the degree of K is less than $m - f$. We then have

$$\lim_{N\to\infty} \frac{A(KF + B, M, N)}{N} = q^{-m}. \tag{3.2}$$

Furthermore, since F divides M, we have

$$A(B, F, N) = \sum_K A(KF + B, M, N), \tag{3.3}$$

where the summation is over all K of degree $< m - f$. The number of such K is q^{m-f}. Therefore, from (3.2) and (3.3), we have for any B,

$$\lim_{N\to\infty} \frac{A(B, F, N)}{N} = q^{m-f}q^{-m} = q^{-f}.$$

Hence, θ is u.d. mod F. ∎

EXAMPLE 3.1. If F does not divide M, then there exists a sequence θ that is u.d. mod M but is not u.d. mod F. Let m and f be the degrees of M and F, respectively. Let (R_i), $i = 1, 2, \ldots, q^m$, be a complete residue system mod M with $R_1 = M$ and $0 \le \text{degree } R_i < m$ for $i = 2, 3, \ldots, q^m$. Let θ be the periodic sequence $R_1, \ldots, R_{q^m}, R_1, \ldots, R_{q^m}, \ldots$. Then θ is clearly u.d. mod M. However, θ is not u.d. mod F. For if $m < f$, then there is no $R_i \equiv 0$ (mod F), and so, our assertion is true. Second, let $f \le m$. Then all the R_i of degree $< f$ are $\not\equiv 0(\text{mod } F)$. For each z, $f \le z < m$, the $(q - 1)q^z$ elements R_i of degree z are divided evenly among the q^f residue classes mod F, each residue class containing $(q - 1)q^{z-f}$ such R_i. Since $R_1 = M \not\equiv 0(\text{mod } F)$, the total number of R_i, $1 \le i \le q^m$, such that $R_i \equiv 0(\text{mod } F)$ is equal to

$$\sum_{z=f}^{m-1} (q - 1)q^{z-f} = q^{m-f} - 1.$$

However, if θ were u.d. mod F, there would have to be q^{m-f} such R_i. ∎

EXAMPLE 3.2. If a sequence θ is both u.d. mod M and u.d. mod F, where M and F are relatively prime, then θ need not be u.d. mod MF. Consider the following example. Let $\alpha_1, \ldots, \alpha_q$ be the elements of $GF(q)$ listed in any fixed order, and let θ be the periodic sequence $\alpha_1, \ldots, \alpha_q, \alpha_1, \ldots, \alpha_q, \ldots$. Now take $M = x$ and $F = x + 1$. Then M and F are relatively prime, θ is u.d. mod M and u.d. mod F, but obviously not u.d. mod MF since the residues of degree 1 are not in this sequence. ■

A residue of a polynomial A mod M can be written as an expression of the form $\alpha_{m-1}x^{m-1} + \alpha_{m-2}x^{m-2} + \cdots + \alpha_0$ with $\alpha_i \in GF(q)$ for $0 \leq i \leq m - 1$. Let $\omega_1, \ldots, \omega_r$ be a fixed basis for the vector space $GF(q)$ over $GF(p)$. Then $\alpha_{m-1} = a_1\omega_1 + \cdots + a_r\omega_r$ with $a_j \in GF(p)$ for $1 \leq j \leq r$. Define

$$e(A, M) = e^{2\pi i a_1/p},$$

where we identify, of course, $GF(p)$ with $\mathbb{Z}/p\mathbb{Z}$.

THEOREM 3.2. The sequence (A_n) in $\mathbf{\Phi}$ is u.d. mod M if and only if

$$\lim_{N\to\infty} \frac{1}{N} \sum_{n=1}^{N} e(A_nC, M) = 0 \qquad \text{for all } C \not\equiv 0 \pmod{M}.$$

PROOF. The notion of u.d. mod M is identical with the notion of u.d. in the residue class ring $\mathbf{\Phi}/M\mathbf{\Phi}$, considered as an additive abelian group in the discrete topology. In the light of Corollary 1.2 of Chapter 4, it will suffice to show that the characters of the above group are precisely the functions $\psi_C(A) = e(AC, M)$, where C runs through a complete residue system mod M. It is easily seen that each ψ_C is a character of $\mathbf{\Phi}/M\mathbf{\Phi}$. Now, let $C \not\equiv 0 \pmod{M}$, so without loss of generality we may take C to be of the form

$$C = \gamma_r x^r + \cdots + \gamma_0, \qquad 0 \leq r < m, \quad \gamma_r \neq 0.$$

With $A = \omega_1\gamma_r^{-1}x^{m-1-r}$ we get then $e(AC, M) \neq 1$. If $C \not\equiv D \pmod{M}$, then the preceding argument shows the existence of an $A \in \mathbf{\Phi}$ with $e(A(C - D), M) \neq 1$, or $e(AC, M) \neq e(AD, M)$. Thus, distinct polynomials C and D from a complete residue system mod M yield distinct characters ψ_C and ψ_D. Since the cardinality of the set of characters of $\mathbf{\Phi}/M\mathbf{\Phi}$ is equal to the cardinality of a complete residue system mod M, we are already done. ■

Uniform Distribution in $GF\{q, x\}$

Let $\mathbf{\Phi}' = GF\{q, x\}$ denote the field containing $\mathbf{\Phi} = GF[q, x]$ that consists of all the expressions

$$\alpha = \sum_{i=-\infty}^{m} c_i x^i, \qquad c_i \in GF(q).$$

By definition, m is the degree of α (provided that $c_m \neq 0$). The integer m may be positive, negative, or 0. We write degree $0 = -\infty$. The integral part of α is $[\alpha] = \sum_{i=0}^{m} c_i x^i$, and the fractional part is $\{\alpha\} = \sum_{i=-\infty}^{-1} c_i x^i$. Hence, $[\alpha]$ is a polynomial over $GF(q)$. Obviously we have $[\alpha + \beta] = [\alpha] + [\beta]$ and $\{\alpha + \beta\} = \{\alpha\} + \{\beta\}$.

DEFINITION 3.2. Let (α_n), $n = 1, 2, \ldots$, be a sequence in $\mathbf{\Phi}'$, let β be an arbitrary element in $\mathbf{\Phi}'$, and let N and k be positive integers. Finally, let $A(\beta, k, N)$ be the number of α_n with $1 \leq n \leq N$ such that degree $(\{\alpha_n - \beta\}) < -k$. Then the sequence (α_n) is said to be *u.d. mod 1 in* $\mathbf{\Phi}'$ if

$$\lim_{N\to\infty} \frac{A(\beta, k, N)}{N} = q^{-k} \qquad \text{for all } k \geq 1 \text{ and all } \beta \in \mathbf{\Phi}'.$$

For given $\alpha \in \mathbf{\Phi}'$, let c_{-1} be the coefficient of x^{-1} in the expression for α (set $c_{-1} = 0$ if x^{-1} does not appear in the expression for α). Let $\omega_1, \ldots, \omega_r$ again be a fixed basis for the vector space $GF(q)$ over $GF(p)$. Then $c_{-1} = b_1\omega_1 + \cdots + b_r\omega_r$ with $b_j \in GF(p)$ for $1 \leq j \leq r$. We define

$$e(\alpha) = e^{2\pi i b_1/p}. \tag{3.4}$$

THEOREM 3.3. The sequence (α_n) is u.d. mod 1 in $\mathbf{\Phi}'$ if and only if

$$\lim_{N\to\infty} \frac{1}{N} \sum_{n=1}^{N} e(C\alpha_n) = 0 \qquad \text{for every } C \in \mathbf{\Phi},\ C \neq 0. \tag{3.5}$$

PROOF. The set $\mathbf{\Phi}_0' = \{\alpha \in \mathbf{\Phi}' \colon \text{degree } \alpha < 0\}$ is an additive abelian group. Let $k \geq 1$ be fixed. Then $\mathbf{\Phi}_k' = \{\alpha \in \mathbf{\Phi}' \colon \text{degree } \alpha < -k\}$ is a subgroup of $\mathbf{\Phi}_0'$, and the condition

$$\lim_{N\to\infty} \frac{A(\beta, k, N)}{N} = q^{-k} \qquad \text{for all } \beta \in \mathbf{\Phi}'$$

characterizes the u.d. of $(\{\alpha_n\} + \mathbf{\Phi}_k')$, $n = 1, 2, \ldots$, in the quotient group $\mathbf{\Phi}_0'/\mathbf{\Phi}_k'$, furnished with the discrete topology. As in the proof of Theorem 3.2, one shows that the nontrivial characters of $\mathbf{\Phi}_0'/\mathbf{\Phi}_k'$ are precisely the functions $\psi_C(\alpha + \mathbf{\Phi}_k') = e(C\alpha)$ with $C \in \mathbf{\Phi}$, $C \neq 0$, and degree $C < k$. Letting k run through all positive integers, we arrive at the desired criterion. ■

DEFINITION 3.3. The sequence (α_n) in $\mathbf{\Phi}'$ is said to be *weakly u.d. mod 1 in* $\mathbf{\Phi}'$ if

$$\lim_{t\to\infty} \frac{A(\beta, k, q^t)}{q^t} = q^{-k} \qquad \text{for all } k \geq 1 \text{ and all } \beta \in \mathbf{\Phi}'.$$

DEFINITION 3.4. An element $\xi \in \mathbf{\Phi}'$ is called *irrational* if ξ is not contained in $GF(q, x)$, the collection of all quotients A/B of elements $A, B \in \mathbf{\Phi}$, $B \neq 0$.

THEOREM 3.4. Let $\xi \in \mathbf{\Phi}'$ be irrational. Let (A_n) be a sequence formed by using all the elements of $\mathbf{\Phi}$ and such that degree $A_{n+1} \geq$ degree A_n for $n \geq 1$ and $A_i \neq A_j$ for $i \neq j$. Then the sequence $(A_n\xi)$ is weakly u.d. mod 1 in $\mathbf{\Phi}'$.

PROOF. In obvious analogy with the criterion in (3.5), we have to show

$$\lim_{t\to\infty} q^{-t} \sum_{n=1}^{q^t} e(A_nC\xi) = 0 \qquad \text{for every } C \in \mathbf{\Phi},\ C \neq 0. \tag{3.6}$$

By the construction of (A_n), (3.6) is equivalent to

$$\lim_{t\to\infty} q^{-t} \sum_{\substack{n \\ \text{degree } A_n < t}} e(A_nC\xi) = 0 \qquad \text{for every } C \in \mathbf{\Phi},\ C \neq 0, \tag{3.7}$$

and since ξ is irrational, (3.7) follows from the relation

$$\lim_{t\to\infty} q^{-t} \sum_{\substack{A\in\mathbf{\Phi} \\ \text{degree } A < t}} e(A\alpha) = 0 \qquad \text{for } \alpha \in \mathbf{\Phi}'\backslash\mathbf{\Phi},$$

which in turn is an easy consequence of the fundamental relation

$$\sum_{\substack{A\in\mathbf{\Phi} \\ \text{degree } A < m}} e(A\alpha) = 0 \qquad \text{for } \alpha \in \mathbf{\Phi}' \text{ with degree } \{\alpha\} \geq -m, \tag{3.8}$$

where m is any positive integer. ■

Notes

Definition 3.1 and Theorem 3.1 are from Hodges [1]. The Weyl criterion in Theorem 3.2 was given by Kuipers and Scheelbeek [2]. Earlier Hodges [1] had found a necessary condition. Other Weyl criteria for u.d. mod M have been established by Meijer and Dijksma [1] and Dijksma [2] (see Exercise 3.1). Definitions 3.2, 3.3, and 3.4 and Theorem 3.3 and 3.4 are from Carlitz [2]. The definition of the function $e(\alpha)$ is from Carlitz [1]. Theorem 3.4 was greatly improved by Meijer and Dijksma [1], who showed that with a specific ordering of the sequence (A_n) the sequence $(A_n\xi)$ will even be u.d. mod 1 in $\mathbf{\Phi}'$. The theory of u.d. in $\mathbf{\Phi}$ was further developed by Hodges [1, 2, 3], Dijksma [1, 2, 3], and Meijer and Dijksma [1]. For u.d. mod M with the special polynomial $M = x$, see Gotusso [1] and also Exercises 3.5–3.8. Various interesting results on u.d. in $\mathbf{\Phi}'$ were shown by Hodges [3], Dijksma [2, 3], Meijer and Dijksma [1], and Rhin [2, 3]. Since $\mathbf{\Phi}_0'$ is a compact (additive) group in the topology induced by the degree valuation, the results of Chapter 4 apply to u.d. mod 1 in $\mathbf{\Phi}'$. A good account of the theory of u.d. in both $\mathbf{\Phi}$ and $\mathbf{\Phi}'$ can be found in Dijksma [5]. For metric theorems, see Dijksma [4] (with some improvements in Dijksma [5]), de Mathan [1, 2], and Rhin [1]. An analogue of P.V. numbers in $\mathbf{\Phi}'$ was defined by Bateman and Duquette [1] and further studied by Grandet-Hugot [2, 3]. For results on u.d. in the setting of local fields, we refer to Beer [2] and de Mathan [3].

Exercises

3.1. Let $e(\alpha)$ be the function defined in (3.4). Prove the following Weyl criterion. The sequence (A_n) in $\mathbf{\Phi}$ is u.d. mod M if and only if

$$\lim_{N\to\infty} \frac{1}{N} \sum_{n=1}^{N} e(CM^{-1}A_n) = 0$$

for all $C \in \mathbf{\Phi}$ with $C \neq 0$ and degree $C <$ degree M.

3.2. The sequence (A_n) in $\mathbf{\Phi}$ is u.d. if and only if

$$\lim_{N\to\infty} \frac{1}{N} \sum_{n=1}^{N} e(\alpha A_n) = 0$$

for all "rationals" $\alpha \in GF(q, x)$ with $\alpha \notin \mathbf{\Phi}$.

3.3. Let (α_n) be a sequence in $\mathbf{\Phi}'$, and let $M \in \mathbf{\Phi}$ have positive degree. Prove that if $(M^{-1}\alpha_n)$ is u.d. mod 1 in $\mathbf{\Phi}'$, then $([\alpha_n])$ is u.d. mod M.

3.4. Let (A_n) be a sequence in $\mathbf{\Phi}$ such that $(A_n\xi)$ is u.d. mod 1 in $\mathbf{\Phi}'$ for every irrational $\xi \in \mathbf{\Phi}'$ (see the notes for the existence of such a sequence). Prove that the sequence $([A_n\xi])$ is u.d. for every irrational $\xi \in \mathbf{\Phi}'$.

3.5. Let $b_1, b_2, \ldots, b_q$ be the elements of $GF(q)$. For a sequence (a_n) of elements in $GF(q)$ and $N \geq 1$, let $A(s, N)$ be the number of elements among $a_1, \ldots, a_N$ that are equal to b_s. Show first that (a_n) is u.d. mod $M = x$ if and only if $\lim_{N\to\infty} A(s, N)/N = 1/q$ for $1 \leq s \leq q$. Prove also that (a_n) is u.d. mod x if and only if

$$\lim_{N\to\infty} A(h, N)/A(k, N) = 1$$

for $1 \leq h, k \leq q$. If (a_n) is u.d. mod x, we say that (a_n) is *u.d. in* $GF(q)$.

3.6. If (a_n) is u.d. in $GF(q)$ (see Exercise 3.5), then the same holds for the sequence $(\bar{a}_n)$ where $\bar{a}_n = a_n^{-1}$ for $a_n \neq 0$ and $\bar{a}_n = 0$ for $a_n = 0$.

3.7. Let (a_n) be u.d. in $GF(q)$ (see Exercise 3.5), and let $f(x) = x^k$, $k \geq 1$. Then, $(f(a_n))$ is u.d. in $GF(q)$ if and only if $(k, q - 1) = 1$.

3.8. Let (a_n) be u.d. in $GF(q)$ (see Exercise 3.5). For $f \in \mathbf{\Phi}$, prove that $(f(a_n))$ is u.d. in $GF(q)$ if and only if f is a permutation polynomial, that is, if and only if the mapping φ_f: $a \mapsto f(a)$, $a \in GF(q)$, is a permutation of $GF(q)$.

3.9. See the proof of Theorem 3.3. Prove in detail that the characters of $\mathbf{\Phi}_0'/\mathbf{\Phi}_k'$ are precisely the functions ψ_C with $C \in \mathbf{\Phi}$ and degree $C < k$.

3.10. Give a detailed proof for (3.8).

BIBLIOGRAPHY

Akuliničev, N. M.: [1] On a dynamical system related to the distribution of the fractional parts of second-degree polynomials (Russian), *Dokl. Akad. Nauk SSSR* **143,** 503–505 (1962); Engl. trans., *Soviet Math. Dokl.* **3,** 405–407 (1962).

Amara, M.: [1] Ensembles fermés de nombres algébriques, *Ann. Sci. École Norm. Sup.* (3) **83,** 215–270 (1967).

Amice, Y.: [1] Interpolation *p*-adique, *Bull. Soc. Math. France* **92,** 117–180 (1964). [2] Un théorème de finitude, *Ann. Inst. Fourier* **14,** fasc. 2, 527–531 (1964).

Ammann, A.: [1] Sur une application d'un théorème de calcul intégral à l'étude des répartitions modulo 1, *C. R. Séances Soc. Phys. Hist. Nat. Genève* **64,** 58–61 (1947).

Anzai, H., and Kakutani, S.: [1] Bohr compactifications of a locally compact abelian group II, *Proc. Imp. Acad. Tokyo* **19,** 533–539 (1943).

Armacost, D. L.: [1] Generators of monothetic groups, *Canad. J. Math.* **23,** 791–796 (1971); Remark, *ibid.* **25,** 672 (1973).

Arnol'd, V. I., and Krylov, A. L.: [1] Uniform distribution of points on a sphere and some ergodic properties of solutions of linear ordinary differential equations in a complex region (Russian), *Dokl. Akad. Nauk SSSR* **148,** 9–12 (1963); Engl. trans., *Soviet Math. Dokl.* **4,** 1–5 (1963).

Artémiadis, N., and Kuipers, L.: [1] On the Weyl criterion in the theory of uniform distribution modulo 1, *Amer. Math. Monthly* **76,** 654–656 (1969).

Auerbach, H.: [1] Sur les groupes linéaires bornés (III), *Studia Math.* **5,** 43–49 (1934).

Auslander, L.: [1] Ergodic automorphisms, *Amer. Math. Monthly* **77,** 1–19 (1970).

Auslander, L., and Brezin, J.: [1] Uniform distribution in solvmanifolds, *Advances in Math.* **7,** 111–144 (1971).

Baayen, P. C., and Hedrlín, Z.: [1] The existence of well distributed sequences in compact spaces, *Indag. Math.* **27,** 221–228 (1965).

Baayen, P. C., and Helmberg, G.: [1] On families of equi-uniformly distributed sequences in compact spaces, *Math. Ann.* **161,** 255–278 (1965).

Bachman, G.: [1] *Introduction to p-Adic Numbers and Valuation Theory*, Academic Press, New York-London, 1964.

Bahvalov, N. S.: [1] Approximate computation of multiple integrals (Russian), *Vestnik Moskov. Univ. Ser. Mat. Meh. Astr. Fiz. Him.* **1959,** no. 4, 3–18. [2] An estimate of the main remainder term in quadrature formulae (Russian), *Ž. Vyčisl. Mat. i Mat. Fiz.* **1,** 64–77 (1961); Engl. trans., *U.S.S.R. Comp. Math. and Math. Phys.* **1,** 68–82 (1961).

Bahvalov, N. S., Korobov, N. M., and Čencov, N. N.: [1] The application of number-theoretic nets to numerical analysis problems (Russian), *Proc. Fourth All-Union Math. Congr.* (Leningrad, 1961), Vol. II, pp. 580–587, Izdat. "Nauka", Leningrad, 1964.

Baker, A.: [1] On some diophantine inequalities involving the exponential function, *Canad. J. Math.* **17**, 616–626 (1965).

Baker, R. C.: [1] A diophantine problem on groups. I, *Trans. Amer. Math. Soc.* **150**, 499–506 (1970). [2] A diophantine problem on groups. II, *Proc. London Math. Soc.* (3) **21**, 757–768 (1970). [3] A diophantine problem on groups. III, *Proc. Cambridge Phil. Soc.* **70**, 31–47 (1971). [4] Discrepancy modulo one and capacity of sets, *Quart. J. Math.* (2) **22**, 597–603 (1971). [5] On a theorem of Erdös and Taylor, *Bull. London Math. Soc.* **4**, 373–374 (1972). [6] Slowly growing sequences and discrepancy modulo one, *Acta Arith.* **23**, 279–293 (1973).

Bass, J.: [1] Suites uniformément denses, moyennes trigonométriques, fonctions pseudo-aléatoires, *Bull. Soc. Math. France* **87**, 1–64 (1959). [2] Nombres aléatoires, suites arithmétiques, méthode de Monte-Carlo, *Publ. Inst. Statist. Univ. Paris* **9**, 289–325 (1960). [3] *Les fonctions pseudo-aléatoires*, Mém. Sci. Math., fasc. 153, Gauthier-Villars, Paris, 1962.

Bass, J., and Bertrandias, J.-P.: [1] Moyennes de sommes trigonométriques et fonctions d'autocorrélation, *C. R. Acad. Sci. Paris* **245**, 2457–2459 (1957).

Bass, J., and Couot, J.: [1] Un théorème d'arithmétique et son application à la théorie des moyennes dans l'espace de Hilbert, *C. R. Acad. Sci. Paris* Sér. A **269**, 1194–1197 (1969).

Bass, J., and Guillod, J.: [1] Méthode de Monte-Carlo et suites uniformément denses, *Chiffres* **1**, 151–155 (1958).

Bateman, P. T.: [1] Sequences of mass distributions on the unit circle which tend to a uniform distribution, *Amer. Math. Monthly* **71**, 165–172 (1964).

Bateman, P. T., and Duquette, A. L.: [1] The analogue of the Pisot-Vijayaraghavan numbers in fields of formal power series, *Illinois J. Math.* **6**, 594–606 (1962).

Bauer, W.: [1] Diskrepanz in separablen metrischen Räumen, *Monatsh. Math.*, to appear.

Beer, S.: [1] Zur Theorie der Gleichverteilung im p-adischen, *Sitzgsber. Österr. Akad. Wiss.*, Math.-naturw. Kl., Abt. II, **176**, 499–519 (1967). [2] Über die Diskrepanz von Folgen in bewerteten Körpern, *Manuscripta Math.* **1**, 201–209 (1969). [3] Die Diskrepanz von Differenzenfolgen im p-adischen, *Monatsh. Math.* **76**, 289–294 (1972).

Behnke, H.: [1] Über die Verteilung von Irrationalitäten mod 1, *Abh. Math. Sem. Hamburg* **1**, 252–267 (1922). [2] Zur Theorie der diophantischen Approximationen. I, *Abh. Math. Sem. Hamburg* **3**, 261–318 (1924); II, *ibid.* **4**, 33–46 (1926).

Benzinger, L.: [1] Uniformly distributed sequences on locally compact groups, Ph.D. thesis, Syracuse Univ., 1972.

Beresin, I. S., and Shidkow, N. P.: [1] *Numerische Methoden I*, VEB Deutscher Verlag der Wissenschaften, Berlin, 1970.

Berg, I. D., Rajagopalan, M., and Rubel, L. A.: [1] Uniform distribution in locally compact abelian groups, *Trans. Amer. Math. Soc.* **133**, 435–446 (1968).

Berg, I. D., and Rubel, L. A.: [1] Densities on locally compact abelian groups, *Bull. Amer. Math. Soc.* **74**, 298–300 (1968). [2] Densities on locally compact abelian groups, *Ann. Inst. Fourier* **19**, fasc. 1, 81–107 (1969); Erratum, *ibid.* **19**, fasc. 2, 533 (1969).

Bergström, V.: [1] Beiträge zur Theorie der endlichdimensionalen Moduln und der diophantischen Approximationen, *Medd. Lunds Univ. Mat. Sem.* **2** (1935). [2] Einige Bemerkungen zur Theorie der diophantischen Approximationen, *Fysiogr. Sälsk. Lund. Förh.* **6**, no. 13, 1–19 (1936); also in *Medd. Lunds Univ. Mat. Sem.* **4** (1939).

Berlekamp, E. R., and Graham, R. L.: [1] Irregularities in the distributions of finite sequences, *J. Number Th.* **2,** 152–161 (1970).

Bertrandias, F.: [1] Ensembles remarquables d'adèles algébriques, *Bull. Soc. Math. France, Suppl., Mém.* **4**(1965).

Bertrandias, J.-P.: [1] Calcul d'une intégrale au moyen de la suite $X_n = An$. Évaluation de l'erreur, *Publ. Inst. Statist. Univ. Paris* **9,** 335–357 (1960). [2] Sur le produit de plusieurs fonctions pseudo-aléatoires. Application à la répartition modulo 1, *C. R. Acad. Sci. Paris* **250,** 2498–2500 (1960). [3] Suites pseudo-aléatoires et critères d'équirépartition modulo un, *Compositio Math.* **16,** 23–28 (1964).

Besicovitch, A. S.: [1] On the sum of digits of real numbers represented in the dyadic number system, *Math. Ann.* **110,** 321–330 (1935). [2] The asymptotic distribution of the numerals in the decimal representation of the squares of the natural numbers, *Math. Z.* **39,** 146–156 (1935).

Bésineau, J.: [1] Sur un problème de Gel'fond relatif à la fonction "somme des chiffres," *C. R. Acad. Sci. Paris* Sér. A **272,** 453–456 (1971). [2] Indépendance statistique d'ensembles liés à la fonction "somme des chiffres," *Acta Arith.* **20,** 401–416 (1972).

Best, E.: [1] On sets of fractional dimension. III, *Proc. London Math. Soc.* (2) **47,** 436–454 (1942).

Beyer, W. A.: [1] Hausdorff dimension of level sets of some Rademacher series, *Pacific J. Math.* **12,** 35–46 (1962).

Beyer, W. A., Metropolis, N., and Neergaard, J. R.: [1] Statistical study of digits of some square roots of integers in various bases, *Math. Comp.* **24,** 455–473 (1970). [2] The generalized serial test applied to expansions of some irrational square roots in various bases, *Math. Comp.* **24,** 745–747 (1970).

Billingsley, P.: [1] *Ergodic Theory and Information*, Wiley, New York, 1965. [2] *Convergence of Probability Measures*, Wiley, New York, 1968.

Billingsley, P., and Topsøe, F.: [1] Uniformity in weak convergence, *Z. Wahrscheinlichkeitsth.* **7,** 1–16 (1967).

Binder, C.: [1] Über einen Satz von de Bruijn und Post, *Sitzgsber. Österr. Akad. Wiss.*, Math.-naturw. Kl., Abt. II, **179,** 233–251 (1970).

Bird, R. S.: [1] Integers with given initial digits, *Amer. Math. Monthly* **79,** 367–370 (1972).

Blanchard, A.: [1] *Initiation à la théorie analytique des nombres premiers*, Dunod, Paris, 1969.

Blum, J. R., and Hanson, D. L.: [1] On invariant probability measures. I, *Pacific J. Math.* **10,** 1125–1129 (1960); II, *ibid.* **11,** 63–71 (1961).

Blum, J. R., and Mizel, V. J.: [1] On a theorem of Weyl and the ergodic theorem, *Z. Wahrscheinlichkeitsth.* **20,** 193–198 (1971). [2] A generalized Weyl equidistribution theorem for operators, with applications, *Trans. Amer. Math. Soc.* **165,** 291–307 (1972).

Bohl, P.: [1] Über ein in der Theorie der säkulären Störungen vorkommendes Problem, *J. reine angew. Math.* **135,** 189–283 (1909).

Bonnesen, T., and Fenchel, W.: [1] *Theorie der konvexen Körper*, Erg. Math. Grenzgeb., Band 3, Heft 1, Julius Springer, Berlin, 1934.

Borel, E.: [1] Les probabilités dénombrables et leurs applications arithmétiques, *Rend. Circ. Mat. Palermo* **27,** 247–271 (1909). [2] *Leçons sur la théorie des fonctions*, 2nd ed., Gauthier-Villars, Paris, 1914.

Borevich, Z. I., and Shafarevich, I. R.: [1] *Number Theory*, Academic Press, New York-London, 1966.

Boyd, D. W.: [1] Transcendental numbers with badly distributed powers, *Proc. Amer. Math. Soc.* **23**, 424–427 (1969).

Brezin, J.: [1] Applications of nilmanifold theory to diophantine approximations, *Proc. Amer. Math. Soc.* **33**, 543–547 (1972).

Brown, J. L., Jr., and Duncan, R. L.: [1] A generalized Weyl criterion, *Duke Math. J.* **35**, 699–705 (1968). [2] Modulo one uniform distribution of the sequence of logarithms of certain recursive sequences, *Fibonacci Quart.* **8**, no. 5, 482–486 (1970). [3] Asymptotic distribution of real numbers modulo one, *Amer. Math. Monthly* **78**, 367–372 (1971). [4] Modulo one uniform distribution of certain Fibonacci-related sequences, *Fibonacci Quart.* **10**, no. 3, 277–280 (1972).

Buck, R. C.: [1] The measure theoretic approach to density, *Amer. J. Math.* **68**, 560–580 (1946).

Bundgaard, S.: [1] Über die Werteverteilung der Charaktere abelscher Gruppen, *Danske Vid. Selsk.*, mat.-fys. Medd., **14**, 1–29 (1936). [2] Eine Verallgemeinerung des Kronecker-Weylschen Satzes über diophantische Approximation, *9^e Congr. Math. Scand.* **1938**, 219–224.

Burkard, R. E.: [1] Zur Theorie der Gleichverteilung modulo Δ, Thesis, Univ. of Vienna, 1967. [2] Zur gleichmässigen Gleichverteilung von Polynomwerten modulo Δ, *Sitzgsber. Österr. Akad. Wiss.*, Math.-naturw. Kl., Abt. II, **177**, 395–406 (1969).

Callahan, F. P.: [1] Density and uniform density, *Proc. Amer. Math. Soc.* **15**, 841–843 (1964).

Cantor, D. G.: [1] On sets of algebraic integers whose remaining conjugates lie in the unit circle, *Trans. Amer. Math. Soc.* **105**, 391–406 (1962). [2] On the elementary theory of diophantine approximation over the ring of adeles I, *Illinois J. Math.* **9**, 677–700 (1965). [3] Irrational power series, *Indag. Math.* **27**, 777–786 (1965).

Carlitz, L.: [1] The singular series for sums of squares of polynomials, *Duke Math. J.* **14**, 1105–1120 (1947). [2] Diophantine approximation in fields of characteristic p, *Trans. Amer. Math. Soc.* **72**, 187–208 (1952).

Carlson, D. L.: [1] Good sequences of integers, Thesis, Univ. of Colorado, 1971.

Carroll, F. W.: [1] On some classes of noncontinuable analytic functions, *Trans. Amer. Math. Soc.* **94**, 74–85 (1960). [2] Some properties of sequences, with an application to noncontinuable power series, *Pacific J. Math.* **24**, 45–50 (1968).

Carroll, F. W., and Kemperman, J. H. B.: [1] Noncontinuable analytic functions, *Duke Math. J.* **32**, 65–84 (1965).

Cassels, J. W. S.: [1] Some metrical theorems in Diophantine approximation. I, *Proc. Cambridge Phil. Soc.* **46**, 209–218 (1950). [2] Some metrical theorems of Diophantine approximation. II, *J. London Math. Soc.* **25**, 180–184 (1950). [3] Some metrical theorems in Diophantine approximation. III, *Proc. Cambridge Phil. Soc.* **46**, 219–225 (1950). [4] Some metrical theorems of Diophantine approximation. IV, *Indag. Math.* **12**, 14–25 (1950). [5] A theorem of Vinogradoff on uniform distribution, *Proc. Cambridge Phil. Soc.* **46**, 642–644 (1950). [6] An extension of the law of the iterated logarithm, *Proc. Cambridge Phil. Soc.* **47**, 55–64 (1951). [7] On a paper of Niven and Zuckerman, *Pacific J. Math.* **2**, 555–557 (1952). [8] A new inequality with application to the theory of diophantine approximation, *Math. Ann.* **126**, 108–118 (1953). [9] *An Introduction to Diophantine Approximation*, Cambridge Tracts in Math. and Math. Physics, no. 45, Cambridge Univ. Press, London, 1957. [10] On a problem of Steinhaus about normal numbers, *Colloq. Math.* **7**, 95–101 (1959).

Cavior, S. R.: [1] Uniform distribution of polynomials modulo *m*, *Amer. Math. Monthly* **73,** 171–172 (1966).

Chabauty, C.: [1] Sur la répartition modulo 1 de certaines suites *p*-adiques, *C. R. Acad. Sci. Paris* **231,** 465–466 (1950).

Champernowne, D. G.: [1] The construction of decimals normal in the scale of ten, *J. London Math. Soc.* **8,** 254–260 (1933).

Chandrasekharan, K.: [1] *Introduction to Analytic Number Theory*, Grundlehren d. math. Wiss., Band 148, Springer-Verlag, Berlin-Heidelberg-New York, 1968.

Chauvineau, J.: [1] Sur la répartition modulo 1 de certaines fonctions périodiques, *C. R. Acad. Sci. Paris* **252,** 4090–4092 (1961). [2] Suites continûment et périodiquement réparties modulo 1, *C. R. Acad. Sci. Paris* **256,** 839–841 (1963). [3] Sur la répartition en valuation *p*-adique, *C. R. Acad. Sci. Paris* **259,** 3907–3909 (1964). [4] Complément au théorème métrique de Koksma dans $\mathbb{R}$ et dans $\mathbb{Q}_p$, *C. R. Acad. Sci. Paris* **260,** 6252–6255 (1965). [5] Sur la *C*-équirépartition modulo 1 en valuation *p*-adique, *C. R. Acad. Sci. Paris* Sér. A **262,** 557–559 (1966). [6] Sur la répartition dans $\mathbb{R}$ et dans $\mathbb{Q}_p$, *Acta Arith.* **14,** 225–313 (1968).

Christol, G.: [1] Equirépartition dans les séries formelles, Séminaire Delange-Pisot-Poitou, 11e année (1969/70), Exp. 4.

Chui, C. K.: [1] Concerning rates of convergence of Riemann sums, *J. Approximation Th.* **4,** 279–287 (1971).

Chung, K. L.: [1] An estimate concerning the Kolmogoroff limit distribution, *Trans. Amer. Math. Soc.* **67,** 36–50 (1949).

Ciesielski, Z., and Kesten, H.: [1] A limit theorem for the fractional parts of the sequence $\{2^k t\}$, *Proc. Amer. Math. Soc.* **13,** 596–600 (1962).

Cigler, J.: [1] Asymptotische Verteilung reeller Zahlen mod 1, *Monatsh. Math.* **64,** 201–225 (1960). [2] Der individuelle Ergodensatz in der Theorie der Gleichverteilung mod 1, *J. reine angew. Math.* **205,** 91–100 (1960). [3] Ein gruppentheoretisches Analogon zum Begriff der normalen Zahl, *J. reine angew. Math.* **206,** 5–8 (1961). [4] Ziffernverteilung in δ-adischen Brüchen, *Math. Z.* **75,** 8–13 (1961). [5] Hausdorffsche Dimensionen spezieller Punktmengen, *Math. Z.* **76,** 22–30 (1961). [6] Folgen normierter Masse auf kompakten Gruppen, *Z. Wahrscheinlichkeitsth.* **1,** 3–13 (1962). [7] Über eine Verallgemeinerung des Hauptsatzes der Theorie der Gleichverteilung, *J. reine angew. Math.* **210,** 141–147 (1962). [8] The fundamental theorem of van der Corput on uniform distribution and its generalizations, *Compositio Math.* **16,** 29–34 (1964). [9] Some applications of the individual ergodic theorem to problems in number theory, *Compositio Math.* **16,** 35–43 (1964). [10] Methods of summability and uniform distribution mod 1, *Compositio Math.* **16,** 44–51 (1964). [11] Stark gleichverteilte Folgen in kompakten Gruppen, *Monatsh. Math.* **70,** 430–436 (1966). [12] A characterization of well-distributed sequences, *Compositio Math.* **17,** 263–267 (1966). [13] Some remarks on the distribution mod 1 of tempered sequences, *Nieuw Arch. voor Wisk.* (3) **16,** 194–196 (1968). [14] On a theorem of H. Weyl, *Compositio Math.* **21,** 151–154 (1969). [15] Well-distributed sequences, *Bull. Soc. Math. France, Suppl., Mém.* **25,** 39–43 (1971).

Cigler, J., and Helmberg, G.: [1] Neuere Entwicklungen der Theorie der Gleichverteilung, *Jber. Deutsch. Math.-Verein.* **64,** 1–50 (1961).

Cigler, J., and Volkmann, B.: [1] Über die Häufigkeit von Zahlenfolgen mit gegebener Verteilungsfunktion, *Abh. Math. Sem. Hamburg* **26,** 39–54 (1963).

Cohen, P. J.: [1] On a conjecture of Littlewood and idempotent measures, *Amer. J. Math.* **82,** 191–212 (1960).

Colebrook, C. M.: [1] The Hausdorff dimension of certain sets of nonnormal numbers, *Michigan Math. J.* **17,** 103–116 (1970).

Colebrook, C. M., and Kemperman, J. H. B.: [1] On non-normal numbers, *Indag. Math.* **30,** 1–11 (1968).

Coles, W. J.: [1] On a theorem of van der Corput on uniform distribution, *Proc. Cambridge Phil. Soc.* **53,** 781–789 (1957).

Colombeau, J.-F.: [1] Ensembles normaux et ensembles normaux en sens large, *C. R. Acad. Sci. Paris* Sér. A **269,** 270–272 (1969).

Conroy, H.: [1] Molecular Schrödinger equation. VIII: A new method for the evaluation of multidimensional integrals, *J. Chemical Phys.* **47,** 5307–5318 (1967).

Cooke, R. G.: [1] *Infinite Matrices and Sequence Spaces*, Macmillan, London, 1950.

Copeland, A. H.: [1] Admissible numbers in the theory of probability, *Amer. J. Math.* **50,** 535–552 (1928).

Copeland, A. H., and Erdös, P.: [1] Note on normal numbers, *Bull. Amer. Math. Soc.* **52,** 857–860 (1946).

Corrádi, K. A., and Kátai, I.: [1] On the theory of multiplicative functions, *Ann. Univ. Sci. Budapest. Eötvös Sect. Math.* **9,** 147–155 (1966).

Couot, J.: [1] Applications des suites $m\theta$ à l'intégration numérique, *C. R. Acad. Sci. Paris* Sér. A **264,** 183–186 (1967). [2] Applications des suites $m\boldsymbol{\theta}$ à l'intégration multiple sur le tore, *C. R. Acad. Sci. Paris* Sér. A **266,** 131–134 (1968). [3] Convergence sûre de méthodes arithmétiques de Monte-Carlo pour l'inversion d'un opérateur linéaire, *C. R. Acad. Sci. Paris* Sér. A **267,** 799–802 (1968). [4] Ergodicité et critères topologiques d'équirépartition d'une famille d'éléments d'un groupe, *C. R. Acad. Sci. Paris* Sér. A **272,** 1045–1048 (1971).

Csillag, P.: [1] Über die gleichmässige Verteilung nichtganzer positiver Potenzen mod 1, *Acta Litt. Sci. Szeged* **5,** 13–18 (1930).

Cugiani, M.: [1] Successioni uniformemente distribuite nei domini p-adici, *Ist. Lombardo Accad. Sci. Lett. Rend.* A **96,** 351–372 (1962).

Darling, D. A.: [1] The Kolmogorov-Smirnov, Cramér-von Mises tests, *Ann. Math. Statist.* **28,** 823–838 (1957).

Davenport, H.: [1] Note on irregularities of distribution, *Mathematika* **3,** 131–135 (1956). [2] On a theorem of P. J. Cohen, *Mathematika* **7,** 93–97 (1960). [3] Note on an irrational power series, *Proc. Amer. Math. Soc.* **17,** 1–5 (1966).

Davenport, H., and Erdös, P.: [1] Note on normal decimals, *Canad. J. Math.* **4,** 58–63 (1952).

Davenport, H., Erdös, P., and LeVeque, W. J.: [1] On Weyl's criterion for uniform distribution, *Michigan Math. J.* **10,** 311–314 (1963).

Davenport, H., and LeVeque, W. J.: [1] Uniform distribution relative to a fixed sequence, *Michigan Math. J.* **10,** 315–319 (1963).

Davis, P. J., and Rabinowitz, P.: [1] Some Monte Carlo experiments in computing multiple integrals, *Math. Tables Aids Comput.* **10,** 1–8 (1956). [2] *Numerical Integration*, Blaisdell, Waltham, Mass., 1967.

De Bruijn, N. G., and Erdös, P.: [1] Sequences of points on a circle, *Indag. Math.* **11,** 14–17 (1949).

De Bruijn, N. G., and Post, K. A.: [1] A remark on uniformly distributed sequences and Riemann integrability, *Indag. Math.* **30,** 149–150 (1968).

Decomps-Guilloux, A.: [1] Ensembles d'éléments algébriques dans les adèles, *C. R. Acad. Sci. Paris* **261,** 1929–1931 (1965). [2] Généralisation des nombres de Salem aux adèles, *Acta Arith.* **16,** 265–314 (1969).

Delange, H.: [1] Sur la distribution des valeurs de certaines fonctions arithmétiques, *Colloque sur la Théorie des Nombres*, Centre belge de recherches mathématiques, Brussels, 1955, pp. 147–161. [2] Sur la distribution des entiers ayant certaines propriétés, *Ann. Sci. École Norm. Sup.* (3) **73,** 15–74 (1956). [3] On a theorem of Erdös, *Scripta Math.* **23,** 109–116 (1957). [4] On some arithmetical functions, *Illinois J. Math.* **2,** 81–87 (1958). [5] Un théorème sur les fonctions arithmétiques multiplicatives et ses applications, *Ann. Sci. École Norm. Sup.* (3) **78,** 1–29 (1961). [6] Sur les fonctions arithmétiques multiplicatives, *Ann. Sci. École Norm. Sup.* (3) **78,** 273–304 (1961). [7] Sur la distribution des fractions irréductibles de dénominateur n ou de dénominateur au plus égal à x, *Hommage au Professeur Lucien Godeaux*, Centre belge de recherches mathématiques, Brussels, 1968, pp. 75–89. [8] On integral-valued additive functions, *J. Number Th.* **1,** 419–430 (1969). [9] On the distribution modulo 1 of additive functions, *J. Indian Math. Soc.* **34,** 215–236 (1970). [10] Quelques résultats nouveaux sur les fonctions additives, *Bull. Soc. Math. France, Suppl., Mém.* **25,** 45–53 (1971). [11] Sur la distribution des valeurs des fonctions additives, *C. R. Acad. Sci. Paris* Sér. A **275,** 1139–1142 (1972).

De Mathan, B.: [1] Sur un théorème métrique d'équirépartition mod 1 dans un corps de séries formelles sur un corps fini, *C. R. Acad. Sci. Paris* Sér. A **265,** 289–291 (1967). [2] Théorème de Koksma dans un corps de séries formelles sur un corps fini, *Séminaire Delange-Pisot-Poitou*, 9ᵉ année (1967/68), Exp. 4. [3] Approximations diophantiennes dans un corps local, *Bull. Soc. Math. France, Suppl., Mém.* **21** (1970). [4] Un problème métrique d'approximation diophantienne, *Bull. Soc. Math. France* **99,** 369–385 (1971). [5] Un critère de non-eutaxie, *C. R. Acad. Sci. Paris* Sér. A **273,** 433–436 (1971).

Dennis, J. S.: [1] The distribution of sequences in Abelian groups, *Proc. Cambridge Phil. Soc.* **67,** 1–11 (1970).

Descovich, J.: [1] Zur Theorie der Gleichverteilung auf kompakten Räumen, *Sitzgsber. Österr. Akad. Wiss.*, Math.-naturw. Kl., Abt. II, **178,** 263–283 (1969).

Diamond, H. G.: [1] The distribution of values of Euler's phi function, *Proc. Symp. Pure Math.*, Vol. XXIV, American Math. Society, Providence, R.I., 1973, pp. 63–75.

Dijksma, A.: [1] The measure theoretic approach to uniform distribution of sequences in $GF[q,x]$, *Mathematica* (Cluj) **11,** 221–240 (1969). [2] Uniform distribution of polynomials over $GF\{q,x\}$ in $GF[q,x]$, Part I, *Indag. Math.* **31,** 376–383 (1969). [3] Uniform distribution of polynomials over $GF\{q,x\}$ in $GF[q,x]$, Part II, *Indag. Math.* **32,** 187–195 (1970). [4] Metrical theorems concerning uniform distribution in $GF[q,x]$ and $GF\{q,x\}$, *Nieuw Arch. voor Wisk.* (3) **18,** 279–293 (1970). [5] Uniform distribution in $GF\{q,x\}$ and $GF[q,x]$, Thesis, Technische Hogeschool Delft, 1971.

Dijksma, A., and Meijer, H. G.: [1] Note on uniformly distributed sequences of integers, *Nieuw Arch. voor Wisk.* (3) **17,** 210–213 (1969).

Donoghue, W. F., Jr.: [1] *Distributions and Fourier Transforms*, Academic Press, New York, 1969.

Donsker, M.: [1] Justification and extension of Doob's heuristic approach to the Kolmogorov-Smirnov theorems, *Ann. Math. Statist.* **23,** 277–281 (1952).

Doob, J. L.: [1] Heuristic approach to the Kolmogorov-Smirnov theorems, *Ann. Math. Statist.* **20,** 393–403 (1949).

Dowidar, A. F.: [1] A note on the generalized uniform distribution (mod 1), *J. Natur. Sci. and Math.* **11,** 185–189 (1971).

Dowidar, A. F., and Petersen, G. M.: [1] The distribution of sequences and summability, *Canad. J. Math.* **15,** 1–10 (1963).

Dress, F.: [1] Sur l'équirépartition de certaines suites $(x\lambda_n)$, *Acta Arith.* **14,** 169–175 (1968). [2] Intersection d'ensembles normaux, *J. Number Th.* **2,** 352–362 (1970).

Dress, F., and Mendès France, M.: [1] Caractérisation des ensembles normaux dans Z, *Acta Arith.* **17,** 115–120 (1970).

Drewes, A.: [1] Diophantische Benaderingsproblemen, Thesis, Vrije Univ. Amsterdam, 1945.

Ducray, S.: [1] Normal sequences, *J. Univ. Bombay* (N. S.) **31,** Sect. A, 1–4 (1962/63).

Dudley, U.: [1] Oscillating sequences modulo one, *Trans. Amer. Math. Soc.* **126,** 420–426 (1967).

Duncan, R. L.: [1] An application of uniform distributions to the Fibonacci numbers, *Fibonacci Quart.* **5,** no. 2, 137–140 (1967).

Dupain, Y., and Lesca, J.: [1] Répartition des sous-suites d'une suite donnée, *Acta Arith.* **23,** 307–314 (1973).

Dutka, J.: [1] The square root of 2 to 1,000,000 decimals, *Math. Comp.* **25,** 927–930 (1971).

Dvoretzky, A., Kiefer, J., and Wolfowitz, J.: [1] Asymptotic minimax character of the sample distribution function and of the classical multinomial estimator, *Ann. Math. Statist.* **27,** 642–669 (1956).

Eckmann, B.: [1] Über monothetische Gruppen, *Comment. Math. Helv.* **16,** 249–263 (1943/44).

Eggleston, H. G.: [1] The fractional dimension of a set defined by decimal properties, *Quart. J. Math.* (1) **20,** 31–36 (1949). [2] Sets of fractional dimensions which occur in some problems of number theory, *Proc. London Math. Soc.* (2) **54,** 42–93 (1951). [3] *Convexity*, Cambridge Tracts in Math. and Math. Physics, no. 47, Cambridge Univ. Press, London, 1958.

Elliott, P. D. T. A.: [1] On the limiting distribution of additive functions (mod 1), *Pacific J. Math.* **38,** 49–59 (1971). [2] The Riemann zeta function and coin tossing, *J. reine angew. Math.* **254,** 100–109 (1972). [3] On the distribution of arg $L(s, \chi)$ in the half-plane $\sigma > \frac{1}{2}$, *Acta Arith.* **20,** 155–169 (1972). [4] On distribution functions (mod 1): Quantitative Fourier inversion, *J. Number Th.* **4,** 509–522 (1972).

Ennola, V.: [1] On the distribution of fractional parts of sequences. I, II, III, *Ann. Univ. Turku.* Ser. A I, no. 91, 92 (1966), no. 97 (1967). [2] On metric diophantine approximation, *Ann. Univ. Turku.* Ser. A I, no. 113 (1967).

Erdös, P.: [1] On the smoothness of the asymptotic distribution of additive arithmetical functions, *Amer. J. Math.* **61,** 722–725 (1939). [2] On the uniform distribution of the roots of certain polynomials, *Ann. of Math.* (2) **43,** 59–64 (1942). [3] On the distribution function of additive functions, *Ann. of Math.* (2) **47,** 1–20 (1946). [4] On the strong law of large numbers, *Trans. Amer. Math. Soc.* **67,** 51–56 (1950). [5] Some results on diophantine approximation, *Acta Arith.* **5,** 359–369 (1959). [6] Some unsolved problems, *Magyar Tud. Akad. Mat. Kutató Int. Közl.* **6,** 221–254 (1961). [7] Problems and results on diophantine approximations, *Compositio Math.* **16,** 52–65 (1964).

Erdös, P., and Davenport, H.: [1] A theorem on uniform distribution, *Magyar Tud. Akad. Mat. Kutató Int. Közl.* **8,** 3–11 (1963).

Erdös, P., and Gál, I. S.: [1] On the law of the iterated logarithm. I, II, *Indag. Math.* **17,** 65–76, 77–84 (1955).

Erdös, P., Kac, M., van Kampen, E. R., and Wintner, A.: [1] Ramanujan sums and almost periodic functions, *Studia Math.* **9,** 43–53 (1940).

Erdös, P., and Koksma, J. F.: [1] On the uniform distribution modulo 1 of lacunary sequences, *Indag. Math.* **11,** 79–88 (1949). [2] On the uniform distribution mod 1 of sequences $\{f(n, \theta)\}$, *Indag. Math.* **11,** 299–302 (1949).

Erdös, P., and Lorentz, G. G.: [1] On the probability that n and $g(n)$ are relatively prime, *Acta Arith.* **5,** 35–44 (1958).

Erdös, P., and Rényi, A.: [1] A probabilistic approach to problems of Diophantine approximation, *Illinois J. Math.* **1,** 303–315 (1957).

Erdös, P., and Taylor, S. J.: [1] On the set of points of convergence of a lacunary trigonometric series and the equidistribution properties of related sequences, *Proc. London Math. Soc.* (3) **7,** 598–615 (1957).

Erdös, P., and Turán, P.: [1] On the uniformly dense distribution of certain sequences of points, *Ann. of Math.* (2) **41,** 162–173 (1940). [2] On interpolation. III, *Ann. of Math.* (2) **41,** 510–553 (1940). [3] On a problem in the theory of uniform distribution. I, II, *Indag. Math.* **10,** 370–378, 406–413 (1948). [4] On the distribution of roots of polynomials, *Ann. of Math.* (2) **51,** 105–119 (1950).

Erdös, P., and Wintner, A.: [1] Additive arithmetical functions and statistical independence, *Amer. J. Math.* **61,** 713–721 (1939).

Faĭnleĭb, A. S.: [1] Distribution of values of Euler's function (Russian), *Mat. Zametki* **1,** 645–652 (1967); Engl. trans., *Math. Notes* **1,** 428–432 (1967). [2] A generalization of Esseen's inequality and its application in probabilistic number theory (Russian), *Izv. Akad. Nauk SSSR,* Ser. Mat. **32,** 859–879 (1968); Engl. trans., *Math. of the USSR—Izvestija* **2,** 821–844 (1968).

Feller, W.: [1] On the Kolmogoroff-Smirnoff limit theorems, *Ann. Math. Statist.* **19,** 177–189 (1948). [2] *An Introduction to Probability Theory and Its Applications,* Vol. II, Wiley, New York, 1966.

Fleischer, W.: [1] Sätze zur Theorie der Gleichverteilung von Folgen auf kompakten Räumen, *Sitzgsber. Österr. Akad. Wiss.*, Math.-naturw. Kl., Abt. II, **176,** 521–545 (1967). [2] Das Wienersche Mass einer gewissen Menge von Vektorfunktionen, *Monatsh. Math.* **75,** 193–197 (1971). [3] Eine Diskrepanzabschätzung für stetige Funktionen, *Z. Wahrscheinlichkeitsth.* **23,** 18–21 (1972).

Florek, K.: [1] Une remarque sur la répartition des nombres $n\xi$ (mod 1), *Colloq. Math.* **2,** 323–324 (1951).

Forman, W., and Shapiro, H. N.: [1] An arithmetic property of certain rational powers, *Comm. Pure Appl. Math.* **20,** 561–573 (1967).

Fortet, R.: [1] Sur une suite également répartie, *Studia Math.* **9,** 54–70 (1940).

Franel, J.: [1] Question 1260, *L'Intermédiaire Math.* **5,** 77 (1898). [2] Question 1547, *L'Intermédiaire Math.* **6,** 149 (1899). [3] A propos des tables des logarithmes, *Vierteljschr. Naturforsch. Ges. Zürich* **1917,** 286–295. [4] Les suites de Farey et les problèmes des nombres premiers, *Nachr. Ges. Wiss. Göttingen,* Math.-phys. Kl., **1924,** 198–201.

Franklin, J. N.: [1] On the equidistribution of pseudo-random numbers, *Quart. Appl. Math.* **16,** 183–188 (1958). [2] Deterministic simulation of random processes, *Math.*

Comp. **17,** 28–59 (1963). [3] Equidistribution of matrix-power residues modulo one, *Math. Comp.* **18,** 560–568 (1964).

Fréchet, M.: [1] *Généralités sur les probabilités. Éléments aléatoires,* 2nd ed., Gauthier-Villars, Paris, 1950.

Furstenberg, H.: [1] *Stationary Processes and Prediction Theory,* Ann. Math. Studies, no. 44, Princeton Univ. Press, Princeton, N.J., 1960. [2] Strict ergodicity and transformations of the torus, *Amer. J. Math.* **83,** 573–601 (1961). [3] Disjointness in ergodic theory, minimal sets, and a problem in Diophantine approximation, *Math. Systems Theory* **1,** 1–50 (1967).

Furstenberg, H., Keynes, H. B., and Shapiro, L.: [1] Prime flows in topological dynamics, *Israel J. Math.* **14,** 26–38 (1973).

Gaal, S. A.: [1] *Point Set Topology,* Academic Press, New York-London, 1964.

Gabai, H.: [1] On the discrepancy of certain sequences mod 1, *Indag. Math.* **25,** 603–605 (1963). [2] On the discrepancy of certain sequences mod 1, *Illinois J. Math.* **11,** 1–12 (1967).

Gál, I. S., and Gál, L.: [1] The discrepancy of the sequence $\{(2^n x)\}$, *Indag. Math.* **26,** 129–143 (1964).

Gál, I. S., and Koksma, J. F.: [1] Sur l'ordre de grandeur des fonctions sommables, *C. R. Acad. Sci. Paris* **227,** 1321–1323 (1948). [2] Sur l'ordre de grandeur des fonctions sommables, *Indag. Math.* **12,** 192–207 (1950).

Galambos, J.: [1] Distribution of arithmetical functions. A survey, *Ann. Inst. Henri Poincaré,* Sect. B, **6,** no. 4, 281–305 (1970). [2] A generalization of a theorem of Borel concerning the distribution of digits in dyadic expansions, *Amer. Math. Monthly* **78,** 774–779 (1971). [3] The Hausdorff dimension of sets related to g-expansions, *Acta Arith.* **20,** 385–392 (1972). [4] Probabilistic theorems concerning expansions of real numbers, *Periodica Math. Hungar.* **3,** 101–113 (1973).

Gallagher, P. X.: [1] Metric simultaneous diophantine approximation, *J. London Math. Soc.* **37,** 387–390 (1962). [2] Metric simultaneous diophantine approximation. II, *Mathematika* **12,** 123–127 (1965).

Gaposhkin, V. F.: [1] Lacunary series and independent functions (Russian), *Uspehi Matem. Nauk* **21,** no. 6, 3–82 (1966); Engl. trans., *Russian Math. Surveys* **21,** no. 6, 1–82 (1966).

Gel'fond, A. O.: [1] A common property of number systems (Russian), *Izv. Akad. Nauk SSSR,* Ser. Mat. **23,** 809–814 (1959).

Gel'fond, A. O., and Linnik, Yu. V.: [1] *Elementary Methods in the Analytic Theory of Numbers,* M.I.T. Press, Cambridge, Mass., 1966.

Gerl, P.: [1] Punktfolgen auf Kurven und Flächen, *Monatsh. Math.* **67,** 401–432 (1963). [2] Einige metrische Sätze in der Theorie der Gleichverteilung mod 1, *J. reine angew. Math.* **216,** 50–66 (1964). [3] Zur Gleichverteilung auf Flächen, *J. reine angew. Math.* **216,** 113–122 (1964). [4] Konstruktion gleichverteilter Punktfolgen, *Monatsh. Math.* **69,** 306–317 (1965). [5] Eine Bemerkung zur gleichmässigen Gleichverteilung mod 1, *Monatsh. Math.* **70,** 106–110 (1966). [6] Einige mehrdimensionale Sätze zur gleichmässigen Gleichverteilung, *Indag. Math.* **30,** 151–161 (1968). [7] Relative Gleichverteilung in lokalkompakten Räumen, *Math. Z.* **121,** 24–50 (1971). [8] Relative Gleichverteilung in lokalkompakten Räumen II, *Monatsh. Math.* **75,** 410–422 (1971). [9] Gleichverteilung auf der Kugel, *Arch. der Math.* **24,** 203–207 (1973).

Gillet, A.: [1] Sur la répartition modulo un des multiples d'un nombre réel, *C. R. Acad. Sci.*

Paris **261,** 4946–4947 (1965). [2] Sur la répartition modulo 1 des multiples d'un nombre réel, *C. R. Acad. Sci. Paris* Sér. A **264,** 223–225 (1967).

Goldstein, L. J.: [1] Analogues of Artin's conjecture, *Trans. Amer. Math. Soc.* **149,** 431–442 (1970).

Good, I. J.: [1] Normal recurring decimals, *J. London Math. Soc.* **21,** 167–169 (1946).

Gotusso, L.: [1] Successioni uniformemente distribuite in corpi finiti, *Atti Sem. Mat. Fis. Univ. Modena* **12,** 215–232 (1962/63).

Graham, R. L., and van Lint, J. H.: [1] On the distribution of $n\theta$ modulo 1, *Canad. J. Math.* **20,** 1020–1024 (1968).

Grandet-Hugot, M.: [1] Études de certaines suites $\{\lambda\alpha^n\}$ dans les adèles, *Ann. Sci. École Norm. Sup.* (3) **83,** 171–185 (1966). [2] Une propriété des "nombres de Pisot" dans un corps de séries formelles, *C. R. Acad. Sci. Paris* Sér. A **265,** 39–41 (1967); errata, *ibid.*, p. 551. [3] Éléments algébriques remarquables dans un corps de séries formelles, *Acta Arith.* **14,** 177–184 (1967/68).

Grassini, E.: [1] Successioni regolarmente equidistribuite negli spazi topologici, *Ist. Lombardo Accad. Sci. Lett. Rend.* A**99,** 950–962 (1965).

Haber, S.: [1] On a sequence of points of interest for numerical quadrature, *J. Res. Nat. Bur. Standards,* Sect. B **70,** 127–136 (1966). [2] Sequences of numbers that are approximately completely equidistributed, *J. Assoc. Comput. Machin.* **17,** 269–272 (1970). [3] Numerical evaluation of multiple integrals, *SIAM Rev.* **12,** 481–526 (1970). [4] Experiments on optimal coefficients, in S. K. Zaremba (Ed.), *Applications of Number Theory to Numerical Analysis,* Academic Press, New York, 1972, pp. 11–37.

Haber, S., and Osgood, C. F.: [1] On a theorem of Piatetsky-Shapiro and approximation of multiple integrals, *Math. Comp.* **23,** 165–168 (1969). [2] On the sum $\sum \langle n\alpha\rangle^{-t}$ and numerical integration, *Pacific J. Math.* **31,** 383–394 (1969).

Hahn, F. J.: [1] On affine transformations of compact abelian groups, *Amer. J. Math.* **85,** 428–446 (1963).

Hahn, H.: [1] *Theorie der reellen Funktionen,* 1. Band, 1. Aufl., Julius Springer, Berlin, 1921.

Halmos, P. R.: [1] *Measure Theory,* Van Nostrand, Princeton, N. J., 1950. [2] *Lectures on Ergodic Theory,* Math. Soc. Japan, Tokyo, 1956.

Halmos, P. R., and Samelson, H.: [1] On monothetic groups, *Proc. Nat. Acad. Sci. U.S.A.* **28,** 254–258 (1942).

Halmos, P. R., and von Neumann, J.: [1] Operator methods in classical mechanics II, *Ann. of Math.* (2) **43,** 332–350 (1942).

Halter-Koch, F.: [1] Abgeschlossene Mengen algebraischer Zahlen, *Abh. Math. Sem. Hamburg* **38,** 65–79 (1972).

Halton, J. H.: [1] On the efficiency of certain quasi-random sequences of points in evaluating multi-dimensional integrals, *Numer. Math.* **2,** 84–90 (1960); Erratum, *ibid.*, p. 196. [2] The distribution of the sequence $\{n\xi\}$ $(n = 0, 1, 2, \ldots)$, *Proc. Cambridge Phil. Soc.* **61,** 665–670 (1965). [3] A retrospective and prospective survey of the Monte Carlo method, *SIAM Rev.* **12,** 1–63 (1970). [4] Estimating the accuracy of quasi-Monte Carlo integration, in S. K. Zaremba (Ed.), *Applications of Number Theory to Numerical Analysis,* Academic Press, New York, 1972, pp. 345–360.

Halton, J. H., and Zaremba, S. K.: [1] The extreme and L^2 discrepancies of some plane sets, *Monatsh. Math.* **73,** 316–328 (1969).

Hammersley, J. M.: [1] Monte Carlo methods for solving multivariable problems, *Ann. New York Acad. Sci.* **86,** 844–874 (1960).

Hammersley, J. M., and Handscomb, D. C.: [1] *Monte Carlo Methods*, Methuen, London, 1964.

Hanson, H. A.: [1] Some relations between various types of normality of numbers, *Canad. J. Math.* **6,** 477–485 (1954).

Hardy, G. H.: [1] On double Fourier series, and especially those which represent the double zeta-function with real and incommensurable parameters, *Quart. J. Math.* (1) **37,** 53–79 (1906). [2] *Divergent Series*, Oxford Univ. Press, London, 1949.

Hardy, G. H., and Littlewood, J. E.: [1] Some problems of Diophantine approximation. III: The fractional part of $n^k\theta$, *Acta Math.* **37,** 155–191 (1914). [2] Some problems of Diophantine approximation. VI: The series $\Sigma e(\lambda_n)$ and the distribution of the points $(\lambda_n\alpha)$, *Proc. Nat. Acad. Sci. U.S.A.* **3,** 84–88 (1917). [3] Some problems of Diophantine approximation: The lattice points of a right-angled triangle. I., *Proc. London Math. Soc.* (2) **20,** 15–36 (1922). [4] Some problems of Diophantine approximation: The lattice points of a right-angled triangle. II., *Abh. Math. Sem. Hamburg* **1,** 212–249 (1922). [5] Some problems of Diophantine approximation: A series of cosecants, *Bull. Calcutta Math. Soc.* **20,** 251–266 (1930). [6] Notes on the theory of series XXIV: A curious power series, *Proc. Cambridge Phil. Soc.* **42,** 85–90 (1946).

Hardy, G. H., and Wright, E. M.: [1] *An Introduction to the Theory of Numbers*, 4th ed., Clarendon, Oxford, 1960.

Hartman, S.: [1] Une généralisation d'un théorème de M. Ostrowski sur la répartition des nombres mod 1, *Ann. Soc. Polon. Math.* **22,** 169–172 (1949). [2] Sur une méthode d'estimation des moyennes de Weyl pour les fonctions périodiques et presque périodiques, *Studia Math.* **12,** 1–24 (1951). [3] Über die Abstände von Punkten $n\xi$ auf der Kreisperipherie, *Ann. Soc. Polon. Math.* **25,** 110–114 (1952). [4] Remarks on equidistribution on non-compact groups, *Compositio Math.* **16,** 66–71 (1964).

Hartman, S., and Hulanicki, A.: [1] Sur les ensembles denses de puissance minimum dans les groupes topologiques, *Colloq. Math.* **6,** 187–191 (1958).

Hartman, S., Marczewski, E., and Ryll-Nardzewski, C.: [1] Théorèmes ergodiques et leurs applications, *Colloq. Math.* **2,** 109–123 (1951).

Hartman, S., and Ryll-Nardzewski, C.: [1] Zur Theorie der lokal kompakten abelschen Gruppen, *Colloq. Math.* **4,** 157–188 (1957).

Haselgrove, C. B.: [1] A method for numerical integration, *Math. Comp.* **15,** 323–337 (1961).

Haviland, E. K.: [1] On the distribution functions of the reciprocal of a function and of a function reduced mod 1, *Amer. J. Math.* **63,** 408–414 (1941).

Hecke, E.: [1] Über analytische Funktionen und die Verteilung von Zahlen mod. Eins, *Abh. Math. Sem. Hamburg* **1,** 54–76 (1922).

Hedrlín, Z.: [1] On integration in compact metric spaces, *Comm. Math. Univ. Carol.* **2,** 4, 17–19 (1961). [2] Remark on integration in compact metric spaces, *Comm. Math. Univ. Carol.* **3,** 1, 31 (1962).

Helmberg, G.: [1] A theorem on equidistribution in compact groups, *Pacific J. Math.* **8,** 227–241 (1958). [2] Ein Satz über Gleichverteilung in kompakten Gruppen (II), *Monatsh. Math.* **63,** 368–377 (1959). [3] Zerlegungen des Mittelwerts fastperiodischer Funktionen. I, *J. reine angew. Math.* **207,** 31–52 (1961); II, *ibid.* **208,** 1–21 (1961). [4] Eine Familie von Gleichverteilungskriterien in kompakten Gruppen, *Monatsh. Math.* **66,** 417–423 (1962). [5] Abstract theory of uniform distribution, *Compositio Math.* **16,**

72–82 (1964). [6] A class of criteria concerning uniform distribution in compact groups, *Compositio Math.* **16,** 196–203 (1964). [7] Gleichverteilte Folgen in lokal kompakten Räumen, *Math. Z.* **86,** 157–189 (1964). [8] On a convolution of sequences in a compact group, *Indag. Math.* **26,** 266–274 (1964). [9] Einige Diskrepanzabschätzungen für Punktfolgen im $\mathbb{R}^s$, *Monatsh. Math.* **75,** 223–238 (1971).

Helmberg, G., and Paalman-de Miranda, A.: [1] Almost no sequence is well distributed, *Indag. Math.* **26,** 488–492 (1964).

Helson, H., and Kahane, J.-P.: [1] A Fourier method in diophantine problems, *J. Analyse Math.* **15,** 245–262 (1965).

Hewitt, E., and Ross, K. A.: [1] *Abstract Harmonic Analysis*, Vols. I, II, Grundlehren d. math. Wiss., Band 115, 152, Springer-Verlag, Berlin-Göttingen-Heidelberg, 1963, 1970.

Hewitt, E., and Savage, L. J.: [1] Symmetric measures on Cartesian products, *Trans. Amer. Math. Soc.* **80,** 470–501 (1955).

Hewitt, E., and Stromberg, K.: [1] *Real and Abstract Analysis*, Springer-Verlag, New York-Heidelberg-Berlin, 1969.

Hewitt, E., and Zuckerman, H. S.: [1] On a theorem of P. J. Cohen and H. Davenport, *Proc. Amer. Math. Soc.* **14,** 847–855 (1963).

Heyer, H.: [1] *Dualität lokalkompakter Gruppen*, Lecture Notes in Math., Vol. 150, Springer-Verlag, Berlin-Heidelberg-New York, 1970.

Hiergeist, F. X.: [1] Some generalizations and extensions of uniformly distributed sequences, Thesis, Univ. of Pittsburgh, 1964.

Hill, J. D.: [1] The Borel property of summability methods, *Pacific J. Math.* **1,** 399–409 (1951). [2] Remarks on the Borel property, *Pacific J. Math.* **4,** 227–242 (1954).

Hlawka, E.: [1] Zur formalen Theorie der Gleichverteilung in kompakten Gruppen, *Rend. Circ. Mat. Palermo* (2) **4,** 33–47 (1955). [2] Über einen Satz von van der Corput, *Arch. der Math.* **6,** 115–120 (1955). [3] Folgen auf kompakten Räumen, *Abh. Math. Sem. Hamburg* **20,** 223–241 (1956). [4] Normal gleichverteilte Folgen auf kompakten Räumen, *Anz. Österr. Akad. Wiss.*, Math.-naturw. Kl., **94,** 94–96 (1957). [5] Zum Hauptsatz der Theorie der Gleichverteilung, *Anz. Österr. Akad. Wiss.*, Math.-naturw. Kl., **94,** 313–317 (1957). [6] Folgen auf kompakten Räumen. II, *Math. Nachr.* **18,** 188–202 (1958). [7] Zur Theorie der diophantischen Approximationen, *Anz. Österr. Akad. Wiss.*, Math.-naturw. Kl., **95,** 41–48 (1958). [8] Erbliche Eigenschaften in der Theorie der Gleichverteilung, *Publ. Math. Debrecen* **7,** 181–186 (1960). [9] Über *C*-Gleichverteilung, *Ann. Mat. Pura Appl.* (IV) **49,** 311–326 (1960). [10] Cremonatransformation von Folgen modulo 1, *Monatsh. Math.* **65,** 227–232 (1961). [11] Über die Diskrepanz mehrdimensionaler Folgen mod 1, *Math. Z.* **77,** 273–284 (1961). [12] Funktionen von beschränkter Variation in der Theorie der Gleichverteilung, *Ann. Mat. Pura Appl.* (IV) **54,** 325–333 (1961). [13] Lösung von Integralgleichungen mittels zahlentheoretischer Methoden I, *Sitzgsber. Österr. Akad. Wiss.*, Math.-naturw. Kl., Abt. II, **171,** 103–123 (1962). [14] Zur angenäherten Berechnung mehrfacher Integrale, *Monatsh. Math.* **66,** 140–151 (1962). [15] Geordnete Schätzfunktionen und Diskrepanz, *Math. Ann.* **150,** 259–267 (1963). [16] Discrepancy and uniform distribution of sequences, *Compositio Math.* **16,** 83–91 (1964). [17] Uniform distribution modulo 1 and numerical analysis, *Compositio Math.* **16,** 92–105 (1964). [18] Trigonometrische Interpolation bei Funktionen von mehreren Variablen, *Acta Arith.* **9,** 305–320 (1964). [19] Mathematische Modelle zur kinetischen Gastheorie, *Sitzgsber. Österr. Akad. Wiss.*, Math.-naturw. Kl., Abt. II, **174,** 287–307 (1965). [20] Geometrie der Zahlen und trigonometrische Interpolation bei Funktionen von mehreren Variablen, in *Les Tendances*

Géométriques en Algèbre et Théorie des Nombres, Édition du Centre National de la Recherche Scientifique, Paris, 1966, pp. 83–86. [21] Mathematische Modelle der kinetischen Gastheorie III, *Sitzgsber. Österr. Akad. Wiss.*, Math.-naturw. Kl., Abt. II, **178**, 1–12 (1969). [22] Interpolation analytischer Funktionen auf dem Einheitskreis, in P. Turán (Ed.), *Number Theory and Analysis*, Plenum, New York-VEB Deutscher Verlag der Wissenschaften, Berlin, 1969, pp. 97–118. [23] Mathematische Modelle der kinetischen Gastheorie II, *Symposia Mathematica IV (Ist. Naz. di Alta Mat.)*, Academic Press, London-New York, 1970, pp. 81–97. [24] Ein metrischer Satz in der Theorie der *C*-Gleichverteilung, *Monatsh. Math.* **74,** 108–118 (1970). [25] Zur Definition der Diskrepanz, *Acta Arith.* **18,** 233–241 (1971). [26] Discrepancy and Riemann integration, in L. Mirsky (Ed.), *Studies in Pure Mathematics (Papers Presented to Richard Rado)*, Academic Press, New York, 1971, pp. 121–129. [27] Ein metrisches Gegenstück zu einem Satz von W. A. Veech, *Monatsh. Math.* **76,** 436–447 (1972). [28] Über eine Methode von E. Hecke in der Theorie der Gleichverteilung, *Acta Arith.* **24,** 11–31 (1973).

Hlawka, E., and Kreiter, K.: [1] Lösung von Integralgleichungen mittels zahlentheoretischer Methoden II, *Sitzgsber. Österr. Akad. Wiss.*, Math.-naturw. Kl., Abt. II, **172,** 229–250 (1963).

Hlawka, E., and Kuich, W.: [1] Geordnete Schätzfunktionen und Diskrepanz II, *Sitzgsber. Österr. Akad. Wiss.*, Math.-naturw. Kl., Abt. II, **174,** 235–286 (1965).

Hlawka, E., and Mück, R.: [1] A transformation of equidistributed sequences, in S. K. Zaremba (Ed.), *Applications of Number Theory to Numerical Analysis*, Academic Press, New York, 1972, pp. 371–388. [2] Über eine Transformation von gleichverteilten Folgen II, *Computing* **9,** 127–138 (1972).

Hlawka, E., and Niederreiter, H.: [1] Diskrepanz in kompakten abelschen Gruppen I, *Manuscripta Math.* **1,** 259–288 (1969).

Hobson, E. W.: [1] *The Theory of Functions of a Real Variable and the Theory of Fourier's Series*, Vol. I, 3rd ed., Cambridge Univ. Press, London, 1927.

Hodges, J. H.: [1] Uniform distribution of sequences in $GF[q, x]$, *Acta Arith.* **12,** 55–75 (1966). [2] Uniform distribution of polynomial-generated sequences in $GF[q, x]$, *Ann. Mat. Pura Appl.* (IV) **82,** 135–142 (1969). [3] On uniform distribution of sequences in $GF\{q, x\}$ and $GF[q, x]$, *Ann. Mat. Pura Appl.* (IV) **85,** 287–294 (1970). [4] A note on irregularities in distribution of sequences of integers, *Portugaliae Math.* **31,** 171–176 (1972).

Holewijn, P. J.: [1] Contribution to the theory of asymptotic distribution mod 1, Thesis, Technische Hogeschool Delft, 1965. [2] On the uniform distribution of sequences of random variables, *Z. Wahrscheinlichkeitsth.* **14,** 89–92 (1969). [3] Note on Weyl's criterion and the uniform distribution of independent random variables, *Ann. Math. Statist.* **40,** 1124–1125 (1969).

Horn, S., and Schach, S.: [1] An extension of the Hewitt-Savage zero-one law, *Ann. Math. Statist.* **41,** 2130–2131 (1970).

Hsu, L. C.: [1] Note on the numerical integration of periodic functions and of partially periodic functions, *Numer. Math.* **3,** 169–173 (1961).

Hua, L.-K.: [1] Die Abschätzung von Exponentialsummen und ihre Anwendung in der Zahlentheorie, *Enzykl. d. math. Wiss.*, 1. Band, 2. Teil, Heft 13, Teil 1, B. G. Teubner, Leipzig, 1959, pp. 1–123. [2] *Additive Theory of Prime Numbers*, Trans. of Math. Monographs, Vol. 13, American Math. Society, Providence, R.I., 1965.

Hua, L.-K., and Wang, Y.: [1] Remarks concerning numerical integration, *Sci. Record*

(N.S.) **4,** 8–11 (1960). [2] On diophantine approximations and numerical integrations. I, II, *Sci. Sinica* **13,** 1007–1010 (1964). [3] On numerical integration of periodic functions of several variables, *Sci. Sinica* **14,** 964–978 (1965).

Huxley, M. N.: [1] The distribution of Farey points. I, *Acta Arith.* **18,** 281–288 (1971).

Ibragimov, I. A.: [1] On asymptotic distribution of values of certain sums (Russian), *Vestnik Leningrad. Univ.* **15,** no. 1, 55–69 (1960); Engl. trans., *Selected Transl. Math. Statist. Prob.* **5,** 33–48 (1965). [2] The central limit theorem for sums of functions of independent random variables and sums of the form $\sum f(2^k t)$ (Russian), *Teor. Verojatnost. i Primenen.* **12,** 655–665 (1967); Engl. trans., *Theor. Probability Appl.* **12,** 596–607 (1967).

Ilyasov, I.: [1] Summation of composite functions of the Euler function (Russian), *Dokl. Akad. Nauk SSSR* **178,** 529–532 (1968); Engl. trans., *Soviet Math. Dokl.* **9,** 130–134 (1968).

Isbell, J. R.: [1] *Uniform Spaces*, American Math. Society, Providence, R.I., 1964.

Jacobs, K.: [1] Gleichverteilung mod 1, *Selecta Mathematica IV*, Springer-Verlag, Berlin-Heidelberg-New York, 1972, pp. 57–93.

Jager, H.: [1] On decimal expansions, Zahlentheorie-Tagung math. Forsch.-inst. Oberwolfach 1970, *Ber. math. Forsch.-inst. Oberwolfach* **5,** 67–75 (1971).

Jagerman, D. L.: [1] The autocorrelation function of a sequence uniformly distributed modulo 1, *Ann. Math. Statist.* **34,** 1243–1252 (1963). [2] The autocorrelation and joint distribution functions of the sequences $\{(a/m)j^2\}$, $\{(a/m)(j + \tau)^2\}$, *Math. Comp.* **18,** 211–232 (1964). [3] Some theorems concerning pseudo-random numbers, *Math. Comp.* **19,** 418–426 (1965).

Jessen, B.: [1] The theory of integration in a space of an infinite number of dimensions, *Acta Math.* **63,** 249–323 (1934).

Judin, A. A.: [1] Irregularity in the distribution of sequences (Russian), *Izv. Akad. Nauk. Tadžik. SSR*, Otdel. Fiz.-Tehn. i Him. Nauk **1966,** no. 2, 7–18.

Kac, I. S.: [1] Uniform distribution on a sphere, *Bull. Acad. Polon. Sci.*, Cl. III **5,** 485–486 (1957).

Kac, M.: [1] Sur les fonctions indépendantes (V), *Studia Math.* **7,** 96–100 (1938). [2] On the distribution of values of sums of the type $\sum f(2^k t)$, *Ann. of Math.* (2) **47,** 33–49 (1946). [3] Probability methods in some problems of analysis and number theory, *Bull. Amer. Math. Soc.* **55,** 641–665 (1949). [4] *Statistical Independence in Probability, Analysis and Number Theory*, Carus Math. Monographs, no. 12, Wiley, New York, 1959.

Kac, M., Salem, R., and Zygmund, A.: [1] A gap theorem, *Trans. Amer. Math. Soc.* **63,** 235–243 (1948).

Kac, M., van Kampen, E. R., and Wintner, A.: [1] Ramanujan sums and almost periodic functions, *Amer. J. Math.* **62,** 107–114 (1940).

Kahane, J.-P.: [1] Sur les mauvaises répartitions modulo 1, *Ann. Inst. Fourier* **14,** fasc. 2, 519–526 (1964).

Kahane, J.-P., and Salem, R.: [1] Distribution modulo one and sets of uniqueness, *Bull. Amer. Math. Soc.* **70,** 259–261 (1964).

Kakutani, S.: [1] On cardinal numbers related with a compact abelian group, *Proc. Imp. Acad. Tokyo* **19,** 366–372 (1943).

Kano, T.: [1] Criteria for uniform and weighted uniform distribution mod 1, *Comment. Math. Univ. St. Pauli* **20,** 83–91 (1971).

Karacuba, A. A.: [1] Estimates for trigonometric sums by Vinogradov's method, and some applications (Russian), *Trudy Mat. Inst. Steklov.* **112,** 241–255 (1971); Engl. trans., *Proc. Steklov Inst. Math.* **112,** American Math. Society, Providence, R.I., 1973, pp. 251–265.

Karimov, B.: [1] On the distribution of the fractional parts of certain linear forms in a unit square (Russian), *Izv. Akad. Nauk UzSSR*, Ser. Fiz.-Mat. Nauk, **10,** no. 1, 19–22 (1966). [2] On the question of the number of fractional parts of certain linear forms in a rectangle (Russian), *Izv. Akad. Nauk UzSSR*, Ser. Fiz.-Mat. Nauk, **10,** no. 2, 21–28 (1966).

Kátai, I.: [1] Zur Gleichverteilung modulo Eins, *Ann. Univ. Sci. Budapest. Eötvös Sect. Math.* **7,** 73–77 (1964).

Każdan, D. A.: [1] Uniform distribution in the plane (Russian), *Trudy Moskov. Mat. Obšč.* **14,** 299–305 (1965); Engl. trans., *Trans. Moscow Math. Soc. 1965*, American Math. Society, Providence, R.I., 1967, pp. 325–332.

Kelley, J. L.: [1] *General Topology*, Van Nostrand, New York-Toronto-London, 1955.

Kelly, B. G. A.: [1] Densities and measures of linear sets, *Canad. Math. Bull.* **16,** 61–66 (1973).

Kemperman, J. H. B.: [1] On the asymptotic distribution of a sequence in a compact group, *Notices Amer. Math. Soc.* **5,** no. 2, Abstr. 542–10 (1958). [2] Probability methods in the theory of distributions modulo one, *Compositio Math.* **16,** 106–137 (1964). [3] On the distribution of a sequence in a compact group, *Compositio Math.* **16,** 138–157 (1964). [4] Distributions modulo 1 of slowly changing sequences, *Nieuw Arch. voor Wisk.* (3) **21,** 138–163 (1973).

Kennedy, P. B.: [1] A note on uniformly distributed sequences, *Quart. J. Math.* (2) **7,** 125–127 (1956).

Keogh, F. R., Lawton, B., and Petersen, G. M.: [1] Well distributed sequences, *Canad. J. Math.* **10,** 572–576 (1958).

Keogh, F. R., and Petersen, G. M.: [1] A strengthened form of a theorem of Wiener, *Math. Z.* **71,** 31–35 (1959).

Kesten, H.: [1] Uniform distribution mod 1, *Ann. of Math.* (2) **71,** 445–471 (1960). [2] Uniform distribution mod 1. II, *Acta Arith.* **7,** 355–380 (1961/62). [3] The discrepancy of random sequences $\{kx\}$, *Acta Arith.* **10,** 183–213 (1964/65). [4] On a conjecture of Erdös and Szüsz related to uniform distribution mod 1, *Acta Arith.* **12,** 193–212 (1966/67).

Khintchine, A.: [1] Ein Satz über Kettenbrüche mit arithmetischen Anwendungen, *Math. Z.* **18,** 289–306 (1923). [2] Einige Sätze über Kettenbrüche, mit Anwendungen auf die Theorie der Diophantischen Approximationen, *Math. Ann.* **92,** 115–125 (1924). [3] Über einen Satz der Wahrscheinlichkeitsrechnung, *Fund. Math.* **6,** 9–20 (1924). [4] Über eine Klasse linearer Diophantischer Approximationen, *Rend. Circ. Mat. Palermo* **50,** 170–195 (1926). [5] *Asymptotische Gesetze der Wahrscheinlichkeitsrechnung*, Erg. Math. Grenzgeb., Band 2, Heft 4, Julius Springer, Berlin, 1933. [6] Eine arithmetische Eigenschaft der summierbaren Funktionen, *Mat. Sb.* **41,** 11–13 (1934). [7] *Continued Fractions*, Noordhoff, Groningen, 1963.

Kiefer, J.: [1] On large deviations of the empiric d.f. of vector chance variables and a law of the iterated logarithm, *Pacific J. Math.* **11,** 649–660 (1961).

Kiefer, J., and Wolfowitz, J.: [1] On the deviation of the empiric distribution function of vector chance variables, *Trans. Amer. Math. Soc.* **87,** 173–186 (1958).

Knapowski, S.: [1] Über ein Problem der Gleichverteilungstheorie, *Colloq. Math.* **5,** 8–10 (1957).

Knichal, V.: [1] Dyadische Entwicklungen und Hausdorffsches Mass, *Mém. Soc. Roy. Sci. Bohême*, Cl. des Sciences **1933,** no. 14, 1–18. [2] Dyadische Entwicklungen und Hausdorffsches Mass, *Čas. Mat. Fys.* **65,** 195–209 (1936).

Knopp, K.: [1] *Theorie und Anwendung der unendlichen Reihen*, 5. Aufl., Grundlehren d. math. Wiss., Band 2, Springer-Verlag, Berlin-Göttingen-Heidelberg-New York, 1964.

Knuth, D. E.: [1] Construction of a random sequence, *Nordisk Tidskr. Informations-Behandling* **5,** 246–250 (1965). [2] *The Art of Computer Programming*, Vol. 2: *Seminumerical Algorithms*, Addison-Wesley, Reading, Mass., 1969.

Koksma, J. F.: [1] Over stelsels Diophantische ongelijkheden, Thesis, Groningen, 1930 [2] Asymptotische verdeling van reële getallen modulo 1. I, II, III, *Mathematica* (Leiden) **1,** 245–248; **2,** 1–6, 107–114 (1933). [3] Ein mengentheoretischer Satz über die Gleichverteilung modulo Eins, *Compositio Math.* **2,** 250–258 (1935). [4] *Diophantische Approximationen*, Erg. Math. Grenzgeb., Band 4, Heft 4, Julius Springer, Berlin, 1936. [5] Über die Diskrepanz (mod 1) und die ganzzahligen Lösungen gewisser Ungleichungen, *Indag. Math.* **3,** 36–41 (1941). [6] Een algemeene stelling uit de theorie der gelijkmatige verdeeling modulo 1, *Mathematica* B (Zutphen) **11,** 7–11 (1942/43). [7] Eenige integralen in de theorie der gelijkmatige verdeeling modulo 1, *Mathematica* B (Zutphen) **11,** 49–52 (1942/43). [8] On decimals, *Nieuw Arch. voor Wisk.* (2) **21,** 242–267 (1943). [9] On a definite integral in the theory of uniform distribution, *Nieuw Arch. voor Wisk.* (2) **23,** 40–54 (1949). [10] An arithmetical property of some summable functions, *Indag. Math.* **12,** 354–367 (1950). [11] Some theorems on diophantine inequalities, *Math. Centrum Amsterdam*, Scriptum no. 5, 1950. [12] A Diophantine property of some summable functions, *J. Indian Math. Soc.* **15,** 87–96 (1951). [13] On a certain integral in the theory of uniform distribution, *Indag. Math.* **13,** 285–287 (1951). [14] Estimations de fonctions à l'aide d'intégrales de Lebesgue, *Bull. Soc. Math. Belg.* **6,** 4–13 (1953/54). [15] Sur les suites $(\lambda_n x)$ et les fonctions $g(t) \in L^{(2)}$, *J. de Math. Pure Appl.* **35,** 289–296 (1956). [16] The theory of asymptotic distribution modulo one, *Compositio Math.* **16,** 1–22 (1964).

Koksma, J. F., and Salem, R.: [1] Uniform distribution and Lebesgue integration, *Acta Sci. Math.* (Szeged) **12B,** 87–96 (1950).

Kolmogorov, A.: [1] *Grundbegriffe der Wahrscheinlichkeitsrechnung*, Erg. Math. Grenzgeb., Band 2, Heft 3, Julius Springer, Berlin, 1933. [2] Sulla determinazione empirica di una legge di distribuzione, *Giorn. Ist. Ital. Attuari* **4,** 83–91 (1933).

König, D., and Szücs, A.: [1] Mouvement d'un point abandonné à l'intérieur d'un cube, *Rend. Circ. Mat. Palermo* **36,** 79–90 (1913).

Korobov, N. M.: [1] On functions with uniformly distributed fractional parts (Russian), *Dokl. Akad. Nauk SSSR* **62,** 21–22 (1948). [2] Some problems on the distribution of fractional parts (Russian), *Uspehi Matem. Nauk* **4,** no. 1, 189–190 (1949). [3] On sums of fractional parts (Russian), *Dokl. Akad. Nauk SSSR* **67,** 781–782 (1949). [4] Normal periodic systems and a question on the sums of fractional parts (Russian), *Uspehi Matem. Nauk* **5,** no. 3, 135–136 (1950). [5] Concerning some questions of uniform distribution (Russian), *Izv. Akad. Nauk SSSR*, Ser. Mat., **14,** 215–238 (1950). [6] Normal periodic systems and their applications to the estimation of sums of fractional parts (Russian), *Izv. Akad. Nauk SSSR*, Ser. Mat., **15,** 17–46 (1951); Engl. trans., *Amer. Math. Soc. Transl.* (2) **4,** 31–58 (1956). [7] Fractional parts of exponential functions (Russian), *Trudy Mat. Inst. Steklov.* **38,** 87–96 (1951). [8] Some many dimensional problems of the theory of Diophantine approximations (Russian), *Dokl. Akad. Nauk SSSR* **84,** 13–16 (1952). [9] On normal periodic systems (Russian), *Izv. Akad. Nauk SSSR*, Ser. Mat., **16,** 211–216 (1952). [10] On a question of diophantine inequalities,

Compt. Rend. Premier Congrès Hongrois, Budapest, 1952, pp. 259–262. [11] Unimprovable estimates on trigonometric sums with exponential functions (Russian), *Dokl. Akad. Nauk SSSR* **89,** 597–600 (1953). [12] Multidimensional problems of the distribution of fractional parts (Russian), *Izv. Akad. Nauk. SSSR*, Ser. Mat., **17,** 389–400 (1953). [13] Numbers with bounded quotient and their applications to questions of Diophantine approximation (Russian), *Izv. Akad. Nauk SSSR*, Ser. Mat., **19,** 361–380 (1955). [14] On completely uniform distribution and conjunctly normal numbers (Russian), *Izv. Akad. Nauk SSSR*, Ser. Mat., **20,** 649–660 (1956). [15] Approximate calculation of multiple integrals with the aid of methods in the theory of numbers (Russian), *Dokl. Akad. Nauk SSSR* **115,** 1062–1065 (1957). [16] The approximate computation of multiple integrals (Russian), *Dokl. Akad. Nauk SSSR* **124,** 1207–1210 (1959). [17] Properties and calculation of optimal coefficients (Russian), *Dokl. Akad. Nauk. SSSR* **132,** 1009–1012 (1960); Engl. trans., *Soviet Math. Dokl.* **1,** 696–700 (1960). [18] Bounds on trigonometric sums involving uniformly distributed functions (Russian), *Dokl. Akad. Nauk SSSR* **133,** 1011–1014 (1960); Engl. trans., *Soviet Math. Dokl.* **1,** 923–926 (1960). [19] Application of number-theoretical sieves to integral equations and interpolation formulas (Russian), *Trudy Mat. Inst. Steklov.* **60,** 195–210 (1961). [20] *Number-theoretical Methods in Approximate Analysis* (Russian), Fizmatgiz, Moscow, 1963. [21] Distribution of fractional parts of exponential functions (Russian), *Vestnik Moskov. Univ.*, Ser. I Mat. Meh., **21,** no. 4, 42–46 (1966). [22] Some problems in the theory of diophantine approximation (Russian), *Uspehi Matem. Nauk* **22,** no. 3, 83–118 (1967); Engl. trans., *Russian Math. Surveys* **22,** no. 3, 80–118 (1967). [23] Trigonometric sums with exponential functions and the distribution of signs in repeating decimals (Russian), *Mat. Zametki* **8,** 641–652 (1970); Engl. trans., *Math. Notes* **8,** 831–837 (1970). [24] On the distribution of digits in periodic fractions (Russian), *Mat. Sb.* **89** (131), 654–670 (1972).

Korobov, N. M., and Postnikov, A. G.: [1] Some general theorems on the uniform distribution of fractional parts (Russian), *Dokl. Akad. Nauk SSSR* **84,** 217–220 (1952).

Kovalevskaja, E. I.: [1] On the joint distribution of fractional parts of polynomials (Russian), *Izv. Akad. Nauk BSSR*, Ser. Fiz.-Mat. Nauk, **1971,** no. 5, 13–23.

Krause, J. M.: [1] Fouriersche Reihen mit zwei veränderlichen Grössen, *Ber. Verh. Sächs. Akad. Wiss. Leipzig*, Math.-naturw. Kl., **55,** 164–197 (1903).

Kruse, A. H.: [1] Estimates of $\sum_{k=1}^{N} k^{-s}\langle kx\rangle^{-t}$, *Acta Arith.* **12,** 229–261 (1967).

Kubilius, J.: [1] *Probabilistic Methods in the Theory of Numbers*, Trans. of Math. Monographs, Vol. 11, American Math. Society, Providence, R.I., 1964.

Kuiper, N. H.: [1] Distribution mod 1 of some continuous functions, *Indag. Math.* **12,** 460–466 (1950).

Kuipers, L.: [1] De asymptotische verdeling modulo 1 van de waarden van meetbare functies, Thesis, Vrije Univ. Amsterdam, 1947. [2] On real periodic functions and functions with periodic derivatives, *Indag. Math.* **12,** 34–40 (1950). [3] Continuous and discrete distribution modulo 1, *Indag. Math.* **15,** 340–348 (1953). [4] Distribution modulo 1 (functions $f(t)$ with bounded $tf'(t)$), *Indag. Math.* **15,** 478–483 (1953). [5] Some remarks on asymptotic distribution functions, *Arch. der Math.* **8,** 104–108 (1957). [6] Continuous distribution modulo 1, *Nieuw Arch. voor Wisk.* (3) **10,** 78–82 (1962). [7] Continuous distribution mod 1 and independence of functions, *Nieuw Arch. voor Wisk.* (3) **11,** 1–3 (1963). [8] Uniform distribution modulo m, *Nieuw Arch. voor Wisk.* (3) **14,** 119–123 (1966). [9] Simple proof of a theorem of J. F. Koksma, *Nieuw Tijdschr. voor Wisk.* **55,** 108–111 (1967). [10] Remark on the Weyl-Schoenberg criterion in the theory of asymptotic distribution of real numbers, *Nieuw Arch. voor Wisk.* (3)

16, 197–202 (1968). [11] Remark on a paper by R. L. Duncan concerning the uniform distribution mod 1 of the sequence of logarithms of the Fibonacci numbers, *Fibonacic Quart.* **7,** no. 5, 465–466, 473 (1969). [12] A remark on a theorem of L. Carlitz, *Mat. Vesnik* **9** (24), 113–116 (1972).

Kuipers, L., and Meulenbeld, B.: [1] Asymptotic *C*-distribution. I, II, *Indag. Math.* **11,** 425–431, 432–437 (1949). [2] Some theorems in the theory of uniform distribution, *Indag. Math.* **12,** 53–56 (1950). [3] New results in the theory of *C*-uniform distribution, *Indag. Math.* **12,** 266–271 (1950). [4] Uniform distribution (mod 1) in sequences of intervals, *Indag. Math.* **12,** 382–392 (1950). [5] On real functions of *n* variables, *Indag. Math.* **14,** 490–497 (1952).

Kuipers, L., and Scheelbeek, P. A. J.: [1] Gleichverteilung in kompakten, topologischen Gruppen, *Indag. Math.* **26,** 261–265 (1964). [2] Uniform distribution of sequences from direct products of groups, *Ann. Sc. Norm. Sup. Pisa* **22,** 599–606 (1968).

Kuipers, L., and Shiue, J.-S.: [1] On the distribution modulo *m* of sequences of generalized Fibonacci numbers, *Tamkang J. Math.* **2,** 181–186 (1971). [2] A distribution property of the sequence of Lucas numbers, *Elemente der Math.* **27,** 10–11 (1972). [3] A distribution property of a linear recurrence of the second order, *Rend. Accad. Naz. Lincei* **52,** 6–10 (1972). [4] A distribution property of the sequence of Fibonacci numbers, *Fibonacci Quart.* **10,** no. 4, 375–376, 392 (1972). [5] Asymptotic distribution modulo *m* of sequences of integers and the notion of independence, *Atti Accad. Naz. Lincei Mem.* (8) **11,** 63–90 (1972).

Kuipers, L., and Stam, A. J.: [1] On a general form of the Weyl criterion in the theory of asymptotic distribution. I, II, *Proc. Japan Acad.* **45,** 530–535, 536–540 (1969).

Kuipers, L., and Uchiyama, S.: [1] Notes on the uniform distribution of sequences of integers, *Proc. Japan Acad.* **44,** 608–613 (1968).

Kuipers, L., and van der Steen, P.: [1] Metric theorems in the theory of asymptotic distribution modulo 1, *Indag. Math.* **28,** 300–310 (1966).

Kulikova, M. F.: [1] A construction problem connected with the distribution of fractional parts of the exponential function (Russian), *Dokl. Akad. Nauk SSSR* **143,** 522–524 (1962); Engl. trans., *Soviet Math. Dokl.* **3,** 422–425 (1962). [2] Construction of a number α whose fractional parts $\{\alpha g^x\}$ are rapidly and uniformly distributed (Russian), *Dokl. Akad. Nauk SSSR* **143,** 782–784 (1962); Engl. trans., *Soviet Math. Dokl.* **3,** 483–485 (1962).

Lacaze, B.: [1] Réunions finies de suites et génération de nombres dont la fonction de répartition est une interpolée linéaire, *C. R. Acad. Sci. Paris* Sér. A **269,** 1000–1002 (1969). [2] Estimation de moyennes et fonctions de répartition de suites d'échantillonnage, *Ann. Inst. Henri Poincaré*, Sect. B, **9,** 145–165 (1973).

Landau, E.: [1] *Diophantische Gleichungen mit endlich vielen Lösungen*, A. Walfisz (Ed.), Deutscher Verlag der Wissenschaften, Berlin, 1959.

Lang, S.: [1] *Introduction to Diophantine Approximations*, Addison-Wesley, Reading, Mass., 1966.

Lawton, B.: [1] A note on well distributed sequences, *Proc. Amer. Math. Soc.* **10,** 891–893 (1959).

Lebesgue, H.: [1] Sur certaines démonstrations d'existence, *Bull. Soc. Math. France* **45,** 132–144 (1917).

Leonov, V. P.: [1] On the central limit theorem for ergodic endomorphisms of compact commutative groups (Russian), *Dokl. Akad. Nauk SSSR* **135,** 258–261 (1960); Engl. trans., *Soviet Math. Dokl.* **1,** 1252–1255 (1960). [2] *Some Applications of Higher*

Semi-invariants to the Theory of Stationary Random Processes (Russian), Izdat. "Nauka," Moscow, 1964.

Lerch, M.: [1] Question 1547, *L'Intermédiaire Math.* **11**, 144–145 (1904).

Lesca, J.: [1] Équirépartition dans un anneau d'adèles, *Séminaire Delange-Pisot-Poitou*, 8ᵉ année (1966/67), Exp. 15. [2] Sur les approximations diophantiennes à une dimension, Thèse Sc. Math., Grenoble, 1968. [3] Démonstration d'une conjecture de P. Erdös, *Acta Arith.* **14**, 425–427 (1968). [4] Suites équiréparties dans un espace localement compact, *L'Enseignement Math.* (2) **17**, 311–328 (1971). [5] Sur la répartition modulo 1 de la suite $n\alpha$, *Acta Arith.* **20**, 345–352 (1972).

Lesca, J., and Mendès France, M.: [1] Ensembles normaux, *Acta Arith.* **17**, 273–282 (1970).

LeVeque, W. J.: [1] Note on a theorem of Koksma, *Proc. Amer. Math. Soc.* **1**, 380–383 (1950). [2] On n-dimensional uniform distribution mod 1, *Michigan Math. J.* **1**, 139–162 (1953). [3] The distribution modulo 1 of trigonometric sequences, *Duke Math. J.* **20**, 367–374 (1953). [4] On uniform distribution modulo a subdivision, *Pacific J. Math.* **3**, 757–771 (1953). [5] *Topics in Number Theory*, 2 vols., Addison-Wesley, Reading, Mass., 1956. [6] On the frequency of small fractional parts in certain real sequences, *Trans. Amer. Math. Soc.* **87**, 237–260 (1958). [7] On the frequency of small fractional parts in certain real sequences. III, *J. reine angew. Math.* **202**, 215–220 (1959). [8] On the frequency of small fractional parts in certain real sequences. II, *Trans. Amer. Math. Soc.* **94**, 130–149 (1960). [9] An inequality connected with Weyl's criterion for uniform distribution, *Proc. Symp. Pure Math.*, Vol. VIII, American Math. Society, Providence, R.I., 1965, pp. 22–30.

Levin, B. V., and Faĭnleĭb, A. S.: [1] Application of some integral equations to problems of number theory (Russian), *Uspehi Matem. Nauk* **22**, no. 3, 119–197 (1967); Engl. trans., *Russian Math. Surveys* **22**, no. 3, 119–204 (1967). [2] Integral limit theorems for certain classes of additive arithmetic functions (Russian), *Trudy Moskov. Mat. Obšč.* **18**, 19–54 (1968); Engl. trans., *Trans. Moscow Math. Soc. 1968*, American Math. Society, Providence, R. I., 1969, pp. 19–57. [3] Multiplicative functions and probabilistic number theory (Russian), *Izv. Akad. Nauk SSSR*, Ser. Mat., **34**, 1064–1109 (1970); Engl. trans., *Math. of the USSR-Izvestija* **4**, 1071–1118 (1970).

Loève, M.: [1] *Probability Theory*, 3rd ed., Van Nostrand, Princeton, N.J., 1963.

Long, C. T.: [1] Note on normal numbers, *Pacific J. Math.* **7**, 1163–1165 (1957). [2] On real numbers having normality of order k, *Pacific J. Math.* **18**, 155–160 (1966).

Loomis, L. H.: [1] *An Introduction to Abstract Harmonic Analysis*, Van Nostrand, New York-Toronto-London, 1953.

Lorentz, G. G.: [1] A contribution to the theory of divergent sequences, *Acta Math.* **80**, 167–190 (1948). [2] Borel and Banach properties of methods of summation, *Duke Math. J.* **22**, 129–141 (1955).

Loynes, R. M.: [1] Some results in the probabilistic theory of asymptotic uniform distribution modulo 1, *Z. Wahrscheinlichkeitsth.* **26**, 33–41 (1973).

Luthar, I. S.: [1] On an application of uniform distribution of sequences, *Colloq. Math.* **8**, 89–93 (1961).

Maak, W.: [1] *Fastperiodische Funktionen*, Grundlehren d. math. Wiss., Band 61, Springer-Verlag, Berlin-Göttingen-Heidelberg, 1950. [2] Der Kronecker-Weylsche Gleichverteilungssatz für beliebige Matrizengruppen, *Abh. Math. Sem. Hamburg* **17**, 91–94 (1951). [3] Verallgemeinerung des Kroneckerschen Approximationssatzes auf Matrizengruppen, *Nachr. Akad. Wiss. Göttingen*, Math. phys. Kl. II, **1965**, 175–180.

Mahler, K.: [1] On the fractional parts of the powers of a rational number. I, *Acta Arith.* **3**,

89–93 (1938). [2] On the fractional parts of the powers of a rational number. II, *Mathematika* **4,** 122–124 (1957). [3] *Lectures on Diophantine Approximations*, Part I, Univ. of Notre Dame, 1961. [4] An unresolved problem on the powers of 3/2, *J. Austral. Math. Soc.* **8,** 313–321 (1968). [5] *Introduction to p-adic Numbers and Their Functions*, Cambridge Tracts in Math., no. 64, Cambridge Univ. Press, London, 1973. [6] Arithmetical properties of the digits of the multiples of an irrational number, *Bull. Austral. Math. Soc.* **8,** 191–203 (1973). [7] On a class of diophantine inequalities, *Bull. Austral. Math. Soc.* **8,** 247–259 (1973).

Maisonneuve, D.: [1] Recherche et utilisation des 'bons treillis.' Programmation et résultats numériques, in S. K. Zaremba (Ed.), *Applications of Number Theory to Numerical Analysis*, Academic Press, New York, 1972, pp. 121–201.

Marstrand, J. M.: [1] On Khinchin's conjecture about strong uniform distribution, *Proc. London Math. Soc.* (3) **21,** 540–556 (1970).

Maxfield, J. E.: [1] A short proof of Pillai's theorem on normal numbers, *Pacific J. Math.* **2,** 23–24 (1952). [2] Normal k-tuples, *Pacific J. Math.* **3,** 189–196 (1953).

Meijer, H. G.: [1] Irrational power series, *Indag. Math.* **25,** 682–690 (1963). [2] Uniform distribution of g-adic integers, *Indag. Math.* **29,** 535–546 (1967). [3] The discrepancy of a g-adic sequence, *Indag. Math.* **30,** 54–66 (1968). [4] On uniform distribution of integers and uniform distribution mod. 1, *Nieuw Arch. voor Wisk.* (3) **18,** 271–278 (1970). [5] On a distribution problem in finite sets, *Indag. Math.* **35,** 9–17 (1973).

Meijer, H. G., and Dijksma, A.: [1] On uniform distribution of sequences in $GF[q, x]$ and $GF\{q, x\}$, *Duke Math. J.* **37,** 507–514 (1970).

Meijer, H. G., and Niederreiter, H.: [1] On a distribution problem in finite sets, *Compositio Math.* **25,** 153–160 (1972).

Meijer, H. G., and Sattler, R.: [1] On uniform distribution of integers and uniform distribution mod. 1, *Nieuw Arch. voor Wisk.* (3) **20,** 146–151 (1972).

Méla, J.-F.: [1] Suites lacunaires de Sidon, ensembles propres et points exceptionnels, *Ann. Inst. Fourier* **14,** fasc. 2, 533–538 (1964).

Mendès France, M.: [1] Nombres normaux et fonctions pseudo-aléatoires, *Ann. Inst. Fourier* **13,** fasc. 2, 91–104 (1963). [2] Dimension de Hausdorff. Application aux nombres normaux, *Séminaire Delange-Pisot*, 5e année (1963/64), Exp. 6. [3] A set of nonnormal numbers, *Pacific J. Math.* **15,** 1165–1170 (1965). [4] Nombres normaux. Applications aux fonctions pseudo-aléatoires, *J. Analyse Math.* **20,** 1–56 (1967). [5] Deux remarques concernant l'équirépartition des suites, *Acta Arith.* **14,** 163–167 (1968). [6] Nombres transcendants et ensembles normaux, *Acta Arith.* **15,** 189–192 (1969). [7] La réunion des ensembles normaux, *J. Number Th.* **2,** 345–351 (1970). [8] Suites et sous-suites équiréparties modulo 1, *Journées arithmétiques françaises* (Univ. Provence, 1971). [9] Les suites à spectre vide et la répartition (mod 1), *J. Number Th.* **5,** 1–15 (1973).

Meulenbeld, B.: [1] On the uniform distribution of the values of functions of n variables, *Indag. Math.* **12,** 59–65 (1950).

Meyer, Y.: [1] Nombres algébriques et répartition modulo 1, *C. R. Acad. Sci. Paris* Sér. A **268,** 25–27 (1969). [2] Nombres algébriques, nombres transcendants et équirépartition modulo 1, *Acta Arith.* **16,** 347–350 (1970). [3] *Nombres de Pisot, Nombres de Salem, et Analyse Harmonique*, Lecture Notes in Math., Vol. 117, Springer-Verlag, Berlin-Heidelberg-New York, 1970. [4] Nombres transcendants et répartition modulo 1, *Bull. Soc. Math. France, Suppl., Mém.* **25,** 143–149 (1971). [5] *Algebraic Numbers and Harmonic Analysis*, North-Holland, Amsterdam, 1972.

Miklavc, A.: [1] Elementary proofs of two theorems on the distribution of numbers $n\theta$ (mod 1), *Proc. Amer. Math. Soc.* **39,** 279–280 (1973).

Mikolás, M.: [1] On a problem of Hardy and Littlewood in the theory of diophantine approximations, *Publ. Math. Debrecen* **7,** 158–180 (1960).

Mineev, M. P.: [1] A Diophantine equation involving an exponential function and its application to the study of an ergodic sum (Russian), *Izv. Akad. Nauk SSSR,* Ser. Mat., **22,** 585–598 (1958). [2] A metric theorem on trigonometric sums with rapidly increasing functions (Russian), *Uspehi Matem. Nauk* **14,** no. 3, 169–171 (1959).

Monna, A. F.: [1] Sur une transformation simple des nombres p-adiques en nombres réels, *Indag. Math.* **14,** 1–9 (1952).

Mordell, L. J.: [1] Irrational power series, *Proc. Amer. Math. Soc.* **12,** 522–526 (1961). [2] The series $\sum a_n/(1 - xe^{2n\pi i\alpha})$, *J. London Math. Soc.* **38,** 111–116 (1963). [3] Irrational power series. II, *Acta Arith.* **11,** 181–188 (1965). [4] Irrational power series. III, *Proc. Amer. Math. Soc.* **16,** 819–821 (1965).

Moskvin, D. A.: [1] On the distribution of the fractional parts of a sequence which is more general than the exponential function (Russian), *Izv. Vysš. Učebn. Zaved. Mat.* **1970,** no. 12, 72–77.

Mück, R.: [1] Über ein Problem in der Theorie der C-Gleichverteilung, *J. reine angew. Math.* **222,** 201–206 (1966).

Mück, R., and Philipp, W.: [1] Distances of probability measures and uniform distribution mod 1, to appear.

Muhutdinov, R. H.: [1] A Diophantine equation with a matrix exponential function (Russian), *Dokl. Akad. Nauk SSSR* **142,** 36–37 (1962); Engl. trans., *Soviet Math. Dokl.* **3,** 28–29 (1962). [2] A central limit theorem for ergodic sums (Russian), *Dokl. Akad. Nauk UzSSR* **1967,** no. 2, 3–7. [3] The asymptotic distribution of sums with gaps (Russian), *Izv. Akad. Nauk UzSSR,* Ser. Fiz.-Mat. Nauk, **12,** no. 4, 15–21 (1968).

Müller, G.: [1] Sätze über Folgen auf kompakten Räumen, *Monatsh. Math.* **67,** 436–451 (1963).

Murdoch, B. H.: [1] A note on well-distributed sequences, *Canad. J. Math.* **17,** 808–810 (1965).

Muromskiĭ, A. A.: [1] Series of cosecants with monotone coefficients (Russian), *Vestnik Moskov. Univ.,* Ser. I Mat. Meh., **24,** no. 4, 52–60 (1969).

Nagasaka, K.: [1] On Hausdorff dimension of non-normal sets, *Ann. Inst. Statist. Math.* **23,** 515–522 (1971).

Narkiewicz, W.: [1] On distribution of values of multiplicative functions in residue classes, *Acta Arith.* **12,** 269–279 (1967).

Neville, E. H.: [1] The structure of Farey series, *Proc. London Math. Soc.* (2) **51,** 132–144 (1949).

Newman, M.: [1] Irrational power series, *Proc. Amer. Math. Soc.* **11,** 699–702 (1960).

Niederreiter, H.: [1] Diskrepanz in kompakten abelschen Gruppen II, *Manuscripta Math.* **1,** 293–306 (1969). [2] Almost-arithmetic progressions and uniform distribution, *Trans. Amer. Math. Soc.* **161,** 283–292 (1971). [3] Distribution of sequences and induced orders, *Nieuw Arch. voor Wisk.* (3) **19,** 210–219 (1971). [4] On the distribution of pseudo-random numbers generated by the linear congruential method, *Math. Comp.* **26,** 793–795 (1972). [5] Methods for estimating discrepancy, in S. K. Zaremba (Ed.), *Applications of Number Theory to Numerical Analysis,* Academic Press, New York, 1972, pp. 203–236. [6] A distribution problem in finite sets, in S. K. Zaremba (Ed.),

Applications of Number Theory to Numerical Analysis, Academic Press, New York, 1972, pp. 237–248. [7] Distribution of Fibonacci numbers mod 5^k, *Fibonacci Quart.* **10,** no. 4, 373–374 (1972). [8] On a class of sequences of lattice points, *J. Number Th.* **4,** 477–502 (1972). [9] On the existence of uniformly distributed sequences in compact spaces, *Compositio Math.* **25,** 93–99 (1972). [10] Discrepancy and convex programming, *Ann. Mat. Pura Appl.* (IV) **93,** 89–97 (1972). [11] On a number-theoretical integration method, *Aequationes Math.* **8,** 304–311 (1972). [12] Uniform distribution of lattice points, *Proc. Number Theory Conference* (Boulder, Colo., 1972), pp. 162–166. [13] Application of diophantine approximations to numerical integration, in C. F. Osgood (Ed.), *Diophantine Approximation and Its Applications*, Academic Press, New York, 1973, pp. 129–199. [14] Metric theorems on the distribution of sequences, *Proc. Symp. Pure Math.*, Vol. XXIV, American Math. Society, Providence, R.I., 1973, pp. 195–212. [15] Zur quantitativen Theorie der Gleichverteilung, *Monatsh. Math.* **77,** 55–62 (1973). [16] The distribution of Farey points, *Math. Ann.* **201,** 341–345 (1973). [17] Well distributed sequences with respect to systems of convex sets, to appear. [18] On the distribution of pseudo-random numbers generated by the linear congruential method. II, to appear. [19] Quantitative versions of a result of Hecke in the theory of uniform distribution mod 1, to appear.

Niederreiter, H., and Philipp, W.: [1] On a theorem of Erdös and Turán on uniform distribution, *Proc. Number Theory Conference* (Boulder, Colo., 1972), pp. 180–182. [2] Berry-Esseen bounds and a theorem of Erdös and Turán on uniform distribution mod 1, *Duke Math. J.*, **40,** 633–649 (1973).

Nienhuys, J. W.: [1] Not locally compact monothetic groups I, II, *Indag. Math.* **32,** 295–310, 311–326 (1970); Corrections, *ibid.* **33,** 59 (1971).

Niven, I.: [1] *Irrational Numbers*, Carus Math. Monographs, no. 11, Wiley, New York, 1956. [2] Uniform distribution of sequences of integers, *Trans. Amer. Math. Soc.* **98,** 52–61 (1961). [3] *Diophantine Approximations*, Interscience Tracts, Vol. 14, Wiley, New York, 1963. [4] Uniform distribution of sequences of integers, *Compositio Math.* **16,** 158–160 (1964).

Niven, I., and Zuckerman, H. S.: [1] On the definition of normal numbers, *Pacific J. Math.* **1,** 103–109 (1951).

Olivier, M.: [1] Sur le développement en base g des nombres premiers, *C. R. Acad. Sci. Paris* Sér. A **272,** 937–939 (1971).

O'Neil, P. E.: [1] A new criterion for uniform distribution, *Proc. Amer. Math. Soc.* **24,** 1–5 (1970).

Ostrowski, A.: [1] Bemerkungen zur Theorie der Diophantischen Approximationen I, II, III, *Abh. Math. Sem. Hamburg* **1,** 77–98, 250–251 (1922); **4,** 224 (1926). [2] Mathematische Miszellen. IX. Notiz zur Theorie der Diophantischen Approximationen, XVI. Zur Theorie der linearen Diophantischen Approximationen, *Jber. Deutsch. Math.-Verein.* **36,** 178–180 (1927); **39,** 34–46 (1930).

Paganoni, L.: [1] Sulla equivalenza di due famiglie di insiemi di fronte alla equidistribuzione, *Ist. Lombardo Accad. Sci. Lett. Rend.* A **102,** 412–420 (1968).

Parry, W.: [1] On the β-expansion of real numbers, *Acta Math. Acad. Sci. Hungar.* **11,** 401–416 (1960).

Parthasarathy, K. R.: [1] *Probability Measures on Metric Spaces*, Academic Press, New York, 1967.

Pathiaux, M.: [1] Répartition modulo 1 de la suite $(\lambda\alpha^n)$, *Séminaire Delange-Pisot-Poitou*, 11e année (1969/70), Exp. 13.

Peck, L. G.: [1] On uniform distribution of algebraic numbers, *Proc. Amer. Math. Soc.* **4**, 440–443 (1953).

Perron, O.: [1] *Die Lehre von den Kettenbrüchen*, 3. Aufl., B. G. Teubner, Stuttgart, 1954.

Petersen, G. M.: [1] Almost convergence and uniformly distributed sequences, *Quart. J. Math.* (2) **7**, 188–191 (1956). [2] *Regular Matrix Transformations*, McGraw-Hill, London, 1966. [3] On the structure of well distributed sequences. IV, *Indag. Math.* **29**, 128–131 (1967). [4] On the structure of well distributed sequences. V, *Indag. Math.* **29**, 229–233 (1967).

Petersen, G. M., and McGregor, M. T.: [1] On the structure of well distributed sequences, *Nieuw Arch. voor Wisk.* (3) **11**, 64–67 (1963). [2] On the structure of well distributed sequences. II, *Indag. Math.* **26**, 477–487 (1964). [3] On the structure of well distributed sequences. III, *Indag. Math.* **28**, 42–48 (1966).

Petersen, G. M., and Zame, A.: [1] Summability properties for the distribution of sequences, *Monatsh. Math.* **73**, 147–158 (1969).

Petersen, K.: [1] On a series of cosecants related to a problem in ergodic theory, *Compositio Math.*, **26**, 313–317 (1973).

Petersen, K., and Shapiro, L.: [1] Induced flows, *Trans. Amer. Math. Soc.* **177**, 375–390 (1973).

Peyerimhoff, A.: [1] *Lectures on Summability*, Lecture Notes in Math., Vol. 107, Springer-Verlag, Berlin-Heidelberg-New York, 1969.

Pham Phu Hien: [1] Fonctions admettant une répartition asymptotique des valeurs, *C. R. Acad. Sci. Paris* Sér. A **267**, 803–806 (1968).

Philipp, W.: [1] Ein metrischer Satz über die Gleichverteilung mod 1, *Arch. der Math.* **12**, 429–433 (1961). [2] Über einen Satz von Davenport-Erdös-LeVeque, *Monatsh. Math.* **68**, 52–58 (1964). [3] An *n*-dimensional analogue of a theorem of H. Weyl, *Compositio Math.* **16**, 161–163 (1964). [4] Some metrical theorems in number theory, *Pacific J. Math.* **20**, 109–127 (1967). [5] Ein zentraler Grenzwertsatz mit Anwendungen auf die Zahlentheorie, *Z. Wahrscheinlichkeitsth.* **8**, 185–203 (1967). [6] Das Gesetz vom iterierten Logarithmus mit Anwendungen auf die Zahlentheorie, *Math. Ann.* **180**, 75–94 (1969); Corrigendum, *ibid.* **190**, 338 (1971). [7] Some metrical theorems in number theory. II, *Duke Math. J.* **37**, 447–458 (1970); Errata, *ibid.*, p. 788. [8] *Mixing Sequences of Random Variables and Probabilistic Number Theory*, Mem. Amer. Math. Soc., no. 114, American Math. Society, Providence, R.I., 1971. [9] Empirical distribution functions and uniform distribution mod 1, in C. F. Osgood (Ed.), *Diophantine Approximation and Its Applications*, Academic Press, New York, 1973, pp. 211–234. [10] Limit theorems for lacunary series and uniform distribution mod 1, to appear.

Pillai, S. S.: [1] On normal numbers, *Proc. Indian Acad. Sci.*, Sect. A, **10**, 13–15 (1939). [2] On normal numbers, *Proc. Indian Acad. Sci.*, Sect. A, **12**, 179–184 (1940).

Pisot, C.: [1] Eine merkwürdige Klasse ganzer algebraischer Zahlen, *J. reine angew. Math.* **209**, 82–83 (1962).

Pisot, C., and Salem, R.: [1] Distribution modulo 1 of the powers of real numbers larger than 1, *Compositio Math.* **16**, 164–168 (1964).

Pjateckiĭ-Šapiro, I. I.: [1] On the laws of distribution of the fractional parts of an exponential function (Russian), *Izv. Akad. Nauk SSSR*, Ser. Mat., **15**, 47–52 (1951). [2] On a generalization of the notion of uniform distribution of fractional parts (Russian), *Mat. Sb.* **30** (72), 669–676 (1952). [3] Fractional parts and some questions of the theory of trigonometric series (Russian), *Uspehi Matem. Nauk* **8**, no. 3, 167–170

(1953). [4] On the distribution of the fractional parts of the exponential function (Russian), *Moskov. Gos. Ped. Inst. Uč. Zap.* **108,** 317–322 (1957).

Polosuev, A. M.: [1] A multidimensional case of unimprovable estimates of trigonometric sums with exponential functions (Russian), *Dokl. Akad. Nauk SSSR* **104,** 186–189 (1955). [2] On a problem concerned with a uniform distribution of a system of functions (Russian), *Dokl. Akad. Nauk SSSR* **122,** 346–348 (1958). [3] Uniform distribution of a system of functions representing a solution of linear first order difference equations (Russian), *Dokl. Akad. Nauk SSSR* **123,** 405–406 (1958). [4] On a problem of uniform distribution of a system of functions (Russian), *Vestnik Moskov. Univ.*, Ser. I Mat. Meh., **1960,** no. 2, 21–32. [5] On the uniformity of the distribution of a system of functions forming the solution of a system of linear finite-difference equations of first order. I. Criterion for uniformity of distribution, II. Construction of a uniformly distributed system (Russian), *Vestnik Moskov. Univ.*, Ser. I Mat. Meh., **1960,** no. 5, 29–39, no. 6, 20–25. [6] An unimprovable estimate of a multidimensional trigonometric sum with exponential functions (Russian), *Vestnik Moskov. Univ.*, Ser. I Mat. Meh., **25,** no. 1, 9–16 (1970).

Pólya, G., and Szegö, G.: [1] *Aufgaben und Lehrsätze aus der Analysis I, II*, 3. Aufl., Grundlehren d. math. Wiss., Band 19, Springer-Verlag, Berlin-Göttingen-Heidelberg-New York, 1964.

Pontryagin, L. S.: [1] *Topological Groups*, 2nd ed., Gordon and Breach, New York, 1966.

Popken, J.: [1] Irrational power series, *Indag. Math.* **25,** 691–694 (1963).

Post, K. A.: [1] Integration in locally compact spaces by means of uniformly distributed sequences, Thesis, Technische Hogeschool Eindhoven, 1967. [2] A note on disjoint unions of basis sets in a topological measure space, *Nieuw Arch. voor Wisk.* (3) **18,** 151–153 (1970).

Postnikov, A. G.: [1] On the distribution of the fractional parts of the exponential function (Russian), *Dokl. Akad. Nauk SSSR* **86,** 473–476 (1952). [2] Estimation of an exponential trigonometric sum (Russian), *Izv. Akad. Nauk SSSR*, Ser. Mat., **20,** 661–666 (1956). [3] A criterion for testing the uniform distribution of an exponential function in the complex domain (Russian), *Vestnik Leningrad. Univ.* **12,** no. 13, 81–88 (1957). [4] A test for a completely uniformly distributed sequence (Russian), *Dokl. Akad. Nauk SSSR* **120,** 973–975 (1958). [5] Arithmetic modeling of random processes (Russian), *Trudy Mat. Inst. Steklov.* **57** (1960). [6] The frequency with which the fractional part of an exponential function lies in a given interval (Russian), *Uspehi Matem. Nauk* **16,** no. 3, 201–205 (1961). [7] Dynamical systems in number theory (Russian), *Proc. Fourth All-Union Math. Congr.* (Leningrad, 1961), Vol. II, Izdat. "Nauka," Leningrad, 1964, pp. 124–131. [8] Ergodic problems in the theory of congruences and of diophantine approximations (Russian), *Trudy Mat. Inst. Steklov.* **82** (1966); Engl. trans., *Proc. Steklov Inst. Math.* **82,** American Math. Society, Providence, R.I., 1967.

Postnikov, A. G., and Pjateckiĭ-Šapiro, I. I.: [1] On Bernoulli-normal sequences of symbols (Russian), *Izv. Akad. Nauk SSSR*, Ser. Mat., **21,** 501–514 (1957). [2] A Markov sequence of symbols and a normal continued fraction (Russian), *Izv. Akad. Nauk SSSR*, Ser. Mat., **21,** 729–746 (1957).

Postnikova, L. P.: [1] On a theorem of N. M. Korobov (Russian), *Dokl. Akad. Nauk SSSR* **134,** 42–43 (1960); Engl. trans., *Soviet Math. Dokl.* **1,** 1019–1020 (1960). [2] On the concept of the collective of Mises (Russian), *Proc. Sixth All-Union Conf. Theory Prob. and Math. Statist.* (Vilnius, 1960), pp. 75–76. [3] Sequences behaving like random ones (Russian), *Proc. Fourth All-Union Math. Congr.* (Leningrad, 1961), Vol. II, Izdat. "Nauka," Leningrad, 1964, pp. 131–134. [4] Fluctuations in the distribution of fractional parts (Russian), *Dokl. Akad. Nauk SSSR* **161**, 1282–1284 (1965); Engl. trans., *Soviet Math. Dokl.* **6**, 597–600 (1965). [5] Quantitative form of Borel's problem (Russian), *Acta Arith.* **21**, 235–250 (1972).

Rademacher, H.: [1] *Lectures on Elementary Number Theory*, Blaisdell, New York-Toronto-London, 1964.

Raikov, D. A.: [1] On some arithmetical properties of summable functions (Russian), *Mat. Sb.* **1** (43), 377–384 (1936).

Rajagopalan, M.: [1] Structure of monogenic groups, *Illinois J. Math.* **12**, 205–214 (1968).

Rajagopalan, M., and Rotman, J. J.: [1] Monogenic groups, *Compositio Math.* **18**, 155–161 (1967).

Ranga Rao, R.: [1] Relations between weak and uniform convergence of measures with applications, *Ann. Math. Statist.* **33**, 659–680 (1962).

Rauzy, G.: [1] Caractérisation des ensembles normaux, *Bull. Soc. Math. France* **98**, 401–414 (1970). [2] Normalité de Q^*, *Acta Arith.* **19**, 43–47 (1971). [3] Fonctions entières et répartition modulo 1, *Bull. Soc. Math. France* **100**, 409–415 (1972).

Rees, D.: [1] Note on a paper by I. J. Good, *J. London Math. Soc.* **21**, 169–172 (1946).

Rényi, A.: [1] Representations for real numbers and their ergodic properties, *Acta Math. Acad. Sci. Hungar.* **8**, 477–493 (1957). [2] *Wahrscheinlichkeitsrechnung*, Hochschulbücher f. Math., Band 54, VEB Deutscher Verlag der Wissenschaften, Berlin, 1962.

Révész, P.: [1] *The Laws of Large Numbers*, Academic Press, New York-London, 1968.

Rhin, G.: [1] Quelques résultats métriques dans un corps de séries formelles sur un corps fini, *Séminaire Delange-Pisot-Poitou*, 9e année (1967/68), Exp. 21. [2] Généralisation d'un théorème de I. M. Vinogradov à un corps de séries formelles sur un corps fini, *C. R. Acad. Sci. Paris* Sér. A **272**, 567–569 (1971). [3] Répartition modulo 1 dans un corps de séries formelles sur un corps fini, *Dissertationes Math.* **95** (1972). [4] Sur la répartition modulo 1 des suites $f(p)$, *Acta Arith.* **23**, 217–248 (1973).

Richtmyer, R. D.: [1] *On the evaluation of definite integrals and a quasi-Monte Carlo method based on properties of algebraic numbers*, Report LA-1342, Los Alamos Sci. Lab., Los Alamos, N.M., 1951. [2] *A non-random sampling method based on congruences for Monte-Carlo problems*, AEC Research and Development Rep. NYO-8674, AEC Computing and Applied Mathematics Center, New York Univ., New York, 1958.

Richtmyer, R. D., Devaney, M., and Metropolis, N.: [1] Continued fraction expansions of algebraic numbers, *Numer. Math.* **4**, 68–84 (1962).

Riesz, F.: [1] Sur la théorie ergodique, *Comment. Math. Helv.* **17**, 221–239 (1944/45).

Rimkevičiūtė, L.: [1] Concerning uniform distribution of points on an interval (Lithuanian; Russian summary), *Vilniaus Valst. Univ. Mokslo Darbai Mat. Fiz.* **8**, 49–56 (1958).

Rizzi, B.: [1] Sulle successioni uniformemente dense, *Giorn. Ist. Ital. Attuari* **31**, 57–62 (1970).

Rjaben'kiĭ, V. S.: [1] Tables and interpolation of a certain class of functions (Russian), *Dokl. Akad. Nauk SSSR* **131**, 1025–1027 (1960); Engl. trans., *Soviet Math. Dokl.* **1**, 382—384 (1960). [2] A way of obtaining difference schemes and the use of number-theoretic nets for the solution of the Cauchy problem by the method of finite differences (Russian), *Trudy Mat. Inst. Steklov.* **60**, 232–237 (1961).

Robbins, H.: [1] Equidistribution of sums of random variables, *Proc. Amer. Math. Soc.* **4**, 786–799 (1953).

Rodosskiĭ, K. A.: [1] On some trigonometric sums (Russian), *Mat. Sb.* **60** (102), 219–234 (1963).

Rohlin, V. A.: [1] On endomorphisms of compact commutative groups (Russian), *Izv. Akad. Nauk SSSR*, Ser. Mat., **13**, 329–340 (1949).

Rolewicz, S.: [1] Some remarks on monothetic groups, *Colloq. Math.* **13**, 27–28 (1964).

Roos, P.: [1] Iterierte Resttransformationen von Zahldarstellungen, *Z. Wahrscheinlichkeitsth.* **4**, 45–63 (1965).

Roos, P., and Arnold, L.: [1] Numerische Experimente zur mehrdimensionalen Quadratur, *Sitzgsber. Österr. Akad. Wiss.*, Math.-naturw. Kl., Abt. II, **172**, 271–286 (1963).

Roth, K. F.: [1] On irregularities of distribution, *Mathematika* **1**, 73–79 (1954). [2] Rational approximations to algebraic numbers, *Mathematika* **2**, 1–20 (1955).

Rubel, L. A.: [1] Uniform distribution in locally compact groups, *Comment. Math. Helv.* **39**, 253–258 (1965). [2] Uniform distribution and densities on locally compact abelian groups, *Symposia on Theoretical Physics and Mathematics*, Vol. 9 (Madras, 1968), Plenum Press, New York, 1969, pp. 183–189.

Rudin, W.: [1] *Fourier Analysis on Groups*, Interscience, New York-London, 1962. [2] An arithmetic property of Riemann sums, *Proc. Amer. Math. Soc.* **15**, 321–324 (1964).

Ryll-Nardzewski, C.: [1] Sur les suites et les fonctions également réparties, *Studia Math.* **12**, 143–144 (1951).

Šahov, Yu. N.: [1] On calculating the eigenvalues of a multidimensional symmetric kernel using number-theoretic nets (Russian), *Ž. Vyčisl. Mat. i Mat. Fiz.* **3**, 988–997 (1963); Engl. trans., *U.S.S.R. Comp. Math. and Math. Phys.* **3**, 1350–1362 (1963). [2] The calculation of integrals of increasing multiplicity (Russian), *Ž. Vyčisl. Mat. i Mat. Fiz.* **5**, 911–916 (1965); Engl. trans., *U.S.S.R. Comp. Math. and Math. Phys.* **5**, no. 5, 184–192 (1965). [3] Some bounds in the construction of Bernoulli-normal sequences of signs (Russian), *Mat. Zametki* **10**, 501–510 (1971); Engl. trans., *Math. Notes* **10**, 724–730 (1971). [4] On the construction of Markov-normal sequences of signs (Russian), *Moskov. Gos. Ped. Inst. Uč. Zap.* **375**, 143–155 (1971).

Šalát, T.: [1] A remark on normal numbers, *Rev. Roumaine Math. Pures Appl.* **11**, 53–56 (1966). [2] Zu einigen Fragen der Gleichverteilung (mod 1), *Czechoslovak Math. J.* **18**, 476–488 (1968).

Salem, R.: [1] Power series with integral coefficients, *Duke Math. J.* **12**, 153–172 (1945). [2] Uniform distribution and capacity of sets, *Medd. Lunds Univ. Mat. Sem. Suppl.* **1952**, 193–195. [3] *Algebraic Numbers and Fourier Analysis*, Heath Math. Monographs, Boston, Mass., 1963.

Saltykov, A. I.: [1] Tables for computing multiple integrals by the method of optimal coefficients (Russian), *Ž. Vyčisl. Mat. i Mat. Fiz.* **3**, 181–186 (1963); Engl. trans., *U.S.S.R. Comp. Math. and Math. Phys.* **3**, 235–242 (1963).

Sanders, J. M.: [1] Verdelingsproblemen bij gegeneraliseerde duale breuken, Thesis, Vrije Univ. Amsterdam, 1950.

Šarygin, I. F.: [1] The use of number-theoretic methods of integration in the case of non-periodic functions (Russian), *Dokl. Akad. Nauk SSSR* **132**, 71–74 (1960); Engl. trans., *Soviet Math. Dokl.* **1**, 506–509 (1960). [2] A lower estimate for the error of quadrature formulas for certain classes of functions (Russian), *Ž. Vyčisl. Mat. i Mat. Fiz.* **3**, 370–376 (1963); Engl. trans., *U.S.S.R. Comp. Math. and Math. Phys.* **3**, 489–497 (1963).

Schmidt, K.: [1] Über einen Zusammenhang zwischen gleichverteilten Punkt- und Massfolgen, *J. reine angew. Math.* **244**, 94–96 (1970). [2] Eine Diskrepanz für Massfolgen auf lokalkompakten Gruppen, *Z. Wahrscheinlichkeitsth.* **17**, 48–52 (1971). [3] Über die *C*-Gleichverteilung von Massen, *Z. Wahrscheinlichkeitsth.* **17**, 327–332 (1971).

Schmidt, K., and Zinterhof, P.: [1] Über Quadraturformeln auf T^ω, *Computing* **6**, 94–96 (1970).

Schmidt, W. M.: [1] A metrical theorem in diophantine approximation, *Canad. J. Math.* **12**, 619–631 (1960). [2] On normal numbers, *Pacific J. Math.* **10**, 661–672 (1960). [3] Über die Normalität von Zahlen zu verschiedenen Basen, *Acta Arith.* **7**, 299–309 (1962). [4] Normalität bezüglich Matrizen, *J. reine angew. Math.* **214/215**, 227–260 (1964). [5] Metrical theorems on fractional parts of sequences, *Trans. Amer. Math. Soc.*, **110**, 493–518 (1964).
[6] Irregularities of distribution, *Quart. J. Math.* (2) **19**, 181–191 (1968). [7] Irregularities of distribution. II, *Trans. Amer. Math. Soc.* **136**, 347–360 (1969). [8] Irregularities of distribution. III, *Pacific J. Math.* **29**, 225–234 (1969). [9] Irregularities of distribution. IV, *Invent. Math.* **7**, 55–82 (1969). [10] Disproof of some conjectures on Diophantine approximations, *Studia Sci. Math. Hungar.* **4**, 137–144 (1969). [11] Irregularities of distribution. V, *Proc. Amer. Math. Soc.* **25**, 608–614 (1970). [12] Simultaneous approximation to algebraic numbers by rationals, *Acta Math.* **125**, 189–201 (1970). [13] Irregularities of distribution. VI, *Compositio Math.* **24**, 63–74 (1972). [14] Irregularities of distribution. VII, *Acta Arith.* **21**, 45–50 (1972). [15] Irregularities of distribution. VIII, to appear.

Schnabl, R.: [1] Zur Theorie der homogenen Gleichverteilung modulo 1, *Sitzgsber. Österr. Akad. Wiss.*, Math.-naturw. Kl., Abt. II, **172**, 43–77 (1963).

Schoenberg, I.: [1] Über die asymptotische Verteilung reeller Zahlen mod 1, *Math. Z.* **28**, 171–199 (1928). [2] Solution of Problem 5090, *Amer. Math. Monthly* **71**, 332–334 (1964).

Schreiber, J.-P.: [1] Sur les nombres de Chabauty-Pisot-Salem des extensions algébriques de Q_p, *C. R. Acad. Sci. Paris* Sér. A **269**, 71–73 (1969). [2] Une caractérisation des nombres de Pisot-Salem des corps p-adiques Q_p, *Bull. Soc. Math. France, Suppl., Mém.* **19**, 55–63 (1969).

Schreier, J.: [1] Eine Bemerkung über Erzeugende in kompakten Gruppen, *Fund. Math.* **25**, 198–199 (1935).

Schreier, J., and Ulam, S.: [1] Sur le nombre des générateurs d'un groupe topologique compact et convexe, *Fund. Math.* **24**, 302–304 (1935).

Schwarz, W.: [1] Irrationale Potenzreihen, *Arch. der Math.* **13**, 228–240 (1962). [2] Irrationale Potenzreihen II, *Arch. der Math.* **17**, 435–437 (1966).

Schwarz, W., and Wallisser, R.: [1] Über gewisse nichtfortsetzbare Potenzreihen, *Monatsh. Math.* **77**, 63–66 (1973); II, *ibid.*, 251–266.

Schweiger, F.: [1] Metrische Theorie einer Klasse zahlentheoretischer Transformationen, *Acta Arith.* **15**, 1–18 (1968); Corrigendum, *ibid.* **16**, 217–219 (1969). [2] Normalität bezüglich zahlentheoretischer Transformationen, *J. Number Th.* **1**, 390–397 (1969). [3] Metrische Theorie kettenbruchähnlicher Ziffernentwicklungen, Zahlentheorie-Tagung math. Forsch.-inst. Oberwolfach 1970, *Ber. math. Forsch.-inst. Oberwolfach* **5**, 159–172 (1971).

Scoville, R.: [1] Some measure algebras on the integers, *Pacific J. Math.* **34**, 769–779 (1970).

Senge, H. G.: [1] Closed sets of algebraic numbers, *Duke Math. J.* **34**, 307–323 (1967).

Senge, H. G., and Straus, E.: [1] PV-numbers and sets of multiplicity, *Proc. Washington State Univ. Conf. on Number Theory* (Pullman, Wash., 1971), pp. 55–67; also in *Periodica Math. Hungar.* **3**, 93–100 (1973).

Serre, J.-P.: [1] *Abelian l-adic Representations and Elliptic Curves*, Benjamin, New York, 1968.

Shapiro, H. N., and Sparer, G. H.: [1] Composite values of exponential and related sequences, *Comm. Pure Appl. Math.* **25,** 569–616 (1972).

Shapiro, L.: [1] Irregularities of distribution in dynamical systems, *Recent Advances in Topological Dynamics*, Lecture Notes in Math., Vol. 318, Springer-Verlag, Berlin-Heidelberg-New York, 1973, pp. 249–252.

Shiue, J.-S.: [1] On a theorem of uniform distribution of g-adic integers and a notion of independence, *Rend. Accad. Naz. Lincei* **50,** 90–93 (1971).

Shreider, Yu. A.: [1] *The Monte Carlo Method*, Pergamon Press, New York, 1966.

Sierpiński, W.: [1] Sur la valeur asymptotique d'une certaine somme, *Bull. Int. Acad. Polon. Sci.* (Cracovie) A, **1910,** 9–11. [2] On the asymptotic value of a certain sum (Polish), *Rozprawy Wydz. Mat. Przyr. Akad. Um.* **50,** 1–10 (1910). [3] Démonstration élémentaire du théorème de M. Borel sur les nombres absolument normaux et détermination effective d'un tel nombre, *Bull. Soc. Math. France* **45,** 125–132 (1917).

Sigmund, K.: [1] Über Verteilungsmasse von Massfolgen auf kompakten Gruppen, *Compositio Math.* **21,** 299–311 (1969).

Slater, N. B.: [1] The distribution of the integers N for which $\{\theta N\} < \phi$, *Proc. Cambridge Phil. Soc.* **46,** 525–534 (1950). [2] Distribution problems and physical applications, *Compositio Math.* **16,** 176–183 (1964). [3] Gaps and steps for the sequence $n\theta$ mod 1, *Proc. Cambridge Phil. Soc.* **63,** 1115–1123 (1967).

Śliwa, J.: [1] On distribution of values of $\sigma(n)$ in residue classes, *Colloq. Math.* **27,** 283–291 (1973).

Smirnov, N. V.: [1] Sur les écarts de la courbe de distribution empirique (Russian; French summary), *Mat. Sb.* **6** (48), 3–26 (1939).

Smoljak, S. A.: [1] Interpolation and quadrature formulas for the classes W_s^α and E_s^α (Russian), *Dokl. Akad. Nauk SSSR* **131,** 1028–1031 (1960); Engl. trans., *Soviet Math. Dokl.* **1,** 384–387 (1960).

Sobol', I. M.: [1] An accurate error estimate for multidimensional quadrature formulae for the functions of the class S_p (Russian), *Dokl. Akad. Nauk SSSR* **132,** 1041–1044 (1960); Engl. trans., *Soviet. Math. Dokl.* **1,** 726–729 (1960). [2] On the evaluation of multidimensional integrals (Russian), *Dokl. Akad. Nauk SSSR* **139,** 821–823 (1961); Engl. trans., *Soviet Math. Dokl.* **2,** 1022–1025 (1961). [3] An exact estimate of the error in multidimensional quadrature formulae for functions of the classes $\tilde{W}_1$ and $\tilde{H}_1$ (Russian), *Ž. Vyčisl. Mat. i Mat. Fiz.* **1,** 208–216 (1961); Engl. trans., *U.S.S.R. Comp. Math. and Math. Phys.* **1,** 228–240 (1961). [4] On the evaluation of infinite-dimensional integrals (Russian), *Ž. Vyčisl. Mat. i Mat. Fiz.* **1,** 917–922 (1961); Engl. trans., *U.S.S.R. Comp. Math. and Math. Phys.* **1,** 1086–1091 (1961). [5] An integral encountered in quadrature formulae theory (Russian), *Ž. Vyčisl. Mat. i Mat. Fiz.* **6,** 1084–1089 (1966); Engl. trans., *U.S.S.R. Comp. Math. and Math. Phys.* **6,** no. 6, 189–196 (1966). [6] The distribution of points in a cube and the approximate evaluation of integrals (Russian), *Ž. Vyčisl. Mat. i Mat. Fiz.* **7,** 784–802 (1967); Engl. trans., *U.S.S.R. Comp. Math. and Math. Phys.* **7,** no. 4, 86–112 (1967). [7] *Multidimensional Quadrature Formulas and Haar Functions* (Russian), Izdat. "Nauka," Moscow, 1969. [8] Calculation of improper integrals using equidistributed sequences (Russian), *Dokl. Akad. Nauk SSSR* **210,** 278–281 (1973); Engl. trans., *Soviet Math. Dokl.* **14,** 734–738 (1973).

Solodov, V. M.: [1] On the calculation of multiple integrals (Russian), *Dokl. Akad. Nauk SSSR* **127,** 753–756 (1959). [2] On the error involved in a numerical integration (Russian), *Dokl. Akad. Nauk SSSR* **148,** 284–287 (1963); Engl. trans., *Soviet Math. Dokl.* **4,** 85–88 (1963). [3] Integration over regions different from the unit cube

(Russian), *Ž. Vyčisl. Mat. i Mat. Fiz.* **8,** 1334–1341 (1968); Engl. trans., *U.S.S.R. Comp. Math. and Math. Phys.* **8,** no. 6, 198–207 (1968). [4] An application of the method of optimal coefficients to numerical integration (Russian), *Ž. Vyčisl. Mat. i Mat. Fiz.* **9,** 14–29 (1969); Engl. trans., *U.S.S.R. Comp. Math. and Math. Phys.* **9,** no. 1, 14–34 (1969).

Sós, V. T.: [1] On the theory of diophantine approximations. I, *Acta Math. Acad. Sci. Hungar.* **8,** 461–472 (1957). [2] On the distribution mod 1 of the sequence $n\alpha$, *Ann. Univ. Sci. Budapest. Eötvös Sect. Math.* **1,** 127–134 (1958).

Spears, J. L., and Maxfield, J. E.: [1] Further examples of normal numbers, *Publ. Math. Debrecen* **16,** 119–127 (1969).

Stackelberg, O. P.: [1] A uniform law of the iterated logarithm for functions *C*-uniformly distributed mod 1, *Indiana Univ. Math. J.* **21,** 515–528 (1971).

Stapleton, J. H.: [1] On the theory of asymptotic distributions (mod 1) and its extension to abstract spaces, Ph.D. thesis, Purdue Univ., 1957.

Starčenko, L. P.: [1] Construction of sequences jointly normal with a given one (Russian), *Izv. Akad. Nauk SSSR*, Ser. Mat., **22,** 757–770 (1958); Erratum, *ibid.* **23,** 635–636 (1959). [2] The construction of a completely uniformly distributed sequence (Russian), *Dokl. Akad. Nauk SSSR* **129,** 519–521 (1959).

Stark, H. M.: [1] *An Introduction to Number Theory*, Markham, Chicago, 1970.

Steinhaus, H.: [1] Sur les fonctions indépendantes (VI), *Studia Math.* **9,** 121–131 (1940). [2] *One Hundred Problems in Elementary Mathematics*, Basic Books, New York, 1964.

Stoneham, R. G.: [1] The reciprocals of integral powers of primes and normal numbers, *Proc. Amer. Math. Soc.* **15,** 200–208 (1964). [2] A study of 60,000 digits of the transcendental '*e*', *Amer. Math. Monthly* **72,** 483–500 (1965). [3] On (j, ε)-normality in the rational fractions, *Acta Arith.* **16,** 221–238 (1969). [4] A general arithmetic construction of transcendental non-Liouville normal numbers from rational fractions, *Acta Arith.* **16,** 239–254 (1969). [5] On absolute (j, ε)-normality in the rational fractions with applications to normal numbers, *Acta Arith.* **22,** 277–286 (1973). [6] On the uniform ε-distribution of residues within the periods of rational fractions with applications to normal numbers, *Acta Arith.* **22,** 371–389 (1973).

Stoyantsev, V. T.: [1] Indeterminate methods of integration with a finite number of feasible methods (Russian), *Ž. Vyčisl. Mat. i Mat. Fiz.* **9,** 1235–1246 (1969); Engl. trans., *U.S.S.R. Comp. Math. and Math. Phys.* **9,** no. 6, 1–16 (1969).

Stromberg, K.: [1] A note on the convolution of regular measures, *Math. Scand.* **7,** 347–352 (1959). [2] Probabilities on a compact group, *Trans. Amer. Math. Soc.* **94,** 295–309 (1960).

Sudan, G.: [1] Sur le problème du rayon réfléchi, *Rev. Roumaine Math. Pures Appl.* **10,** 723–733 (1965).

Supnick, F., Cohen, H. J., and Keston, J. F.: [1] On the powers of a real number reduced modulo one, *Trans. Amer. Math. Soc.* **94,** 244–257 (1960).

Surányi, J.: [1] Über die Anordnung der Vielfachen einer reellen Zahl mod 1, *Ann. Univ. Sci. Budapest. Eötvös Sect. Math.* **1,** 107–111 (1958).

Swierczkowski, S.: [1] On successive settings of an arc on the circumference of a circle, *Fund. Math.* **46,** 187–189 (1958).

Szüsz, P.: [1] On a problem in the theory of uniform distribution (Hungarian), *Compt. Rend. Premier Congrès Hongrois*, Budapest, 1952, pp. 461–472. [2] Über die Verteilung der Vielfachen einer komplexen Zahl nach dem Modul des Einheitsquadrats, *Acta Math. Acad. Sci. Hungar.* **5,** 35–39 (1954). [3] Lösung eines Problems von Herrn Hartman, *Studia Math.* **15,** 43–55 (1955). [4] Über die metrische Theorie der diophantischen Approximationen. II, *Acta Arith.* **8,** 225–241 (1962/63).

Teghem, J.: [1] Sur un type d'inégalités diophantiennes, *Indag. Math.* **1,** 31–41 (1939). [2] Sur des applications de certaines estimations de sommes trigonométriques, *Colloque sur la Théorie des Nombres*, Centre belge de recherches mathématiques, Brussels, 1955, pp. 183–204.

Thorp, E., and Whitley, R.: [1] Poincaré's conjecture and the distribution of digits in tables, *Compositio Math.* **23,** 233–250 (1971).

Tijdeman, R.: [1] Note on Mahler's $\frac{3}{2}$-problem, *Norsk. Vid. Selsk. Skrifter* **16,** 1–4 (1972). [2] On a distribution problem in finite and countable sets, *J. Combinatorial Theory* Ser. A **15,** 129–137 (1973).

Titchmarsh, E. C.: [1] *The Theory of the Riemann Zeta-Function*, Clarendon Press, Oxford, 1951.

Topsøe, F.: [1] On the connection between P-continuity and P-uniformity in weak convergence, *Teor. Verojatnost. i Primenen.* **12,** 279–288 (1967); also in *Theor. Probability Appl.* **12,** 241–250 (1967). [2] *Topology and Measure*, Lecture Notes in Math., Vol. 133, Springer-Verlag, Berlin-Heidelberg-New York, 1970.

Tsuji, M.: [1] On the uniform distribution of values of a function mod. 1, *Proc. Imp. Acad. Tokyo* **19,** 66–69 (1943). [2] On the uniform distribution of numbers mod. 1, *J. Math. Soc. Japan* **4,** 313–322 (1952).

Tyan, M. M.: [1] The distribution of values of Euler's function $\varphi(n)$ (Russian), *Litovsk. Mat. Sb.* **6,** 105–119 (1966).

Uchiyama, M., and Uchiyama, S.: [1] A characterization of uniformly distributed sequences of integers, *J. Fac. Sci. Hokkaidô Univ.*, Ser. I, **16,** 238–248 (1962).

Uchiyama, S.: [1] On the uniform distribution of sequences of integers, *Proc. Japan Acad.* **37,** 605–609 (1961). [2] A note on a theorem of J. N. Franklin, *Math. Comp.* **20,** 139–140 (1966). [3] A note on the uniform distribution of sequences of integers, *J. Fac. Sci. Shinshu Univ.* **3,** 163–169 (1968).

Ungar, P.: [1] The theorem of Coles on uniform distribution, *Comm. Pure Appl. Math.* **20,** 609–618 (1967).

Usol'cev, L. P.: [1] An additive problem with an increasing number of terms with the exponential function (Russian), *Izv. Vysš. Učebn. Zaved. Mat.* **1967,** no. 3, 96–104. [2] Limit theorems for the uniform distribution of certain sequences (Russian), *Izv. Vysš. Učebn. Zaved. Mat.* **1967,** no. 4, 112–121. [3] A construction problem connected with the uniform distribution of the fractional parts of exponential functions (Russian), *Izv. Vysš. Učebn. Zaved. Mat.* **1967,** no. 12, 75–83.

Van Aardenne-Ehrenfest, T.: [1] Proof of the impossibility of a just distribution of an infinite sequence of points over an interval, *Indag. Math.* **7,** 71–76 (1945). [2] On the impossibility of a just distribution, *Indag. Math.* **11,** 264–269 (1949).

Van Dantzig, D.: [1] Über topologisch homogene Kontinua, *Fund. Math.* **15,** 102–125 (1930). [2] Nombres universels ou ν!-adiques avec une introduction sur l'algèbre topologique, *Ann. Sci. École Norm. Sup.* (3) **53,** 275–307 (1936). [3] Zur topologischen Algebra III, *Compositio Math.* **3,** 408–426 (1936).

Van de Lune, J.: [1] On the distribution of a specific number-theoretical sequence, *Math. Centrum Amsterdam*, Report ZW 1969-004 (1969).

Vanden Eynden, C. L.: [1] The uniform distribution of sequences, Ph.D. thesis, Univ. of Oregon, 1962.

Van der Corput, J. G.: [1] Zahlentheoretische Abschätzungen, *Math. Ann.* **84**, 53–79 (1921). [2] Zahlentheoretische Abschätzungen mit Anwendung auf Gitterpunktprobleme. I, *Math. Z.* **17**, 250–259 (1923); II, *ibid.* **28**, 301–310 (1928). [3] Neue zahlentheoretische Abschätzungen. I, *Math. Ann.* **89**, 215–254 (1923). [4] Neue zahlentheoretische Abschätzungen. II, *Math. Z.* **29**, 397–426 (1929). [5] Diophantische Ungleichungen I. Zur Gleichverteilung modulo Eins, *Acta Math.* **56**, 373–456 (1931). [6] Diophantische Ungleichungen II. Rhythmische Systeme A, B, *Acta Math.* **59**, 209–328 (1932). [7] Verteilungsfunktionen I–VIII, *Proc. Akad. Amsterdam* **38**, 813–821, 1058–1066 (1935); **39**, 10–19, 19–26, 149–153, 339–344, 489–494, 579–590 (1936).

Van der Corput, J. G., and Pisot, C.: [1] Sur la discrépance modulo un, *Indag. Math.* **1**, 143–153, 184–195, 260–269 (1939).

Veech, W. A.: [1] A Kronecker-Weyl theorem modulo 2, *Proc. Nat. Acad. Sci. U.S.A.* **60**, 1163–1164 (1968). [2] Strict ergodicity in zero dimensional dynamical systems and the Kronecker-Weyl theorem mod 2, *Trans. Amer. Math. Soc.* **140**, 1–33 (1969). [3] Some questions of uniform distribution, *Ann. of Math.* (2) **94**, 125–138 (1971). [4] Well distributed sequences of integers, *Trans. Amer. Math. Soc.* **161**, 63–70 (1971).

Verbickiĭ, I. L.: [1] Estimate of the remainder term in the Kronecker-Weyl approximation theorem (Russian), *Teor. Funkciĭ Funkcional. Anal. i Priložen Vyp.* **5**, 84–98 (1967).

Ville, J.: [1] *Étude critique de la notion de collectif*, Monographies des Probabilités, fasc. 3, Gauthier-Villars, Paris, 1939.

Vinogradov, I. M.: [1] On the fractional parts of integral polynomials (Russian), *Izv. Akad. Nauk SSSR* **20**, 585–600 (1926). [2] On the number of fractional parts of a polynom lying in a given interval, *Mat. Sb.* **1** (43), 3–8 (1936). [3] The representation of odd numbers as a sum of three primes (Russian), *Dokl. Akad. Nauk SSSR* **15**, 291–294 (1937). [4] On an estimate of trigonometric sums with prime numbers (Russian), *Izv. Akad. Nauk SSSR*, Ser. Mat., **12**, 225–248 (1948). [5] *The Method of Trigonometrical Sums in the Theory of Numbers*, Interscience, London-New York, 1954. [6] Trigonometric sums involving values of a polynomial (Russian), *Izv. Akad. Nauk SSSR*, Ser. Mat., **21**, 145–170 (1957). [7] On the distribution of the fractional parts of the values of a polynomial (Russian), *Izv. Akad. Nauk SSSR*, Ser. Mat., **25**, 749–754 (1961). [8] On the distribution of systems of fractional parts of values of several polynomials (Russian), *Izv. Akad. Nauk SSSR*, Ser. Mat., **26**, 793–796 (1962).

Visser, C.: [1] The law of nought-or-one in the theory of probability, *Studia Math.* **7**, 143–149 (1938).

Volkmann, B.: [1] Über Klassen von Mengen natürlicher Zahlen, *J. reine angew. Math.* **190**, 199–230 (1952). [2] Über Hausdorffsche Dimensionen von Mengen, die durch Zifferneigenschaften charakterisiert sind. I-VI, *Math. Z.* **58**, 284–287 (1953); **59**, 247–254, 259–270, 425–433 (1953/54); **65**, 389–413 (1956); **68**, 439–449 (1958). [3] On uniform distribution and the density of sum sets, *Proc. Amer. Math. Soc.* **8**, 130–136 (1957). [4] Gewinnmengen, *Arch. der Math.* **10**, 235–240 (1959). [5] On uniform distribution and the density of sets of lattice points, *Compositio Math.* **16**, 184–185 (1964). [6] On non-normal numbers, *Compositio Math.* **16**, 186–190 (1964). [7] Über extreme Anormalität bei Ziffernentwicklungen, *Math. Ann.* **190**, 149–153 (1970).

Von Mises, R.: [1] Über Zahlenfolgen, die ein kollektiv-ähnliches Verhalten zeigen, *Math. Ann.* **108**, 757–772 (1933). [2] *Mathematical Theory of Probability and Statistics*, Academic Press, New York-London, 1964.

Von Neumann, J.: [1] Gleichmässig dichte Zahlenfolgen (Hungarian; German summary), *Mat. Fiz. Lapok* **32,** 32–40 (1925); also in *Collected Works,* Vol. I, Pergamon Press, Oxford, 1961, pp. 58–66.

Walfisz, A.: [1] *Weylsche Exponentialsummen in der neueren Zahlentheorie,* Math. Forschungsberichte, Band 16, VEB Deutscher Verlag der Wissenschaften, Berlin, 1963.

Wall, D. D.: [1] Normal numbers, Thesis, Univ. of California, 1949.

Wallisser, R.: [1] Einige Sätze über nichtfortsetzbare Potenzreihen, *Math. Z.* **113,** 61–67 (1970). [2] Eine Bemerkung über irrationale Werte und Nichtfortsetzbarkeit von Potenzreihen mit ganzzahligen Koeffizienten, *Colloq. Math.* **23,** 141–144 (1971).

Warnock, T. T.: [1] Computational investigations of low-discrepancy point sets, in S. K. Zaremba (Ed.), *Applications of Number Theory to Numerical Analysis,* Academic Press, New York, 1972, pp. 319–343.

Weil, A.: [1] *L'intégration dans les groupes topologiques et ses applications,* Act. Sci. Ind. 869, 1145, Hermann, Paris, 1939, 1951.

Wendel, J. G.: [1] Haar measure and the semigroup of measures on a compact group, *Proc. Amer. Math. Soc.* **5,** 923–929 (1954).

Weyl, H.: [1] Über die Gibbssche Erscheinung und verwandte Konvergenzphänomene, *Rend. Circ. Mat. Palermo* **30,** 377–407 (1910); also in *Gesammelte Abhandlungen,* Band I, Springer-Verlag, Berlin-Heidelberg-New York, 1968, pp. 321–353. [2] Über ein Problem aus dem Gebiete der diophantischen Approximationen, *Nachr. Ges. Wiss. Göttingen,* Math.-phys. Kl., **1914,** 234–244; also in *Gesammelte Abhandlungen,* Band I, Springer-Verlag, Berlin-Heidelberg-New York, 1968, pp. 487–497. [3] Sur une application de la théorie des nombres à la mécanique statistique et la théorie des perturbations, *L'Enseignement Math.* **16,** 455–467 (1914); also in *Gesammelte Abhandlungen,* Band I, Springer-Verlag, Berlin-Heidelberg-New York, 1968, pp. 498–510. [4] Über die Gleichverteilung von Zahlen mod. Eins, *Math. Ann.* **77,** 313–352 (1916); also in *Gesammelte Abhandlungen,* Band I, Springer-Verlag, Berlin-Heidelberg-New York, 1968, pp. 563–599; and in *Selecta Hermann Weyl,* Birkhäuser Verlag, Basel-Stuttgart, 1956, pp. 111–147.

White, B. E.: [1] Complete minimum discrepancy distributions for one through six points in the unit square, Technical Report #118, Computer Sciences Dept., Univ. of Wisconsin, 1971. [2] The L^2 discrepancy of the Roth sequence in $[0, 1]^2$ for an arbitrary number of. points, Technical Report #170, Computer Sciences Dept., Univ. of Wisconsin, 1973.

Wiener, N.: [1] The quadratic variation of a function and its Fourier coefficients, *J. Math. Phys.* **3,** 72–94 (1924).

Wintner, A.: [1] On the cyclical distribution of the logarithms of the prime numbers, *Quart. J. Math.* (1) **6,** 65–68 (1935).

Yudin, A.: [1] New proof of Erdös-Turán's theorem (Russian), *Litovsk. Mat. Sb.* **9,** 839–848 (1969).

Zame, A.: [1] The distribution of sequences modulo 1, *Canad. J. Math.* **19,** 697–709 (1967). [2] On the measure of well-distributed sequences, *Proc. Amer. Math. Soc.* **18,** 575–579 (1967). [3] On the equivalence of two modes of distribution of a sequence, *Monatsh. Math.* **72,** 157–167 (1968). [4] Almost convergence and well-distributed sequences, *Canad. J. Math.* **20,** 1211–1214 (1968). [5] On normal sets of numbers, *Acta Arith.* **20,** 147–154 (1972). [6] On a problem of Narkiewicz concerning uniform distributions of sequences of integers, *Colloq. Math.* **24,** 271–273 (1972).

Zane, B.: [1] Uniform distribution modulo *m* of monomials, *Amer. Math. Monthly* **71,** 162–164 (1964).

Zaremba, S. K.: [1] Good lattice points, discrepancy, and numerical integration, *Ann. Mat. Pura Appl.* (IV) **73,** 293–317 (1966). [2] The mathematical basis of Monte Carlo and quasi-Monte Carlo methods, *SIAM Rev.* **10,** 303–314 (1968). [3] Some applications of multidimensional integration by parts, *Ann. Polon. Math.* **21,** 85–96 (1968). [4] Good lattice points in the sense of Hlawka and Monte Carlo integration, *Monatsh. Math.* **72,** 264–269 (1968). [5] A quasi-Monte Carlo method for computing double and other multiple integrals, *Aequationes Math.* **4,** 11–22 (1970). [6] A remarkable lattice generated by Fibonacci numbers, *Fibonacci Quart.* **8,** no. 2, 185–198 (1970). [7] La discrépance isotrope et l'intégration numérique, *Ann. Mat. Pura Appl.* (IV) **87,** 125–136 (1970). [8] Sur la discrépance des suites aléatoires, *Z. Wahrscheinlichkeitsth.* **20,** 236–248 (1971). [9] La méthode des 'bon treillis' pour le calcul numérique des intégrales multiples, in S. K. Zaremba (Ed.), *Applications of Number Theory to Numerical Analysis*, Academic Press, New York, 1972, pp. 39–119. [10] Good lattice points modulo primes and composite numbers, in C. F. Osgood (Ed.), *Diophantine Approximation and Its Applications*, Academic Press, New York, 1973, pp. 327–356. [11] Good lattice points modulo composite numbers, *Monatsh. Math.*, to appear.

Zaretti, A.: [1] Equivalenza di due definizioni di successioni equidistribuite nei domini *P*-adici, *Atti Sem. Mat. Fis. Univ. Modena* **13,** 111–118 (1964). [2] Sopra un teorema di B. Eckmann riguardante successioni uniformemente distribuite, *Ist. Lombardo Accad. Sci. Lett. Rend.* A **100,** 99–108 (1966).

Zeller, K., and Beekmann, W.: [1] *Theorie der Limitierungsverfahren*, 2. Aufl., Erg. Math. Grenzgeb., Band 15, Springer-Verlag, Berlin-Heidelberg-New York, 1970.

Zinterhof, P.: [1] Einige zahlentheoretische Methoden zur numerischen Quadratur und Interpolation, *Sitzgsber. Österr. Akad. Wiss.*, Math.-naturw. Kl., Abt. II, **177,** 51–77 (1969).

Zlebov, E. D.: [1] Pisot-Vijayaraghavan numbers and basic units of algebraic fields (Russian), *Vesci Akad. Navuk BSSR*, Ser. Fiz.-Mat. Navuk, **1966,** no. 4, 110–112.

Zygmund, A.: [1] *Trigonometric Series*, Vol. 1, Cambridge Univ. Press, London, 1968.

LIST OF SYMBOLS AND ABBREVIATIONS

Together with each symbol or abbreviation the definition and/or page reference is given.

Set Theory:

$A \backslash B$	set-theoretic difference of A and B
A'	(in Chapters 3 and 4), complement of A in underlying set, 174
$A \times B$	Cartesian product of A and B
A^{∞}	Cartesian product of denumerably many copies of A
$\pi_{j \in J} A_j$	Cartesian product of family of sets
$\dot{\subseteq}$	313
card A	cardinality of A
$\mathbf{a}$	cardinality of denumerable set
$\mathbf{c}$	cardinality of the continuum
$f \circ g$	composite mapping $f(g(.))$
c_M	characteristic function of M
$\mathbb{Z}$	integers
$\mathbb{Z}^{+}$	positive integers
$\mathbf{Q}$	rational numbers
$\mathbf{R}$	real numbers
$\mathbf{C}$	complex numbers
$\mathbb{Z}^s$	s-dimensional lattice points
$\mathbf{R}^s$	s-dimensional euclidean space
I	unit interval, 1
$\bar{I}$	closed unit interval, 2
I^s	s-dimensional unit cube, 47
$\bar{I}^s$	closed s-dimensional unit cube, 48

Analysis:

$(x_n), (x_k)$	sequences
(s_{jk})	double sequence, 18
$\lim_{x\to a-0} f(x), f(a-0)$	left-hand side limit of f at a
$\lim_{x\to a+0} f(x), f(a+0)$	right-hand side limit of f at a
$\lim_{M,N\to\infty} f(M,N)$	limit of $f(M,N)$ as $\min(M,N)\to\infty$
$f = O(g)$	Landau symbol: f/g bounded
$f = o(g)$	Landau symbol: $f/g\to 0$
$f = \Omega(g)$	Landau symbol: $f \neq o(g)$, 105
$\Delta f(n), \Delta x_n, \Delta f(t)$	difference operators, 13, 28, 80
$\Delta^k x_n, \Delta^k f(t)$	difference operators of higher order, 29, 86
$\Delta_j, \Delta_{j_1\ldots j_p}$	147
$\Delta_j^*, \Delta_{j_1\ldots j_p}^*$	148
$V(f)$	variation of f on $[0, 1]$, 143
$V^{(k)}(f)$	variation of f on $[0, 1]^k$ in the sense of Vitali, 147
λ	Lebesgue measure (unless stated otherwise)
Re z	real part of complex number z
Im z	imaginary part of complex number z
arg z	argument of complex number z, 230
$\bar{z}$	conjugate of complex number z
$\sqrt{x}$	nonnegative square root of x
$\log_k x$	iterated logarithm function, 24
$B_k(x)$	Bernoulli polynomial, 24
exp (t)	$e^{2\pi it}$ for real t
Exp (t)	e^t for real t
$\Vert\mathbf{h}\Vert$	maximum norm of lattice point $\mathbf{h}$, 49
$r(\mathbf{h})$	116
$P^{(q)}(\mathbf{g})$	162
(C, r)	Cesàro means, 62
(H, r)	Hölder means, 62
(E, s)	Euler method, 63
(R, p_n)	simple Riesz means, 63
F	(in Chapter 3, Section 4), summation method of almost convergence, 215
$F_{\mathbf{A}}$	summation method of A-almost convergence, 217
$\Vert\mathbf{A}\Vert$	(in Chapter 3), norm of matrix method $\mathbf{A}$, 207

Number Theory:

$[x]$	integral part of real number x, 1
$\{x\}$	fractional part of real number x, 1
$[x]_\Delta$	4

$\{x\}_\Delta$	4
$[\mathbf{x}]$	integral part of $\mathbf{x} \in \mathbf{R}^s$, 47
$\{\mathbf{x}\}$	fractional part of $\mathbf{x} \in \mathbf{R}^s$, 47
$[x]_m$	316
$\{x\}_m$	316
$\langle t \rangle$	distance from t to the nearest integer, 121
$[a_0, a_1, a_2, \ldots]$	simple continued fraction, 122
$a \mid b$	a divides b
$a \equiv b \pmod{m}$	a congruent to b modulo m
(a, b)	greatest common divisor of a and b
$[a, b]$	least common multiple of a and b
$\phi(n)$	Euler phi function
$\mu(n)$	Moebius function
$\sum_{d \mid n}$	sum over the positive divisors d of n

Uniform Distribution:

(x_n), (x_k)	sequences
u.d.	uniform distribution, uniformly distributed
u.d. mod 1	uniform distribution modulo 1, uniformly distributed modulo 1, 1
A-u.d. mod 1	A-uniform distribution modulo 1, A-uniformly distributed modulo 1, 61
c.u.d. mod 1	continuous uniform distribution modulo 1, continuously uniformly distributed modulo 1, 78
u.d. mod Δ	uniform distribution modulo Δ, uniformly distributed modulo Δ, 4
u.d. mod m	uniform distribution modulo m, uniformly distributed modulo m, 305
c.u.d. mod m	continuous uniform distribution modulo m, continuously uniformly distributed modulo m, 316
k-u.d.	uniformly distributed of order k, 320
w.d. mod 1	well distributed modulo 1, 40
d.f.	distribution function, 56
d.f. (mod 1)	distribution function modulo 1, 53
a.d.f. (mod 1)	asymptotic distribution function modulo 1, 53
A-a.d.f. (mod 1)	A-asymptotic distribution function modulo 1, 60
$A(E;N)$, $A(E;N;\omega)$	counting function, 1, 47, 175, 319
$A(E;N,k)$	40, 48
$A(j,m,N)$	counting function modulo m, 305
$A^*(j,m,N)$	306
$A(\alpha,k,N)$	(in Chapter 5, Section 2), 320
$A(B,M,N)$	(in Chapter 5, Section 3), 326
$A(\beta,k,N)$	(in Chapter 5, Section 3), 329

$A([a,b);M,N)$	(in Chapter 1, Section 2), counting function for double sequences, 18
$A_b(a;N), A_b(a;N;\alpha)$	69
$A_b(B_k;N), A_b(B_k;N;\alpha)$	69
$T(a, b), T(\mathbf{a}, \mathbf{b})$	78, 83
$C(P;n)$	251
$D_N, D_N(x_1,\ldots,x_N)$	discrepancy, 88
$D_N^*, D_N^*(x_1,\ldots,x_N)$	discrepancy, 90
$D_N, D_N(\mathbf{x}_1,\ldots,\mathbf{x}_N)$	multidimensional discrepancy, 93
$D_N^*, D_N^*(\mathbf{x}_1,\ldots,\mathbf{x}_N)$	multidimensional discrepancy, 93
$D_N(\omega), D_N^*(\omega)$	discrepancy of infinite sequence (or of finite sequence containing at least N terms), 89, 90, 93
$J_N, J_N(\mathbf{x}_1,\ldots,\mathbf{x}_N)$	isotropic discrepancy, 93
$J_N(\omega)$	isotropic discrepancy of infinite sequence (or of finite sequence containing at least N terms), 94
$D_N^{(p)}$	L^p discrepancy, 97
$D_N(\omega;f)$	discrepancy with respect to distribution function, 90
$R_N(x)$	remainder function, 107
M_N	maximal deviation, 189
P. V. number	Pisot-Vijayaraghavan number, 36
$(x_n) * (y_n)$	convolution of sequences, 260
$\chi_k(\alpha)$	320

Algebra, Linear Algebra:

$<\mathbf{x}, \mathbf{y}>$	standard inner product in $\mathbf{R}^s$, 48
$\mathrm{sp}\,(\mathscr{V})$	linear subspace spanned by $\mathscr{V}$, 172, 221
$\mathbf{E}$	identity matrix
$\lVert\mathbf{A}\rVert$	norm of complex square matrix, 222
$\mathrm{tr}\,(\mathbf{A})$	trace of complex square matrix, 223
$\overline{\mathbf{A}}$	conjugate of complex square matrix, 222
$\mathbf{A}^T$	transpose of complex square matrix, 222
$(\mathbf{A}\mid\mathbf{B})$	inner product of complex square matrices, 223
$\mathbf{A}\oplus\mathbf{B}$	Kronecker product of matrices, 225
$\mathbf{GL}(k)$	general linear group of rank k over $\mathbf{C}$, 222
$\mathbf{U}(k)$	unitary group of rank k over $\mathbf{C}$, 222
$GF(q)$	finite field of q elements, 326
$\Phi, GF[q,x]$	ring of polynomials over $GF(q)$, 326
$GF(q,x)$	field of rational functions over $GF(q)$, 329
$\Phi', GF\{q,x\}$	completion of $GF(q, x)$ with respect to degree valuation, 328
Φ'_k	329
$A \equiv B \pmod{M}$	congruence in $GF[q, x]$, 326
$[\alpha]$	integral part of $\alpha \in GF\{q, x\}$, 329
$\{\alpha\}$	fractional part of $\alpha \in GF\{q, x\}$, 329

$e(A, M)$	328
$e(\alpha)$	329

Group Theory, Topological Groups:

e	identity element, 220
a^{-1}	inverse of a
aM, Ma	left (right) translate of M by a, 221
M^{-1}	set of inverses of elements of M, 221
$G \times H$	direct product of (topological) groups G and H, 233
$\prod_{j \in J} G_j$	direct product of family of (topological) groups
$\prod^*_{j \in J} G_j$	weak direct product of family of groups, 233
μ	(in Chapter 4), Haar measure (unless stated otherwise), 221
$\nu(D)$	259
$\mathscr{A}(G)$	Banach algebra of almost-periodic functions on G, 297
$M(f)$	mean value of almost-periodic function f, 297
$\hat{G}$	dual group (character group) of G, 232
$\hat{G}^p$	set of periodic characters of G, 288
$\bar{G}$	(in Chapter 4, Section 5), Bohr compactification of G, 289
$\bar{G}^p$	periodic compactification of G, 289
$\hat{x}$	232
$A(\hat{G}, H)$	annihilator of H in $\hat{G}$, 232
$\mathbb{Z}$	additive group of integers in discrete topology
$\mathbb{Z}^s$	additive group of s-dimensional lattice points in discrete topology
$\mathbb{Z}(p^\infty)$	quasi-cyclic p-group, 293
$\mathbb{Z}_p$	additive group of p-adic integers in p-adic topology, 321
$\mathbf{Q}_d$	additive group of rational numbers in discrete topology, 283
$\mathbf{Q}_p$	additive group of p-adic numbers in $p$adic topology, 321
$\mathbf{R}$	additive group of real numbers in euclidean topology
$\mathbf{R}^s$	additive group of s-dimensional euclidean space in euclidean topology
$\mathbf{R}^0$	$\{e\}$, 233
$\mathbf{R}_d$	additive group of real numbers in discrete topology, 273
T	one-dimensional circle group, 224
T^s	s-dimensional circle group, 268
T_d	one-dimensional circle group in discrete topology, 273
G_0	universal compact monothetic group, 273
$\mathrm{GL}(k)$	general linear group of rank k over $\mathbf{C}$ in topology of entrywise convergence, 222

U(k)	unitary group of rank k over C in topology of entrywise convergence, 222
$S_k(\alpha)$	320
$\mathbf{R}_p^{(h)}$	322
$\mathbf{R}_p^k$	322
$\mathbf{R}_p$	322

Measure Theory:

λ	Lebesgue measure (unless stated otherwise)
μ	(in Chapter 3), nonnegative regular normed Borel measure, 171
μ	(in Chapter 4), Haar measure (unless stated otherwise), 221
μ_x, ϵ_x	normed point measure at x, 179, 258
$\bar{\mu}$	outer measure induced by μ, 198
μ_∞	(complete) product measure induced by μ on product of denumerably many copies of given measure space, 182
$T^{-1}\mu$	measure defined by $(T^{-1}\mu)(.) = \mu(T^{-1}.)$, 179
$\lambda * \nu$	convolution of measures, 258
$\int f d\mu$	286
$M(f)$	mean value of almost-periodic function f, 297
$\mathcal{M}^+(G)$	set of nonnegative regular normed Borel measures on G, 258
$\mathcal{R}$	(in Chapter 5, Section 1), 313
$\mathcal{A}$	(in Chapter 5, Section 2), 319
$\mathcal{B}$	(in Chapter 5, Section 2), 324

Topology, Functional Analysis:

int M	interior of M, 94
$\overline{M}$	closure of M, 94
∂M	topological boundary of M, 174
M'	(in Chapters 3 and 4), complement of M in underlying space, 174
c_M	characteristic function of M
X^∞	Cartesian product of denumerably many copies of space X in product topology, 182
w(X)	weight of space X, 227
$\mathcal{U}$	uniformity, 194
$U \circ V$	194
R mod 1	170
T	one-sided shift (unless stated otherwise), 183, 216

$\mathscr{B}(X)$	Banach algebra of bounded real-valued Borel-measurable functions on X under supremum norm, 171
$\mathscr{R}(X)$	Banach algebra of real-valued continuous functions on X under supremum norm, 171
$\mathscr{C}(X)$	Banach algebra of complex-valued continuous functions on X under supremum norm, 172
$\mathscr{A}(G)$	Banach algebra of almost-periodic functions on G, 297
$L^2[0, 1]$	Banach space of square-integrable functions on $[0, 1]$, 25
$L^1(\mu)$	Banach space of μ-integrable functions, 270
$\|f\|$	supremum norm of $f \in \mathscr{B}(X)$, 171
$\|\sigma\|$	supremum norm of bounded sequence σ of real numbers, 216
$\|\mathbf{A}\|$	norm of complex square matrix, 222
$(\mathbf{A} \mid \mathbf{B})$	inner product of complex square matrices, 223
■	end of proof, end of example

AUTHOR INDEX

SUBJECT INDEX

A CATALOG OF SELECTED

DOVER BOOKS

IN SCIENCE AND MATHEMATICS

Mathematics

FUNCTIONAL ANALYSIS (Second Corrected Edition), George Bachman and Lawrence Narici. Excellent treatment of subject geared toward students with background in linear algebra, advanced calculus, physics and engineering. Text covers introduction to inner-product spaces, normed, metric spaces, and topological spaces; complete orthonormal sets, the Hahn-Banach Theorem and its consequences, and many other related subjects. 1966 ed. 544pp. 6⅛ x 9¼. 0-486-40251-7

ASYMPTOTIC EXPANSIONS OF INTEGRALS, Norman Bleistein & Richard A. Handelsman. Best introduction to important field with applications in a variety of scientific disciplines. New preface. Problems. Diagrams. Tables. Bibliography. Index. 448pp. 5⅜ x 8½. 0-486-65082-0

VECTOR AND TENSOR ANALYSIS WITH APPLICATIONS, A. I. Borisenko and I. E. Tarapov. Concise introduction. Worked-out problems, solutions, exercises. 257pp. 5⅜ x 8¼. 0-486-63833-2

AN INTRODUCTION TO ORDINARY DIFFERENTIAL EQUATIONS, Earl A. Coddington. A thorough and systematic first course in elementary differential equations for undergraduates in mathematics and science, with many exercises and problems (with answers). Index. 304pp. 5⅜ x 8½. 0-486-65942-9

FOURIER SERIES AND ORTHOGONAL FUNCTIONS, Harry F. Davis. An incisive text combining theory and practical example to introduce Fourier series, orthogonal functions and applications of the Fourier method to boundary-value problems. 570 exercises. Answers and notes. 416pp. 5⅜ x 8½. 0-486-65973-9

COMPUTABILITY AND UNSOLVABILITY, Martin Davis. Classic graduate-level introduction to theory of computability, usually referred to as theory of recurrent functions. New preface and appendix. 288pp. 5⅜ x 8½. 0-486-61471-9

ASYMPTOTIC METHODS IN ANALYSIS, N. G. de Bruijn. An inexpensive, comprehensive guide to asymptotic methods–the pioneering work that teaches by explaining worked examples in detail. Index. 224pp. 5⅜ x 8½ 0-486-64221-6

APPLIED COMPLEX VARIABLES, John W. Dettman. Step-by-step coverage of fundamentals of analytic function theory–plus lucid exposition of five important applications: Potential Theory; Ordinary Differential Equations; Fourier Transforms; Laplace Transforms; Asymptotic Expansions. 66 figures. Exercises at chapter ends. 512pp. 5⅜ x 8½. 0-486-64670-X

INTRODUCTION TO LINEAR ALGEBRA AND DIFFERENTIAL EQUATIONS, John W. Dettman. Excellent text covers complex numbers, determinants, orthonormal bases, Laplace transforms, much more. Exercises with solutions. Undergraduate level. 416pp. 5⅜ x 8½. 0-486-65191-6

RIEMANN'S ZETA FUNCTION, H. M. Edwards. Superb, high-level study of landmark 1859 publication entitled "On the Number of Primes Less Than a Given Magnitude" traces developments in mathematical theory that it inspired. xiv+315pp. 5⅜ x 8½. 0-486-41740-9